Methods in Cellular Imaging

THE AMERICAN PHYSIOLOGICAL SOCIETY
METHODS IN PHYSIOLOGY SERIES

Methods in
Cellular
Imaging

Edited by
AMMASI PERIASAMY, Ph.D.

Director
W. M. Keck Center for Cellular Imaging
Professor of Biology and Biomedical Engineering
University of Virginia
Charlottesville, Virginia

Published for the American Physiological Society by
OXFORD
UNIVERSITY PRESS
2001

OXFORD

UNIVERSITY PRESS

Oxford   New York

Athens   Auckland   Bangkok   Bogotá   Buenos Aires   Calcutta
Cape Town   Chennai   Dar es Salaam   Delhi   Florence   Hong Kong   Istanbul
Karachi   Kuala Lumpur   Madrid   Melbourne   Mexico City   Mumbai
Nairobi   Paris   São Paulo   Shanghai   Singapore   Taipei   Tokyo   Toronto   Warsaw

and associated companies in
Berlin   Ibadan

Library of Congress Cataloging-in-Publication Data
Methods in cellular imaging / edited by Ammasi Periasamy.
p. cm.– (American Physiological Society's methods in physiology series)
Includes bibliographical references and index.
ISBN 0-19-513936-4
1. Fluorescence microscopy.   2. Cytology–Methodology.
3. Microscopy.
I. Periasamy, Ammasi.
II. Methods in physiology series.
QH212.F55 M48 2001   570'.28'2–dc21   00-062378

9 8 7 6 5 4 3 2

Printed in Hong Kong
on acid-free paper

Preface

This book is different in many important ways from the other books currently available on cellular imaging techniques. The introductions to the various sections indicate many of these differences. In a single volume we have addressed the basics of fluorescence, the different types of fluorophores used for cellular labeling, the detectors used for cellular imaging, and state-of-the-art light microscopy imaging systems as they are applied to biological and clinical research. Each chapter covers fundamental concepts and principles, methodology, and biological applications. The chapters are grouped according to the light microscopy imaging methods for live cell imaging and their contents are summarized in the introduction to each section. This should make the methods in cellular imaging easier to understand for all readers but especially for graduate students, postdoctoral fellows and scientists who are new to state-of-the-art light microscopy imaging systems.

This is the first book to address the many technological developments in multiphoton excitation (MPE) fluorescence microscopy for deep tissue imaging since its invention. In addition to MPE, the book also includes many other advanced techniques such as spectral imaging, laser tweezers, total internal reflection, high-resolution atomic force microscopy, and bioluminescence imaging.

The section on fluorescence resonance energy transfer (FRET) and fluorescence lifetime imaging (FLIM) is a unique feature in this book. It describes the physics of FRET and the basics of different light microscopic FRET techniques to provide high spatial and temporal resolution with the goal of localizing and quantitating the protein–protein interactions in live specimens. The chapters in this section provide many possible combinations of conventional and green fluorescent protein fluorophores for FRET with appropriate filter combinations and spectral configurations.

The book also provides step-by-step information on the preparation of digital light microscope images for use at conferences, in graphic manipulation, and for websites.

I would like to extend special thanks to all of the contributors who spent their time in preparing the valuable manuscripts to be included in this book. I also want to thank Mr. Jeffrey House, Oxford University Press, for his constant feedback and timely answers to all of my questions. This book is dedicated to my parents and teachers—particularly Profs. B. M. Sivaram, Brain Herman, and David L. Brautigan, who chiseled this hard rock in a different way into a scientist. My special thanks to Melanie Stenz, Hal Noakes, and Masilamani Elangovan for their valuable suggestions in preparing this book. Most of all I want to thank my wife Anandhi and my children Tamil (Anush), Monica, and Andrew Rajan for supporting me as I worked on this book.

I would like to thank the W. M. Keck Foundation, the Academic Enhancement Program of the University of Virginia, and the following organizations for their generous support of this book:

     Bio-Rad Laboratories Life Sciences Group
     Carl Zeiss, Inc.
     Chroma Technology Corporation
     Coherent Laser Group
     Nikon, Inc.
     Omega Optical, Inc.

Charlottesville, Va.   A. P.

Contents

Part IV Other Advanced Methods in Cellular Imaging

Introduction, 343

Contributors

DANIEL AXELROD, Ph.D.
Department of Physics
Biophysical Research Division
University of Michigan
Ann Arbor, Michigan

MAGARIDA BARROSO, Ph.D.
Department of Biology
University of Virginia
Charlottesville, Virginia

PHILIPPE I. H. BASTIAENS, Ph.D.
European Molecular Biology Laboratory
Heidelberg, Germany

KEITH BERLAND, Ph.D.
Department of Physics
Emory University
Atlanta, Georgia

DAVID L. BRAUTIGAN, Ph.D.
Center for Cell Signaling
University of Virginia Health
*   Sciences Center*
Charlottesville, Virginia

CHRISTOF BUEHLER, Ph.D.
Paul Scherrer Institut
Villigen PSI, Switzerland

WAYNE E. CASCIO, M.D.
Department of Medicine
School of Medicine
University of North Carolina
Chapel Hill, North Carolina

VICTORIA CENTONZE, Ph.D.
Department of Cellular & Structural Biology
University of Texas Health Science Center at
*   San Antonio*
San Antonio, Texas

MARIAH L. COLENO, Ph.D.
Laser Microbeam and Medical Program
Beckman Laser Institute and Medical
*   Clinic*
University of California
Irvine, California

LANCE DAVIDSON, Ph.D.
Department of Cell Biology
University of Virginia Health Sciences Center
Charlottesville, Virginia

RICHARD N. DAY, Ph.D.
Departments of Medicine and Cell
*   Biology*
National Science Foundation Center for
*   Biological Timing*
University of Virginia Health Sciences
*   Center*
Charlottesville, Virginia

KEES DE GRAUW, Ph.D.
University of Utrecht
Utrecht, the Netherlands

JAMES N. DEMAS, Ph.D.
Department of Chemistry
University of Virginia
Charlottesville, Virginia

ALBERTO DIASPRO, Ph.D.
INFM and Department of Physics
University of Genoa
Genoa, Italy

MARY E. DICKINSON, Ph.D.
Biological Imaging Center
Beckman Institute
California Institute of Technology
Pasadena, California

CHEN-YUAN DONG, Ph.D.
Department of Physics
National Taiwan University
Taipei, Taiwan, Republic of China

MICHAEL R. DUCHEN, MRCP, Ph.D.
Life Sciences Imaging Consortium and
    Mitochondrial Biology Group
Department of Physiology
University College London
London, United Kingdom

ANDREW K. DUNN, Ph.D.
Nuclear Magnetic Resonance Center
Massachusetts General Hospital
Harvard Medical School
Charlestown, Massachusetts

MASILAMANI ELANGOVAN, M.S.
W.M. Keck Center for Cellular Imaging
Department of Biology
University of Virginia
Charlottesville, Virginia

STEVEN P. ELMORE, Ph.D.
Department of Cell Biology and Anatomy
    School of Medicine
University of North Carolina
Chapel Hill, North Carolina

DANIEL L. FARKAS, Ph.D.
Department of Bioengineering
University of Pittsburgh and
    Center for Light Microscope Imaging and
    Biotechnology
Carnegie Mellon University
Pittsburgh, Pennsylvania

SCOTT E. FRASER, Ph.D.
Biological Imaging Center
Beckman Institute
California Institute of Technology
Pasadena, California

TODD FRENCH, Ph.D.
Molecular Devices Corporation
Synnyvale, California

MUHAMMED GAD, Ph.D.
Ain Shams University
Faculty of Sciences,
    Microbiology Department
Abbasia, Cairo, Egypt

HANS C. GERRITSEN, Ph.D.
University of Utrecht
Utrecht, the Netherlands

MICHAEL E. GEUSZ, Ph.D.
Department of Biological Sciences
Bowling Green State University
Bowling Green, Ohio

GERALD GORDON, Ph.D.
Department of Cell Biology and Anatomy
University of North Carolina
Chapel Hill, North Carolina

ENRICO GRATTON, Ph.D.
Laboratory of Fluorescence Dynamics
Department of Physics
University of Illinois at Urbana-
    Champaign
Urbana, Illinois

WILLIAM H. GUILFORD, Ph.D.
Department of Biomedical Engineering
University of Virginia
Charolottesville, Virginia

PETER HAGGIE, Ph.D.
Cardiovascular Research Institute
Department of Medicine and Physiology
University of California
San Francisco, California

IAN S. HARPER, Ph.D.
Microscopy and Imaging Research Facility
School of Biomedical Sciences
Monash University, Clayton
Victoria, Australia

BRIAN HERMAN, Ph.D.
Department of Cellular & Structural Biology
Mail Code 7762
University of Texas Health Science Center at
    San Antonio
San Antonio, Texas

LILY HSU, M.S.
Department of Mechanical
Engineering
Massachusetts Institute of Technology
Cambridge, Massachusetts

ATSUSHI IKAI, Ph.D.
Tokyo Institute of Technology
Tokyo, Japan

JAKE JACOBSON, Ph.D.
Life Sciences Imaging Consortium and
Mitochondrial Biology Group
Department of Physiology
University College London
London, United Kingdom

JULIE KEELAN, Ph.D.
Life Sciences Imaging Consortium and
Mitochondrial Biology Group
Department of Physiology
University College London
London, United Kingdom

RAYMOND KELLER, Ph.D.
Department of Biology
University of Virginia
Charlottesville, Virginia

KI H. KIM, M.S.
Department of Mechanical
Engineering
Massachusetts Institute of Technology
Cambridge, Massachusetts

KARSTEN KÖNIG, Ph.D.
Laser Microscopy Division
Institute of Anatomy II
Friedrich Schiller University
Jena, Germany

JOHN J. LEMASTERS, M.D., Ph.D.
Cell and Molecular Imaging
Facility
Department of Cell Biology and
Anatomy
School of Medicine
University of North Carolina
Chapel Hill, North Carolina

NUPAM MAHAJAN, Ph.D.
Lineberger Cancer Center
University of North Carolina
Chapel Hill, North Carolina

BARRY R. MASTERS, Ph.D.
Department of Mechanical
Engineering
Massachusetts Institute of Technology
Cambridge, Massachusetts

MART MOJET, Ph.D.
Life Sciences Imaging Consortium and
Mitochondrial Biology Group
Department of Physiology
University College London
London, United Kingdom

BARABRA J. MULLER-BORER, Ph.D.
Department of Medicine
School of Medicine
University of North Carolina
Chapel Hill, North Carolina

HAROLD L. NOAKES , JR.
National Science Foundation Center for
Biological Timing
Department of Biology
University of Virginia
Charlottesville, Virginia

AMMASI PERIASAMY, (EDITOR)
W. M. Keck Center for Cellular Imaging
Department of Biology
University of Virginia
Charlottesville, Virginia

TING QIAN, M.D., Ph.D
Department of Cell Biology and Anatomy
School of Medicine
University of North Carolina
Chapel Hill, North Carolina

PETER T. C. SO, Ph.D.
Department of Mechanical Engineering
Massachusetts Institute of Technology
Cambridge, Massachusetts

KENNETH R. SPRING, D.M.D., Ph.D.
Laboratory of Kidney and Electrolyte
Metabolism
National Heart, Lung and Blood
Institute
National Institutes of Health
Bethesda, Maryland

ANTHONY SQUIRE, Ph.D.
European Molecular Biology Laboratory
Heidelberg, Germany

DONNA R. TROLLINGER, Ph.D.
Department of Molecular and Cell Biology
University of California
Davis, California

BRUCE J. TROMBERG, Ph.D.
Laser Microbeam and Medical Program
Beckman Laser Institute and Medical Clinic
University of California
Irvine, California

OLGA VERGUN, Ph.D.
Department of Pharmacology
University of Pittsburgh
Pittsburgh, Pennsylvania

ALAN S. VERKMAN, M.D., Ph.D.
Cardiovascular Research Institute
Departments of Medicine and Physiology
University of California
San Francisco, California

PETER J. VERVEER, Ph.D.
European Molecular Biology Laboratory
Heidelberg, Germany

LAKSHMANAN VETRIVEL, Ph.D.
Cardiovascular Research Institute
Department of Medicine and Physiology
University of California
San Francisco, California

VINCENT P. WALLACE, Ph.D.
Laser Microbeam and Medical Program
Beckman Laser Institute and Medical
  Clinic
University of California
Irvine, California

HORST WALLRABE, B.A.
Department of Biology
University of Virginia
Charlottesville, Virginia

WATT W. WEBB, Sc.D.
Developmental Resource of Biophysical
  Imaging and Optoelectronics
School of Applied and Engineering Physics
Cornell University
Ithaca, New York

WARREN R. ZIPFEL, Ph.D.
Developmental Resource of Biophysical
  Imaging and Optoelectronics
School of Applied and Engineering Physics
Cornell University
Ithaca, New York

I

BASICS OF FLUORESCENCE, FLUOROPHORES, MICROSCOPY, AND DETECTORS

INTRODUCTION

The ability to study the development, organization, and function of unicellular and higher organisms and to investigate structures and mechanism on the micron size has allowed scientists to better grasp the often misunderstood relationship between microscopic and macroscopic behavior. Further, the microscope preserves temporal and spatial relationships that are frequently lost in traditional biochemical techniques and gives two- or three-dimensional resolution that other laboratory methods cannot.

The benefits of fluorescence microscopy techniques are numerous. Some advantages of these techniques are the inherent specificity and sensitivity of fluorescence, the high temporal, spatial, and three-dimensional resolution that is possible, and the enhancement of contrast resulting from detection of an absolute rather than relative signal. Additionally, the plethora of well-described spectroscopic techniques providing different types of information and the commercial availability of fluorescent probes, many of which exhibit an environment- or analytic-sensitive response, broaden the range of possible applications in the biomedical sciences. Recent advancements in available light sources, detection systems, data acquisition methods, and image enhancement, analysis, and display methods have further broadened the applications in which fluorescence microscopy can be successfully implemented for various biological studies.

Chapter 1 explains the basic principles of fluorescence excitation and emission, and various parameters involved, such as absorption, emission, quenching, quantum yield, anisotropy, and photobleaching. This helps with understanding the fluorescence labeling, imaging and quantification of the cellular events. Chapter 2 explains the kind of fluorophore one should use for measuring different ion concentration, membrane potential, and pH. This chapter provides not only information on how to load the cells with different fluorophore molecules but also a list of probes used for different cellular function studies with excitation and emission wavelengths. Appropriate selection of detectors for acquiring the fluorescence signals is an important aspect of the fluorescence microscopy. Chapter 3 describes the characteristics of detectors, such as sensitivity, signal-to-noise (S/N) ratio, and types of detectors, which includes charge-coupled device (CCD) cameras, photodiodes, and photomultiplier tubes (PMT). The different cameras that are selected for different microscopic techniques are described in Chapter 4. In this chapter, the author clearly addresses how the biologist or physiologist can choose a microscope to monitor or image the cellular activity. The authors provided enough information to follow the microscope terminology and their respective definitions, and information regarding where to go for different parts of the imaging system, such as lenses, cameras, and software.

The wide-field microscope image includes information above and below the focal plane. The information inside the cell could be obtained by digital deconvolution of the optical sectioned images. The digital deconvolution method will not help to monitor the real time cell signaling, and so Chapter 5 describes the laser scanning confocal microscope, which would reject the out-of-focus information by using a pinhole in front of the detector. Chapter 5 explains the basic principles involved in confocal microscopy and how to implement the ratio imaging using confocal microscopy for calcium, pH, and membrane potential in cardiac myocytes. Chapter 6 focuses mainly one aspect of the biological application, which is mitochondrial signaling.

Chapter 7 describes the fluorescence recovery after photobleaching (FRAP) method and its instrumentation and data acquisition for biological samples using a conventional fluorescence microscopy configuration. This FRAP method provides fundamental information about the dynamics of biologically important solutes and macromolecules in cell membrane and aqueous compartments. It is important to understand in this modern e-commerce society, how to present the digital image acquired from light microscopy. Chapter 8 provides step-by-step information on how to use Adobe Illustrator, how to convert the file to different formats such as Tiff and JPEG, how to use Macromedia Fireworks, which helps for graphic manipulation, and how to present your microscopic images on your website.

1

Basics of Fluorescence

Keith Berland

1. Introduction

Fluorescence imaging has become a very powerful tool in modern biological investigations. Its strength and popularity are largely based on the exquisite sensitivity and specificity of fluorescence measurements. The wide availability of high-quality fluorescence probes and labeling protocols has facilitated the spread of applications for studying molecular and cellular systems throughout numerous fields of biological and biomedical investigation.

In the most basic implementations of fluorescence microscopy, one simply purchases a " fluorescence microscope" and starts acquiring pictures. Impressive images can be generated using standard commercial systems, with hardly any thought or energy spent considering the underlying fluorescence processes or imaging parameters. On the other hand, there is much to be gained from a deeper understanding of basic fluorescence properties. At the very least, this will ensure proper technique to optimize image quality, and it will often open the door to much more sophisticated applications of fluorescence (many of which are discussed elsewhere in this book). A thorough understanding of the underlying fluorescence processes is particularly valuable in gathering and correctly interpreting quantitative information regarding molecular and cellular processes.

With this in mind, this chapter provides a brief introduction to the fundamentals of molecular fluorescence. The main goals are to define the basic parameters that characterize fluorescence emission and to briefly discuss their importance in fluorescence imaging. Readers interested in gaining a deeper understanding of fluorescence phenomena, the history of their discovery, and their applications in cellular imaging are referred to the excellent sources contained in the suggested reading list at the end of this chapter and to the references they provide.

2. Basics of Fluorescence

Fluorescence is one of the many different luminescence processes in which molecules emit light. *Fluorescence* refers to the light emitted during the rapid relaxation of fluorescent molecules following excitation by light absorption. To

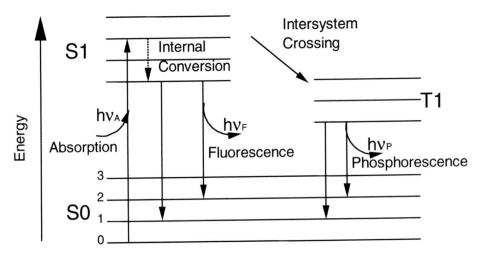

FIGURE 1–1 A Jablonski diagram representing the energy levels for a fluorescent molecule and several important transitions. The states S0 and S1 represent the ground and first singlet electronic states, while T1 is a triplet electronic state. There may of course be important higher excited states for a given molecule, but for simplicity they are ignored here. The closely spaced states within each electronic level (numbered 0, 1, 2, and 3 in the ground state) represent vibrational energy levels of the molecule. Planck's constant ($h = 6.63 \times 10^{-34}$ Joule s) relates a photon's energy, E, to its frequency, ν, as $E = h\nu$. Note: The wavelength, λ, and frequency, ν, are related by the speed of light, $c = 3 \times 10^8$ m/s, with $c = \lambda\nu$. Thus, the photon energy can also be calculated as $E = hc/\lambda$. A convenient mnemonic for calculating the approximate photon energy (in electron volts, eV) is that $E \approx 1234/\lambda$, where λ is the wavelength in nanometers.

introduce the basic properties of fluorescence, it is best to begin by considering the energy level structure of typical fluorescent molecules.

The characteristic molecular states and relaxation processes involved in fluorescence emission are illustrated in Figure 1–1, in what is referred to as a *Jablonski diagram*. S0 and S1 are singlet electronic states, and T1 is the lowest energy triplet state for the molecule. Singlet and triplet states are distinguished by the orientation of the electron spins.* The closely spaced levels within each electronic state represent the vibrational levels of the molecule. Very often the vibrational levels are similar for the different electronic states. Due to the quantum mechanical nature of the underlying processes, energy is absorbed or emitted in discrete units (i.e., photons). A single photon has energy $h\nu$, as defined in the figure caption. The transitions shown in the diagram are discussed in the following sections.

2.1 Absorption

At room temperature, a fluorescent molecule will normally occupy the ground state (S0) of both the electronic and vibrational energy levels. Before any fluorescence emission can occur, an energy source, namely, an absorbed photon

*Singlet states have *antiparallel* electron spins, while triplet states have *parallel* electron spins.

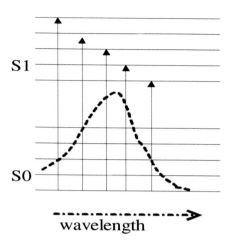

FIGURE 1–2 This schematic diagram shows how the transitions to various vibrational levels (arrows) of the molecule combine to define the absorption spectrum of the molecule (dashed line). As usual, the height of the dashed curve represents the relative strength of the transition at each excitation wavelength. A similar collection of transitions from S1 back to S0 will define the emission spectrum.

of energy hv_A, is required to excite the molecule. The energy of the absorbed photon will lift the molecule to higher level electronic and vibronic energy states. One particular absorption transition is shown in Figure 1–1, but there will generally be a whole range of allowed transitions with differing energy/frequency that corresponds to the different vibrational energy levels of the molecule, as shown in Figure 1–2. Some of the transitions are more favorable than others, as represented by the dashed curve.* Taken together, they make up the absorption spectrum for the molecule. Some practical details regarding fluorescence excitation are discussed in Section 3.1.

2.2 Fluorescence and Molecular Relaxation Pathways

Following photon absorption, an excited molecule will quickly relax to the ground vibrational level of S1. This fast vibrational relaxation process is called *internal conversion,* and it typically occurs on the picosecond (10^{-12} s) timescale. One important consequence of the rapid internal conversion is that all subsequent relaxation pathways proceed from the lowest vibrational level of S1. Other details of this fast relaxation process are typically not very important for fluorescence imaging applications, and thus will not be discussed here. Upon reaching the ground vibrational state of the excited state, there are three main pathways by which the molecule may depopulate the S1 state: fluorescence emission, nonradiative relaxation, and intersystem crossing to the triplet state.

---

*Discussions regarding the transition rates and selection rules for molecular excitation and emission spectra can be found in many introductory quantum physics and physical chemistry texts (e.g., Eisberg R. and R. Resnick. *Quantum Physics of Atoms, Molecules, Solids, Nuclei, and Particles,* 2nd Ed. New York: John Wiley & Sons, 1985).

2.2.1 Fluorescence emission

Fluorescence is generated when a molecule occupying the singlet electronic excited state relaxes to the ground state via spontaneous photon emission, an incoherent radiative process. There are many different parameters that are important in characterizing fluorescence emission as introduced below. We begin with two important spectral properties of fluorescence; the Stokes' shift and the shape of the emission spectrum.

2.2.1.1 The Stokes' shift. From the discussion thus far, one can clearly see that the energy of any photon emitted via fluorescence is less than the energy of the originally absorbed photon due to the energy lost to internal conversion. Fluorescence emission will thus have a redder color (i.e., longer wavelength and thus lower energy photons) than the original excitation source. This red shift is called the *Stokes' shift*. The actual extent of the red shift depends on the particular molecule and solvent. As an example Figure 1–3 shows the Stokes' shift of ~25 nm for rhodamine-123 in methanol. Other molecular processes including solvent relaxation and chemical reactions may also contribute to the Stokes' shift of a given probe.

It is worth noting that the Stokes' shift is very important for the extremely high sensitivity of fluorescence imaging measurements. The red shift allows the use of optical filters, which can block the excitation source from reaching the detector so that fluorescence detection (unlike absorption) is measured against a very low background.

2.2.1.2 The emission spectrum. Figure 1–1 shows just one of the characteristic transitions involved in fluorescence emission. The complete emission spectrum is made up of multiple transitions from the lowest vibrational state of S1 back to the various ground-state vibrational levels. Like absorption, some fluorescence transitions will be more favorable than others, and together they determine the

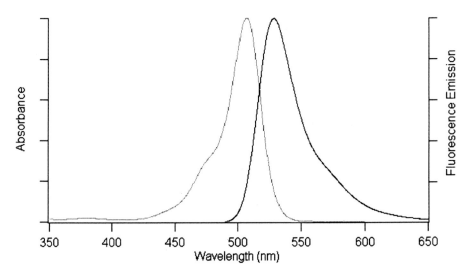

FIGURE 1–3 Rhodamine-123 in methanol. The absorption spectrum is the left wave, and the fluorescence emission spectrum is the right wave. [Courtesy of Molecular Probes, Inc.]

overall emission spectrum. Fluorescent molecules in solution typically have spectra similar to those in Figure 1–3, which shows both the absorption and the emission spectra for rhodamine-123 in methanol.

Several interesting features of the fluorescence spectrum are worth noting: (*1*) The spectrum is typically independent of the excitation energy (i.e., wavelength). This is a consequence of rapid internal conversion because fluorescence transitions initiate from the lowest vibrational state of S1 regardless of the initial excited state. (*2*) For many molecules, the vibrational energy level spacing is quite similar for the ground and excited electronic states. To the extent that this is true, the fluorescence spectrum of a particular molecule will resemble the mirror image of its absorption spectrum because the same transitions are most favorable for both absorption and emission. (*3*) Detailed vibrational structure is generally lost for fluorophores in solution. Spectral information can still, however, provide important information regarding fluorescence probes and/or their environment (e.g., see Section 2.5).

2.2.2 Quenching and nonradiative relaxation

For any given molecular system there exist at least some mechanisms that allow an excited molecule to relax back to the ground state without any associated fluorescence emission. These nonradiative transitions may be due to intramolecular or intermolecular relaxation pathways, and are said to quench the fluorescence. Quenching pathways compete with the fluorescence relaxation pathway and will thus reduce (or in some cases eliminate) fluorescence emission. Some examples of nonradiative relaxation mechanisms include resonance energy transfer, excited state chemical reactions, static quenching, and collisional quenching. Resonance energy transfer is a process in which an excited state molecule (the "donor") transfers its energy to a nearby molecule (the "acceptor") through near-field electromagnetic interaction. Static quenching occurs when a fluorophore forms a nonfluorescent complex with another molecule. Finally, collisional quenching refers to the loss of the fluorophore's excited state energy upon colliding with a quenching agent. Molecular oxygen is a strong fluorescence quencher, as are metals.

While reduced fluorescence efficiency is generally undesirable, the nonradiative relaxation mechanisms can be very important in fluorescence imaging. In particular the sensitivity of these processes to the environment surrounding the fluorophore, such as through complex formation or interaction with solvent molecules, can be exploited in imaging applications to probe the local environment. For example, resonance energy transfer imaging is a powerful tool for detecting interaction and/or co-localization of different molecular species (Clegg, 1995; Wu and Brand, 1994). Similarly, the sensitivity of certain fluorophores to various quenching agents can serve to determine the concentration (or absence) of the quenching molecule, such as molecular oxygen.*

2.2.3 Intersystem crossing and phosphorescence

Intersystem crossing occurs when an excited state molecule relaxes to the triplet excited state. The triplet state is distinguished from singlet states by the

---

*The Molecular Probes, Inc., Handbook (Haugland, 2000) contains numerous examples of fluorescent probes, which are sensitive to various environmental factors and other biomolecules.

opposite orientation of an electron spin. Molecules that reach the triplet state relax via photon emission or nonradiative relaxation. The radiative transition from the triplet state is called *phosphorescence*. The characteristic timescale for phosphorescence relaxation is much longer than that for fluorescence emission (see Section 2.3.1). Phosphorescence is generally weaker than fluorescence as well because nonradiative transitions often dominate the triplet state relaxation.

As in the case of fluorescence quenching, intersystem crossing to the triplet state reduces the fluorescence yield. In addition, the molecule may spend a substantial period of time (microseconds and longer) in the triplet state, during which time it is unavailable to generate fluorescence photons. Under normal illumination conditions, a significant steady-state population of triplet state molecules can build up even for fluorescent molecules with modest probability for intersystem crossing. This triplet state saturation can lead to substantially reduced fluorescence signal levels (Tsien and Waggoner, 1995).

2.3 Fluorescence Lifetime and Quantum Yield

We next consider the characteristic timescales of the molecular transitions introduced earlier. Because the main interest here is fluorescence, we shall simplify the model as shown in Figure 1–4. The excitation rate, W_A, will be discussed later in Section 3.1.1. Gamma (Γ) is the rate coefficient for the S1 to S0 transition via fluorescence emission. The "nonfluorescent" rate, k_{nf}, represents all transitions from the state S1 that do not result in fluorescence emission. It includes nonradiative relaxation mechanisms and intersystem crossing, and its value is the sum of the rate coefficients for each contributing process. Two important fluorescence parameters, the fluorescence lifetime and the quantum yield, can now be defined in terms of these rate constants. The contribution of k_1 will be ignored because the rapid relaxation by internal conversion occurs on the picosecond timescale versus the typical nanosecond timescale for fluorescence relaxation. This approach simply reflects the earlier statement that all emission occurs from the lowest vibrational state of S1.

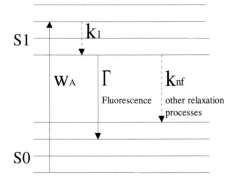

FIGURE 1–4 A simplified Jablonski diagram emphasizing transition rates.

2.3.1 Fluorescence lifetime

The fluorescence lifetime, τ, is a measure of the average time a molecule will spend in the excited state before relaxing back to the ground state (by either of the two pathways shown). From the model it is easily seen that

$$\tau = \frac{1}{\Gamma + k_{nf}} \tag{1}$$

The fluorescence lifetime should not be confused with the actual time any given molecule spends in the excited state, as fluorescence is a random process. Rather (for this simple model), a population of molecules simultaneously excited to S1 will decay exponentially back to the ground state with time constant τ (i.e., $\propto \exp[-t/\tau]$). Alternatively, the same exponential factor is proportional to the probability that a single excited molecule will remain in state S1 a time t after being excited. The lifetime of many common probes is on the order of a few nanoseconds (e.g., fluorescein isothiocyanate at pH 7.2 has a lifetime of 3.9 ns; Draaijer et al., 1995), but can vary over a wide range depending on the specific fluorescent molecule, the solvent/environment, and other factors. Occasionally it is useful to consider a quantity referred to as the *natural lifetime,* which refers to the lifetime in the absence of any nonfluorescent relaxation pathways (i.e., $k_{nf} = 0$). Thus, the natural lifetime is given by $\tau_n = 1/\Gamma$

Measuring the lifetime can be quite useful in fluorescence imaging (Draaijer et al., 1995; French et al., 1998; Lackowicz and Szmacinski, 1996; Wang et al., 1996; also see Chapters 15–19). First, it provides an additional contrast parameter so that molecules with the same spectra (color) but different lifetimes can be distinguished. The lifetime also depends on the local environment, and lifetime imaging can provide a sensitive measure of local conditions (e.g., pH and ion concentration; see Chapter 18). Because the lifetime is an intrinsic property of a fluorescent molecule, the results of such measurements are independent of probe concentration. This can be an important advantage in cells and tissue, where probe concentration may not be uniform. Lifetime measurements are also less susceptible to misinterpretation due to photobleaching than are intensity measurements.

2.3.2 Quantum yield

The quantum yield of a fluorescent molecule is a measure of its fluorescence efficiency, defined as the fraction of all excited molecules that relax by fluorescence emission. One can also validly interpret the quantum yield as the probability that a given excited molecule will produce a fluorescence photon. The quantum yield is defined in terms of the rate constants as

$$Q = \frac{\Gamma}{\Gamma + k_{nf}} \tag{2}$$

It is apparent from this definition that the quantum yield would reach its maximum value of unity only when fluorescence is the single allowed relaxation process. Fluorescent molecules commonly used as probes in microscopy have quantum yields ranging from very low (<0.05) to near unity. High quantum yield is generally desirable in most imaging applications. On the other hand, because nonfluorescent relaxation pathways (and the value of k_{nf}) can be very sensitive to the local environment, the quantum efficiency can sometimes be used as a probe of the environment near the fluorescent molecule. For example, the intercalating dye ethidium bromide is very weakly fluorescent when not bound to DNA and highly fluorescent (high quantum yield) once bound (Haugland, 2000). Ethidium fluorescence is thus highly sensitive to the presence of nucleic acids and can be used to locate and quantify DNA in fluorescence images.

2.4 Anisotropy

The dipole transition moments involved in both molecular absorption and fluorescence have spatial orientation. When a sample is illuminated with polarized light, those molecules whose absorption dipole orientation is aligned with the direction of the optical electric field are preferentially excited. Excitation with polarized light, therefore, photoselects a subpopulation of a randomly oriented sample, resulting in an oriented excited state population.* As a result, the fluorescence emission will also be partially polarized. Two different parameters, anisotropy (r) and polarization (P), are commonly used to measure the degree of polarization. They are defined in terms of the fluorescence intensity polarized parallel ($\parallel$) and perpendicular ($\perp$) to the excitation polarization:

$$r = \frac{I_\parallel - I_\perp}{I_\parallel + 2I_\perp}, \qquad P = \frac{I_\parallel - I_\perp}{I_\parallel + I_\perp} \qquad (3)$$

If necessary, one can easily convert between anisotropy and polarization.† Fluorescence anisotropy is often a more natural quantity to work with, although a discussion of the reasons are beyond the scope of this discussion.

For a randomly oriented population of molecules with parallel excitation and emission dipoles, the maximum anisotropy is 0.4 because photoselection will only result in a partially oriented system. Any reorientation of the transition dipoles before fluorescence emission will reduce the anisotropy toward zero. Anisotropy can thus be used as a measure of molecular rotations provided that the rotational motion is substantial on the timescale of the fluorescence lifetime.

2.5 Environmental Influence on Fluorescence Properties

As discussed earlier, fluorescence parameters like absorption and emission spectra, intensity, lifetime, and quantum yield are often very sensitive to the local

---

*There is actually a distribution of orientations, weighted by $\cos^2(\theta)$, where theta is the angle between the molecular dipole orientation and the polarization direction.

†$r = \dfrac{2P}{3 - P} \cdot P = \dfrac{3r}{2 + r}$. These are easily derived from Equation 3.

environment of the fluorescent molecule. Some of the factors that can alter the fluorescence properties of a given fluorophore include interaction with other fluorophores, the solvent, or ions in solution. Molecular associations and complex formation may also influence the fluorescence emission. The environmental sensitivity of fluorescence can be very important in imaging applications. Known environmental effects on specific fluorescent molecules can be quantified in order to measure important biological parameters, such as ion concentration (e.g., calcium), pH, and the presence of molecular interactions. As discussed earlier, lifetime imaging and energy transfer imaging are two useful methods for exploiting the environmental sensitivity. Ratio imaging is another example of this type of application (Tsien and Poenie, 1986; Bright et al., 1989). Figure 1–5 shows one specific example of a probe used for ratio imaging in which the molecular spectra are dramatically shifted as the pH of the solvent varies. Measurements of the spectra can thus be used to image local pH throughout the system under study.

To correctly interpret quantitative fluorescence measurements, it can be fairly important to consider potential shifts in various fluorescence parameters, even when they are not expected. For example, in a homogeneous system fluorescence intensity is proportional to molecular concentration. In the heterogeneous environment often encountered in biological imaging, however, intensity measurements interpreted in terms of probe concentration could, in some cases, actually reflect local pH or the presence of a quenching agent. These possibilities should always be considered when analyzing quantitative data.

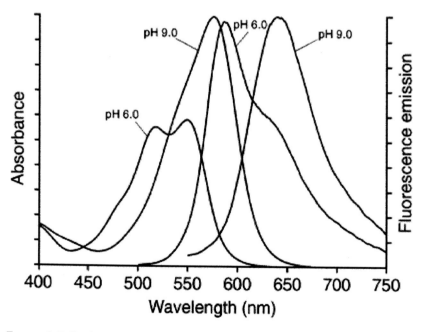

FIGURE 1–5 Environmental effects on fluorescence properties. The shift in the absorption and emission spectra of carboxy SNARF-1 for two different buffers (pH 6 and 9) is shown. The spectra undergo dramatic spectral shifts as a function of pH. [Courtesy of Molecular Probes, Inc.]

2.6 Photobleaching

Unfortunately, fluorescent molecules are often not terribly robust, and bleaching is almost always an issue in fluorescence imaging. Upon repeated excitation, fluorescent molecules will undergo irreversible chemical reactions after which they are no longer fluorescent. Photobleaching is highly dependent on the illumination level, although the chemical mechanisms of the bleaching process are not well understood. One of the known bleaching mechanisms is due to interaction with molecular oxygen. Multiphoton absorption and excited state absorption can also play a role in the bleaching of some molecular systems. The average number of emission cycles that occur before photobleaching depends on both the molecule in question and the environment. Some molecules may bleach after emitting only a few photons, while other more robust probes can emit millions of fluorescence photons before bleaching. Choosing the most stable fluorophores, when possible, is important for optimizing fluorescence image quality and measured signal-to-noise (S/N) ratios. Unfortunately, there are very few data available on the bleaching characteristics of most common fluorophores. The lack of data forces some degree of trial and error when choosing the best probes for a given system or environment.

There are a number of other strategies for reducing the effects of bleaching in imaging applications. The general idea behind most effective approaches is to increase the fluorescence detection sensitivity while simultaneously reducing the illumination level. By maximizing the detection efficiency it should become possible to collect images with good S/N ratios over much longer time periods than without this optimization. This approach probably does not increase the total number of photons emitted per fluorescent molecule (Hirschfeld, 1976) before bleaching. Rather, the S/N ratio is maintained by making more efficient use of the photons that are emitted. With modern highly sensitive instrumentation, even single molecules can be detected (Ambrose et al., 1999). There are a number of important factors that should be considered in order to maximize the collection efficiency. First, it is important to use high numerical aperture lenses to efficiently collect fluorescence, as discussed in Section 3.2. Dichroic mirrors and filters should be carefully chosen to match the spectrum of each fluorophore. In addition, using electronic detectors with high quantum efficiency (high sensitivity) can greatly enhance imaging results. It is also important to consider the spectral response of various cameras and photomultiplier tubes because their sensitivity to different wavelengths can vary greatly even within the visible spectrum. Finally, use of a shutter to block illumination when not acquiring data will also significantly reduce photobleaching.

Besides working to increase the overall measurement sensitivity, it is sometimes possible to chemically reduce the photobleaching rate. For example, because bleaching is often highly dependent on oxygen concentration, various methods to remove oxygen from the sample can be very effective in stabilizing fluorescence emission. This assumes that the reduced oxygen concentration and the method used to remove the oxygen are not otherwise detrimental to the system under investigation. A number of chemical agents that can help reduce photobleaching are also available, although their effects on the system under study are not always benign.

It is worth noting that while bleaching can impede imaging applications, it can also be a great advantage. For example, in fluorescence photobleaching recovery, the molecules within a small region are intentionally bleached (Axelrod et al., 1976). By monitoring the recovery of the fluorescence signal as new molecules diffuse in from unbleached regions of the sample, it is possible to measure diffusion rates of various fluorescent species. This application is discussed in more detail in a later chapter of this volume.

3. A Few Practical Comments on Fluorescence Imaging

The experimental requirements in fluorescence imaging can be broken down into several distinct issues, all of which should be optimized in order to acquire the highest quality fluorescence images. Many of these topics are covered elsewhere in this book, so I will focus on only two topics of practical importance: the illumination conditions and fluorescence collection efficiency.

3.1 Illumination

3.1.1 Quantifying absorption probabilities

The most natural parameter for specifying molecular absorption probability is the absorption cross section, σ_A. The molecular excitation rate, W_A, is given by

$$W_A = \frac{\sigma_A(\lambda)}{A} \frac{I}{h\nu_A} \tag{4}$$

where I is the illumination power (in Watts) and A is the cross-sectional area of the beam where it interacts with the molecule (i.e., I/A is the light flux in Watts/cm^2). The cross section has been written explicitly as a function of wavelength to emphasize the spectral dependence of the absorption process. This equation assumes that the excitation source has a uniform spatial profile. For nonuniform illumination, W_A will have the same spatial profile as the input intensity (for one-photon excitation). The molecular absorption cross section has units of area (e.g., cm^2) and has a simple geometric interpretation. It can be thought of as the effective "size" of the molecule for interacting with the applied radiation. The quantity $(\sigma_A[\lambda])/A$ represents the fraction of photons per second, $I/h\nu_A$, incident upon the illuminated area A that will be absorbed by a molecule inside the beam area.

As noted earlier, the absorption cross section is different for each fluorophore, and it is also affected by the solvent, interaction with other molecules, excitation wavelength, and other factors. A maximum cross section on the order of 10^{-16} cm^2 is characteristic of some of the most common fluorescence probes. For example, fluorescein in water (at high pH) excited at 488 nm has a cross section of 3×10^{-16} cm^2.

Many readers will perhaps be more familiar with the molar extinction coefficient, ε, units of liter Mole^{-1} cm^{-1}, also often written as M^{-1}cm^{-1}), which is another common measure of molecular absorption. The Beer-Lambert law

$$I = I_0 \, 10^{-\varepsilon c l} \tag{5}$$

relates the transmitted intensity, I, to the initial intensity I_0 for a beam passing through a path length l of an absorbing solution with molar concentration c. The extinction coefficient and absorption cross section are related by

$$\sigma_A = 3.8 \times 10^{-21} \varepsilon \tag{6}$$

for cross sections in cm^2 and ε in liters/Mole-cm. In imaging applications, the cross section is a more natural quantity to work with because the extinction path is not always easily defined for many common imaging geometries. In addition, thinking in terms of cross sections allows one to more easily consider what is happening on a per-molecule basis.

3.1.2 Choosing excitation wavelength and power

There are two main concerns in choosing excitation wavelength and power in fluorescence imaging. The first is to maximize the measured S/N ratio over the required image exposure time, including the effects of bleaching. The second is to minimize photodamage and thus the invasiveness of the measurement. To minimize photodamage and photobleaching, it is usually best to limit the power reaching the sample as much as possible. On the other hand, fluorescence signals are very often much weaker than desired, and it becomes necessary to increase the excitation power to improve signal levels.

3.1.2.1 Choosing excitation wavelength. Because the molecular excitation rate depends only on the product of the cross section and the illumination level, and the emission spectrum is independent of the excitation wavelength, one may conclude that the choice of excitation wavelength is not terribly important. To a certain extent this is reasonable, and exciting at wavelengths with a low absorption cross section can be compensated by increasing illumination. Varying the excitation wavelength can provide a valuable option for applications in which it is necessary to excite (or not excite) multiple fluorophores within the sample. This is particularly relevant in cellular imaging where it can be helpful to avoid exciting autofluorescent species. On the other hand, the measured S/N ratios are usually best optimized by working near the peak of the absorption spectrum in order to minimize the illumination levels. In most cases this will also help reduce photodamage to the sample, although not always because the degree of photodamage and photobleaching is wavelength dependent.

3.1.2.2 Saturation. The discussion thus far has assumed that saturation and other such effects are negligible, which is not always accurate. On average, a fluorescent molecule can emit no more than $1/\tau$ photons per second. As the excitation rate approaches this limiting fluorescence rate, the excitation rate will deviate from its linear dependence on input power. Thus, turning up the power will no longer proportionally increase the fluorescence output. Illumination above the saturation level will increase photobleaching, photodamage, and background noise from scattered light without improving the measured signal level. Therefore, it is often valuable to do a quick calculation in order to choose illumination levels for optimum image quality. Such a calculation depends only on the fluorescent molecule's absorption cross section and illumination levels, but not the probe concentration. For example, fluorescein has a lifetime of $\sim 4 \times 10^{-9}$ s and thus a maximum emission rate of $\sim 2.5 \times 10^8$ s^{-1}. At the peak absorption cross section of $\sim 3 \times 10^{-16}$ cm^2 (for illumination at 488 nm), light focused to a power density (I/A) of $\sim 3.5 \times 10^5$ W/cm^2 (8.6×10^{23} photons/cm^2-s)* yields an excitation rate of 2.6×10^8 s^{-1}. This is certainly near saturation, and greater illumination would not likely be very helpful. Three other points are worth noting. First, because the cross section and lifetime will be affected by the environment, this type of calculation can only be used as a rough guide to illumination levels. Second, most fluorophores will rapidly bleach under such high illumination, and the peak emission rates calculated will not likely be sustainable for very long, typically much less than a second! Finally, if there is substantial intersystem crossing, triplet state saturation may also be an important factor (Tsien and Waggoner, 1995).

Saturation levels may also be determined experimentally by comparing fluorescence signal levels with illumination intensity. This approach is complicated by collection of out-of-focus light and will thus work best with thin sample sections or using confocal detection.

3.2 Fluorescence Collection Efficiency

Because fluorescence is emitted in all directions, it is impossible for a standard lens to collect all the light and send it to a detector. The collection efficiency for an objective lens is a function of its numerical aperture (NA); higher NA lenses collect more light. Provided one assumes fluorescence emission is spatially isotropic, the collection efficiency (CE) of an objective lens (for an emitter at the focal point) is given by

$$\text{CE} = \frac{1 - \cos(\theta)}{2} \tag{7}$$

*The actual power density depends on the profile of the illumination. For uniform (wide-field) illumination, one would simply divide the total power by the illuminated area. For laser illumination with a gaussian beam profile, the beam waist ($1/e^2$ radius) is $\omega_0 \sim \lambda / \pi A$, where A is the numerical aperture of the lens. The power density in the center of the beam at the focal plane is then $2I/\pi\omega_0$, with I expressed in Watts.

The angle theta is determined from the NA of the lens by $n \sin(\theta) = NA$, where n is the index of refraction ($n = 1$ in air, $n \simeq 1.5$ for common immersion oils). Thus, for example, an air lens with NA = 0.9 will collect about 28% of the fluorescence. Note that most of the light will never make it through the lens to the detector, but high NA lenses collect *substantially* more light than low NA lenses.

A number of other factors will further reduce the overall light level reaching the detector, such as the transmission efficiency of dichroic mirrors, lenses, and filters in the optical path. The detector itself will also have varied efficiency for different colors of light. Both of these issues are discussed in detail in Chapters 2–4 of this volume. All of these factors must be taken into account to determine the overall efficiency of fluorescence detection. Surprisingly, by the time all factors are accounted for, a total efficiency of a few percent is quite good!

4. SUMMARY

The aim of this chapter has been to introduce the fundamental concepts and parameters that characterize molecular fluorescence. It is hoped that an understanding of this information will aid researchers in correctly applying fluorescence imaging technology and in interpreting experimental results. The concepts introduced can also serve as the foundation for understanding the more sophisticated applications of fluorescence measurements, and readers are encouraged to consult the literature to further explore these topics.

REFERENCES

Ambrose, W. P., P. M. Goodwin, J. H. Jett, A. V. Orden, J. H. Werner, and R. A. Keller. Single molecule fluorescence spectroscopy at ambient temperature. *Chem. Rev. (Washington, D.C.)* 99:2929–2956, 1999.

Axelrod, D., D. E. Koppel, J. Schlessinger, E. Elson, and W. W. Webb. Mobility measurement by analysis of fluorescence photobleaching recovery kinetics. *Biophys. J.* 16:1055–1069, 1976.

Bright, G., G. W. Fisher, J. Rugowsta, and D. L. Taylor. Fluorescence ratio imaging microscopy. In: *Methods in Cell Biology.* Vol. 30, edited by D. L. Taylor and Y. L. Wang. San Diego: Academic Press, 1989, pp. 157–192.

Clegg, R. M. Fluorescence resonance energy transfer. *Curr. Opin. Biotechnol.* 6:103–110, 1995.

Draaijer, A., R. Sanders, and H. C. Gerritsen. Fluorescence lifetime imaging, a new tool in confocal microscopy. In *Handbook of Biological Confocal Microscopy*, 2nd Ed., edited by J. B. Pawley. New York: Plenum Press, 1995, pp. 491–505.

French, T., P. T. C. So, C. Y. Dong, K. M. Berland, and E. Gratton. Fluorescence lifetime imaging techniques for microscopy. In: *Video Microscopy*, edited by G. Sluder and D. E. Wolf. New York: Academic Press, 1998, pp. 277–304.

Haugland, R. P. *Handbook of Fluorescent Probes and Research Chemicals.* Available at: http://www. probes. com/handbook. 2000.

Hirschfeld, T. Quantum efficiency independence of the time integrated emission from a fluorescent molecule. *Appl. Optics* 15:3135–3139, 1976.

Lackowicz, J. R., and H. Szmacinski. Imaging applications of time-resolved fluorescence spectroscopy. In: *Fluorescence Imaging Spectroscopy and Microscopy*, edited by X. F. Wang and B. Herman. New York: John Wiley & Sons, 1996, pp. 273–304.

Tsien, R. Y., and M. Poenie. Fluorescence ratio imaging: A new window into intracellular ionic signaling. *Trends Biochem. Sci.* 11:450–455, 1986.

Tsien, R. Y., and A. Waggoner. Fluorophores for confocal microscopy. In: *Handbook of Biological Confocal Microscopy,* 2nd Ed., edited by J. B. Pawley. New York: Plenum, 1995, pp. 267–279.

Wang, X. F., A. Periasamy, P. Wodnicki, G. Gordon, and B. Herman. Time-resolved fluorescence lifetime imaging microscopy: Instrumentation and biomedical applications. In: *Fluorescence Imaging Spectroscopy and Microscopy,* edited by X. F. Wang and B. Herman. New York: John Wiley & Sons, 1996, pp. 273–304.

Wu, P., and L. Brand. Resonance energy transfer: Methods and applications. *Anal. Biochem.* 218:1–13, 1994.

SUGGESTED READINGS

Brand, L., and M. L. Johnson. *Fluorescence Spectroscopy* (Methods in Enzymology, Vol. 278). New York: Academic Press, 1997.

Cantor, C. R., and P. R. Schimmel. *Biophysical Chemistry Part II: Techniques for the Study of Biological Structure and Function.* New York: W. H. Freeman and Co., 1980.

Haugland, R. P. *Handbook of Fluorescent Probes and Research Chemicals.* Available at: http://www.probes.com/handbook. 2000.

Herman, B., and H. J. Tanke. *Fluorescence Microscopy,* 2nd Ed. New York: Springer-Verlag, 1998.

Lakowicz, J. R. *Principles of Fluorescence Spectroscopy,* 2nd Ed. New York: Plenum, 1999.

Lakowicz, J. R. *Topics in Fluorescence Spectroscopy: Techniques* (Vol. 1, 1991); *Principles* (Vol. 2, 1991); Biochemical Applications (Vol. 3, 1992); *Probe Design and Chemical Sensing* (Vol. 4, 1994); *Nonlinear and Two-Photon Induced Fluorescence* (Vol. 5, 1997). New York: Plenum.

Pawley, J. B. *Handbook of Biological Confocal Microscopy.* New York: Plenum, 1995.

Rost, F. W. D. *Fluorescence Microscopy,* Vols. 1, 2. New York: Cambridge University Press, 1992.

Taylor, D. L., and Y. L. Wang. *Fluorescence Microscopy of Living Cells in Culture,* Parts A and B (Methods in Cell Biology, Vols. 29, 30). New York: Academic Press, 1989.

Wang, X. F., and B. Herman. *Fluorescence Imaging Spectroscopy and Microscopy.* New York: Wiley, 1996.

2

Fluorophores and Their Labeling Procedures for Monitoring Various Biological Signals

Ian S. Harper

1. Introduction

Fluorescence microscopy occupies a unique position in the biological and biomedical sciences where fluorescence probe specificity and sensitivity can provide important information regarding the biochemical, biophysical, and structural status of cells and tissues. The continuing development of fluorescent probes, or fluorophores, in conjunction with the strong emergence over the past two decades of confocal and multiphoton microscopy (and specialized applications such as fluorescene recovery after photobleaching [FRAP], fluorescence resonance energy transfer [FRET], and fluorescence lifetime imaging [FLIM]) has been a major contributor to our understanding of dynamic processes in cells and tissue.

A detailed discussion of all currently available fluorophores and their characteristics is not within the scope of this chapter, and for this the reader is referred to recent reviews (Kasten, 1999; Tsien and Waggoner, 1995) and to the highly recommended *Handbook of Fluorescent Probes and Research Chemicals* (Haugland, 1999). Additional help with selection of specific probes made on the basis of their spectral properties can be found on Bio-Rad's fluorochrome website (http://fluorochrome. biorad.com). Fluorescent probes and reagents are available from Amersham Pharmacia Biotech, Calbiochem, Fluka, Jackson ImmunoResearch Laboratories, Molecular Probes, Polysciences, Serotec, Sigma-Aldrich, and others. Fluorescent proteins (such as green fluorescent protein [GFP]) and GFP vectors are available from Clontech Laboratories, Quantum Biotechnologies, and Life Technologies.

The approach in this chapter is to emphasize the general principles of fluorophore selection and loading for the investigation of various biological signals and to illustrate the use of multiple-labeling procedures through a few specific examples. The most commonly used probes and their spectral characteristics are listed to facilitate choosing dyes for multiple-label experiments (Tables 2–1 through 2–4). The fluorescent labels currently available for protein and antibody conjugation are also included (Table 2–1) for completeness and because it is common for investigators to chemically fix and stain tissue at the end of experiments for subsequent immunofluorescence.

A feature of fluorescence imaging that is frequently exploited is the ability to monitor a number of specific probes simultaneously, thus allowing co-localization or multiparameter imaging of structures within cells or tissues. Naturally, this requires that the probes can be distinctly resolved based on excitation and emission spectra. Moreover, the ability to introduce multiple probes into live specimens presents an opportunity to investigate biological function and activity in a unique and dynamic way. The fluorophores and methods described in this chapter are relevant to both conventional (wide-field) fluorescence microscopy and confocal fluorescence microscopy, and much of this information (with the exception fluorophore excitations) is also applicable to multiphoton microscopy.

TABLE 2–1 Fluorescent probes for conjugation to antibodies and other reagents

Emission Range	Fluorophore	Excitation	Emission
Blue	AMCA (aminomethylcourmarin)	350	450
	Cascade Blue	400	420
Green	Alexa 488	491	515
	Cy2 (cyanine)	489	505
	FITC (fluorescein isothiocyanate)	494	520
	FluoroGreen	490	520
	BODIPY FL	505	513
	Oregon Green 488	496	524
	Oregon Green 514	511	530
Yellow	Eosin	524	544
	Alexa 532	531	554
Orange	BODIPY TMR	543	569
	Alexa 546	553	569
	TRITC (tetrarhodamine isothiocyanate)	554	576
	TMR (tetramethylrhodamine)	555	580
Orange–red	Cy3 (indocarbocyanine)	552	568
	R-phycoerythrin	565	578
	Lissamine rhodamine	570	590
	Alexa 568	573	596
	Cy3.5	578	593
	Alexa 594	585	610
	BODIPY TR	592	618
	Texas Red	595	615
Red–far red	Allophycocyanin	650	671
	Cy5 (indodicarbocyanine)	650	670
	Cy5.5	678	698
	Cy7	747	768

Note: The fluorophores presented in Tables 2–1 through 2–4 should be considered as a guide only and are not comprehensive. Peak excitation and emission wavelengths may vary slightly because the data are derived from multiple sources and vary according to different procedures and apparatuses.

2. Selection of Fluorophores

Because there now exists a wide array of fluorescence probes for most imaging applications, one of the problems may not be to find a probe, but rather to find the optimal probe for a given application. Choice of the most appropriate probe will be determined in many cases by both the instrumentation (most commonly, excitation source and emission filters) and the application. The basic requirements of the ideal probe are that it should (1) specifically associate with the target molecule, structure, or cell; (2) be nontoxic; (3) be very bright (i.e., have a high quantum efficiency); (4) not photobleach; and (5) not dissociate or leach out of the cell. Cost of the fluorophore may also be a consideration, particularly if it is to be used in large amounts as for bulk samples or a long work-up is required to optimize a method.

Possibly the most important parameter for the selection of a label, after target specificity, is that of compatibility with the excitation source and emission filters available on the microscope. Mercury lamps have peak emissions at 366, 405, 436, 546, and 578 nm and a strong background continuum, whereas xenon lamps have a strong flat spectrum from about 350 to 700 nm with several intense peaks in the 700 to 1000 nm range. In contrast, lasers emit monochromatic light of very high intensity, and various lasers are chosen for their specific emission characteristics. The most common lasers for confocal microscopy are generally Ar ion (457, 488, and 514 nm), green HeNe (543 nm), mixed gas ArKr (457, 488, 568, with or without 647 nm), yellow HeNe (594 nm), and red HeNe (633 nm). Either the ArKr or a combination, such as the Ar ion (488 nm) plus green and red HeNe lasers, will provide good sources for triple label imaging because the excitation lines are well separated. For the ultraviolet region, Ar ion lasers provide lines at 351 and 364 nm, Kr ion lasers at 413 nm, and HeCd lasers at 325 and 442 nm.

It is important to choose fluorophores that will be maximally excited by the lamp or laser line, thereby requiring less gain on the detector to achieve a reasonable image and hence less background noise. Narrow-band, high-efficiency interference filters are preferable for filtering the emission and will also mean that there is less resultant bleed-through between channels. Although it is possible to reduce bleed-through by image processing techniques (Carlsson and Mossberg, 1992), this is not advocated. It is more sensible to select fluorophores and systems for optimal excitation, sensitivity, and spectral separation. The past few years have seen the introduction of fluorophores aimed specifically at the laser-based flow cytometry and confocal microscopy markets where multiple labeling of specimens is common. Triple fluorophore labeling with excitation in the blue, green/yellow, and red regions can be accomplished with little overlap between regions. For example, probes like Oregon Green 488, Alexa 568, and Cy5 (Table 2–1) can now be employed with good spectral separation and should be preferred, particularly when multiple labels over the visible light spectrum are required. Figure 2–1 (see also Color Figure 2–1) illustrates Hela cells imaged with an ArKr laser (488, 568, and 647 nm) after the cytoskeleton was labeled with antibodies conjugated to Oregon Green 488 and Cy5; the nucleus is labeled with propidium iodide. There is clear separation of these fluorescence emission spectra that can be checked by single-line excitation. Most of the automated multiuser confocal systems now also

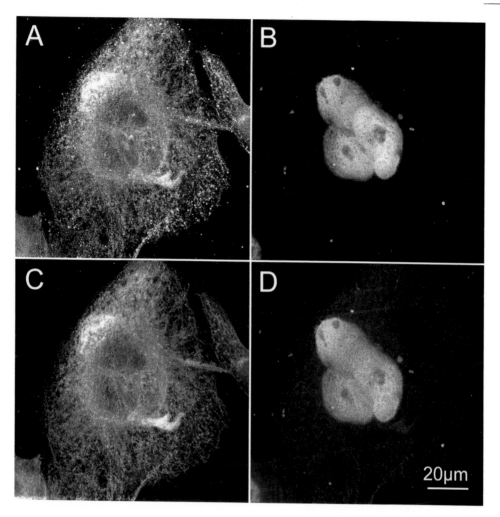

FIGURE 2–1 Triple labeling of Hela cervical carcinoma cells transfected with to overexpress the protein ISG60. *A,* Alexa 488–linked secondary antibody staining (488/530 ± 30 nm excitation/emission). *B,* Propidium iodide (568/590 ± 30 nm ex./em.). *C,* Cy5-linked anti-alpha-tubulin antibody (647/LP680 nm ex./em.). *D,* RGB color overlay produced from placing *A, B,* and *C* into green, red, and blue color planes, respectively. This clearly demonstrates co-localization (in cyan) of the two antibodies. LP, long pass.

provide sequential scanning routines that will scan specific channels individually and then rapidly merge resultant images to provide a double- or triple-channel image (see Chapter 5).

Although many probes are not entirely selective for a specific organelle like $DiOC_6(3)$ for endoplasmic reticulum (ER) (Terasaki and Reese, 1992) or calcein for mitochondria (Minamikawa et al., 1999), careful selection and testing of probe concentrations, loading conditions, and strategies (such as selective quenching after loading) can be effectively used to locate the fluorescence to the area of interest. This is discussed briefly in the following section and in more detail in Chapter 6.

Finally, selection of a probe should also take into account the possible need for localization, even at an ultrastructural level, following chemical fixation. Many probes are now available in a modified form that can be cross-linked by formaldehyde or glutaraldehyde, commonly used in histology and electron microscopy. Modifications include the addition of amino groups (particularly lysine) to dextran and the addition of thiol-reactive chloromethyl groups to a range of probes that are known as Tracker probes (e.g., MitoTracker, CellTracker). Probes like propidium iodide and acridine orange, which intercalate with DNA, may also be retained after fixation. Apart from direct observation of postfixation fluorescence, fluorescence can be tracked after fixation by photoconversion to a diaminobenzidine (DAB) product that is observable with light microscopy and, after further processing, with electron microscopy (Sandell and Masland, 1998). Fluorophores that have been shown to precipitate DAB when irradiated with light at their absorption peak, include acridine orange, DAPI, DiI, DiO, $DiOC_{18}(3)$, and their derivatives eosin, ethidium bromide, fluorescein isothiocyanate (FITC), lucifer yellow, propidium iodide, and tetrarhodamine isothiocyanate (TRITC).

3. LOADING AND LABELING LIVE CELLS

Methods for loading living cells with fluorophores differ depending on the number of parameters, some of which can be modified and manipulated to enhance the amount and specificity of loading or to reduce possible toxicity or photo injury. The exact loading conditions depend on the probe being used and the cellular or molecular target. A recent review by Cullander (1999) discusses probe selection and handling, so the following discussion is restricted to labeling procedures to target specific organelles, cells, or pathways.

In general, lipid-soluble dyes like the carbocyanines (DiI, DiD, and DiO) will associate with cell membranes. They may diffuse through the plasma membrane and stain all interior membranes, or they may remain predominantly in the plasma membrane itself. On the other hand, nonlipophilic and free acid forms of fluorophores like Fura-2, Fluo-3, and BCECF will require special loading procedures such as electroporation, scrape-loading, microinjection, hypotonic shock, or osmotic lysis following pinocytotic loading (Haugland, 1999). One of the most common methods of introducing fluorophores into the cell is through incubation with the acetoxymethyl (AM) ester form of the probe. Although this is one of the more variable procedures, it does provide potentially exploitable options for improving selectivity of fluorophore loading.

The main factors that affect fluorophore loading are the fluorophore concentration, incubation time, temperature, and agitation. In addition, the surface area and mass of material, particularly when larger amounts of tissue are being labeled, should also be taken into consideration because effective dye concentration will become reduced as dye diffuses into the tissue. Labeling individual cells (either in suspension or attached to a coverslip) probably represents the most favorable situation for AM ester loading, as most of the parameters can be controlled with relative ease, and loading strategies can be systematically tested. As tissue samples increase in size, a more complex situation arises because since both dye loading and microscopical resolution may attenuate with increasing distance from the periphery of the tissue. In this case, dye loading through perfusion should be attempted, as is done in whole organs like the heart (Delnido et al., 1998) or liver.

3.1 Loading Acetoxymethyl Ester Forms

Loading AM forms is comparatively simple, relying on the fact that the AM derivatives of fluorophores have ester moieties that mask the carboxy groups of the probes, making the compound permeant. Once inside the cell, endogenous esterases remove the ester groups, thus converting the probe to a nonpermeant salt or acid form that remains trapped within the cytosol. An additional advantage for some probes (e.g., calcein) is that they are not fluorescent in their AM form and hence do not contribute to an extracellular fluorescent signal. Successful loading of the cell with fluorophore depends on a combination of factors that influence the final destination of the probe. Generally, mammalian cells and tissues can be loaded with solutions of about 1–10 μM for 10–30 min at temperatures ranging from ambient to 37°C. Under these conditions, it is estimated that intracellular concentrations of hydrolyzed probe may reach 30–50 μM. Changing these parameters tends to influence the final pattern of fluorophore distribution (see Fig. 2–2), and successful results depend very much on careful evaluation of the contribution of these parameters for each tissue or cell type.

For fluorophores used to report ion fluxes in the cytosol, it is usual to load for a shorter time (i.e. 10–20 min) at 37°C, to ensure that most of the probe is hydrolyzed in the cytosol. Longer periods and/or lower temperatures tend to increase the organelle loading (Fig. 2–2). Even under conditions favoring cytosolic loading, a significant proportion of the fluorophore might gain entry to other organelles (Moore et al., 1990) and be a source of significant error (Scanlon et al., 1987), so careful evaluation of fluorophore distribution is required. Several possibilities exist for evaluating the contribution of fluorescence from noncytosolic compartments, including preferential quenching of cytosolic fluorescence using Mn^{2+} or Co^{2+} (Griffiths et al., 1997; Minamikawa et al., 1999) or differential permeabilization (Lemasters et al., 1994; Chapter 5). An alternative approach is to extract nonspecific fluorescence based on the distribution of a second specific probe. Thus, Fluo-3 loaded intracellularly in a nonspecific fashion can still be used to detect calcium fluxes in mitochondrial and cytosolic compartments based on the distribution of mitochondrially loaded tetramethylrhodamine methyl ester (TMRM) (Chacon et al., 1996; Ohata et al., 1998).

3.2 Leakage and Hydrolysis of AM Acetoxymethyl Esters

Leakage of hydrolyzed AM derivatives is an acknowledged problem, but it is usually encountered only after several hours and is more prevalent in some tissues than in others. Several remedies are used to combat this. First, some probes have been modified by the addition of chloromethyl moieties that covalently bind to intracellular thiols and help retain the probes within the cells. These probes are known as

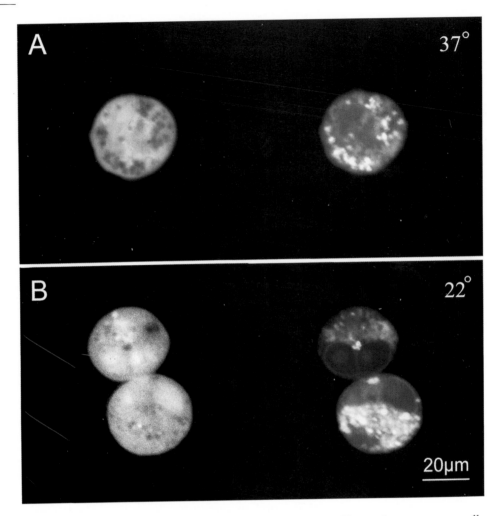

FIGURE 2–2 Effect of temperature on loading AM derivatives. Chinese hamster ovary cells were loaded for 30 min with 1 μM calcein AM (left images) and 500 nM tetramethyl-rhodamine methyl ester (TMRM) (right images) at 37°C or 22°C. At 37°C hydrolysis of the AM proceeds maximally and calcein is retained predominantly in the cytoplasm, with mitochondria (high TMRM fluorescence) excluding calcein (A). At the lower temperature, calcein AM has been able to diffuse into mitochondria and nuclei (B).

CellTracker probes and persist for hours and even days within labeled cells. A second possibility to enhance probe retention in cells is to use, in millimolar amounts, an inhibitor of organic anion transport, such as probenecid or sulfinpyrazone (Bootman et al., 1996; Griffiths et al., 1997). These inhibitors all attenuate, rather than completely prevent, dye leakage, and in our laboratories have also been effectively used to reduce loading variation in chinese hamster ovary cells. Anion transport inhibitors are also effective in reducing probe sequestration in cells (Di Virgilio et al., 1990), which may be a major cause of reduced and punctate loading in some cells.

On the other hand, probe leakage from the cytosol may be exploited to advantage. Trollinger et al. (1997) utilized a regime of cold loading for Rhod-2 (4°C for 1 h) followed by an extended warm incubation period (several hours at 37°C) during which the cytoplasmic concentration of the probe decreased substantially, leaving a predominantly mitochondrial distribution. This facilitated analysis of mitochondrial calcium in a probe, which is not completely selective for mitochondria.

Although most probes that are loaded in the form of an AM ester are also available as carboxylic (free) acids, it is also possible to chemically hydrolyze AM esters in the laboratory using KOH. This is useful for checking hydrolysis rates and to produce free acids in small amounts as needed. To do this, a small amount of the ester is dissolved in dimethylsulfoxide to make a 1 mM stock, diluted to 0.5 mM with methanol, and then hydrolyzed for several hours (preferably overnight) by the addition of a drop of 2M KOH (for a precise method, see http://www.probes.com/media/pis/g002.pdf).

4. Fluorophores for Live Cell Imaging and Signaling

4.1 Cell Viability

Cell viability can be monitored in several ways, but the most common procedure is to monitor probes that cross the cell membrane in injured, permeabilized, or ruptured cells. Probes such as propidium iodide, ethidium homodimer-1, and Sytox Green will only enter the cell after significant membrane damage, at which time they bind to nucleic acid and label the nuclei intensely (see Table 2–3, later). These probes may be added in micromolar amounts to experimental media and buffers to allow continuous monitoring of viability. While leakage of a cytosolic located dye would occur during or after cell death, leakage may also occur in viable cells. Thus, it is common to determine the relative incidence of cell death using a combination of live/dead cell stains, such as calcein and ethidium homodimer-1, Sybr1-14, propidium iodide, or others (Decherchi et al., 1997).

4.2 General Morphology and Structure

Most of the probes used to stain specific organelles can be co-loaded with a general cytoplasmic or membrane probe, such as calcein or $DiOC_6(3)$, which then acts as a more general stain for the whole cell body. These probes can also be used with optical sectioning techniques to reconstruct the entire cell and provide shape and volume information over time (see Chapter 6).

4.3 Specific Organelles

Some permeant probes used for specific organelles are listed in Table 2–2.

4.3.1 Nucleus

There are a number of fluorescent dyes that are membrane permeant and that will localize in the nucleus of living cells. These include DAPI and Hoechst 33258, which are excitable by ultraviolet light, and hexidium iodide. Used at low

TABLE 2–2 Fluorophores used for microscopy of live cells/tissue: Permeant probes for specific organelles

Parameter	Fluorophore	Excitation	Emission
Nucleus	DAPI	358	461
	Hoechst 33258	350	461
	Acridine orange	500	526
	Hexidium iodide	518	600
	Syto dyes	Various	
	7-Aminoactinomycin D	546	647
Mitochondria	MitoFluor Green	489	517
	MitoTracker Green FM	490	516
	MitoTracker Orange CMTMRos	551	576
	MitoTracker Orange CM-H$_2$TM	551	576
	MitoTracker Red CMXRos	578	599
	MitoTracker Red CM-H$_2$XRos	578	599
	Rhodamine-123	507	529
	Rhodamine-6G	528	551
	DiOC$_6$ (3)	484	519
	Rhodamine B hexyl ester	556	578
Mitochondrial membrane potential	TMRE (tetramethylrhodamine methyl ester)	549	574
	TMRM (tetramethylrhodamine ethyl ester)	548	573
	JC-1	510	527, 590
Golgi	NBD C$_6$–ceramide	466	536
	BODIPY FL C$_5$–ceramide	505	512
	BODIPY FL sphingomyelin	505	511
	BODIPY FL Brefeldin A	505	511
	BODIPY FL cerebroside	505	511
Endoplasmic reticulum (ER)	ER-Tracker Blue-White DPX	374	575
	DiOC$_6$ (3)	484	519
	Rhodamine B hexyl ester	556	578
Lysosomes	Fluorescent dextrans	Various	
	LysoTracker Green	504	511
	LysoTracker Yellow	534	551
	LysoTracker Red	577	590
	LysoSensor dyes	Various	
Vacuoles	FM1-43	510	626
	MDY-64 (yeast)	456	505
	Carboxy DCFDA	490	520
Plasma membrane	FM1-43	510	626
	FM4-64	515	640
	RH414	532	716

concentrations, these dyes are nontoxic and can be used to study chromatin dynamics. The Syto dyes (e.g., Syto 11 to 16) stain nucleic acids in general, and, while they will differentiate nuclei, there is also a variable (sometimes high) amount of cytoplasmic staining. Other nuclear probes, such as 7-aminoactinomycin D (7AAD) and acridine orange may also give variable results, but it is usually worth the effort to optimize staining conditions. Although acridine orange binds to both DNA and RNA, it is possible to differentiate between these, as the emission of the

DNA-bound form is predominantly green and there is a significant red shift to 650 nm when it associates with RNA (Kasten, 1999).

Injection of fluorescent-labeled DNA precursors, such as Cy5-dUTP, has been effectively used to image dynamics of chromosome assembly and movement (Manders et al., 1999). Another widely used approach is to target nuclear structures, such as histones or centromeres using GFP–fusion protein constructs (Shelby et al., 1996).

4.3.2. Mitochondria and mitochondrial membrane potential

A fairly wide range of probes is available for localization of mitochondria (Table 2–2), including the short-chained carbocyanine dyes such as $DiOC_6(3)$, DASPMI, rhodamine-6G, and rhodamine-123 and the newer generation MitoTracker and MitoFluor probes. The cationic probes, rhodamine-123 and tetramethylrhodamines (TMRM and TMRE), and MitoFluor all show intense select staining for mitochondria, but are not retained after cell death or fixation. In contrast, the thiol-reactive chloromethyl moieties of the Tracker probes aid their retention in the organelle, even after aldehyde fixation. MitoTracker Green will accumulate in mitochondria, regardless of mitochondrial membrane potential, and is therefore considered to be a good general stain for mitochondria. In contrast, $DiOC_6(3)$ and the other lipophilic carbocyanine dyes stain most intracellular compartments, but selective mitochondrial staining can be achieved at low concentrations in several cell types.

MitoTracker Orange and Red are tetramethylrosamines that are used as potentiometric indicators because they redistribute across the mitochondrial membrane based on membrane potential (Mathur et al. 2000). There is, however, some indication that retention of the Tracker probes occurs largely due to binding of the chloromethyl moieties to SH groups, so the results should be carefully assessed (Scorrano et al., 1999). Another useful potentiometric probe is JC-1, which is a green fluorescent cyanine dye that remains in the monomeric form in low concentrations or at low membrane potential and becomes aggregated in aqueous concentrations higher than 0.1 μM or at higher potentials. The so-called J-aggregate form is red shifted, and hence ratiometric measurement with 490 nm excitation and 530/590 nm emission has been found to be a sensitive indicator of mitochondrial membrane potential (Di Lisa et al., 1995; Reers et al., 1995).

Tetramethylrhodamine methyl and ethyl esters (TMRM and TMRE, respectively) are rapidly taken up by living cells in a manner that depends on membrane potential (Farkass et al., 1989; Ehrenberg et al., 1988) and can be calibrated to show membrane potential in live cells (Chacon et al., 1994). The latter dyes are cationic redistribution dyes with a nernstian accumulation in organelles, and hence the ratio of fluorescence intensities measured in two compartments separated by a membrane can be used to quantitatively monitor mitochondrial membrane potential. We have carried out TMR loading by incubating cells briefly (10–30 min) in about 500 nM TMRM, with 50–150 nM added to all subsequent buffers during subsequent imaging and measurement (Harper et al., 1993; Lemasters et al., 1994). Higher concentrations of TMRM and TMRE are reported to inhibit respiration; therefore, some investigators prefer to use concentrations as low as 30 nM.

Figure 2–3 shows the use of TMRM in tracking mitochondrial membrane potential changes in conjunction with cytosolic Ca^{2+} in a cardiac myocyte experiencing anoxia. A more detailed review of functional imaging of mitochondria is presented in Chapter 6.

Several reduced forms of mitochondrial probes, which require oxidation before becoming fluorescent, are available for probing mitochondrial oxidation

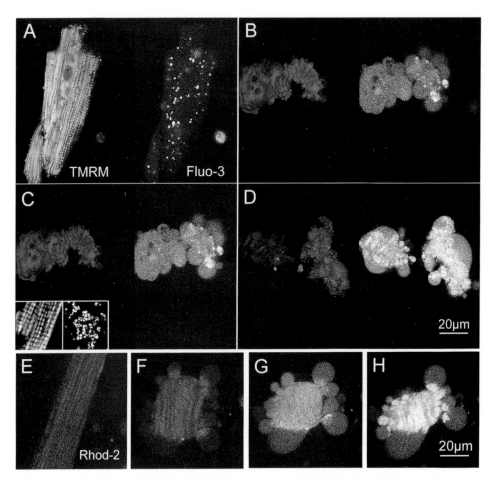

FIGURE 2–3 An example of the use of multiple fluorophores to label cardiac myocytes for investigating physiology during normoxic and anoxic conditions. A–D, Myocytes were labeled with TMRM and Fluo-3 (A) to follow mitochondrial membrane potential and cytosolic calcium, respectively. The cells were then subjected to chemical anoxia (B, 90 min; C, 98 min) followed by 15 min of re-oxygenation (D). Note the punctate fluorescence in A, which is generally undesirable and probably corresponds to autosomal/lysosomal uptake of incompletely solubilized probe. Note also that TMRM allows the inspection of both cell morphology and mitochondrial membrane potential in these particular cells and depicts the loss of mitochondrial membrane potential (decreased TMRM intensity) before a major increase in cytoslic calcium. Inset, Comparison of TMRM-labeled mitochondria imaged in the intact myocyte (left) or after isolation (right). E–H, Rhod-2 AM loading of cardiac myocyte used to follow mitochondrial calcium increase during anoxia E–H, 0, 47, 55, and 60 min chemical anoxia, respectively.

status and the presence of reactive oxygen species. These include dihydrorhodamine-123, dihydrorhodamine-6G, dihydrotetramethylrosamine, MitoTracker Red CM-H$_2$Xros, and MitoTracker Orange CM-H$_2$TMRos (Poot et al., 1996).

4.3.3 Endoplasmic reticulum and Golgi

Most probes suitable for staining of the ER and Golgi are either lipids or chemical inhibitors of protein movement, such as Brefeldin A. Fluorescent ceramide or sphingolipid analogues are most commonly used for live cell imaging (Bai and Pagano, 1997; Pagano and Chen, 1998), whereas the fluorescent-conjugated lectins concanavalin A (Con A) and wheat germ agglutinin (WGA), which bind to the ER and Golgi apparatus, respectively, have been more useful for localization after aldehyde fixation. ER-Tracker Blue-White DPX is a highly selective and photostable stain for the ER in live cells. Unlike the conventional ER stain DiOC$_6$(3), ER-Tracker Blue-White DPX does not stain mitochondria, and staining at low concentrations does not appear to be toxic to cells.

Fluorescent Brefeldin A (either as the green-fluorescent BODIPY FL or the red-orange-fluorescent BODIPY 558/568) has been used to selectively localize ER and Golgi apparatus in several different cell lines (Deng et al., 1995). The probes do not generally give an identical staining pattern, as there is some distortion of intracellular trafficking of protein through the ER and Golgi (Cole et al., 2000). Brefeldin A also alters the morphology of endosomes and lysosomes, so these probes should be used with appropriate controls.

4.4 Ionic Environment

The importance of measuring intracellular ions is reflected in the number of fluorescent probes that are available for the principle ions Ca^{2+}, Na$^+$, and H$^+$ (Table 2–3). Choice of indicators depends on whether intensometric or ratiometric analysis is required and on fluorescence characteristics, binding affinity, or sensitivity range.

Intensometric probes, such as Fluo-3 (see Fig. 2–3), Oregon Green BAPTA-1, Calcium Green, and Sodium Green are single excitation/emission wavelength dyes that provide valuable data regarding relative intensities and that can also be calibrated (Morgan and Thomas, 1999; Simpson, 1999). In contrast, ratiometric indicators may be dual emission (e.g., Indo-1 for calcium, SNARF-1 for pH) or dual excitation (e.g., Fura-2 for calcium, BCECF for pH) and provide the ability to calibrate and exclude artifacts of loading, leaching, photobleaching, and movement. Fluorescent spectra are published for most indicators and are important data to have because the excitation wavelength significantly affects the photochemistry and emission spectra for probes. For example, carboxy SNARF can be excited by 488, 514, 530, or 568 nm light, yielding substantially different spectra (Blank et al., 1992; Chacon et al., 1994; Haugland, 1996).

Although combinations are not widely used, it is also possible to use combinations of two dyes to effect a ratiometric dual emission approach, for example, Fura Red and Fluo-3 (Lipp and Niggli, 1993; Schild et al., 1994). In this instance great care must be taken to ensure that there is similar loading and intracellular distribution of the dye. This can be overcome by use of dextrans double

TABLE 2–3 Fluorophores used for microscopy of live cells/tissue: Cellular activities

Parameter	Fluorophore	Excitation	Emission
Cell morphology			
Shape/volume	Calcein (AM) acetoxymethyl ester	495	520
	$DiOC_6$ (3)	484	501
Cell viability			
Live	Most AM forms		
	FDA, CFDA, DCFDA	490	520
	Calcein AM	495	520
Dead (nuclear)	Sytox Green	504	523
	TO-PRO-1, TOTO-1	513	530
	Ethidium homodimer-1	528	617
	Propidium iodide	530	615
	TO-PRO-3	631	661
Ion physiology			
Ca^{2+}	Indo-1	350	405/485
	Fura-2	335/362	505
	Fluo-3	506	526
	Fluo-4	494	516
	Fura Red	458, 488	597
	Calcium Green	506	526
	Calcium Orange	550	580
	Calcium Crimson	590	620
	Oregon Green 488 BAPTA	488	520
	Rhod-2	553	576
	X-Rhod-1	580	602
pH	BCECF	439, 490	530
	Carboxy SNARF-1	488–568	580/640
	Carboxy SNAFL-2	514	546/630
Na^+	SBFI	340	410/590
	Sodium Green	488	535
K^+	PBFI	340/380	505
Cl^-	SPQ	344	443
	Bis-DMXPQ	365	450/565
Mg^{2+}	Mag-Indo-1	343	405/485
Cyclic adenosine monophosphate (cAMP)	cAMP Fluorosensor	488	520/580
Reactive oxygen species (probes are nonfluorescent until oxidized)			
Dihydrorhodamine-123		507	529
Dihydrorhodamine-6G		528	551
Tetramethylrosamine		578	599
$H_2DCFDA/DCFH$ (2′7′-dichlorofluorescin)		488	530

Note: Fluorophores with two wavelengths for excitation or emission are dual excitation or dual emission ratiometric indicators.

conjugated with Calcium Green-1 and a calcium-insensitive probe, such as Texas Red, provided that the cells or tissues are able to be loaded by microinjection, endocytosis, or scrape loading. It is also possible to carry out simultaneous imaging of Ca^{2+} and pH by co-loading with two indicators and with measurements at multiple wavelengths (Wiegman et al., 1993; Opas, 1997; Martinezzaguilan et al., 1996), facilitating correction of pH-sensitive calcium indicators under conditions in which changes in both pH and calcium occur.

On the other hand, dextrans conjugated to pH indicators like BCECF, SNARF, and SNAFL (pK$_a$ 7.0–7.8 range) can be used to follow uptake into acidic organelles as their fluorescence is either quenched or significantly shifted under pH 5.5. Accurate measurement of pH in the region of 2–3 should be done with dextrans conjugated to probes such as Oregon Green, Rhodol Green, DM-NERF, or Cl-NERF, which have a pK$_a$ of 3.8–5.6.

Several reports are available for further details of methods and protocols for imaging Ca^{2+} (Hayes et al., 1996; Stricker and Whitaker, 1999; Schild 1996), pH (House, 1994; Weinlich et al., 1997; Opas, 1997; Cody et al., 1993), Na^+ (Harootunian et al, 1989; Satoh et al., 1991; Jung et al., 1992), and Cl^-.

Reactive oxygen species (Ros; including singlet oxygen, superoxides, and hydroxy radicals) can be detected using a number of probes (Tables 2–2 and 2–3) that react with Ros and in the process are converted from a nonfluorescent form (dichlorofluorescin, DCFH) to a fluorescent form (dichlorofluorescein, DCF) by oxidation. Recently DCFH has been used as a marker for oxidative stress in cardiac muscle (Swift and Sarvazyan, 2000). Reduced forms of mitochondrial probes, such as dihydrorhodamine-123 and MitoTracker Red CM-H$_2$XRos are also available for Ros detection.

4.5 Tracers for Membranes and Cells

A number of lipophillic fluorophores are available to facilitate probing of membrane and lipid dynamics, endocytosis and intracellular lipid trafficking, and cell fusions (Table 2–4). These fluorophores can also be used to examine neuronal development and morphology and to track cells.

Fluorescently conjugated particles, bacteria (live or dead), latex spheres, and dextrans have all been used to study endocytotic uptake and intracellular movement. FM1-43 and similar styryl dyes have proved to be useful probes for membrane trafficking because they reversibly stain membranes, are impermeable to membranes, and are more fluorescent when bound to membranes than when in solution. FM1-43 has been used to study endocytosis and trafficking and recycling in synaptic nerve terminals (Cousin and Robinson, 1999). Selective quenching of background (non-endocytosed) membrane signals can be achieved by bathing in another fluorophore, sulforhodamine (Pyle et al., 1999). Similarly, trypan blue can be used to quench extracellular fluorescence after internalization of FITC–dextran. Ligands can also be labeled and then used to study receptor location, density, internalization, and trafficking. This has been effectively done using fluoresceinylated or BOPIDY-labeled neurotensin, somatostatin, and mu- and delta-selective opioid peptides in live cells and tissue slices (Beaudet et al., 1998).

TABLE 2–4 Fluorophores used for tracking live cells or cellular processes

Parameter	Fluorophore	Excitation	Emission
Tracking cells	CellTracker Blue (CMHC)	372	470
	Lucifer yellow (and derivatives)	428	533
	CellTracker Green (CMFDA)	492	516
	CFDA succinimidyl ester (CFSE)	494	525
	CellTracker Yellow-Green (CMEDA)	524	544
	CellTracker Orange (CMTMR)	541	565
	CellTracker CM-DiI	553	570
	MitoTracker probes		
Lipophilic tracers	CellTracker CM-DiI	553	570
	DiI	549	565
	DiO	484	501
	DiD	644	655
	PKH26	551	567
	PKH2	490	504
Endocytosis	Fluorescent bacteria, spores, dextrans or beads		
	FM1-43	510	626
	FM4-64	515	640
	$PKH_2 6$	551	567
	PKH_2	490	504
Receptor and peptide distribution and trafficking	Labeled Fluo peptides		
	BODIPY 576/589 Neurotensin	576	589
	Fluo-NPY (neuropeptide Y), Fluo-GHRH (growth hormone-releasing hormone), Fluo-glucagon, etc.	494	520
Fluorescent proteins	Wt Gfp	395 (470)	509 (540)
	GFPuv	395	509
	EBFP	380	440
	SgBFP	387	450
	ECFP	434	477
	SgGFP	474	509
	S65T	488	507
	EGFP	488	507
	EYFP	514	527
	DsRed	558	583

Pagano and coworkers have pioneered the use of lipid analogues to study membrane dynamics and trafficking in living cells (see Pagano and Chen, 1998). NBD and BODIPY-conjugated lipids and lipid analogues have proved powerful in tracking the intracellular distribution and movement of lipids. Also, because emission characteristics change with fluorophore–lipid packing, these lipids may be used to indicate selective internalization and possibly lipid sorting (Chen et al., 1997).

Probes for tracking cells may be cell impermeant, binding irreversibly to the membrane (PKH2, PH26, CM-DiI); cell impermeant, which requires loading by physical means like microinjection (fluorescent-labeled dextrans, lucifer yellow); or membrane permeant (chloromethyl CellTracker probes). Lucifer yellow has long been a favorite probe for studying neuronal morphology and pathways (Belichenko and Dahlstrom, 1995) because it contains a carbohydrazide group that is linked to intracellular molecules during aldehyde fixation. Long-chain carbo-cyanines, such as DiI, DiO, and DiD, have similarly been used to trace cells, cell contacts, and membrane cell continuity, even though they are not cell imperme-ant. These dyes diffuse readily through the cell membrane, but are hydrophobic (Vercelli et al., 2000). CM-DiI has been used as a marker in lymphocyte migration and can be recovered after 40 h (Andrade et al., 1996). It is better retained than DiI after fixation, and, because the probe has increased solubility in aqueous solu-tions, it is an attractive choice for labeling cells in solution.

CellTracker Green (CMFDA) on the other hand, labels the cytosol and has been used to label mast cells and then track them from the blood circulation to the brain (Silverman et al., 2000). We have used CMFDA to label rat platelets, which are then reintroduced to the circulation of an animal with carotid artery injury. Fixation and dissection of the vessel allows subsequent analysis of morphology and size of thrombus formation (Color Fig. 2–4), which, because of retention of the fluorophore in the fixed tissue, can be carried out after a few days in storage.

PKH2 and PKH26, which bind irreversibly to the membrane, have been used extensively for tracking cells up to 60 days with subsequent analysis by flow cytometry (Beavis and Pennline, 1994) and histology (Ford et al., 1996). It is impor-tant, however, to distinguish between fluorophores that are used to track cells by recovering labeled cells at an end point and those that allow repeated or continu-al imaging in live cells. Thus, although PKH2 and PKH26 are suitable for recov-ering cells after several days, repetitive imaging over 5 min is sufficient to cause phototoxicity (Oh et al., 1999). Thus, it is advisable to include appropriate controls and probes to monitor cell viability when dynamic imaging is required.

4.6 Fluorescent Proteins

Naturally fluorescent proteins, such as the GFP family, are now a proven and powerful tool for molecular and cellular biology. The GFP sequenced gene can be used as a marker for gene expression in many cell types, and new variants with specific ionic sensitivities or constructs show potential for use as probes for Ca^{2+}, pH, or cAMP.

Green fluorescent protein is an extremely stable protein of 238 amino acids found naturally in photogenic cells of the jellyfish *Aquorea victoria*. The intrinsic fluorescence of the protein is due to a unique chromophore that is formed post-translationally in the protein by cyclization and oxidation of residues 65–67, Ser-Tyr-Gly (Tsien, 1998). Green fluorescent protein fluorescence, which depends on oxidation of the chromophore, has major and minor excitation peaks at 395 and 474 nm, emits green light with a peak at 509 nm, and persists after formaldehyde-or ethanol-based fixativation.

Using recombinant DNA technology, the coding sequence for GFP can be spliced with that of other proteins to create fluorescent fusion proteins. These fusion proteins can be expressed in vivo and their distribution and movement studied. When GFP expression is placed under the control of a specific promoter or DNA regulatory sequence, the fusion protein becomes a reporter of transcriptional activity.

Wild-type GFP is not optimal for some reporter gene applications, as it has a low extinction coefficient and low expression level, and so several optimized variants have been developed and are now available commercially. These include SuperGlow (sg) GFP, sgBFP, enhanced GFP (EGFP), blue EBFP, and yellow EYFP (Table 2–4). Recently, a red fluorescent protein (DsRed) was isolated from *Discosoma striata* (a relative of the IndoPacific sea anenome), thus extending the range of fluorescent proteins for use in coexpression or FRET studies. These variants have a significantly increased extinction coefficient, making them 4–35 times brighter. Also, various mutations of their gene sequence with codon alterations improve expression in mammalian or other cells. Using specific fusion vectors, the probe can be targeted to specific organelles, cells, or tissues in vivo or in vitro, which is useful for dynamic imaging of cellular structure.

Apart from their use in direct imaging methods, GFPs are also common in applications exploiting the resonance energy transfer (FRET) between two GFPs of different colors (see Chapter 17). When two fluorophores are in close proximity (<100 Å apart) and when the emission of one (the donor) overlaps with the excitation of a second (the acceptor), donor emission may directly excite the acceptor. FRET is thus an indicator of fluorophore proximity at a molecular level and can be used to study the spatiotemporal dynamics of proteins in living cells, such as receptor–ligand interactions. Fluorescent protein pairs suitable for FRET are GFP–BFP and ECFP–EYFP and of course between these and other fluorophores. Further details outlining FRET principles and use of GFP are given in Chapter 17.

REFERENCES

Andrade, W., T. J. Seabrook, M. G. Johnston, and J. B. Hay. The use of the lipophilic fluorochrome CM-DiI for tracking the migration of lymphocytes. *J. Immunol. Methods.* 194: 181–189, 1996.

Bai, J., and R. E. Pagano. Measurement of spontaneous transfer and transbilayer movement of BODIPY-labeled lipids in lipid vesicles. *Biochemistry* 36:8840–8848, 1997.

Beaudet, A., D. Nouel, T. Stroh, F. Vandenbulcke, C. Dal-Farra, and J. P. Vincent. Fluorescent ligands for studying neuropeptide receptors by confocal microscopy. *Braz. J. Med. Biol. Res.* 31:1479–1489, 1998.

Beavis, A. J., and K. J. Pennline. Tracking of murine spleen cells in vivo—detection of PKH26-labeled cells in the pancreas of non-obese diabetic (NOD) mice. *J. Immunol. Methods* 170:57–65, 1994.

Belichenko, P. V., and A. Dahlstrom. Confocal laser scanning microscopy and 3-D reconstructions of neuronal structures in human brain cortex. *Neuroimage* 2:201–207, 1995.

Blank, P. S., H. S. Silverman, O. Y. Chung, B. A. Hogue, M. D. Stern, R. G. Hansford, E. G. Lakatta, and M. C. Capogrossi. Cytosolic pH measurements in single cardiac myocytes using carboxy-seminaphthorhodafluor-1. *Am. J. Physiol.* 263:H276–H284, 1992.

Bootman, M. D., K. W. Young, J. M. Young, R. B. Moreton, and M. J. Berridge. Extracellular calcium concentration controls the frequency of intracellular calcium spiking independently of inositol 1,4,5-triphsophate production in HeLa cells. *Biochem. J.* 314:347–354, 1996.

Carlsson, K., and K. Mossberg. Reduction of cross-talk between fluorescent labels in scanning laser microscopy. *J. Microsc.* 167:23–37, 1992.

Chacon, E., H. Ohata, I. S. Harper, D. R. Trollinger, B. Herman, and J. J. Lemasters. Mitochondrial free calcium transients during excitation–contraction coupling in rabbit cardiac myocytes. *FEBS Lett.* 382:31–36, 1996.

Chacon, E., J. M. Reece, A.-L. Nieminen, G. Zabhrebelski, B. Herman, and J. J. Lemasters. Distribution of electrical potential, pH, free Ca^{2+}, and volume inside cultured adult rabbit myocytes during chemical hypoxia: A multi-parameter digitized confocal microscopic study. *Biophys. J.* 66:942–952, 1994.

Chen, C. S., O. C. Martin, and R. E. Pagano. Changes in the spectral properties of a plasma membrane analogue during the first seconds of endocytosis in living cells. *Biophys. J.* 72:37–50, 1997.

Cody, S. H., P. N. Dubbin, A. D. Beischer, N. D. Duncan, J. S. Hill, A. H. Kaye, and D. A. Williams. Intracellular pH mapping with SNARF-1 and confocal microscopy. 1. A quantitative technique for living tissues and isolated cells. *Micron* 24:573–580, 1993.

Cole, L., D. Davies, G. J. Hyde, and A. E. Ashford. ER-tracker dye and BODIPY-brefeldin A differentiate the endoplasmic reticulum and Golgi bodies from the tubular–vacuole system in living hyphae of *Pisolithus tinctorius. J. Microsc.* 197:239–249, 2000.

Cousin, M. A., and P. J. Robinson. Mechanisms of synaptic vesicle recycling illuminated by fluorescent dyes. *J. Neurochem.* 73:2227–2239, 1999.

Cullander, C. Fluorescent probes for confocal microscopy. In: *Methods in Molecular Biology,* edited by S. W. Paddock. Totowa, NJ: Humana Press, 1999, pp. 59–73.

Decherchi, P., P. Cochard, and P. Gauthier. Dual staining assessment of Schwann cell viability within whole peripheral nerves using calcein-AM and ethidium homodimer. *J. Neurosci. Methods* 71:205–213, 1997.

Delnido, P. J., P. Glynn, P. Buenaventura, G. Salama, and A. P. Koretsky. Fluorescence measurement of calcium transients in perfused rabbit heart using Rhod-2. *Am. J. Physiol.* 43:H728–H741, 1998.

Deng, Y. P., J. R. Bennink, H. C. Kang, R. P. Haugland, and J. W. Yewdell. Fluorescent conjugates of brefeldin A selectively stain the endoplasmic reticulum and Golgi complex of living cells. *J. Histochem. Cytochem.* 43:907–915, 1995.

Di Lisa, F., P. S. Blank, R. Colonna, G. Gambassi, H. S. Silverman, M. D. Stern, and R. G. Hansford. Mitochondrial membrane potential in single living adult rat cardiac myocytes exposed to anoxia or metabolic inhibition. *J. Physiol. Lond.* 486:1–13, 1995.

Di Virgilio, F., T. H. Steinberg, and S. C. Silverstein. Inhibition of Fura-2 sequestration and secretion with organic anion transport blockers. *Cell Calcium* 11:57–62, 1990.

Ehrenberg, B., V. Montana, M. D. Wei, J. P. Wuskell, and L. M. Loew. Membrane potential can be determined in individual cells from the nernstian distribution of cationic dyes. *Biophys. J.* 53:785–794, 1988.

Farkas, D. L., M. Wei, P. Febbroriello, J. H. Carson, and L. M. Loew. Simultaneous imaging of cell and mitochondrial membrane potential. *Biophys. J.* 56:1053–1069, 1989.

Ford, J. W., T. H. Welling, J. C. Stanley, and L. M. Messina. PKH26 and I-125-PKH95—Characterization and efficacy as labels for in vitro and in vivo endothelial cell localization and tracking. *J. Surg. Res.* 62:23–28, 1996.

Griffiths, E. J., M. D. Stern, and H. S. Silverman. Measurement of mitochondrial calcium in single living cardiomyocytes by selective removal of cytosolic Indo 1. *Am. J. Physiol.* 273:C37–C44, 1997.

Harootunian, A. T., J. P. Y. Kao, B. K. Eckert, and R. Y. Tsien. Fluorescence ratio imaging of cytosolic free Na in individual fibroblasts and lymphocytes. *J. Biol. Chem.* 264:19458–19467, 1989.

Harper, I. S., J. M. Bond, E. Chacon, J. M. Reece, B. Herman, and J. J. Lemasters. Inhibition of Na^+/H^+ exchange preserves viability, restores mechanical function, and prevents the pH paradox in reperfusion injury to rat neonatal myocytes. *Basic Res. Cardiol.* 88:430–442, 1993.

Haugland, R. P. *Handbook of Fluorescent Probes and Research Chemicals,* 7th Ed. Eugene: Molecular Probes, Inc., 1999.

Hayes, A., S. H. Cody, and D. A. Williams. Measuring cytosolic Ca^{2+} in cells with fluorescent probes: An aid to understanding cell pathophysiology. *Int. Rev. Exp. Pathol.* 36:197–212, 1996.

House, C. R. Confocal ratio-imaging of intracellular pH in unfertilized mouse oocytes. *Zygote* 2:37–45, 1994.

Jung, D. W., L. M. Apel, and G. P. Brierley. Transmembrane gradients of free Na^+ in isolated heart mitochondria estimated using a fluorescent probe. *Am. J. Physiol.* 262:C1047–C1055, 1992.

Kasten, F. H. Introduction to fluorescent probes: Properties, history and applications. In: *Fluorescent and Luminescent Probes for Biological Activity: A Practical Guide to Technology for Quantitative Real-Time Analysis,* edited by W. T. Mason. Cambridge: Academic Press, 1999, pp. 17–39.

Lemasters, J. J., A. L. Nieminen, E. Chacon, J. M. Bond, I. Harper, J. M. Reece, and B. Herman. Single cell microscopic techniques for studying toxic injury. *Methods Toxicol.* 1:238–455, 1994.

Lipp, P., and E. Niggli. Ratiometric confocal Ca^{2+}-measurements with visible wavelength indicators in isolated cardiac myocytes. *Cell Calcium* 14:359–372, 1993.

Manders, E. M. M., H. Kimura, and P. R. Cook. Direct imaging of DNA in living cells reveals the dynamics of chromosome formation. *J. Cell Biol.* 144:813–821, 1999.

Martinezzaguilan, R., M. W. Gurule, and R. M. Lynch. Simultaneous measurement of intracellular pH and Ca^{2+} in insulin-secreting cells by spectral imaging microscopy. *Am. J. Physiol.* 39:C1438–C1446, 1996.

Mathur, A., Y. Hong, B. K. Kemp, A. A. Barrientos, and J. D. Erusalimsky. Evaluation of fluorescent dyes for the detection of mitochondrial membrane potential changes in cultured cardiomyocytes. *Cardiovasc. Res.* 46:126–138, 2000.

Minamikawa, T., A. Sriratana, D. A. Williams, D. N. Bowser, J. S. Hill, and P. Nagley. Chloromethyl-X-rosamine (MitoTracker Red) photosensitizes mitochondria and induces apoptosis in intact human cells. *J. Cell Sci.* 112:2419–2430, 1999.

Moore, E. D. W., P. L. Becker, K. E. Fogarty, D. A. Williams, and F. S. Fay. Ca^{2+} imaging in single living cells: Theoretical and practical issues. *Cell Calcium* 11:157–179, 1990.

Morgan, A. J., and A. P. Thomas. Single cell and subcellular measurement of intracellular Ca^{2+} concentration ($[Ca^{2+}]_{(i)}$). *Calcium Signal. Protocol.* 114:93–123, 1999.

Oh, D. J., G. M. Lee, K. Francis, and B. O. Palsson. Phototoxicity of the fluorescent membrane dyes PKH2 and PKH26 on the human hematopoietic KG1a progenitor cell line. *Cytometry* 36:312–318, 1999.

Ohata, H., E. Chacon, S. A. Tesfai, I. S. Harper, B. Herman, and J. J. Lemasters. Mitochondrial Ca^{2+} transients in cardiac myocytes during the excitation–contraction cycle: Effects of pacing and hormonal stimulation. *J. Bioenerg. Biomembr.* 30:207–222, 1998.

Opas, M. Measurement of intracellular pH and pCa with a confocal microscope. *Trends Cell Biol.* 7:75–80, 1997.

Pagano, R. E., and C. S. Chen. Use of BODIPY-labeled sphingolipids to study membrane traffic along the endocytic pathway. *Ann. N. Y. Acad. Sci.* 845:152–160, 1998.

Poot, M., Y. Z. Zhang, J. A. Kramer, K. S. Wells, L. Jones, D. K. Hanzel, A. G. Lugade, V. L. Singer, and R. P. Haugland. Analysis of mitochondrial morphology and function with novel fixable fluorescent stains. *J. Histochem. Cytochem.* 44:1363–1372, 1996.

Pyle, J. L., E. T. Kavalali, S. Choi, and R. W. Tsien. Visualization of synaptic activity in hippocampal slices with FM1-43 enabled by fluorescence quenching. *Neuron* 24:803–808, 1999.

Reers, M., S. T. Smiley, C. Mottolahartshorn, A. Chen, M. Lin, and L. B. Chen. Mitochondrial membrane potential monitored by JC-1 dye. *Methods Enzymol.* 260:406–417, 1995.

Sandell, J. H., and R. H. Masland. Photoconversion of some fluorescent markers to a diaminobenzidine product. *J. Histochem. Cytochem.* 36:555–559, 1988.

Satoh, H., H. Hayashi, N. Noda, H. Terada, A. Kobayashi, Y. Yamashita, T. Kawai, M. Hirano, and N. Yamazaki. Quantification of intracellular free sodium ions by using a new fluorescent indicator, sodium-binding benzofuran isophthalate in guinea pig myocytes. *Biochem. Biophys. Res. Commun.* 175:611–616, 1991.

Scanlon, M., D. A. Williams, and F. S. Fay. A Ca^{2+}-insensitive form of Fura-2 associated with polymorphonuclear leukocytes. Assessment and accurate Ca^{2+} measurement. *J. Biol. Chem.* 262:6308–6312, 1987.

Schild, D. Laser scanning microscopy and calcium imaging. *Cell Calcium* 19:281–296, 1996.

Schild, D., A. Jung, and H. A. Schultens. Localization of calcium entry through calcium channels in olfactory receptor neurones using a laser scanning microscope and the calcium indicator dyes Fluo-3 and Fura-Red. *Cell Calcium* 15:341–348, 1994.

Scorrano, L., V. Petronilli, R. Colonna, F. Di Lisa, and P. Bernardi. Chloromethy-ltetramethylrosamine (Mitotracker Orange) induces the mitochondrial permeability transition and inhibits respiratory complex I. Implications for the mechanism of cytochrome c release. *J. Biol. Chem.* 274:24657–24663, 1999.

Shelby, R. D., K. M. Hahn, and K. F. Sullivan. Dynamic elastic behavior of alpha-satellite DNA domains visualized in situ in living human cells. *J. Cell Biol.* 135:545–557, 1996.

Silverman, A. J., A. K. Sutherland, M. Wilhelm, and R. Silver. Mast cells migrate from blood to brain. *J. Neurosci.* 20:401–408, 2000.

Simpson, A. W. M. Fluorescent measurement of $[Ca^{2+}]_{(c)}$—Basic practical considerations. *Calcium Signal. Protocol.* 114:3–30, 1999.

Stricker, S. A., and M. Whitaker. Confocal laser scanning microscopy of calcium dynamics in living cells. *Microsc. Res. Tech.* 46:356–369, 1999.

Swift, L. M., and N Sarvazyan. Localization of dichlorofluorescin in cardiac myocytes: Implications for assessment of oxidative stress. *Am. J. Physiol.* 278:H982–H990, 2000.

Terasaki, M., and T. S. Reese. Characterization of endoplasmic reticulum by co-localization of BiP and dicarbocyanine dyes. *J. Cell Sci.* 101:315–322, 1992.

Tsien, R. Y. The green fluorescent protein. *Annu. Rev. Biochem.* 67:509–544, 1998.

Tsien, R. Y., and A. Waggoner. Fluorophores for confocal microscopy: Photophysics and photochemistry. In: *Handbook of Biological Confocal Microscopy,* 2nd Ed., edited by J. B. Pawley, 1995, pp. 267–279.

Vercelli, A., M. Repici, D. Garbossa, and A. Grimaldi. Recent techniques for tracing pathways in the central nervous system of developing and adult mammals. *Brain Res. Bull.* 51:11–28, 2000.

Weinlich, M., U. Heydasch, M. Starlinger, and R. K. Kinne. Intracellular pH-measurements in rat duodenal mucosa in vitro using confocal laserscan microscopy. *Z. Gastroenterol.* 35:263–270, 1997.

Wiegmann, T. B., L. W. Welling, D. M. Beatty, D. E. Howard, S. Vamos, and S. J. Morris. Simultaneous imaging of intracellular $[Ca^{2+}]$ and pH in single MDCK and glomerular epithelial cells—Special communication. *Am. J. Physiol.* 265:C1184–C1190, 1993.

3

Detectors for Fluorescence Microscopy

Kenneth R. Spring

1. Introduction

The field of fluorescence microscopy has developed from one dependent on film to one in which electronic images are the typical output. One of the critical components in fluorescence microscopy is the imaging device because it determines whether the fluorescence may be detected, the relevant structures resolved, or the dynamics of a process visualized (Aikens et al., 1988; Becker, 1996; Bookman, 1990; Bright and Taylor, 1986; Herman, 1998; Periasamy and Herman, 1994). The range of light detection methods and the wide variety of imaging devices currently available make the selection process a difficult and confusing one. This chapter is intended to aid the microscopist in understanding the basics of light detection and the process of selecting a suitable detector for an application.

2. Parameters for Characterizing Detectors

Electronic imaging sensor performance may be described by spectral sensitivity, quantum efficiency, spatial resolution, uniformity, signal-to-noise ratio, dynamic range, and response speed (Spring and Smith, 1987; Spring and Lowy, 1988; Tsay et al., 1990). Spectral sensitivity is the detector signal as a function of the wavelength of the incident light. It is often expressed in terms of the quantum efficiency (QE), which is the percentage of incident photons that are detected.

Limiting spatial resolution is commonly determined from the minimum separation required for discrimination between two high-contrast objects (e.g., white points or lines on a black background). Contrast is an important factor in resolution, as high-contrast objects (e.g., black and white lines) are more easily resolved than low-contrast objects (e.g., adjacent gray lines).

More informative measures of the spatial resolution of an electronic detector are the modulation transfer function (MTF) and the contrast transfer function (CTF). Both show the magnitude of the detector response as a function of spatial frequency. The CTF is determined from the detector response to a series of black and white bars that become narrower and closer together. Each pair of bars is essentially a square wave with 100% contrast. The MTF is similarly determined from the image of 60% contrast sine waves of increasing spatial frequency.

The uniformity of electronic detectors depicts several variables: gain variations across the sensor, regional differences in noise, regional differences in the efficiency of sampling (often termed *shading*), and spatial variation in the efficiency of light collection or transmission (Inoué and Spring, 1997; Shotton, 1993; Sluder and Wolf, 1998).

Electronic detectors are often compared by their signal-to-noise (SN) ratio, a measure of the variation of a signal that indicates the confidence with which the magnitude of the signal can be estimated. Light has an inherent noise component arising from the stochastic nature of the photon flux that is equal to the square root of the signal. Noise also derives from a variety of other sources and in electronic devices can often be reduced by lowering the operating temperature. The thermally sensitive noise arising in electronic devices in the absence of light is termed *dark noise*.

Intrascene dynamic range is derived from the maximum and minimum intensities that can be simultaneously detected in the same field of view. Interscene dynamic range is the range of intensities that can be accommodated when the detector gain, integration time, lens aperture, or other variables are adjusted for differing fields of view. Dynamic range and S/N ratio should not be confused. Dynamic range is often calculated to be the maximum signal that can be accumulated divided by the noise associated with the reading of that signal.

The response speed of an electronic detector is described by its lag. Lag is the fraction of the previous image that carries over into the next image after a prescribed time interval.

3. Electronic Detection of Light

3.1 Detectors with No Spatial Discrimination

Light detectors without spatial discrimination are exemplified by the photomultiplier tube and photodiode. Both devices employ a photosensitive surface that captures incident photons and generates electronic charges that are sensed and amplified.

3.2 Photomultiplier Tube

Photomultiplier tubes (PMTs) are widely used in confocal microscopes, as well as in spectrometers. Photons impinge on a photocathode and liberate electrons that are accelerated toward an electron multiplier composed of a series of curved plates, known as *dynodes*. The output from the dynode chain is a current proportional to the number of photons striking the photocathode and to the voltage drops along the dynodes. Spectral sensitivity depends on the chemical composition of the photocathode. The best devices often incorporate gallium-arsenide and are sensitive from 300 to 800 nm (Fig. 3–1). The peak QE of PMTs has recently

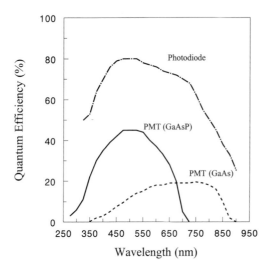

FIGURE 3–1 Spectral sensitivity curves for three electronic detectors showing quantum efficiency as a function of the illumination wavelength. The solid curve shows the response of a photomultiplier tube (PMT) with a photocathode composed of gallium–arsenide–phosphide (GaAsP); the dashed curve shows a similar device with a gallium–arsenide photocathode; the dash–dot curve shows the response of a silicon photodiode with an ultraviolet–transparent window.

been improved to about 40% (GaAsP curve in Fig. 3–1). Photomultiplier tube photocathodes are not uniformly sensitive, and typically the photons are spread over the entire entrance window rather than over one region. Because PMTs do not store charge and respond to changes in input light fluxes within a few nanoseconds, they can be used for the detection and recordings of extremely fast events. These devices also have a huge dynamic range (10^5) over which electrical current output still accurately reflects the photon flux. Finally, the S/N ratio is very high in scientific-grade PMTs because the dark current is extremely low (it can be further reduced by cooling) and the gain may be greater than 10^6.

3.3 Photodiodes

Silicon photodiodes also respond rapidly to light by the generation of a current, but do so without the gain that comes from the electron multiplication of the PMT. Photodiodes have a relatively flat response over the entire visible spectrum (Fig. 3–1) with high QE (80%–90%). The uniformity of the photosensitive surface is excellent, and the dynamic range and response speeds of these devices are among the highest of any light detector. However, they have noise, which is mostly thermal, and consequently exhibit relatively poor S/N ratios under photon-limited conditions.

Photodiodes that incorporate limited gain have been developed (i.e., avalanche photodiodes) and have been utilized in some confocal and wide-field fluorescence microscopes. Although they have up to a 300-fold gain, they exhibit significant dark noise even when cooled to 0°C.

3.4 Area Detectors

3.4.1 Tube-type detectors

Tube-type detectors are exemplified by the vidicon tube camera, in which the photosensitive surface is "read out" by a scanning electron beam. In vidicons, the photosensitive surface stores charge rather than liberating electrons as in a photocathode. Photons captured by the photosensor alter its electrical resistance at their site of impact, and the current of the scanning beam flows more readily through these sites, generating a signal. Because vidicon tube sensors have been supplanted by solid-state detectors and are largely of historical significance, they are not considered in detail here (see references for additional information).

3.4.2 Solid-state detectors

Solid-state detectors consist of a dense matrix of photodiodes incorporating charge storage regions. Several variations on the basic model are commercially available, including the charge-coupled device (CCD), the charge-injection device (CID), and the complimentary metal-oxide semiconductor (CMOS) detector. In each of these detectors, a silicon diode photosensor (often denoted as a *pixel*) is coupled to a charge storage region that is, in turn, connected to an amplifier that reads out the quantity of accumulated charge. In the CID and CMOS detectors, each individual photosensor has an amplifier associated with it, and the signal is output in parallel. Although techniques for charge storage by silicon photodetectors were known for many years before the development of the CCD, a suitable mechanism for the systematic read out of this stored charge needed to be devised before the device became a reality. In a CCD, there is typically only one amplifier at the corner of the entire array, and the stored charge is sequentially transferred to a serial register and then to an output node adjacent to the read-out amplifier.

Because the CCD is presently the most widely used detector for fluorescence microscopy, its performance is considered in detail. Distinctions will be made, where appropriate, between the two classes of CCD cameras: consumer grade and scientific grade. It is important to point out that, although all electronic detectors are analog devices that generate electrical currents or charges, cameras with an internal digitizer have been recently denoted as digital cameras because they do not have an analog signal output.

Some CCD cameras used in scientific applications are operated at room temperature, while others are cooled. A 20°C decrease reduces the dark current of the CCD tenfold. Because the charge storage wells do not fill with thermally generated dark noise during the integration period, longer exposures are possible. Cooled cameras for scientific use may be designated for slow-scan use because their frame rate is less than that of a standard video camera.

A video-rate camera in the United States reads the stored charge and outputs a video field every 16.7 ms to conform to the recommended standards (denoted RS-170 or RS-330), in which 30 video frames are produced per second, with each frame consisting of two interlaced fields (the European standard format requires 50 fields per second with a field every 20 ms). Video exploits the lag in our visual system by generating images at a rate faster than the critical flicker frequency. Each video field contains 50% of the information in an entire frame. Each of the video fields is obtained in sequence to result in a 16.7 ms time difference between each odd or even scan line in the complete image. If the output of a video-rate camera is stopped and light is allowed to fall on the CCD for a prolonged period, the first two video fields produced contain all of the information accumulated during the integration period.

Interlacing the two video fields to produce a complete video frame was a clever solution to the engineering problem resulting from the bandwidth limitations of the electronics and signal transmission and reception components available at the time of the development of television. Today much higher frequency amplifiers and associated electronics allow the production, storage, and subsequent displays at frame rates of up to 1000/s without interlace. Such progressive-scan cameras produce a continuous scan from the top to the bottom of the image. This does not mean that the top lines were obtained before the bottom ones; rather, the devices first integrate the photon flux over the entire sensor and then rapidly displace the accumulated charge to a charge storage and transfer region that is protected from further illumination.

Two CCD designs are commonly used to achieve such rapid transfers: the interline-transfer CCD and the frame-transfer CCD. The interline-transfer CCD incorporates charge transfer channels immediately adjacent to each photodiode so that the accumulated charge can be efficiently and rapidly shifted into them. Interline-transfer CCDs can be electronically shuttered by altering the voltages at the photodiode so that the generated charges are injected into the substrate rather than being shifted to the transfer channels.

The frame-transfer CCD uses a two-part sensor in which the top half is covered by a light-tight mask and is used as a storage region. Light is allowed to fall on the uncovered portion of the sensor, and the accumulated charge is then rapidly shifted into the masked storage region. While the signal is being integrated on the light-sensitive portion of the sensor, the stored charge is read out. A disadvantage of this design is charge smearing during the transfer from the light-sensitive to the masked regions of the CCD, but this can often be compensated for.

The spectral sensitivity of the CCD differs from that of a simple silicon photodiode detector because the CCD surface has channels used for charge transfer. These structures absorb the shorter wavelengths and reduce the blue sensitivity of the device. A typical spectral sensitivity curve for a consumer or scientific-grade CCD is shown by the solid line in Figure 3–2, where it should be noted that the peak QE of 40% is markedly below that of an individual silicon photodiode. Recently, the transparency of the channels has been increased with substantial improvement in blue-green sensitivity of some scientific-grade CCDs (Blue Plus curve in Fig. 3–2). The losses due to the channels are completely eliminated in the back-illuminated CCD. In this design, light falls onto the back of the CCD in a region that has been thinned by etching until it is transparent. The resultant spectral sensitivity curve, also shown in Figure 3–2, illustrates the high QE that can be realized. Back-thinning, however, results in a delicate, relatively expensive sensor that to date has only been employed in scientific-grade CCD cameras.

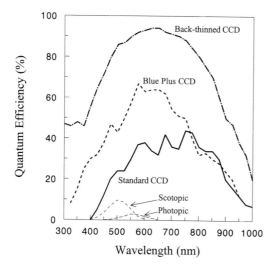

FIGURE 3–2 Spectral sensitivity for similar charge-coupled device (CCD) sensors in three different configurations. The performances of a standard, front-illuminated interline transfer sensor and the same sensor equipped with transparent channels (Blue Plus) are shown. The use of back-thinning to improve performance is also illustrated. The curves at the bottom show the performance of the human eye for both scotopic and photopic vision.

The resolution of a CCD is a function of the number of photodiodes and their size relative to the projected image. Charge-coupled device arrays of 1000 × 1000 sensors are now commonplace in scientific-grade video cameras. The trend in consumer- and scientific-grade CCD manufacturing is for the sensor size to decrease, resulting in cameras with photodiodes as small as 4 × 4 aem being available in the consumer market. A typical MTF curve for a CCD camera with 6.7 μm pixels is shown in Figure 3–3.

Adequate resolution of an object can only be achieved if at least two samples are made for each resolvable unit. Many users prefer three samples per resolvable unit to ensure sufficient sampling. In the case of the epifluorescence microscope, the resolvable unit from the Abbe diffraction limit at a wavelength of 550 nm and a 1.4 numerical aperture (NA) lens is 0.22 μm. If a ×100 lens is used, the projected size of a diffraction-limited spot on the face of the CCD would be 22 μm. A sensor size of 11 × 11 μm would just allow the optical and electronic resolution to be matched, with a 7 × 7 μm sensor size preferred. Although small sensors in a CCD improve spatial resolution, they also limit the dynamic range of the device.

The charge storage capacity of a CCD is proportional to the size of the individual photodiode, such that the maximum number stored is about 1000 times the cross-sectional area of each photodiode. Thus, a CCD with 7 × 7 μm photodiodes should have a maximum charge storage capacity (a full-well capacity) of 49,000 electrons (or holes). A hole is the region of the silicon from where the electron came and constitutes an equally valid and usable measure of detected photons. The measure of electrons will be predominantly used, even though many CCDs read out the number of holes generated rather than electrons. Because CCDs

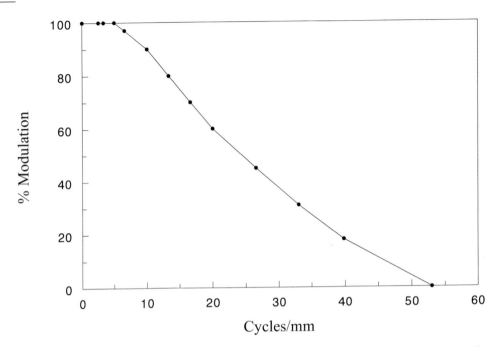

FIGURE 3–3 Modulation transfer function curve for a typical CCD camera equipped with a 1 K × 1 K sensor with 6.7 μm pixels. The spatial frequency of 60% contrast sine waves projected onto the sensor surface is on the abscissa and the resultant modulation percentage on the ordinate. The limiting resolution is normally defined as the 3% modulation level.

do not have inherent gain, one electron–hole pair is accumulated for each detected photon. The dynamic range of a CCD is typically defined as the full-well capacity divided by the camera noise. The camera noise is the sum, in quadrature, of the dark and read-out noise. Recent improvements in CCD design have greatly diminished dark charge to negligible levels and reduced read-out noise to about 10 electrons per pixel. Even room temperature cameras may have such a low dark signal that it can be ignored for integration periods of 10 s or less. Cooling further reduces the dark signal and permits much longer integration (up to several hours) without significant dark charge accumulation. Thus, the dynamic range of a 49,000 electron full-well capacity CCD with 10 electrons of read-out noise and negligible dark noise is about 4900, corresponding to a 12-bit range (i.e., 2^{12}).

A CCD with a 49,000 electron full-well capacity has a maximum achievable S/N ratio of about 220 (i.e., the square root of 49,000). Of course, camera noise would add, in quadrature, to the photon statistical noise and reduce the maximum S/N ratio below this value. A simple estimate of the S/N ratio of any homogeneous region in an image may be made from the average intensity of the region of interest divided by the standard deviation of the intensities of that region.

Cameras for consumer use often have a rectangular format CCD with an aspect ratio of 4/3. This means that the height of the image will be 3/4 of the width to conform to video standards based on our landscape view of the world. Indeed, the newest generation of consumer-grade products designed for HDTV employs a 16/9 aspect ratio. Scientific imaging, on the other hand, is best carried with a square image made up of square pixels, as they are better suited to digital image processing.

Charge-coupled device sensor uniformity is generally very good, with less than a 10% variation in gain between photodiodes. Shading may, however, be introduced into the image from a CCD camera because of inefficiencies in charge transfer. The operation of a CCD requires that each packet of charge underlying a photodiode be transferred to the read-out amplifier. This transfer is accomplished by a series of parallel and serial shifts that displace rows of charge along the chip toward the corner containing the read-out amplifier. If the read-out amplifier is in the upper right-hand corner of a 1000×1000 sensor CCD, the charge from the photosensor nearest to that corner will have to be shifted only once upward toward the amplifier (a parallel shift) and once rightward (a serial shift) to reach the amplifier. On the other hand, the charge from the photodiode in the lower left-hand corner will have to be shifted upward 1000 times and rightward 1000 times to be read out. If the transfer efficiency is 99.9% for each shift, only 13.5% of the charge accumulated by the lower left photodiode would remain after the requisite 2000 shifts. This charge loss would make the lower left corner much darker than the upper right and would also tend to blur or smear that region of the image because of admixture of charges from adjacent photodiodes.

Slow-scan CCD cameras increase charge transfer efficiency by cooling the CCD and slowing the transfer rate. The high-speed charge transfer required in video-rate CCD cameras necessitates a different strategy. They adjust the read-out amplifier gain to compensate for the charge lost from each row by sampling extra pixels outside of the image area. The additional gain required for the lower rows inevitably increases the noise in the highly corrected regions of the sensor.

Slow-scan CCD cameras permit some control over the read-out rate, as well as the size of the pixel that constitutes a sensor. Video-rate CCD cameras are simpler and do not allow such control. Slowing the read out usually reduces the amplifier noise associated with reading the charge, which is a beneficial situation when the photon flux is very low and the signal can be produced relatively slowly (e.g., in a second or two rather than in 33 ms). Scientific-grade CCD cameras usually offer two or more read-out rates so that speed may be traded off against noise.

The size of a pixel in a scientific-grade CCD may be increased by binning, a process in which the charge from a cluster of adjacent photodiodes is pooled and treated as if it came from a larger detector. In binning, several shifts of charge to the serial register and output node storage regions occur before read out. The extent of binning depends on how many shifts occur before the stored charge is read, the only limitation being the charge storage capacity of the serial register, usually twice that of one of the photodiodes, or of the output node, usually three times that of a photodiode. The maximum charge storage capacities of the serial register and output node are not a concern in most fluorescence microscopy applications because binning is employed when light levels are very low and few photons are detected. Binning enables the investigator to trade spatial resolution for sensitivity.

Slow-scan CCD cameras also allow region-of-interest read out. This means that a selected portion of the image can be displayed, as the remainder of the accumulated charge is discarded. The framing rate generally increases in proportion to the reduction in the size of the detected area. For example, a CCD with a sensor size of 1000 × 1000 and an output rate of ten frames/s can produce 100 frames/s if the read-out region is reduced to 100 × 100 diodes. By trading off field-of-view and framing rate, an investigator can adjust to a wider range of experimental circumstances than would be possible with a fixed framing rate video camera.

4. LOW-LIGHT-LEVEL IMAGING OF FLUORESCENCE

Because of the problems of photodestruction of fluorophores in the presence of oxygen (bleaching) and limitations on the numbers of fluorophores that can be in a single region, a variety of sensitive electronic detectors are used in fluorescence microscopy. Only about 5%–10% of the emitted light from an excited fluorophore is collected and transferred to the sensor in a typical epifluorescence microscope. There are two approaches to capturing as much of this limited light flux as possible: integration by a slow-scan CCD, as previously, and capture on a video-rate or progressive-scan CCD camera. The general finding is that a cooled, slow-scan CCD camera always produces a higher S/N ratio than an intensified CCD, provided that sufficient integration time is available.

Image intensifiers were developed for military use to enhance night vision. They have an input photocathode followed by a microchannel plate electron multiplier and a phosphorescent output screen. In the latest generation of these devices, the photocathode, while similar to that in photomultiplier tubes, has a higher QE (up to 50%) in the blue-green end of the spectrum. The gain of the microchannel plate is adjustable over a wide range with a typical maximum of about 80,000; that is, a detected photon at the input leads to a pulse of 80,000 photons from the phosphor screen. The phosphor matches the spectral sensitivity of the eye and is often not ideal for a CCD. Resolution of an intensified CCD depends on both the intensifier and the CCD, but is usually limited by the intensifier microchannel plate geometry to about 75% of that of the CCD alone. The latest versions of image intensifiers (denoted Blue Plus Gen III or Gen IV) employ smaller microchannels (6 μm diameter) and better packing geometry than in the past, with a resultant increase in resolution and elimination of the chicken-wire fixed-pattern noise that plagued earlier devices.

Image intensifiers have a reduced intrascene dynamic range compared with slow-scan CCD cameras. It is difficult to obtain more than a 256-fold intensity range (i.e., 8 bits) from an intensified CCD camera. Intensifier gain may be rapidly and reproducibly changed to accommodate variations in scene brightness, thereby increasing the interscene dynamic range. Indeed, because image intensifiers can be rapidly gated (turned off or on in a few nanoseconds), relatively bright objects can be visualized by a reduction in the on time. A gated, variable gain intensified CCD camera is commercially available with a 12 order of magnitude dynamic range. Gated, intensified CCD cameras are required for most time-resolved fluorescence microscopy applications because the detector must be turned on and off in nanoseconds or its gain rapidly modulates in synchrony with the light source.

Thermal noise from the photocathode and electron multiplication noise from the microchannel plate reduce the S/N ratio in an intensified CCD camera to below that of a slow-scan CCD. The contribution of these components to the noise created by the statistical nature of the photon flux depends on the gain of the device and the temperature of the photocathode. Generally, a reduction of the gain of the intensification stage is employed to limit the noise, although intensified CCD cameras are available with a cooled photocathode.

Intensified CCD cameras have a very fast response limited by the time constant of the output phosphor, and often the CCD camera read out is the slowest step in image acquisition. Because of the low light fluxes emanating from the fluorophores in living cells, intensified CCD cameras are frequently employed to study dynamic events and for ratio imaging of ion-sensitive fluorophores. The simultaneous or near-simultaneous acquisition of two images at different excitation or emission wavelengths is required for ratio imaging. Intensified CCD cameras have the requisite speed and sensitivity for this process.

A hybrid of the image intensifier and the CCD camera is the recently introduced electron-bombarded CCD. In this device, photons are detected by a photocathode similar to that in an image intensifier. The released electrons are accelerated across a gap and impact on the backside of a CCD. These energetic electrons generate multiple charges in the CCD, resulting in a modest gain of a few hundred. The advantages of this device over a cooled, slow-scan CCD are the additional gain and accompanying speed; the main disadvantages are the lower QE of the photocathode and diminished dynamic range. Compared with an intensified CCD, the electron-bombarded CCD usually has higher spatial resolution and a better S/N ratio at moderate light levels. The limited gain adjustment range and modest low-light-level detection capability, however, make the electron-bombarded CCD the solid-state equivalent of the outmoded SIT (silicon intensifier target) camera.

5. ELECTRONIC VERSUS VISUAL DETECTION

How do our eyes compare with electronic detectors? Figure 3–2 shows spectral sensitivity curves for the eye, corresponding to photopic and scotopic vision, arising from the cones and rods, respectively. Peak sensitivity is in the green (photopic at 555 nm and scotopic at 507 nm), with a maximum QE of 3% for photopic vision and 10% for scotopic. Our spatial resolution is not uniform because the cones are not evenly distributed. The highest density occurs in the fovea, where the distance between cones is about 1.5 μm, giving us 5–6 μm limiting spatial resolution on the retina. Under achromatic (black and white) constant illumination conditions, visual intrascene dynamic range is only about 50-fold (i.e., 6 bits). Our visual pigment, rhodopsin, exhibits little thermal noise, and the minimum detectable signal after dark adaptation is about 100–150 photons at the pupil or about 10–15 photons at the retina. The S/N ratio for the eye at the visual detection limit is about 3/1. Lag is about 20 ms at high light levels and about 100 ms in dim illumination.

Compared with our eyes, a scientific-grade CCD camera has a broader spectral sensitivity, much higher quantum efficiency, greater integration capability, more uniformity, better intrascene dynamic range (i.e., more "bits"), comparable or higher S/N ratio, but lower spatial resolution. When matched against our visual system, low-light-level cameras have a much wider spectral range, less lag, and far greater sensitivity and resolution under photon-limited conditions.

6. Choosing the Appropriate Camera for an Application

No single detector will meet all requirements in fluorescence microscopy, so the investigator is often forced to compromise. In addition, the choice is made difficult because the slow-scan cameras are getting faster and the video-rate cameras are often cooled.

When time is the critical parameter, intensified cameras are generally the better choice. If the event under investigation is rapid, but can be precisely triggered, a slow-scan CCD operating in a burst or high-speed mode may be suitable. However, when the event is not readily predictable and the specimen must be monitored continuously at low incident light flux, the intensified CCD is the detector of choice. For this reason, single-molecule fluorescence studies have often employed intensified CCD cameras.

When time is available for image integration, a slow-scan CCD camera will usually outperform an intensified camera in all areas, due mainly to its higher QE and lower noise. Cooling always improves camera performance, although the difference may not be noticeable when the integration time is a few seconds or less and the digitization level is 10–12 bits or less. For applications involving digital deconvolution, the detector of choice is a cooled, scientific-grade, slow-scan camera capable of producing a high-resolution, 14–16 bit image. Some of the latest slow-scan CCDs, however, have such small pixels that the integration period must be limited to avoid saturation of the wells, and the dynamic range and peak S/N ratio may be no better than those of an intensified CCD.

Two types of color CCD cameras are used for scientific applications: a single CCD with a wavelength selection filter or a three-sensor camera. Both use filters to produce red, green, and blue versions of the field-of-view. The single sensor camera utilizes a filter wheel or liquid-crystal tunable filter to acquire the red, green, and blue images in sequence. The three-sensor camera has a beam-splitting prism and trim filters that enable each sensor to image the appropriate color and to acquire all three images simultaneously. Invariably, color cameras are less sensitive than their monochrome counterparts because of the additional beam-splitting and wavelength selection components. In some applications, particularly immunofluorescence, being able to capture multiple wavelengths simultaneously offsets the loss of sensitivity. In addition, some color cameras achieve a higher resolution by offsetting each CCD slightly, thereby increasing the sampling frequency.

Recent improvements in the performance of CMOS cameras herald a potentially important future role for these devices in fluorescence microscopy. The CMOS camera has an amplifier and digitizer associated with each photodiode in an integrated on-chip format. The result is an inexpensive, compact, versatile detector combining the virtues of silicon detection without the problems of charge transfer. The CMOS sensors allow gain manipulation of individual photodiodes, region-of-interest read-out, high-speed sampling, electronic shuttering, and exposure control. They have extraordinary dynamic range, as well as an ideal format for the computer interface. Until recently, they suffered from high fixed-pattern noise associated with switching and sampling artifacts, but these problems are now being solved rapidly. It is likely that they will replace CCD cameras in a number of scientific applications in the near future.

7. DISPLAY OF ELECTRONIC IMAGES

Although we have primarily been concerned with detectors, the display is what we see. Both cathode ray tube (CRT) and liquid-crystal display (LCD) monitors are used in fluorescence microscopy. Two defining factors are most noticeable when an image is observed on a monitor: resolution and color fidelity. Resolution is determined by the pixel geometry, while color fidelity depends on the phosphors or filters in the display.

In a color CRT monitor the pixel is composed of a triad of dots on the display phosphor with characteristic red, green, and blue colors (an RGB monitor). The size of the dots and the spacing of the triads are determined by the scanning beam geometry and a mask placed close to the interior surface of the phosphor. The larger the tube, the higher the resolution of the display.

In an LCD display, the pixel size is determined by the size of the liquid-crystal cells, with three cells needed for each RGB triad. Other solid-state monitors, such as plasma displays, similarly need a triad of cells for each pixel. The difficulties of such large-scale integration have limited both the size and resolution of these monitors. Color is determined by absorption filters rather than by phosphors as in the CRT monitor.

Although our ability to detect small differences in gray scale is rather poor, we have remarkable color discrimination capabilities. A lack of color fidelity for CRT displays occurs because the range of phosphors is limited, particularly in the blue-green end of the spectrum. In the case of LCD monitors, color fidelity is reduced by inadequate display brightness, slow pixel response, and the quality of the color filters employed. Finally, limitations in ink, paper, and transfer technology often result in a print of an image that differs in color from that seen on the monitor, as well as through the microscope (see Chapter 7).

Thus, for a variety of reasons, the electronic image is never identical to what we perceive, and we must understand these differences to make proper use of the information obtained.

REFERENCES

Aikens, R., D. Agard, and J. Sedat. Solid state imagers for optical microscopy. *Methods Cell. Biol.* 29: 291–313, 1988.

Becker, P. L. Quantitative fluorescence measurements. In: *Fluorescence Imaging Spectroscopy and Microscopy, Chemical Analysis,* Vol. 137, edited by X. F. Wang and B. Herman. New York: John Wiley & Sons, 1996, pp. 1–29.

Bookman, R. Temporal response characterization of video cameras. In: *Optical Microscopy for Biology,* edited by B. Herman and K. Jacobson. New York: Wiley-Liss, 1990, pp. 235–250.

Bright, G. R., and D. L. Taylor. Imaging at low light level in fluorescence microscopy. In: *Applications of Fluorescence in the Biomedical Sciences,* edited by D. L. Taylor, F. Lanni, A. S. Waggoner, R. F. Murphy, and R. R. Birge. New York: Alan R. Liss, 1986, pp. 257–288.

Herman, B. *Fluorescence Microscopy,* 2nd Ed., New York: Springer-Verlag, 1998.

Inoué S., and K. R. Spring. *Video Microscopy: The Fundamentals.* New York: Plenum, 1997.

Periasamy, A., and B. Herman. Computerized fluorescence microscopic vision in the biomedical sciences. *J. Comput. Assist. Microsc.* 6:1–26, 1994.

Shotton, D. *Electronic Light Microscopy.* New York: Wiley-Liss, 1993.

Sluder, G., and D. E. Wolf. *Methods in Cell Biology,* Vol. 56, *Video Microscopy.* San Diego: Academic Press, 1998.

Spring, K. R., and R. J. Lowy. Characteristics of low light level television cameras. *Methods Cell Biol.* 29:269–289, 1988.

Spring, K. R., and P. D. Smith. Illumination and detection systems for quantitative fluorescence microscopy. *J. Microsc.* 147:265–278, 1987.

Tsay, T. -T., R. Inman, B. Wray, B. Herman, and K. Jacobson. Characterization of low-light-level cameras for digitized video microscopy. *J. Microsc.* 160:141–159, 1990.

4

Basics of a Light Microscopy Imaging System and Its Application in Biology

Lance Davidson and Raymond Keller

1. Introduction

Advances in microscopy technique have driven the field of biology forward—from the invention of the microscope to the development of phase contrast (Zernike, 1958), differential interference contrast (Allen et al., 1969), epifluorescence (Ploem, 1967), laser scanning confocal (Minsky, 1957), and now multiphoton confocal microscopy (Denk et al., 1990). As the imaging technology evolves (Inoué and Spring, 1997), the biologist works to prepare fixed samples (Presnell and Schreibman, 1997) with enhanced contrast using fluorescent probes (Haugland, 1994) and living samples with green fluorescent protein (Heim et al., 1995).

The purpose of this chapter is twofold; our first goal is to introduce the basic principles of microscope systems, important components useful to the biologist, and discuss how they work for and against the investigator. Our second goal is to provide several examples of different imaging systems put together to meet both the general and specific imaging needs in our own laboratory. (Table 4–1 lists some useful Internet websites for further information.)

The very first question to ask yourself is "why does my laboratory need a microscope?" A well-defined answer to this question will enable you to select a system that will serve your needs both immediately as well as into the near future. Setting priorities will also help to identify where to spend money and where the budget can be tightened.

The biologist works to "construct" a sample with the most contrast possible. The dominant theme in all microscopy techniques is the need to increase contrast between different tissues, different cells, and different structures. In our laboratory, we take considerable effort to enhance the contrast within the sample with fluorescence and visible contrast agents or by microsurgically exposing important cells to the microscope "eye." Various microscopes and techniques serve to enhance, visualize, and record contrast inherent in the sample. Without a suitable sample other accessories, such as motorized stages, focus motors, shutters, filter wheels, cameras, and heated stages, are worthless. With a suitable sample, however, these accessories can allow computer-controlled image acquisition, thus augmenting the capabilities of the biologist.

TABLE 4–1 Useful Internet resources*

General Non-Commercial

Microscopy Primer: Molecular Expressions Introduction to Microscopy (http:// micro.magnet.fsu. edu/primer);

MicroWorld Guide to Online Microscopy Resources: http:// www.mwrn.com/

Microscopy listserv: To subscribe, send the message "Subscribe Microscopy YourUserID@Your Domain" to Listserver@MSA.Microscopy.Com. To unsubscribe send the message "Unsubscribe Microscopy YourUserID@YourDomain" to listserver@msa.microscopy.com

Confocal listserv: To subscribe by e-mail send the message "subscribe confocal your full name" to listserv@ubvm.cc.buffalo.edu. Features discussions of confocal microscopy as well as generally relevant fluorescence microscopy issues

NIH-Image listserv: Send the message "subscribe NIH-IMAGE first-name last-name" to join and "unsubscribe NIH-IMAGE" to be removed from the list to listproc@soils.umn.edu. Features discussion and troubleshooting of NIH-Image software and general image acquisition issues

sci.techniques.microscopy: Scientific microscopy newsgroup

Imaging Systems

Compix: http://www.cimaging.net/. Supplier of Simple PCI and C-Imaging Systems image acquisition systems

Improvision: http://www.improvision.com/. Supplier of the Openlab imaging system

Inovision: http://www.inovis.com/. Supplier of the Isee Analytical Imaging System

Media Cybernetics: http://www.mediacy.com/. Supplier of Image-Pro Plus imaging system

NIH-Image: http://rsb.info.nih.gov/nih-image/. Free software for the acquisition and morphometric analysis of 8 bit images. Can be used with Data Translation (http:// www.datx.com) or Scion Corp. (http://www.scioncorp.com/) frame grabber cards and analog video. Spin-off programs such as Object-Image and Scion PC-Image have been extended from NIH-Image to carry out nondestructive morphometrics and to run on Microsoft Windows–capable computers, respectively

QED Imaging: http://www.qedimaging.com/. Supplier of camera, microscope, and time-lapse "plug-ins" for Adobe Photoshop

Scanalytics: http://www.iplab.com/. Supplier of the IPLab imaging system

Universal Imaging: http://www.image1.com/. Supplier of the Metamorph Imaging System

Cameras

Cohu: http://www.cohu-cameras.com /. Supplier of video cameras

Dage: http://www.dagemti.com/. Supplier of video and digital cameras, including the DC330

Diagnostic Instruments: http://www.diaginc.com/. Supplier of the Spot camera and video couplers for every microscope. Excellent site when considering attaching camera X to microscope Y

Hamamatsu: http://usa.hamamatsu.com/. Supplier of video and digital cameras, including the ORCA series

Nikon: http://www.nikon.com/. Supplier of digital cameras and microscopes (see below), including the Coolpix 990, an excellent consumer-grade color digital camera easily adaptable for microscopy

Optronics: http://www.optronics.com/. Supplier of video and digital cameras, including the Magnafire

Roper Scientific: http://www.prinst.com/. Supplier of digital cameras including the microMax (formerly Princeton Instruments) and PXL (formerly Photometrics)

TABLE 4–1 *(Continued)*

Filters

Chroma Technology: http://www.chroma.com/. Manufacturer of filters and blocks for a wide variety of fluorescent probes and microscopes. Site is not searchable but has extensive technical library available for downloading as well as spectra for filters

Omega Optical: http://www.omegafilters.com/. Manufacturer of filters and blocks for a wide variety of fluorescent probes and microscopes. Searchable site with an extensive database of technical notes and spectra for available filters

Heated Stages, Shutters, Filter Wheels, XYZ Stages, and Other Supplies

Bioptechs: http://www.bioptechs.com/. Supplier of perfusion devices, open and closed cell chambers, heated stages, as well as objective heaters

Diagnostic Instruments: http://www.diaginc.com/. See entry under Cameras

Harvard Apparatus: http://www.harvardapparatus.com/. Supplier of Leiden chambers and temperature-controlled "open" chambers for microperfusion and patch clamping

Ludl Electronic Products: http://www.ludl.com. Supplier of the Bioprecision line of filter wheels, XY stage, and focus controllers

Prior Scientific: http://www.prior.com/. Supplier of XYZ controlled stages

Sutter Instrument: http://www.sutter.com. Supplier of the Lambda-series filter wheels and controllers

Vincent Associates: http://www.uniblitz.com/. Suppliers of UniBlitz shutters and controllers

Microscope Companies

Leica: http://www.leica.com. On-line product information and detailed specifications. A fairly shallow site that is "searchable" but does not have much information

Nikon: http://www.nikon.com. On-line product information and detailed specifications. Much information is distributed into a "frequently asked questions" (FAQ) format. Although the site is not searchable, the "search" page does bring up an extensive table of contents

Olympus: http://www.olympus.com. On-line product information and detailed specifications; has FAQ feature. The site is searchable with a large knowledge database

Zeiss: http://www.zeiss.com. On-line product information. Good general glossary. Very nice data sheet describing notations on objectives. The site is searchable with a large knowledge database

*Information current as of spring 2001.

2. COMMON ILLUMINATION TECHNIQUES

The simplest microscope has four major components: the light source, the condenser, the objective, and the eyepiece. These are still the most important elements, even in the most advanced microscopes. Proper alignment of the lamp and maintenance of Köhler illumination with the condenser (Spencer, 1982) are essential to illuminate the sample evenly with intense light. The next aim of the microscopist is to collect as much of the light that interacts with the sample as possible. Using bright-field illumination techniques, contrast in the image is the result of differences in light absorption in the sample, as well as the diffraction patterns generated by these light-absorbing objects. If a bright-field image is not satisfactory, the microscopist can use methods to enhance other forms of contrast already present in the sample.

2.1 Dark-Field and Low-Angle Illumination

Two other techniques for enhancing the contrast of absorbing or scattering features in a sample are dark-field and low-angle illumination. Dark-field illumination uses a special condenser that focuses a hollow cone of light on the sample in an oblique angle. After the sample is illuminated, the direct illumination is blocked by a mask placed in or just behind the objective (Goldstein, 1999). Only light that is scattered by the sample is allowed through the objective. This technique can illuminate exceedingly small objects below the limit of resolution, as well as thin, nearly transparent samples. Once the sample is thick or translucent, low-angle illumination can be used. Typically a fiberoptic illuminator (a single-source, dual-source, or ring illuminator) shines on the sample from an off-axis, oblique angle, enhancing contrast between parts of the sample that scatter the oblique light differently. Both of these techniques can be useful to enhance contrast due to changes in absorption or scattering in a sample.

2.2 Phase Contrast

Phase contrast microscopy was developed to observe areas with either *different thicknesses* or *different refractive indices* within a sample (Fig. 4–1A). The modern principle requires that an annulus mask be placed in the condenser and that a light-absorbing annulus etched into a glass plate be placed in the objective. The annulus mask allows a single wavefront to interact with the sample, which either advances or retards the speed of the wave as it passes through. The etched annulus, or phase plate, causes light that passes without interacting with the sample to undergo destructive interference. Light waves passing through thicker regions of the plate are retarded compared with those passing through the annulus. If recombined, the two such waves would destructively interfere. If waves that have passed through the annulus interact with a sample that has a different index of refraction, they too will be slowed. This shift keeps them from destructively interfering and increases the contrast of the sample over that of the dark background.

Use of phase contrast is typically limited for several reasons: (1) changes in the index of refraction in samples are typically small, (2) matching phase "rings" are required in the condenser and objective, and (3) phase "halo" results from diffraction from small point-like objects in the sample. Even with these limitations, however, phase contrast techniques do an excellent job of enhancing the contrast between structures with differing refractive indices.

2.3 Differential Interference Contrast Microscopy

Differential interference contrast (DIC; or Nomarski) microscopy was developed as an enhanced phase contrast technique and produces contrast wherever there are *changes* in the index of refraction in a sample (Fig. 4–1B). Differential interference

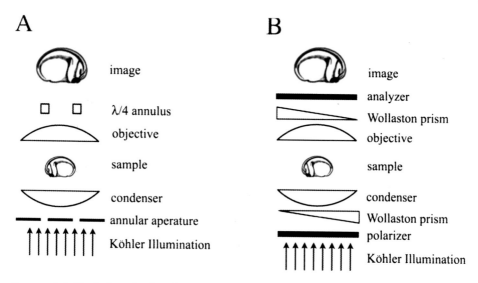

A

image

λ/4 annulus

objective

sample

condenser

annular aperature

Köhler Illumination

B

image

analyzer

Wollaston prism

objective

sample

condenser

Wollaston prism

polarizer

Köhler Illumination

FIGURE 4–1 Optical paths for phase contrast microscopy (*A*) and differential interference contrast microscopy (*B*). Proper illumination begins with Köhler illumination.

contrast uses a pair of polarizers and Wollaston prisms. These elements in the condenser allow only one orientation of polarized illuminating light to pass through. All light that passes straight through without being changed in polarization by interfaces in the sample are rejected. Only light that "glances off" the sample in a manner in which glare is generated from reflected light at the beach or along the motorway is allowed to pass. In this way DIC enhances contrast of interfacial features of the sample. Several factors such as dust or a poor-quality condenser can severely degrade the quality of both phase and DIC images.

2.4 Epifluorescence

Dark-field, phase contrast, and DIC are used to collect images of thin, nearly transparent samples, whereas bright-field and low-angle illumination can collect images of the surface features of more optically opaque samples. Vital dyes and counterstains in histology (Presnell and Schreibman, 1997) were developed to exaggerate structures in samples, thus extending the capacities of these techniques. More recently, fluorescent probes (lineage tracers, antibodies, proteins, RNA) have been developed to enhance the contrast of cellular and subcellular structures. Evolving epifluorescent microscopy techniques continue to keep pace with these rapidly evolving contrast agents. The principles of the epifluorescence technique are very similar to those of dark-field techniques. The illuminator produces light of a particular wavelength that passes through the objective lens (the objective operating as a condenser). The light then excites a fluorophore in the sample, which emits light with a longer wavelength that passes back through the objective to either the eyepiece or camera/detector (Color Fig. 4–2). Advances in this technique occur in several areas. The first advance is the construction of objectives that pass more excitation light with less distortion and collect more of the light that is emitted. The second advanced technology is the development of advanced filters to prevent scattered excitation wavelengths of light from traveling to the eyepiece, as well as filters that enable researchers to collect information simultaneously from several different structures at once, each with a different fluorophore whose absorption and emission spectra are distinct.

Biologists commonly use one of the above contrast-enhancing techniques; however, numerous other techniques have been and are continuing to be developed using a variety of different optical phenomena (see other chapters in this book). Novel microscopy techniques are frequently built on equipment scavenged from existing microscopes. The remainder of this chapter explores the more common components of these microscopes.

3. COMPOUND MICROSCOPES: CONDENSERS, OBJECTIVES, TUBE LENGTH, AND CAMERA COUPLERS

For phase contrast and DIC microscopy, the positioning of the condenser is extremely important. Optical elements, phase rings, and so forth must match the objective. Likewise, the numerical aperture of the condenser should match or exceed that of the objective. Complex high-resolution work can even require the use of oil between the slide and the condenser.

When faced with the choice of which objective to use, the biologist is faced with a bewildering array of lenses with apparently obtuse nomenclatures (Table 4–2 defines some common terms) that change from one manufacturer to another. The two main functions of an objective is to put the most light onto a sample (i.e., in dark-field or fluorescence mode) and to capture as much light as possible (reflected or emitted) from the sample and deliver that light to the eyepiece or camera with as little spatial and chromatic aberrations as possible.

Computer-based objective design has attempted to meet these needs, but the truth is that, when buying objectives, there is still no better test than a side-by-side comparison. Microscope sales engineers should be happy to provide loaner objectives for you to test. Consider the samples and the magnification you will need. Is the image well enough resolved? Do you have enough working distance to reach those regions of your sample that are far from the coverslip? Do you need dry objectives that might be dim for fluorescence work, but have a long working distance, or oil immersion with its bright fluorescence but little working distance? Perhaps an intermediate multi-immersion objective with moderate fluorescence and enough working distance to get by? Do you need ultraviolet-transmitting optics to see DAPI- or Hoescht-labeled DNA or Fura-2 for calcium imaging? Spend time considering other optical elements, such as a zoom lens or video coupler and whether your condenser is sufficient for DIC or phase contrast.

Another term the biologist might not be familiar with is *infinity corrected*. Infinity-corrected microscope design removes two elements formerly in the optical train and eliminates the need for a projection lens for the camera. The practical effect of these new designs is the creation of a short region within the microscope

TABLE 4–2 Glossary

Aberration A distortion (either spatial or chromatic) of the projected image from the true image (Keller, 1995). Spherical aberration due to mismatches in coverslip thickness, mounting media, and immersion media can sometimes be corrected on objectives with "correction collars"

Achromatic objective An objective designed to bring two specific wavelengths of light to the same focal point

Aperture The opening of the objective or condenser that lets illuminating light onto the sample. It can be reduced for enhanced contrast by a field iris on the light source

Apochromatic objective An objective designed to bring three visible colors of light (red, green, and blue) to a single focal point without spherical aberration

Depth of field The span of focus over which the sample is acceptably resolved. It depends on the objective and the illuminating aperture

Field number (FN) A numerical indicator of the field of view observable through an eyepiece (commonly printed on the eyepiece). An eyepiece with a field number of 25 projects a larger field of view than one with a field number of 23

Fluorite objective An objective made of fluorite or other specialty glass that are better corrected than achromats with larger numerical aperture

Light-gathering power A measure of the amount of light that can be collected from a sample by an objective. Depends on the numerical aperture as well as the ability of the objective to pass the desired wavelength. For instance, objectives vary tremendously in their ability to transmit ultraviolet wavelengths

Numerical aperture (NA) A geometric measure of the light-gathering power of an objective defined by (n sin [$a/2$]), where n is the index of refraction of the material between the sample and the front of the objective and a is angle of the objective aperture

Parfocal The case when the same plane of a sample is in focus in both the eyepiece and the camera. Many camera couplers can be focused separately to bring the system into "parfocality"

Plano objective An objective designed to form an image that is resolved over a large field

Working distance The distance between the front glass of the objective and the coverglass in which the objective may be moved to keep the sample in focus. The specific working distance depends on the immersion oil, the sample mounting media, and the objective

immediately behind the objective where the light path is parallel, neither converging nor diverging. These regions are particularly friendly to the placement of filters and other optical accessories, eliminating changes in tube length and the focal plane induced by such accessories in the past.

The last elements in the optical train are the eyepieces and the camera coupler. Printed on each eyepiece is the magnification and the field number. The field number characterizes the field of view of the eyepiece. Although camera adapters have a fixed magnification, the field of view that they project on the camera depends on the size of the detector used in the camera. A small-sized detector will deliver a smaller field of view than a larger detector for the same magnification. Thus, it is important to match the magnification of the camera coupler to the size of the detector used in the camera. For instance, use of the Hamamatsu ORCA camera with a direct imaging coupler (i.e., no magnification) will result in the projection of a very small field of view onto the CCD. This approach seems appealing because it does not introduce additional glass elements into the optical train. The user might compensate for this small field by using a lower magnification objective, thus lowering image quality and brightness. The counterintuitive use of a magnifying element increases the brightness by focusing more light on the CCD chip and allows the use of more efficient, high numerical aperture objectives.

4. COMPOUND MICROSCOPES: LIGHT SOURCES AND FILTER SELECTION

Compound microscopes can be fitted with a number of different light sources. The most generally useful are tungsten-halogen, xenon, and mercury arc-lamps (or a xenon-mercury combination), each with distinct emission spectra. Xenon and tungsten-halogen lamps produce even illumination from 300 nm into the infrared. Xenon, mercury, and combination arc-lamps provide intense sources of illumination useful for fluorescence microscopy. Although tungsten-halogen lamps do not produce such intense light sources, they can provide suitable illumination for live-cell fluorescence microscopy.

The aim of fluorescence microscopy is to acquire images in which structures are labeled with one or more fluorophores. High-quality images require high signal, low background, spectral separation between different fluorophores, in addition to a satisfactorily resolved optical image. Fluorescence microscopy achieves a high signal-to-background ratio through the use of filters that match both the light source and the absorption and emission spectra of the specific fluorophore used for contrast. Most microscopes combine all the filters into a single optical element, the filter cube or block.

The filter cube consists of three elements: an excitation filter, a beam splitter or dichroic, and an emission filter. The excitation filter typically allows only a limited range of wavelengths to pass (i.e., bandpass) that fall within the absorption spectra of the target fluorophore. The emission filter allows only a limited range of wavelengths, typically excluding the excitation wavelengths and allowing all emitted light above a cut-off wavelength (i.e., longpass). The third element is the dichroic mirror, which passes incoming excitation light from the light source to the sample and simultaneously reflects emitted light into the camera or eyepieces. All three of these elements are generally mounted into a single filter cube that inserts into the optical axis. Filter cubes can be constructed to match single or multiple fluorophores.

The intensity of a fluorescence image and the quality of that image depend in large part on the choice of filters used in the filter cube (see Chapter 2). Two of the most common types are bandpass and longpass filters. Bandpass filters are constructed to transmit a range of wavelengths and block out light of longer or shorter wavelengths. Bandpass filters have a characteristic center and width. The narrowest narrow-bandpass filters only allow a 5 nm window of the incident spectra to pass, while widepass filters pass 50 nm of the incident spectra. The wider the window, the more intense the transmitted light. Longpass filters absorb at short wavelengths and transmit light only above a certain threshold. A widepass excitation filter combined with a longpass emission filter often produces the brightest images. Sample autofluorescence can be minimized, at the expense of image brightness, by using either a narrow-bandpass excitation filter, a narrow-bandpass emission filter, a combination of the two, or a neutral density filter on the excitation.

More complex filters can be obtained with multiple bandpass windows to allow the simultaneous collection of emissions from multiple fluorophores.

Due to the broad absorption and emission spectra of some fluorescent probes, the emission from one fluorophore can occasionally pass through the filter set used to collect the second fluorophore. This condition, known as *bleedthrough*, can be minimized by using either different filter sets or different sets of fluorescent probes (e.g., bleedthrough seen with fluorescein isothiocyanate and rhodamine fluorophores can be reduced if rhodamine is replaced with Texas Red; see Chapter 2). Another strategy to reduce bleedthrough is to excite each fluorophore separately by placing fluorophore-specific excitation filters on a filter wheel and collecting two separate images after they have passed through a common dichroic and emission filter. Another software-based strategy involves digital image "subtraction" of a fraction of the bleeding image from the contaminated image. This strategy should only be used as a last resort once all other efforts have failed and artifacts introduced by such digital manipulations can be controlled for (see Chapters 15 and 17 for bleedthrough correction for fluorescence resonance energy transfer).

Filters can become contaminated and degrade with age and misuse. Fingerprints and spilled media can destroy a filter, as can prolonged exposure to intense illumination. Care should be taken when handling filter blocks and when using fluorescence on inverted microscopes to minimize contamination of the cubes. Additional care should be taken to close shutters or sliders to block incident light from entering the filter cube when it is not in use.

5. An Imaging System

Once the biologist is satisfied with the contrast presented in a single image of a single sample, a large number of experiments come to mind. In our laboratory the immediate response is to gather time-lapse sequences. The simplest system for collecting a time-lapse sequence includes a microscope, a camera, a frame grabber or digital acquisition board, a computer, and software to trigger the acquisition at timed intervals (see Fig. 4–3 and Chapter 3). If the sample requires 15°C, 37°C, or CO_2, a special stage for the sample must be attached to the microscope. Once two identifiable structures can be seen with different fluorescent probes (or one fluorescent probe and DIC), an automated filter wheel or shutter must be added.

If the experiments are accommodating, the biologist's head begins to swim with ideas, and the constraint of collecting merely a single time-lapse sequence at a time becomes burdensome. The solution to this is the XYZ stage positioner, which allows multiple simultaneous time-lapse sequences to be collected at once. Although the fully automated microscope outlined here forms the basis of a complete imaging system, further expansion can still accommodate additional functions like laser ablation or drug perfusion.

After the optical system in the above example, the second most important component is the software package. It is the software program that drives the motors, controls the perfusion chamber, and operates the camera. When installed with an appropriate set of compatible computer-driven accessories, most software

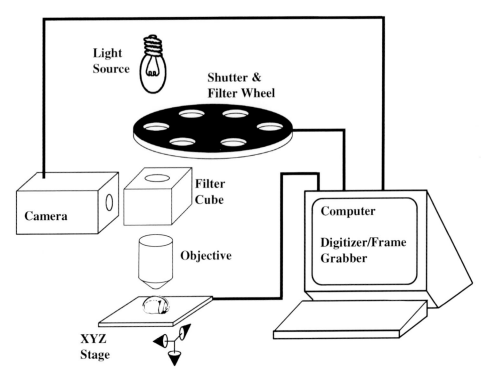

FIGURE 4–3 Basic elements of an imaging system. Three computer-controlled elements are shown in a simple schematic of a computerized imaging system. Images from the camera, either analog or digital, can be captured by either frame grabbing or digitizing hardware in the computer. The combination shutter/filter wheel allows computer-controlled exposure times and selection of excitation filters for the detection of single or multiple fluorophores. The XYZ moveable stage allows for simultaneous time-lapse sequence collection from multiple samples as well as collection of multiple planes of focus for later deconvolution. The computer runs the software for real-time hardware control and image acquisition, as well as postcollection morphometric image analysis.

packages do these tasks equally well. The last task of the software package is to integrate well with the biologist, performing tasks like file management, format conversion, and morphometric analysis. A program that is well liked by laboratory members with programming skills might not be accessible to anyone else. Likewise, the program designed for the lowest common denominator will not be adaptable for more complex tasks. Table 4–1 outlines several of the major software packages and their manufacturers' websites, where up-to-date product information can be found. Imaging system integrators can be found at these companies, as well as through microscope representatives. These people are also important components of the imaging system, as they can provide training, software updates, and technical support.

Stereoscopes can be convenient and easy-to-use alternatives to compound microscopes for low-magnification fluorescence, bright-field, and epiillumination imaging. Modern stereoscopes are "infinity corrected" and can be outfitted with epifluorescence attachments, high-intensity light sources, and digital cameras. Stereoscopes use a single wide-field objective of various low-power magnifications (e.g., ×1.0, ×1.6, ×2.0) with varying amounts of working distance (e.g., 30 mm). For fluorescence, a xenon or mercury arc-lamp is used because of the need for an intense light source. Filter sets can be obtained for a variety of fluorescent probes. The single drawback to the use of the fluorescent stereoscope is its low numerical aperture compared with that used in compound microscope–based fluorescence. The relatively dim signal seen in a fluorescence stereoscope is not generally a limitation when the sample is large and bright. The usefulness and popularity of fluorescent stereoscopes are driving development of hybrid systems that can utilize high magnification and high numerical aperture lenses designed for compound microscopes, turning the stereoscope into something that resembles an upright compound microscope without a substage condenser.

6. PRACTICAL EXAMPLES

This section illustrates how we attempt to meet the microscopy needs of our laboratory (see Color Fig. 4–4 for examples of phase contrast, epifluorescence, and stereoscope-collected images). Mentions of particular manufacturers are not meant as endorsements (or condemnations), but are included for completeness.

Cell movements and patterning in whole frog embryos and explanted embryonic tissues are investigated in our laboratory. We have a large number of uses for microscopy: (*1*) to screen lineage-labeled embryos, (*2*) to screen fluorescent protein–expressing transgenic embryos, (*3*) to document whole-mount RNA in situ patterns in both bright-field and fluorescent wavelengths, (*4*) to document histological sections in bright-field and fluorescent wavelengths, (*5*) to collect low-magnification time-lapse sequences, (*6*) to collect high-magnification phase and low-angle illumination time-lapse sequences, and (*7*) to collect high-magnification low-light time-lapse sequences of lineage-labeled or green fluorescent protein–expressing cells. These are everyday requirements, but we have also used differential interference microscopy (DIC), confocal microscopy, and calcium and pH ratiometric imaging.

6.1 Low-Light Time-Lapse Digital Microscopy

The focus of the work in our laboratory is the tracking of cell movements in whole embryos and tissue explants. Early work used a 16 mm cine-camera driven in time-lapse mode by an electromechanical timer. We currently use an inverted compound microscope (IX70, Olympus) with a digital cooled charge-coupled device (CCD) camera (C4810-95, a.k.a. ORCA, Hamamatsu) driven by image acquisition software (Metamorph version 4.13, Universal Imaging) running on a personal computer (450 MHz Pentium II, Dell). The image acquisition software controls all aspects of image collection, including camera control, shutters and filter wheels (Uniblitz and Sutter Instruments), and XYZ stage positioning (Prior).

Although we have two light sources, a tungsten-halogen lamp and a mercury arc-lamp, we typically use only the halogen because it provides more than enough excitation illumination for fluorescence microscopy of live cells. We use a video coupler to match the size of the image at the camera port to the ORCA camera. This results in some losses of light to the additional optical elements, but the coupler allows us to use higher magnification objectives with higher numerical apertures so that the final image is brighter.

A favorite objective on this microscope is a ×20 (numerical aperture of 0.7) planapo. Large lumpy embryos and tissues require the depth of focus and the reasonable working distance afforded by this relatively low-magnification objective. Samples are cultured in various custom chambers ranging from glass-bottomed petri dishes to low-volume perfusion-capable glass-bottomed acrylic chambers. The ORCA camera is capable of being used with low-angle illumination, as well as in low-light fluorescence mode. We use fiberoptic illuminators (Dolan-Jenner) to follow gross cell and tissue movements.

The ORCA camera is capable of 12 or 16 bit acquisition of 1280 by 1024 pixel images. Used in this mode, a short time-lapse sequence can easily exceed 500 Mb of storage. More typically the ORCA is used in 2×2 binned mode with the resulting 640 by 512 pixel images stored using 8 bit resolution. Image processing and morphometric analysis are carried out off-line with image processing software (Adobe Photoshop, NIH-Image, Object-Image, Scion-Image) on a variety of personal computers (mixed Macs and PCs).

6.2 Screening for Transgenic Embryos and Documenting Gene Expression Patterns

The introduction of exogenous genes creating mutant embryos and the mapping of gene expression patterns in whole embryos are a growing pursuit in developmental biology, and our laboratory is no exception. To assist in the tasks of sorting and documenting fluorescent protein–expressing embryos and documenting whole-mount RNA in situ patterns, we have assembled an imaging system around a fluorescence-equipped stereomicroscope (Olympus SZH11 stereoscope, Kramer Scientific fluorescence attachment). High-resolution color images (Hamamatsu C5810 cooled 3-CCD camera) can be collected from epiilluminated or fluorescent samples using Adobe Photoshop (plug-in provided by Hamamatsu) running on an Apple PowerMac 7500. This system is several years old (newer color cameras are faster and less expensive), but produces high-quality 1024 by 724 pixel images that have completely eliminated the use of film photography.

The tremendous depth of focus of the stereoscope and the placement of the sample at hand level is extremely valuable when orienting and documenting large lumpy embryos. There are, however, several limitations to the use of the stereoscope in epifluorescence mode. The first is imposed by the poor fluorescence performance of the traditional stereoscope. Objectives for the stereoscope typically have very low numerical apertures (less than 0.2) and have not been rigorously optimized for performance in epifluorescence mode. Thus, while a fluorescence-equipped stereoscope is easy to use and effective when combined with sensitive cooled CCD cameras, its efficiency is less than optimal when used with the naked eye. These problems and other design flaws, such as difficult to change filter blocks, are being remedied in second-generation designs.

6.3 Low Magnification Time-Lapse Video Microscopy

"You have to look in order to see something" is a common adage in our laboratory. Tissue movements in the embryo follow queer rules. Simple pictures at the beginning and end tell you little about the intervening movements or even which parts of the tissue at the start contribute to which parts at the end. Skill and mastery of microsurgical techniques require good hands and eyes, as well as the ability to repeat a surgery many times, taking the same tissues at the same stage or varying the surgery in known ways. To follow tissue movements, to demonstrate a difficult microsurgical maneuver, or to check on the consistency of surgeries requires the display of live video and the collection of low-magnification time-lapse sequences. To meet this need we use a stereoscope with a trinocular head (Zeiss SV6), a c-mount, a CCD video camera (Hamamatsu) controlled by a frame grabber (LG3, Scion Corporation) mounted in a personal computer (Apple PowerMac 7600) and running image acquisition software (NIH-Image version 1.62, Wayne Rasband NIMH). We have mounted the computer and camera controller onto a small mobile cart (AnthroCart) that enables the system to be moved from the dissecting station after time-lapse collection for later morphometric analysis. There are numerous limitations to this imaging system (analog video from the CCD and the 8 bit pixel depth restrictions of NIH-Image), but they are far outweighed by the low cost and ease of use.

L. D. is an American Cancer Society Postdoctoral Fellow. R.K. is supported by grants from the National Institutes of Health (NIH-NICHD HD25595).

REFERENCES

Allen, R. D., G. B. David, and G. Nomarski. The Zeiss-Nomarski differential interference equipment for transmitted light microscopy. *Z. Wiss. Mikrosk.* 69:193–221, 1969.

Denk, W., J. H. Strickler, and W. W. Webb. Two-photon laser scanning fluorescence microscopy. *Science* 248:73–76, 1990.

Goldstein, D. J. *Understanding the Light Microscope: A Computer-Aided Introduction.* San Diego: Academic Press, 1999.

Haugland, R. P. *Handbook of Fluorescent Probes and Research Chemicals.* Eugene, OR: Molecular Probes, Inc., 1994.

Heim, R., A. B. Cubitt, and R. Y. Tsien. Improved green fluorescence [letter]. *Nature* 373: 663–664, 1995.

Inoue, S., and K. Spring. *Video Microscopy,* 2nd Ed. New York: Plenum Press, 1997.

Keller, H. E. Objective lenses for confocal microscopy. In: *Handbook of Biological Confocal Microscopy,* edited by J. B. Pawley. New York: Plenum Press, 1995, pp. 111–126.

Minsky, M. Microscopy apparatus. *U.S. Patent 3,013,467,* 1957.

Ploem, J. S. The use of a vertical illuminator with interchangeable dielectric mirrors for fluorescence microscopy with incident light. *Z. Wiss. Mikrosk.* 68:129–142, 1967.

Presnell, J. K., and M. P. Schreibman. *Humason's Animal Tissue Techniques,* 5th Ed. Baltimore: Johns Hopkins University Press, 1997.

Spencer, M. *Fundamentals of Light Microscopy.* Cambridge: Cambridge University Press, 1982.

Zernike, F. The wave theory of microscopic image formation. In: *Concepts of Classical Optics,* edited by J. Strong. San Francisco: W. H. Freeman and Co, 1958, pp. 525–536.

5

Laser Scanning Confocal Microscopy Applied to Living Cells and Tissues

JOHN J. LEMASTERS, TING QIAN, DONNA R. TROLLINGER,
BARBARA J. MULLER-BORER, STEVEN P. ELMORE, AND WAYNE E. CASCIO

1. INTRODUCTION

Single cells often respond to stimuli heterogeneously. Thus, bulk biochemical and physiological measurements can fail to accurately represent the magnitude and time course of individual cellular changes. Responses within single cells may also show spatial heterogeneity. For these reasons, three-dimensionally resolved approaches are needed to study individual cells as they respond to imposed stimuli and stresses. Increasingly, confocal microscopy of parameter-specific fluorophores is the method of choice to assess single cell physiology with high spatial and temporal resolution.

2. OPTICAL PRINCIPLES OF CONFOCAL MICROSCOPY

2.1 Lateral Resolution

The diameter of the smallest self-luminous object resolvable by light microscopy, r, is given by the Abbe equation

$$r = 0.6\, \lambda/NA \qquad (1)$$

where λ is wavelength and NA is the numerical aperture of the objective lens of the microscope. Numerical aperture is defined as $n \sin \alpha$, where n is the index of refraction of the object medium and α is the half angle of light collection of the objective. For a given wavelength, NA characterizes the ability of an objective lens to resolve small objects. For green light (0.54 nm) and an NA 1.4 oil immersion objective lens, lateral resolution is 0.23 μm.

2.2 Superimposition with Thick Specimens

The axial distance in the specimen space that is in focus in a single image is the depth of field. Effective depth of field at high power is 2–3 μm for conventional light microscopes. Because lateral resolving power is much greater than this depth of field, superimposition of features with the specimen degrades overall image quality and lateral resolution. In thicker specimens, superimposition of light from out-of-focus planes causes further image deterioration. In fluorescence microscopy, especially, out-of-focus fluorescence produces diffuse halos around in-focus structures under observation. Confocal microscopy eliminates these artifacts by creating images of very thin optical slices through the specimen. Such confocal images reject light from out-of-focus planes almost completely. The smaller depth of field and the rejection of out-of-focus information achieved by confocal microscopy produce images of remarkable detail and quality.

2.3 Pinhole Principle

In confocal microscopy the microscope objective focuses the illuminating light, which is typically from a laser, to a small spot within the specimen (Fig. 5–1) (Minsky, 1961). At crossover, spot diameter is limited by the NA of the objective lens to about 0.2 μm for a higher power oil immersion lens. The same objective lens then collects light reflected or fluoresced by the specimen. Reflected and fluoresced light then passes through a prism or dichroic mirror to separate it from the illuminating light. Simultaneously, the objective focuses this light onto a small pinhole. Light originating from the in-focus crossover point of the illuminating light beam is focused to a small spot precisely at the pinhole. This in-focus light passes through the pinhole to a photomultiplier beyond.

Light originating from above and below the focal plane forms crossover spots behind and in front of the pinhole, respectively. Consequently, light from planes above and below the plane of focus is spread out at the pinhole and selectively blocked (Fig. 5–1). The simple placement of this pinhole in the objective image plane thus permits selective transmission of in-focus light with rejection of out-of-focus light. This optical design is the basis for creating thin optical slices through thick specimens.

2.4 Image Formation

In laser scanning confocal microscopy, a scan generator moves the excitation light beam from the laser physically across the specimen (White et al., 1987). The scan generator also "descans" fluoresced light so that it may be focused on the stationary pinhole. Light traversing the pinhole strikes a photomultiplier, whose output is stored in computer memory in relation to the x–y position of the laser beam on the specimen. An image is created when this information is displayed on a computer monitor. Together, the laser, dichroic mirror, scan generator, objective lens, pinhole, photomultiplier, and computer form the essential elements of a laser scanning confocal microscope (Fig. 5–1).

2.5 Axial Resolution

The diameter of the pinhole relative to the magnification and NA of the objective lens determines the axial resolution of a confocal microscope. For a given lens, the

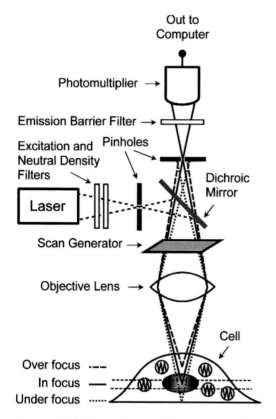

FIGURE 5–1 Scheme of an upright laser scanning confocal microscope. The laser beam passes through neutral density and barrier filters and is focused on a pinhole, which acts as a point source of light. The dichroic mirror reflects the laser beam through the x–y scan generator and objective lens onto the specimen. Fluoresced (or reflected) light passes back through the objective, scan generator, dichroic mirror, and emission barrier filter to be focused on a pinhole placed in the image plane. Light from the in-focus specimen plane passes through the pinhole to a photomultiplier. Light from out-of-focus planes is spread out at the pinhole and does not pass through to the photomultiplier. Photomultiplier output goes to a computer.

smaller the pinhole, the smaller the thickness of the confocal slice. Once the pinhole approximates the size of the diffraction-limited spot (Airy disk) of in-focus light that is projected on the pinhole, however, axial resolution is no longer improved by further decreases of pinhole diameter. Axial resolution at the diffraction limit is given by the following equation:

$$Z = 2n\lambda/(NA)^2 \tag{2}$$

where Z is the diffraction-limited axial resolution (Inoué, 1995).

2.6 Lasers for Confocal Microscopy

Lasers emit one or a few narrow lines of light at a time. The relatively small choice of laser wavelengths constrains the types of fluorophores that can be used for confocal microscopy. Commonly used lasers for confocal microscopy include argon (488 and 514 nm); argon-krypton (488, 568, and 647 nm), helium-cadmium (442 nm), helium-neon (543, 596, and 633 nm), and ultraviolet (UV)-argon (351–364 nm). The argon-krypton laser with well-separated blue, yellow, and red lines has become popular for biological applications, but maintenance costs are high due to frequent laser tube replacement. Multiple lasers can be used instead, such as the relatively inexpensive and durable argon and helium-neon lasers. Multiple lasers or a multiline laser can excite several fluorophores at once. By separating different colors of fluorescence using dichroic and barrier filters, the specific fluorescence of multiple fluorophores may be simultaneously imaged using multiple detectors. Nonconfocal bright-field images may also be simultaneously collected using a transmitted light detector. There are a number of companies (Biorad, Leica, Nikon, Olympus, and Zeiss) that use different kinds of laser combinations and couple them to their scanning head, either fiberoptically or directly. These turnkey laser scanning confocal microscope systems cost $100,000–$500,000 depending on the configuration of the unit.

3. Specimen Preparation

For high-resolution imaging, living cells must first be plated on number $1\frac{1}{2}$ glass coverslips. Such coverslips have a thickness of 170 μm, which provides the proper optical match to nearly all microscope objectives. Unfortunately, cells tend to adhere and grow poorly on glass. To improve adherence, coverslips may be treated with type I collagen, laminin, or polylysine. After coverslips are rinsed in ethanol for sterilization and drying, 1–2 drops (50–100 μl) of rat tail collagen (1 mg/ml in 0.1% acetic acid), laminin (0.1 mg/ml in TRIS-buffered saline), or polylysine (1 mg/ml in saline) are spread out across one surface of the coverslip. After the coverslips are air dried and placed in Petri dishes, cells are plated in the usual fashion. Alternatively, plastic Petri dishes are available with coverslips glued to an opening in their bottoms, but coatings may still be necessary.

To image the cells, the coverslip or Petri dish must be mounted in a special chamber, which can be built in a machine shop or purchased commercially. An open chamber may be used on an inverted microscope, which permits direct access to the cells to add and replace reagents. Closed chambers may be used with both upright and inverted microscopes. If needed, temperature may be regulated by heaters built into the specimen chamber or a stream of tempered air.

4. Minimizing Photodamage

4.1 Relative Dimness of Confocal Images

Because confocal images are formed from light arising from only a thin optical section of a thick specimen, confocal images are much dimmer than wide-field fluorescence images of the same specimen. Accordingly, confocal microscopy requires brighter illumination to generate images of the same brightness as wide-field fluorescence microscopes. Additional light losses compared with wide-field microscopy arise from the greater number of reflective and refractive surfaces in the scan generators of confocal microscopy. Also, photomultipliers used in laser scanning confocal microscopy have a lower quantum efficiency than the best charge-coupled devices (CCDs) used in wide-field fluorescence microscopy (see Chapter 3). Consequently, specimens viewed by confocal microscopy will be exposed to more light than specimens imaged by wide-field fluorescence microscopy and low-light cameras. For this reason, confocal specimens may be vulnerable to photobleaching and photodamage.

4.2 Objective Lenses

Several strategies can minimize and even eliminate photobleaching and photo-damage. Objective lenses of the highest NA should be used to maximize light collection because the efficiency of fluoresced light collection varies with NA^2. For cell monolayers, oil immersion objective lenses with an NA of 1.3–1.4 are preferred. For imaging less than 5 μm into living specimens, oil immersion objective lenses produce sharp and bright confocal images. The images deteriorate, however, as the plane of focus moves deeper than 5 μm into an aqueous specimen. The oil/water interface causes a spherical aberration, which becomes more severe as the specimen plane moves deeper into the aqueous phase (Keller, 1995). This spherical aberration decreases image brightness and lateral resolution. High NA water immersion objective lenses eliminate this aberration and should be used with thick biological specimens. Another potential problem is that most objective lenses are designed for visible light and may not be corrected in the UV and infrared spectrum. Thus, applications with UV or infrared light may require special lenses designed for this purpose.

4.3 Laser Attenuation

One important step in decreasing photobleaching and photodamage is the attenuation of the laser with neutral density filters and a reduced power setting. By increasing photomultiplier gain, good images can often be collected even with attenuations of 300- to 3000-fold. As a rule, lasers should be attenuated to where image quality begins to become unacceptable. Excess attenuation causes a high spatial signal-to-noise ratio, creating a "snowy image" because too few photons are measured from each picture element (pixel). The human eye can easily discern approximately 15 levels of gray. Because the standard deviation of measurement is equal to the square root of the number of photons measured, each pixel must detect 800 photons for a standard deviation of ±3.5%. If this precision is expected for lower gray levels, for example, a pixel intensity of 50, then the highest pixel intensity must represent detection of about 4000 photons.

4.4 Pinhole Diameter

Another strategy to improve sensitivity is to increase pinhole diameter. A larger pinhole transmits more light, which decreases the spatial signal-to-noise ratio without increasing laser illumination. A larger pinhole diminishes axial resolution, but the improvement of sensitivity may be an acceptable trade-off. Such improved sensitivity permits greater laser attenuation to decrease photobleaching and photodamage. For maximum axial resolution, however, pinhole size should be slightly less than the diameter of the diffraction-limited Airy disk that the objective lens projects upon the pinhole. A smaller pinhole diameter will only decrease image brightness without improving axial resolution. Airy disk diameter at the pinhole varies with NA and magnification of the objective lens. Thus, optimal pinhole diameter varies with each objective lens. Many confocal microscope systems employ software to calculate the correct Airy disk diameter and adjust the pinhole accordingly.

4.5 Zoom Magnification

Laser scanning confocal microscopes can "zoom" in continuous increments over about a 10-fold range. Zooming decreases the x and y excursion of the laser beam across the specimen and, hence, increases magnification. With higher zooms, excitation light concentrates into a smaller area of the specimen, which increases photodamage and photobleaching. Hence, lower zoom magnification decreases phototoxicity. In general, the lowest zoom setting should be used that is consistent with the particular needs of an experiment.

A beginner's mistake is to preview samples at high laser excitation and zoom magnification. Good, bright images are obtained initially, but photobleaching and photodamage soon make collection of decent images impossible, even after laser attenuation. Survey images should always be collected judiciously at low illumination and zoom magnification. When an area of interest is found, then zoom magnification and/or laser intensity can be increased to achieve the desired image characteristics.

4.6 Fluorophore Characteristics

The type of fluorophores used also determines the rate at which photodamage and photobleaching occur (see Chapter 2). Rhodamine dyes, for example, are relatively stable in the light beam, whereas fluorescein and acridine orange bleach rapidly and can generate toxic photoproducts. Decreased photobleaching may be achieved by increasing fluorophore concentration because increased fluorophore permits greater laser attenuation for the same amount of fluorescence. Ion- and membrane potential–indicating fluorophores at high concentrations may, however, interfere with the parameter they measure. The biological significance of any interference must be determined empirically.

5. Biological Applications to Living Cells

5.1 Cell Viability and Integrity

Living cells are impermeable to large hydrophilic molecules unless a specific plasma membrane transporter exists. When cells lose viability, their permeability barrier fails. As a consequence, otherwise impermeant molecules cross freely in and out of the cell interior. For example, at loss of cell viability, trypan blue enters cells to stain their nuclei. Similarly, propidium iodide enters nonviable cells and binds to nuclear DNA (Fig. 5–2). This binding to DNA enhances the red fluorescence of propidium iodide, which is easily detected by fluorescence microscopy (Lemasters et al., 1987).

At onset of cell death, cells release soluble intracellular components, including ions, metabolites, and enzymes. Similarly, cells release trapped intracellular fluorophores, such as ester-loaded fluorescein and calcein, after loss of cell viability (Jones and Senft, 1985; Zahrebelski et al., 1995). Fluorescein and calcein are loaded as their neutral, membrane-permeant acetate or acetoxymethyl ester derivatives. After these neutral esters cross into cells, endogenous esterases release the membrane-impermeant free acid forms of these fluorophores into the cytosol. When viability is lost, the trapped cytosolic dye escapes into the extracellular medium, causing cellular fluorescence to abruptly disappear. Conversely, fluorophores in the extracellular medium gain entrance to the cell interior (Color Fig. 5–2). Ideally, cells should retain such fluorophores only until viability is lost. Plasma membrane transporters in many cells, however, translocate anionic fluorophores out to cause a gradual decline of fluorescence from the cytoplasm (Wieder et al., 1993). Thus, retention of fluorophores like calcein may not be as reliable an indicator of cell viability as nuclear exclusion of dyes like propidium iodide.

5.2 Cell Volume and Topography

Cell shape and surface topography are characteristic features of specific cell types. By using calcein to label the intracellular space, serial confocal images through the entire thickness of individual cells may be collected to reconstruct cell volume, shape, and surface topography (Chacon et al., 1994; Zahrebelski et al., 1995). Such reconstructions, or volume renderings, utilize contour-dependent shading to display surface detail, as illustrated for a cultured adult rabbit cardiac myocyte (Fig. 5–3), and they rival scanning electron micrographs in their depiction of surface detail.

Although scanning electron microscopy has greater spatial resolution, confocal microscopy allows consecutive imaging of living cells. This is unlike electron microscopy, which requires chemical fixation, dehydration, and metal coating of specimens. Confocal microscopy also allows serial measurement of cell volume over time.

5.3 Mapping Electrical Potentials with Cationic Fluorophores

The plasma membrane and mitochondrial inner membrane normally maintain negative-inside membrane potentials ($\Delta\Psi$). Membrane-permeant cationic fluorophores, such as rhodamine-123, tetramethylrhodamine methyl ester (TMRM), and tetramethylrhodamine ethyl ester (TMRE), accumulate electrophoretically into negatively charged compartments in response to these membrane potentials (Ehrenberg et al., 1988; Emaus et al., 1986; Johnson et al., 1981). At equilibrium, the accumulation of such permeant fluorophores is related to $\Delta\Psi$ by the Nernst

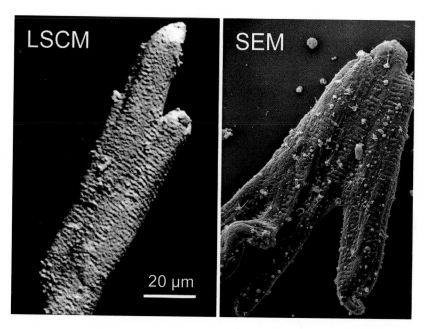

FIGURE 5–3 Confocal volume reconstruction and scanning electron micrograph of cardiac myocytes. On the left is a computer-generated three-dimensional volume reconstruction of a calcein-loaded adult rabbit cardiac myocyte in culture, which was created from serial confocal images collected through the entire thickness of the myocyte. On the right is a scanning electron micrograph of another cardiac myocyte at about the same magnification. [Adapted from Chacon et al., 1994.]

equation:

$$\Delta\Psi = -59 \log F_{in}/F_{out} \qquad (3)$$

where $\Delta\Psi$ is the electrical potential difference and F_{in} and F_{out} are concentrations of fluorophore inside and outside the compartment of interest. From confocal images, intracellular maps of the distribution of electrical potential can be generated using Equation 3. From such maps, $\Delta\Psi$ between the mitochondria and the cytosol and between the cytosol and the extracellular space can be estimated, as illustrated in Figure 5–4 (Chacon et al., 1994).

Typically, plasma membrane $\Delta\Psi$ ranges between −30 and −90 mV, and mitochondrial $\Delta\Psi$ ranges between −120 and −180 mV. Because mitochondria reside inside the cytosol, they are as much as 270 mV more negative than the extracellular space. Thus, mitochondria will accumulate monovalent cationic fluorophores to concentration ratios of up to 30,000 to 1 relative to the extracellular space. Because fluorescence intensity is proportional to concentration, such large concentration gradients cannot be represented using a linear scale of 256 gray levels per pixel (8 bits) of conventional computer memory. Rather, the confocal images must be recorded using 16 bit memory (65,536 gray levels) or a nonlinear logarithmic scale. Alternatively, separate 8 bit images may be collected serially at two different laser illumination intensities. One image is collected at a low laser power

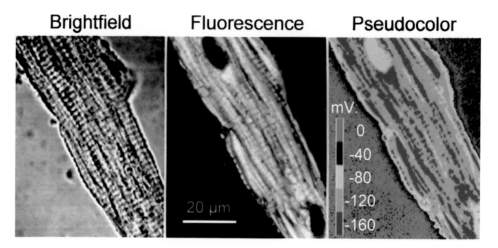

FIGURE 5–4 Distribution of electrical potential in a cardiac myocyte. Red fluorescence of a TMRM-loaded adult rabbit cardiac myocyte was imaged by laser scanning confocal microscopy. The left panel is a non-confocal bright-field image collected with a transmitted light detector. The middle panel is the unprocessed fluorescence image of red TMRM fluorescence. The right panel displays a pseudocolor map of the intracellular distribution of electrical potential. [Adapted from Chacon et al., 1994.]

setting to measure areas with bright fluorescence, typically the mitochondria. A second image of the same field is then recorded at 30- to 100-fold higher laser power. This image records areas with weak fluorescence. Subsequently, these images are combined into a single 16 bit image, taking into account the relative difference in laser intensity. The Nernst equation is then applied on a pixel-by-pixel basis to the combined 16 bit image using average extracellular fluorescence as a reference to generate a two-dimensional map of electrical potential.

Figure 5–4 illustrates a map of $\Delta\Psi$ with respect to the extracellular space in a cultured cardiac myocyte. $\Delta\Psi$ in the nucleus and areas under the sarcolemma is about -80 mV. Because extracellular Ψ is 0, -80 mV represents $\Delta\Psi$ across the plasma membrane. Mitochondria can have a $\Delta\Psi$ of -200 mV relative to the extracellular space. The difference between mitochondrial and cytosolic electrical potential, -120 mV, represents $\Delta\Psi$ across the mitochondrial membrane. Mitochondria typically show heterogeneity of $\Delta\Psi$. This heterogeneity arises, at least in part, from the fact that the thickness of the confocal section is only slightly smaller than mitochondrial diameter. Thus, many confocal sections extend only part way through mitochondria. As a consequence, $\Delta\Psi$ for many mitochondria is underestimated. For this reason, the best overall estimate of mitochondrial $\Delta\Psi$ is the maximal observed $\Delta\Psi$ for mitochondria fully sectioned by the confocal slice (Chacon et al., 1994).

The Nernst equation assumes that fluorophores ideally distribute in response to electrical potential. Some probes, however, show nonideal behavior. For example, low-affinity and high-capacity binding to mitochondria occurs with rhodamine-123 (Emaus et al., 1986). As this binding occurs, quenching of rhodamine-123 fluorescence occurs. Such quenching is concentration dependent and may be decreased by using smaller amounts of fluorophore.

5.4 Artifacts with Potential-Indicating Fluorophores

Electrophoretic uptake of fluorophores may lead to millimolar accumulation into the mitochondrial matrix. Such concentrations can cause metabolic inhibition. Rhodamine-123, for example, inhibits the mitochondrial ATP synthase of oxidative phosphorylation at millimolar concentrations (Emaus et al., 1986). $DiOC_6$, another cationic fluorophore used in both flow cytometry and fluorescence microscopy, inhibits mitochondrial respiration even more strongly (Rottenberg and Wu, 1997). Other fluorophores, such as TMRM and TMRE, can also inhibit respiration at higher concentrations than necessary for confocal imaging (Scaduto and Grotyohann, 1999). As a precaution, however, low fluorophore concentrations (<500 nM) should always be used to minimize quenching and metabolic effects.

JC-1 is another cationic fluorophore used to monitor mitochondrial $\Delta\Psi$ (Reers et al., 1995; Smiley et al., 1991). JC-1 fluoresces green, but after accumulation into mitochondria JC-1 forms so-called J-aggregates that fluoresce red. As the magnitude of mitochondrial $\Delta\Psi$ increases, mitochondrial accumulation of JC-1 increases and more J-aggregates form. Because of increasing J-aggregate formation, the ratio of red (J-aggregate) to green (monomeric) fluorescence becomes a relative indicator of $\Delta\Psi$. By laser scanning confocal microscopy, individual J-aggregates can be visualized inside mitochondria, which gives the false impression that mitochondrial matrix electrical potential is inhomogeneous. JC-1 is used widely to monitor changes of mitochondrial $\Delta\Psi$ by non-confocal microscopy and other fluorescence techniques, especially flow cytometry. Additional information on fluorophores used for mitochondrial imaging can be found in Chapter 6.

5.5 Ion Imaging

Ion-indicating fluorophores may be either microinjected into cells or, more commonly, ester-loaded to trap the ion-sensitive free acid form of the fluorophores inside the cells. In wide-field microscopy, fluorescence of these fluorophores depends not only on the ion being measured but also very strongly on cell shape and thickness. Because confocal microscopy collects an optical slice of uniform thickness, variations due to cell morphology are largely offset, but not entirely because heterogeneous uptake into organelles can lead to heterogeneous intracellular distribution of these fluorophores.

For many fluorophores, a ratioing technique can correct for variations of fluorophore concentration in the light path (Tanasugarn et al., 1984). If ion binding causes a shift of the fluorescence excitation or emission spectrum, then images may be acquired at two different excitation or emission wavelengths, one that increases as the ion of interest increases and another that decreases or stays the same (Fig. 5–5). Background images from cell-free regions representing nonspecific fluorescence must also be collected and subtracted from the experimental images. The next step is to divide the image at one wavelength by the image at the second wavelength on a pixel-by-pixel basis. The resulting ratios are proportional only to ion concentration and not to the amount of fluorophore in the light path. Using a standard curve, ratio values are converted to ion concentrations, which are often displayed using a pseudocolor scale to better highlight small differences. Overall,

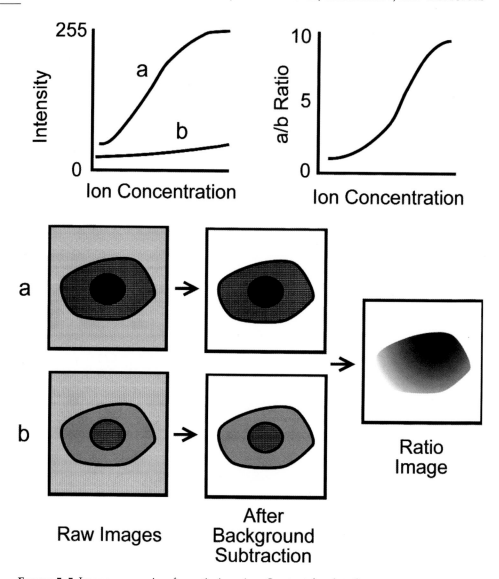

FIGURE 5–5 Image processing for ratio imaging. See text for details.

this ratio procedure corrects for heterogeneity of intracellular fluorophore loading, dye leakage over time, and photobleaching (Fig. 5–5).

5.6 Imaging pH

For laser scanning confocal microscopy, emission wavelengths rather than excitation wavelengths must usually be ratioed because of the very limited number of excitation wavelengths available from laser light sources. To measure pH by confocal microscopy, carboxyseminaphthorhodafluor-1 (SNARF-1) is a useful ratiometric indicator. With 568 nm excitation light from an argon-krypton laser, SNARF-1 emission at 640 nm increases as pH increases, whereas fluorescence at 585 nm remains the same (Chacon et al., 1994; Chapter 2). Thus, pH is proportional to the 640/585 nm ratio. SNARF-1 can also be ratio-imaged using 488 nm excitation light from an argon laser, but the fluorescence is less bright.

For the best results, the SNARF-1 signal should be calibrated in situ because the intracellular environment can affect the pK_a of the dye. For calibration, cells are incubated at different pH values in the presence of 10 μM nigericin and 5 μM valinomycin, ionophores that equilibrate intracellular and extracellular pH. To minimize swelling, NaCl and KCl are replaced with their corresponding gluconate salts. After ester loading of SNARF-1 AM (2–5 μM), SNARF-1 labels both the cytosol and the mitochondria.

Ratio imaging of SNARF-1 fluorescence by confocal microscopy shows heterogeneity of pH within the cytoplasm of both hepatocytes and myocytes (Fig. 5–6). The pH of the cytosol and the nucleus is 7.1–7.2, whereas mitochondrial pH is 7.8–8. These findings show directly that a pH gradient (ΔpH) exists across the mitochondrial inner membrane of 0.6–0.8 pH units. Because mitochondrial $\Delta\Psi$ in these cells is −120 mV, the protonmotive force ($\Delta p = \Delta\Psi - 59\ \Delta$pH) of mitochondria in these intact living cells is −170 to −185, mV in agreement with the

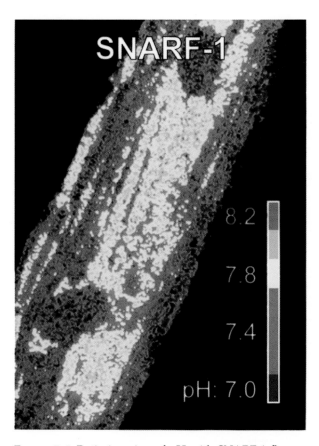

Figure 5–6 Ratio imaging of pH with SNARF-1 fluorescence. A cardiac myocyte was ester-loaded with SNARF-1. Confocal images at emission wavelengths of 585 and 640 nm were collected and ratioed to determine intracellular pH. Areas of high pH correspond to mitochondria. [Adapted from Chacon et al., 1994.]

chemiosmotic hypothesis. SNARF-1 may also be used to measure intracellular and extracellular pH in intact, arterially perfused cardiac ventricular papillary muscle to a depth of 300 μM (Muller-Borer et al., 1998).

Ester loading of fluorophores shows strong temperature dependence (Nieminen et al., 1995; Ohata et al., 1998). For example, incubation of hepatocytes and cardiac myocytes with calcein acetoxymethyl ester at 37°C leads predominantly to calcein uptake into the cytosol and the nucleus (Color Fig. 5–7). Mitochondria exclude the green-fluorescing cytosolic marker and appear as round dark voids in the diffuse cytosolic fluorescence. These voids are positively identified as mitochondria from the uptake of red-fluorescing TMRM. By contrast, cold ester loading at 4°C leads to calcein uptake into both the cytosol and the mitochondria (Color Fig. 5–7). Evidently, the colder temperature inhibits cytosolic esterases to the extent that the neutral esters can diffuse through the cytosol to enter mitochondria. Subsequently, mitochondria esterases hydrolyze the acetoxymethyl group to trap the free acid inside the organelles.

5.7 Free Ca^{2+}

Green-fluorescing Fluo-3 and Fluo-4 and red-fluorescing Rhod-2 are useful fluorophores for nonratiometric imaging of intracellular free Ca^{2+} by confocal microscopy (Haugland, 1999; Minta et al., 1989). Ca^{2+} binding to these dyes causes a 50-fold or greater increase of fluorescence, and the K_d for Ca^{2+} binding is 300–600 nM. Because Ca^{2+} does not shift the fluorescence spectrum of these fluorophores, calibration relies on in situ measurements of maximal and minimal fluorescence after respective additions of a calcium ionophore (e.g., ionomycin or Br-A23187) and EGTA (Harper et al., 1993; Minta et al., 1989). Even without calibration, fluorescence of Fluo-3, Fluo-4, and Rhod-2 provides a sensitive indicator of relative changes of free Ca^{2+}.

5.8 Line Scanning Confocal Microscopy of Ca^{2+} Transients

With nearly all commercial confocal microscopes, collection of two-dimensional images of good signal-to-noise characteristics requires 2–10 s. To improve temporal resolution, the line-scanning mode may be used whereby the x-axis scan is repeated at the same y-axis position to create x versus time images with a time resolution of 25 ms or less (Cheng et al., 1993). For example, Color Figure 5–8 shows simultaneous red and green line-scan images of myocytes co-loaded with green fluorescing Fluo-3 and red fluorescing TMRM. In the x-axis versus time line-scan images, TMRM fluorescence appears as wavy vertical stripes. Each stripe is a single mitochondrion, and each wave in the stripe is cell movement after an electrically stimulated contraction. Fluo-3 fluorescence, in contrast, shows only horizontal bands that correspond to the intracellular Ca^{2+} transients induced by electrical stimulation. This banding occurs both in pixels corresponding to TMRM-labeled mitochondria and in those of the TMRM-unlabeled cytosol. Such results demonstrate that Ca^{2+} transients occur in both the cytosol and the mitochondria during the contractile cycle (Chacon et al., 1996; Ohata et al., 1998).

The line scan in Color Figure 5–8 goes through the nucleus, the cytosol, and both interfibrillar and perinuclear mitochondria. Regions corresponding to the cytosol and the nucleus are unstained by TMRM, whereas interfibrillar and perinuclear mitochondria have bright TMRM labeling. Pixels of intermediate TMRM fluorescence represent the overlap of mitochondrial and cytosolic spaces. Based on this TMRM staining and morphological criteria, pixels can be subdivided into interfibrillar or perinuclear mitochondria (high TMRM fluorescence) and nucleus or cytosol (low TMRM fluorescence). Fluo-3 fluorescence in the pixels of each of the identified cellular compartments is then averaged for each horizontal line in the line scan. By then averaging the repeating transients of Fluo-3 fluorescence, a profile of the Ca^{2+} transient in each compartment is obtained, which shows that the peak change after field stimulation is greatest in the cytosol and least in perinuclear mitochondria (Fig. 5–9). The β-adrenergic agonist isoproterenol increases the magnitude of the Ca^{2+} transient and its subsequent rate of decay. In these experiments, red and green fluorescence are collected simultaneously. Thus movement

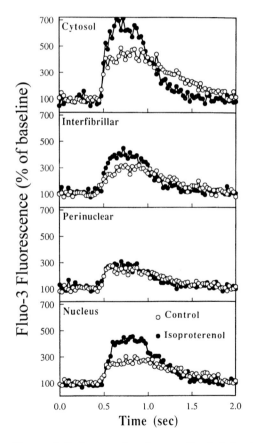

Figure 5–9 Averaged and normalized plots of Fluo-3 fluorescence during electrical stimulation in different regions of a cardiac myocyte. From line-scan images in Color Figure 5.8, TMRM fluorescence was used to identify the cytosol, interfibrillar mitochondria, perinuclear mitochondria, and nucleus. Fluo-3 intensity was then calculated for each region, averaged for the repeated stimulations, normalized as percentage of diastolic intensity, and plotted versus time. [After Ohata et al., 1998.]

during myocyte contraction does not affect the measurement, as would occur if red and green fluorescence were measured sequentially (Chacon et al., 1996; Ohata et al., 1998).

5.9 Ratio Imaging of Free Ca^{2+}

Fluo-3 fluorescence increases after Ca^{2+} binding, but no wavelength shift occurs. Therefore, absolute Ca^{2+} concentrations cannot be estimated by ratio imaging. Indo-1 is a Ca^{2+} indicator whose emission spectrum shifts after Ca^{2+} binding (Grynkiewicz et al., 1985). This property allows Indo-1 to be used for confocal ratio imaging. Indo-1, however, requires excitation with UV light, which necessitates a UV laser and UV-corrected optics. As Ca^{2+} concentration increases, Indo-1 fluorescence increases and decreases, respectively, at emission wavelengths of 405 and 480 nm. After ester loading (2–5 μM Indo-1 AM), Indo-1 distributes into both the cytosol and the mitochondria of cardiac myocytes even at 37°C (Ohata et al., 1998). In unstimulated myocytes, ratio images of Indo-1 fluorescence are uniform. Estimated intracellular free Ca^{2+} concentration is about 200 nM, as based on an ex situ calibration. This may be an overestimation because the binding constant of Indo-1 for Ca^{2+} changes in the intracellular environment (Baker et al., 1994; Bassani et al., 1995; Roe et al., 1990). Hydrogen ions also compete with Ca^{2+} for binding to Indo-1 and other BAPTA-derived probes, and significant changes of the K_d for Ca^{2+} occur when pH falls below about 6.8 (Kawanishi et al., 1991; Lattanzio and Bartschat, 1991). Although intracellular calibration of Indo-1 remains problematic in practice, Indo-1 ratio imaging nonetheless corrects for variable loading of the indicator into various cellular compartments and confirms that Ca^{2+} transients occur in both mitochondria and the cytosol during the excitation-contraction cycle.

5.10 Cold Loading of Ca^{2+} Indicators Followed by Warm Incubation

Ester-loaded fluorophores in the cytosol gradually leak from many cells through an organic anion carrier in the plasma membrane (Wieder et al., 1993). Mitochondria and other organelles release these fluorophores much more slowly. The selective release of fluorophores from the cytosol allows selective mitochondrial localization of ester-loaded fluorophores to be achieved. A two-stage loading protocol is employed. First, cells are ester loaded in the cold to load fluorophores into both the cytosol and the mitochondria. Second, the cells are incubated at 37°C to allow cytosolic but not organellar fluorophores to leak into the extracellular medium (Lemasters et al., 1999; Qian et al., 1999; Trollinger et al., 1997). In this way, mitochondrial fluorophores are retained, but cytosolic fluorophores are lost.

After cold loading and warm incubation with Rhod-2 AM (5 μM), red fluorescence in cardiac myocytes shows a distinctly mitochondrial pattern, comparable with TMRM or rhodamine-123 (Color Fig. 5–10a) (Trollinger et al., 1997, 2000). During electrical stimulation, mitochondrial Rhod-2 fluorescence increases with each stimulation, as evidenced by horizontal bands in the 16 s confocal scans (Color Fig. 5–10b). These fluorescence transients occur in regions showing a mitochondrial pattern of labeling and are inhibited by ruthenium red, a blocker of electrogenic Ca^{2+} uptake by mitochondria (Color Fig. 5–10c). The mitochondrial distribution of Rhod-2 is visualized by placing the myocytes in low Na^+ medium, which increases intracellular Ca^{2+} by reversal of Na^+/Ca^{2+} exchange. Fluorescence now shows the distribution of intracellular Rhod-2 and verifies the mitochondrial, noncytosolic localization of Rhod-2 (Color Fig. 5–10d). After treatment in this way, Rhod-2 fluorescence is punctate instead of diffuse, although mitochondrial morphology is distorted by the Ca^{2+}-induced contracture (Color Fig. 5–10d). Contracture can be prevented by butanedione monoxime (Armstrong and Ganote, 1991), in which case the pattern of punctate Rhod-2 fluorescence after Br-A23187 is virtually identical to that of rhodamine-123 and TMRM-labeled (Trollinger et al., 1997). Importantly, cytosolic areas between mitochondria are dark, showing that Rhod-2 labeling of the cytosol is minimal.

The myocytes in Color Figure 5–10 were simultaneously warm loaded with Fluo-3. Warm ester loading leads to diffuse cytosolic localization of Fluo-3 (Color Fig. 5–10a′). Transients of Fluo-3 fluorescence occur after electrical stimulation, but, unlike the Rhod-2 transients, Fluo-3 transients are not blocked by ruthenium red. Moreover, Fluo-3 florescence increases diffusely throughout the cytosol after the cells are placed in low Na^+ medium (Color Fig. 5–10d′). This contrasts markedly with the punctate distribution of Rhod-2. In comparison with Ca^{2+} measurements using co-loaded TMRM to identify mitochondrial pixels, measurements of mitochondrial Ca^{2+} using indicators localized exclusively to mitochondria has distinct advantages. With TMRM co-loading, pixels with intermediate TMRM fluorescence must be excluded from analysis because such pixels represent areas in the confocal slice where mitochondria and the cytosol overlap. After cold loading of Rhod-2 followed by warm incubation, all the Ca^{2+}-indicating fluorophore is mitochondrial, and every pixel can be used for analysis. The result is increased signal strength and signal-to-noise ratio.

5.11 Artifacts Arising from Lysosomal Localization of Ca^{2+}-Indicating Fluorophores

Lysosomes may be visualized by incubating cells with dextran conjugates, such as rhodamine-dextran, which is endocytosed and concentrated within lysosomes as an indigestible residue (see Color Fig. 5–2). Lysosomes and other acidic organelles, including endosomes and autophagosomes, will also label with fluorescent weak bases, such as LysoTracker Red. LysoTracker Red is uncharged at physiological pH and freely enters acidic organelles. Once inside, however, LysoTracker Red becomes protonated and trapped because the charged form cannot go back across the membrane. Because lysosomes contain esterases, ester-loaded Ca^{2+} indicators also accumulate into acidic endosomal/lysosomal compartments that co-label with LysoTracker Red, as illustrated in Color Figure 5–11 for ester-loaded Fluo-3 (Trollinger et al., 2000). Indeed, lysosomes may contain the majority of total ester-loaded Fluo-3 and Rhod-2 after cold loading and warm incubation.

Confocal microscopes typically record fluorescence intensity in only 256 increments (0–255). Intensity levels exceeding a value of 255 are simply recorded as 255.

Ideally, confocal images should be collected so that no pixel exceeds an intensity value of 254. In the absence of saturation, red fluorescence of cardiac myocytes labeled with Rhod-2 by cold ester loading and warm incubation has no relation to the distribution of green-fluorescing mitochondria labeled with rhodamine-123 (compare Color Fig. 5–12A and 5–12B). Rather, Rhod-2 fluorescence has the same pattern of lysosomal fluorescence as shown in Color Figure 5–11. Under these nonsaturating imaging conditions, no fluorescence transients are observed during electrical stimulation (Fig. 5–12B).

In contrast, when the confocal scan is repeated using 100 times more laser excitation intensity, lysosomal fluorescence becomes saturated and is "clipped" to a gray level intensity of 255. With lysosomal saturation, a mitochondrial pattern of fluorescence emerges in the background (Color Fig. 5–12C). This background fluorescence then shows transients during electrical stimulation (Color Fig. 5–12D). In this way, signals arising from the more weakly labeled mitochondria can be observed despite the bright labeling of a relatively small number of lysosomes. These observations resolve the apparent discrepancy between confocal experiments that describe rapid mitochondrial Ca^{2+} transients after electrical stimulation (Chacon et al., 1996; Ohata et al., 1998; Trollinger et al., 1997, 2000) and studies using microfluorometry of total cell fluorescence that fail to detect mitochondrial transients (DiLisa et al., 1993; Griffiths et al., 1997; Miyata et al., 1991). Lysosomes account for the majority of noncytosolic fluorescence after ester loading despite their small size and number. Consequently, spot microfluorometry of whole cellular fluorescence reports predominantly lysosomal fluorescence under these conditions, whereas confocal imaging distinguishes the different subcellular sources of fluorescence. These experiments underscore the advantages of confocal microscopy to study mitochondrial Ca^{2+} fluxes in intact living cells.

5.12 Mitochondrial Permeability

After cytosolic localization of calcein by warm loading, many stresses cause calcein fluorescence to redistribute into mitochondria with simultaneous loss of the mitochondrial membrane potential (Nieminen et al., 1995). These relatively abrupt changes signify the onset of the mitochondrial permeability transition (MPT). Opening of a high-conductance permeability transition (PT) pore in the mitochondrial inner membrane conducting solutes of molecular weight up to 1500 Daltons causes the MPT and leads quickly to mitochondrial depolarization, uncoupling of oxidative phosphorylation, and large-amplitude mitochondrial swelling (Zoratti and Szabo, 1995). The MPT is now recognized as a crucial event for regulating and promoting both necrotic and apoptotic cell killing (Lemasters et al., 1998).

A pH of less than 7 inhibits the PT pore (Bernardi et al., 1992; Halestrap, 1991), and the naturally occurring acidosis of ischemia prevents onset of the MPT. After reperfusion, however, intracellular pH returns to a normal value of 7.1–7.2, and onset of the MPT may occur, resulting in cell death (Qian et al., 1997). Cyclosporin A is a specific inhibitor of the PT pore and prevents pH-dependent onset of the MPT and cell killing after ischemia/reperfusion.

In contrast, when hepatocytes are reoxygenated at normal pH, mitochondria initially begin to repolarize and reaccumulate TMRM, but soon afterward the mitochondria depolarize and release TMRM (Color Fig. 5–13). Simultaneously, calcein redistributes into the mitochondria. These events directly document the onset of mitochondrial permeablization in the living cells. Subsequently, several minutes after the MPT the cell viability is lost, as indicated by propidium iodide nuclear staining. In the rare cell where the MPT fails to occur, cell killing also does not occur (Color Fig. 5–13). Moreover, reperfusion with cyclosporin A or acidotic pH prevents this mitochondrial permeablization and depolarization (Qian et al., 1997). Thus, confocal microscopy shows that the MPT mediates pH-dependent cell killing after ischemia/reperfusion.

For many cell types, mitochondria have an average diameter of 0.3 μm and cannot be visualized by negative contrast in confocal sections that are three times thicker. To visualize the MPT in such cells, mitochondria can be labeled with calcein using the cold ester loading/warm incubation protocol (Qian et al., 1999). At onset of the MPT, calcein then leaks from the mitochondria into the cytosol. Another approach is to load calcein into both the cytosol and the mitochondria and then quench cytosolic fluorescence with 1 mM $CoCl_2$, a potent quencher of calcein fluorescence (Petronilli et al., 1999). Because Co^{2+} does not readily enter mitochondria, mitochondrial fluorescence is preserved, but it is lost at the onset of the MPT. Co^{2+} has adverse toxic effects, however, especially in longer term experiments (Beyersmann and Hartwig, 1992; Hughes and Barritt, 1989; Seghizzi et al., 1994).

5.13 Reactive Oxygen Species and Cellular Thiols

Reactive oxygen species (ROS), including superoxide, hydrogen peroxide, and hydroxyl radical, are formed both during normal metabolism and as a consequence of various pathological processes. Some probes react with ROS to form highly fluorescent derivatives (Haugland, 1999; Nieminen et al., 1997; Perticarari et al., 1991; Rothe et al., 1988). These include dichlorofluorescin, dihydrorhodamine-123, hydroethidium, and dihydroMitoTracker Red. Confocal microscopy can then be used to determine the formation rate and subcellular location of the ROS-sensing fluorophores. Many of these probes will, however, re-localize within the cell after they react with ROS.

Neutral hydroethidium becomes the hydrophilic ethidium cation, which binds to nuclear DNA with a substantial enhancement of fluorescence. Neutral dihydrorhodamine-123 and dihydroMitoTracker Red become lipophilic cations that accumulate electrophoretically into mitochondria. Thus, the intracellular distribution of these probes may not reflect where the ROS were generated within the cells. However, dichlorofluorescein, which is formed after reaction of nonfluorescent dichlorofluorescin with ROS, remains in the compartment where it is formed, at least as long as membrane permeability barriers remain intact (Nieminen et al., 1997). Glutathione and protein thiols are also important targets of ROS, which can be visualized by fluorescence microscopy. Monochlorobimane reacts with nonprotein thiols (mostly glutathione), and monobromobimane reacts with both protein thiols and nonprotein thiols to form highly fluorescent adducts. These adducts can then be imaged by UV fluorescence microscopy to quantify the intracellular distribution of protein and nonprotein thiols in single cells (Nieminen et al., 1991; Bellomo et al., 1992).

NADH and NADPH emit blue light (450 nm) when excited by near-UV light (340 nm), unlike their oxidized counterparts NAD^+ and $NADP^+$. This fluorescence can be imaged by UV confocal microscopy. NAD(P)H fluorescence is, however, highly quenched in the cytosol. Thus, the majority of all NAD(P)H fluorescence visualized by confocal microscopy is mitochondrial (Nieminen et al., 1997). Similarly, the green fluorescence of oxidized mitochondrial flavin can be used to image the oxidation reduction of flavoproteins in the respiratory chain and elsewhere (Hajnoczky et al., 1995).

6. CONCLUSION

Laser scanning microscopy is a powerful tool to identify and localize specific cellular components with a submicron axial resolution. Unlike electron microscopy, confocal optical microscopy allows nondestructive serial observations of living cells and tissues. Numerous parameter-indicating fluorophores are available for live cell imaging that provide specific information on cellular ion concentration, electrical potential, organelle distribution, viability, membrane permeability, and various specific chemical species. These include markers of cell shape and volume, organelles, membrane potentials, ions, and other chemical constituents. Although not discussed here (see Chapter 2), with the advent of green fluorescent protein and its many mutants, virtually any cellular protein can be fluorescently tagged in living cells. Increasingly, confocal fluorescence microscopy of living cells is an indispensable tool for the molecular biologist, cell biologist, and physiologist.

This work was supported, in part, by grants DK37034, HL27430, AG07218, AA09156, and AG13637 from the National Institutes of Health. The Cell and Molecular Imaging Facility is supported, in part, by grants AA11605 and DK34987 to the Bowles Alcohol Research Center Grant and the Center for Gastrointestinal Biology and Disease.

REFERENCES

Armstrong, S. C., and C. E. Ganote. Effects of 2,3-butanedione monoxime (BDM) on contracture and injury of isolated rat myocytes following metabolic inhibition and ischemia. *J. Mol. Cell Cardiol.* 23:1001–1014, 1991.

Baker, A. J., R. Brandes, J. H. Schreur, S. A. Camacho, and M. W. Weiner. Protein and acidosis alter calcium-binding and fluorescence spectra of the calcium indicator indo-1. *Biophys. J.* 67:1646–1654, 1994.

Bassani, J. W., R. A. Bassani, and D. M. Bers. Calibration of indo-1 and resting intracellular [Ca]$_i$ in intact rabbit cardiac myocytes. *Biophys. J.* 68:1453–1460, 1995.

Bellomo, G., M. Vairetti, L. Stivala, F. Mirabelli, P. Richelmi, and S. Orrenius. Demonstration of nuclear compartmentalization of glutathione in hepatocytes. *Proc. Natl. Acad. Sci. U.S.A.* 89:4412–4416, 1992.

Bernardi, P., S. Vassanelli, P. Veronese, R. Colonna, I. Szabo, and M. Zoratti. Modulation of the mitochondrial permeability transition pore. Effect of protons and divalent cations. *J. Biol. Chem.* 267:2934–2939, 1992.

Beyersmann, D., and A. Hartwig. The genetic toxicology of cobalt. *Toxicol. Appl. Pharmacol.* 115:137–145, 1992.

Chacon, E., H. Ohata, I. S. Harper, D. R. Trollinger, B. Herman, and J. J. Lemasters. Mitochondrial free calcium transients during excitation–contraction coupling in rabbit cardiac myocytes. *FEBS Lett.* 382:31–36, 1996.

Chacon, E., J. M. Reece, A.-L. Nieminen, G. Zahrebelski, B. Herman, and J. J. Lemasters. Distribution of electrical potential, pH, free Ca^{2+}, and volume inside cultured adult rabbit cardiac myocytes during chemical hypoxia: A multiparameter digitized confocal microscopic study. *Biophys. J.* 66:942–952, 1994.

DiLisa, F., G. Gambassi, H. Spurgeon, and R. G. Hansford. Intramitochondrial free calcium in cardiac myocytes in relation to dehydrogenase activation. *Cardiovasc. Res.* 27:1840–1844, 1993.

Ehrenberg, B., V. Montana, M.-D. Wei, J. P. Wuskell, and L. M. Loew. Membrane potential can be determined in individual cells from the Nernstian distribution of cationic dyes. *Biophys. J.* 53:785–794, 1988.

Emaus, R. K., R. Grunwald, and J. J. Lemasters. Rhodamine 123 as a probe of transmembrane potential in isolated rat liver mitochondria: Spectral and metabolic properties. *Biochim. Biophys. Acta* 850:436–448, 1986.

Griffiths, E. J., M. D. Stern, and H. S. Silverman. Measurement of mitochondrial calcium in single living cardiomyocytes by selective removal of cytosolic indo 1. *Am J. Physiol.* 273:C37–C44, 1997.

Grynkiewicz, G., M. Poenie, and R. Y. Tsien. A new generation of Ca^{2+} indicators with greatly improved fluorescence properties. *J. Biol. Chem.* 260:3440–3450, 1985.

Hajnoczky, G., L. D. Robb-Gaspers, M. B. Seitz, and A. P. Thomas. Decoding of cytosolic calcium oscillations in the mitochondria. *Cell* 82:415–424, 1995.

Halestrap, A. P. Calcium-dependent opening of a non-specific pore in the mitochondrial inner membrane is inhibited at pH values below 7. Implications for the protective effect of low pH against chemical and hypoxic cell damage. *Biochem. J.* 278:715–719, 1991.

Harper, I. S., J. M. Bond, E. Chacon, J. M. Reece, B. Herman, and J. J. Lemasters. Inhibition of Na^+/H^+ exchange preserves viability, restores mechanical function, and prevents the pH paradox in reperfusion injury to rat neonatal myocytes. *Bas. Res. Cardiol.* 88:430–442, 1993.

Haugland, R. P. *Handbook of Fluorescent Probes and Research Chemicals,* 7th Ed. Eugene OR. Molecular Probes, Inc., 1999.

Hughes, B. P., and G. J. Barritt. Inhibition of the liver cell receptor-activated Ca^{2+} inflow system by metal ion inhibitors of voltage-operated Ca^{2+} channels but not by other inhibitors of Ca^{2+} inflow. *Biochim. Biophys. Acta* 1013:197–205, 1989.

Inoué, S. Foundations of confocal scanned imaging in light microscopy. In: *Handbook of Biological Confocal Microscopy,* 2nd Ed., edited by J. B. Pawley. New York: Plenum Press, 1995, pp. 1–17.

Johnson, L. V., M. L. Walsh, B. J. Bockus, and L. B. Chen. Monitoring of relative mitochondrial membrane potential in living cells by fluorescence microscopy. *J. Cell Biol.* 88:526–535, 1981.

Jones, K. H., and J. A. Senft. An improved method to determine cell viability by simultaneous staining with fluorescein diacetate–propidium iodide. *J. Histochem. Cytochem.* 33:77–79, 1985.

Kawanishi, T., A.-L. Nieminen, B. Herman, and J. J. Lemasters. Suppression of Ca^{2+} oscillations in cultured rat hepatocytes by chemical hypoxia. *J. Biol. Chem.* 266:20062–20069, 1991.

Keller, H. E. Objective lenses for confocal microscopy. In: *Handbook of Biological Confocal Microscopy,* 2nd Ed., edited by J. B. Pawley. New York: Plenum Press, 1995, pp. 111–126.

Lattanzio, F. A., and D. K. Bartschat. The effect of pH on rate constants, ion selectivity and thermodynamic properties of fluorescent calcium and magnesium indicators. *Biochem. Biophys. Res. Commun.* 177:184–191, 1991.

Lemasters, J. J., J. DiGuiseppi, A.-L. Nieminen, and B. Herman. Blebbing, free Ca^{++} and mitochondrial membrane potential preceding cell death in hepatocytes. *Nature* 325:78–81, 1987.

Lemasters, J. J., A.-L. Nieminen, T. Qian, L. C. Trost, S. P. Elmore, Y. Nishimura, R. A. Crowe, W. E. Cascio, C. A. Bradham, D. A. Brenner, and B. Herman. The mitochondrial permeability transition in cell death: A common mechanism in necrosis, apoptosis and autophagy. *Biochim. Biophys. Acta* 1366:177–196, 1998.

Lemasters, J. J., D. R. Trollinger, T. Qian, W. E. Cascio, and H. Ohata. Confocal imaging of Ca^{2+}, pH, electrical potential and membrane permeability in single living cells. In: *Methods in Enzymology, Green Fluorescent Protein*, Vol 302, edited by P. M. Conn. New York: Academic Press, 1999, pp. 341–358.

Minsky, M., *Microscopy Apparatus*, United States Patent 3,013,467, Dec. 19, 1961 (Filed Nov. 7, 1957).

Minta, A., J. P. Y. Kao, and R. Y. Tsien. Fluorescent indicators for cytosolic calcium based on rhodamine and fluorescein chromophores. *J. Biol. Chem.* 264: 8171–8178, 1989.

Miyata, H., H. S. Silverman, S. J. Sollott, E. G. Lakatta, M. D. Stern, and R. G. Hansford. Measurement of mitochondrial free Ca^{2+} contraction in living single rat cardiac myocytes. *Am. J. Physiol.* 261:H1123–H1134, 1991.

Muller-Borer, B. J., H. Yang, S. A. M. Marzouk, J. J. Lemasters, and W. E. Cascio. pH_i and pH_o at different depths in perfused myocardium measured by confocal fluorescence microscopy. *Am. J. Physiol.* 275:H1937–H1947, 1998.

Nieminen, A.-L., A. M. Byrne, B. Herman, and J. J. Lemasters. Mitochondrial permeability transition in hepatocytes induced by *t*-BuOOH: NAD(P)H and reactive oxygen species. *Am. J. Physiol.* 272:C1286–C1294, 1997.

Nieminen, A.-L., G. J. Gores, T. L. Dawson, B. Herman, and J. J. Lemasters. Mechanisms of toxic injury by $HgCl_2$ in rat hepatocytes studied by multiparameter digitized video microscopy. In: *Optical Microscopy for Biology*, edited by B. Herman and K. Jacobson. New York: Alan R. Liss, 1990, pp. 323–335.

Nieminen, A.-L., A. K. Saylor, S. A. Tesfai, B. Herman, and J. J. Lemasters. Contribution of the mitochondrial permeability transition to lethal injury after exposure of hepatocytes to *t*-butylhydroperoxide. *Biochem. J.* 307: 99–106, 1995.

Ohata, H., E. Chacon, S. A. Tesfai, I. S. Harper, B. Herman, and J. J. Lemasters. Mitochondrial Ca^{2+} transients in cardiac myocytes during the excitation–contraction cycle: Effects of pacing and hormonal stimulation. *J. Bioenerg. Biomembr.* 30:207–222, 1998.

Perticarari, S., G. Presani, M. A. Mangiarotti, and E. Banfi. Simultaneous flow cytometric method to measure phagocytosis and oxidative products by neutrophils. *Cytometry* 12:687–693, 1991.

Petronilli, V., G. Miotto, M. Canton, M. Brini, R. Colonna, P. Bernardi, and F. DiLisa. Transient and long-lasting openings of the mitochondrial permeability transition pore can be monitored directly in intact cells by changes in mitochondrial calcein fluorescence. *Biophys. J.* 76:725–734, 1999.

Qian, T., A.-L. Nieminen, B. Herman, and J. J. Lemasters. Mitochondrial permeability transition in pH-dependent reperfusion injury to rat hepatocytes. *Am. J. Physiol.* 273:C1783–C1792, 1997.

Qian, T., L. C. Trost, and J. J. Lemasters. Quenching or misalignment? Confocal microscopy of onset of the mitochondrial permeability transition in cultured hepatocytes. *Microsc. Microanal.* 5(Suppl. 2):468–469, 1999.

Reers, M., S. T. Smiley, C. Mottola-Hartshorn, A. Chen, M. Lin, and L. B. Chen. Mitochondrial membrane potential monitored by JC-1 dye. *Methods Enzymol.* 260:406–417, 1995.

Roe, M. W., J. J. Lemasters, and B. Herman. Assessment of Fura-2 for measurements of cytosolic free calcium. *Cell Calcium* 11:63–73, 1990.

Rothe, G., A. Oser, and G. Valet. Dihydrorhodamine 123: A new flow cytometric indicator for respiratory burst activity in neutrophil granulocytes. *Naturwissenschaften* 75:354–355, 1988.

Rottenberg, H., and S. Wu. Mitochondrial dysfunction in lymphocytes from old mice: Enhanced activation of the permeability transition. *Biochem. Biophys. Res. Commun.* 240:68–74, 1997.

Scaduto, R. C. Jr., and L. W. Grotyohann. Measurement of mitochondrial membrane potential using fluorescent rhodamine derivatives. *Biophys. J.* 76(1 Pt 1):469–477, 1999.

Seghizzi, P., F. D'Adda, D. Borleri, F. Barbic, and G. Mosconi. Cobalt myocardiopathy: A critical review of literature. *Sci. Total Environ.* 150:105–109, 1994.

Smiley, S. T., M. Reers, C. Mottola-Hartshorn, M. Lin, A. Chen, T. W. Smith, G. D. Steele, Jr., and L. B. Chen. Intracellular heterogeneity in mitochondrial membrane potentials revealed by a J-aggregate–forming lipophilic cation JC-1. *Proc. Natl. Acad. Sci. U.S.A.* 88:3671–3675, 1991.

Tanasugarn, L., P. McNeil, G. T. Reynolds, and D. L. Taylor. Microspectrofluorometry by digital image processing: Measurement of cytoplasmic pH. *J. Cell Biol.* 98:717–724, 1984.

Trollinger, D. R., W. E. Cascio, and J. J. Lemasters. Selective loading of Rhod 2 into mitochondria shows mitochondrial Ca^{2+} transients during the contractile cycle in adult rabbit cardiac myocytes. *Biochem. Biophys. Res. Commun.* 236:738–742, 1997.

Trollinger, D. R., W. E. Cascio, and J. J. Lemasters. Mitochondrial calcium transients in adult rabbit cardiac myocytes: Inhibition by ruthenium red and artifacts caused by lysosomal loading of Ca^{2+}-indicating fluorophores. *Biophys. J.* (in press), 2000.

White, J. G., W. B. Amos, and M. Fordham. An evaluation of confocal versus conventional imaging of biological structures by fluorescence light microscopy. *J. Cell Biol.* 105:41–48, 1987.

Wieder, E. D., H. Hang, and M. H. Fox. Measurement of intracellular pH using flow cytometry with carboxy-SNARF-1. *Cytometry* 14:916–921, 1993.

Zahrebelski, G., A.-L. Nieminen, K. Al-Ghoul, T. Qian, B. Herman, and J. J. Lemasters. Progression of subcellular changes during chemical hypoxia to cultured rat hepatocytes: A laser scanning confocal microscopic study. *Hepatology* 21:1361–1372, 1995.

Zoratti, M., and I. Szabo. The mitochondrial permeability transition. *Biochim. Biophys. Acta* 1241:139–176, 1995.

6

Functional Imaging of Mitochondria Within Cells

MICHAEL R. DUCHEN, JAKE JACOBSON, JULIE KEELAN,
MART H. MOJET, AND OLGA VERGUN

1. INTRODUCTION

Mitochondrial function lies at the heart of cell life and cell death. Mitochondria not only generate adenosine triphosphate (ATP) the major currency used by cells in energy requiring processes, but they also house a range of synthetic enzymes. Mitochondria are also intimately involved in processes of intracellular signaling subtle influences through on the concentration of cytosolic calcium ($[Ca^{2+}]_c$).

The history of our understanding of mitochondrial biology is very instructive. For many years, mitochondria were the preserve of biochemists. Experiments that served to illuminate fundamental processes of mitochondrial biochemistry required preparations of mitochondria isolated from cells and tissues—usually from large relatively homogeneous tissues, such as the liver or the heart—that could then be manipulated under carefully defined conditions. Most physiologists and cell biologists barely disguised the fact that they considered mitochondria boring and on the sidelines of developments in cell signaling that were new and exciting. Only very recently has the fundamental importance of mitochondria in cell signaling and in the control of cell life and cell death become clear to the broader scientific community. Indeed, the integration of mitochondrial function into cell physiology and pathophysiology has only emerged as a topic included in major international scientific meetings over the last few years.

In large measure, this reflects technical advances that have allowed us to study mitochondria in their normal environment—within living cells. The technology primarily engages the application of fluorescence probes combined with fluorimetric and imaging technologies that permit the dynamic study of mitochondrial function within single cells. The rigorous study of mitochondrial potential, redox state, and mitochondrial calcium handling has allowed us to dissect mitochondrial contributions to cell physiology and mitochondrial responses to cell signaling, both physiological and pathophysiological. These issues also serve to exemplify a range of problems in imaging epifluorescence microscopy, and our search for solutions should provide some instructive examples of the application and limitations of this technology.

2. BASIC PRINCIPLES

To understand the principles that define the approach to the study of mitochondria, it is essential to first review the basic principles that govern mitochondrial function. These are illustrated in Figure 6–1. In brief, the supply of substrate, such as pyruvate, to the citric acid cycle (the tricarboxylic acid [TCA] cycle or Krebs' cycle) promotes reduction of nicotinamide adenine dinucleotide (NAD^+) and flavin adenine dinucleotide (FAD) to the reduced forms of NAD^+ (NADH) and FAD ($FADH_2$), respectively. Electrons from NADH or $FADH_2$ are transferred through a series of coupled redox reactions regulated by the components of the respiratory chain eventually to oxygen, which is reduced to generate water. In the process, protons are transferred across the inner mitochondrial membrane into the intermembrane space, thus generating an electrochemical potential for protons, expressed largely as a membrane potential, approximately 150 to 200 mV negative to the cytosol.

This potential lays at the heart of mitochondrial function, providing the driving force for proton influx through the F_1F_0–ATP synthase, an enzyme complex

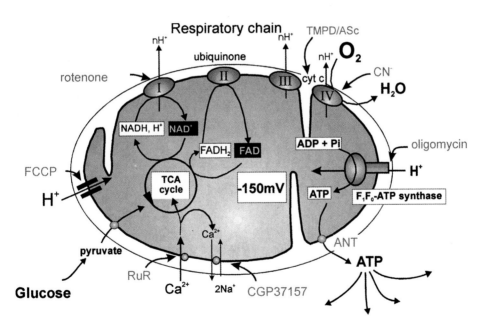

FIGURE 6–1 Illustration of a mitochondrion emphasizing the features of the chemiosmotic basis for oxidative phosphorylation. The sites of action of some of the major biochemical reagents commonly used are indicated. ADP, adenosine diphosphate; ASc, ascorbate; ATP, adenosine triphosphate; CGP37157, Ciba Greigy drug; cyt c, cytochrome c; FAD, flavin adenine dinucleotide; $FADH_2$ reduced form of FAD; FCCP, carbonyl cyanide p-trifluoromethoxyphenylhydrazone; NAD, nicotinamide adenine dinucleotide; NADH, reduced form of NAD; Pi, inorganic phosphate; RuR, ruthenium red; TCA, tricarboxylic acid; TMPD, tetramethyl-p-phenylenediamine. See text for explanation.

that includes a component that is essentially a proton channel. The proton influx drives the ATP synthase to phosphorylate adenosine diphosphate (ADP), generating ATP, which is then transported to the cytosol. The mitochondrial potential, usually referred to as $\Delta\Psi_m$, also provides the driving force for Ca^{2+} uptake into mitochondria. Ca^{2+} is taken up through the uniporter when the $[Ca^{2+}]_c$ is high, moving down an electrochemical potential gradient. Re-equilibration of $[Ca^{2+}]_m$ is then achieved through a $2Na^+/Ca^{2+}$ exchanger in the inner mitochondrial membrane, thought to be electroneutral. This balance of Ca^{2+} between cytosolic and mitochondrial compartments seems to play a key role in the integration of mitochondria into cell function.

The pharmacological or biochemical manipulation of these pathways is central to the investigation of mitochondrial function, and so we have identified the sites of action of some of the major reagents used in Figure 6–1.

The relationship between $\Delta\Psi_m$ and the redox state of the $NADH/NAD^+$ and flavoproteins can be of great experimental value in illuminating the mechanisms underlying mitochondrial contributions or responses to cell physiology and pathophysiology. Although the details of measurement will be discussed later, Figure 6–2

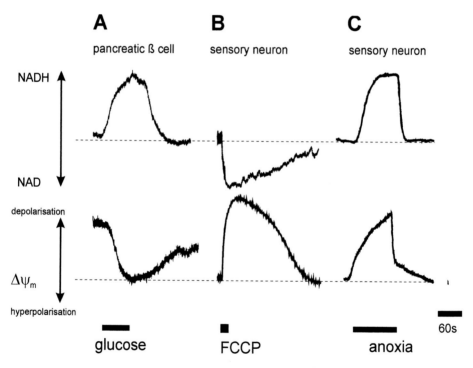

FIGURE 6–2 Changes in NADH autofluorescence and in mitochondrial potential ($\Delta\Psi_m$) in response to increased supply of substrate (A), addition of an uncoupler (B), and inhibition of respiration (C). Note that both A and C are associated with an increase in NADH autofluorescence, but with opposing changes in $\Delta\Psi_m$, while both B and C are associated with a loss of mitochondrial potential, but with opposing changes in autofluorescence. The signals shown come from A, a pancreatic β cell in which the superfusate was switched from 3 to 10 mM glucose; B, a sensory neuron briefly exposed to 1 μM FCCP; and C, a sensory neuron in culture exposed to an anoxic perfusate.

summarizes the changes in these variables with a range of manipulations in an attempt to clarify ways in which mitochondrial potential and redox state will change with each variable. Thus, provision of glucose to a substrate-starved pancreatic β cell activates the TCA cycle, increasing the NADH/NAD$^+$ ratio (Fig. 6–2A). The increased provision of reducing equivalents to the respiratory chain then permits the polarization of $\Delta\Psi_m$ after a delay (Fig. 6–2A), which will in turn serve to increase the rate of ATP production. Collapse of $\Delta\Psi_m$ using an uncoupler, which shuttles protons across the inner membrane and short circuits the membrane, allows the unfettered activation of the respiratory chain, increasing the rate of oxygen consumption, reflected by the increased rate at which NADH is oxidized (Fig. 6–2B). Here, a depolarization of mitochondrial potential is associated with a fall in NADH/NAD$^+$ ratio.

In contrast, inhibition of the respiratory chain (Fig. 6–2C) is also associated with a slower dissipation of $\Delta\Psi_m$, but, as the oxidation of NADH is inhibited, the NADH/NAD$^+$ ratio now increases. The translocation of protons by the respiratory chain stops, and protons leak either through the ATPase or through other leak pathways, which leads to the gradual dissipation of potential (see also below). Interestingly, the apparent rate of change of $\Delta\Psi_m$ under these conditions seems to vary enormously between cell types.

3. Some Technical Considerations

Functional imaging of mitochondria raises a number of technical issues. First, we are trying to resolve signals from small structures within cells. In very flat cells, one can obtain good resolution of fluorescence signals from mitochondria, but as soon as cells are more than a few microns in thickness, optical resolution is limited and either confocal imaging or digital deconvolution is required. This is particularly true for most measurements that involve the partitioning of dyes between compartments, which is discussed later. Thus, the calcium-sensitive indicator Rhod-2 (see later) and the potentiometric probes rhodamine-123, tetramethylrhodamine methyl ester (TMRM), and tetramethylrhodamine ethyl ester (TMRE) all partition electrophoretically between saline and cytosol and then cytosol to mitochondrial matrix. This inevitably tends to leave a significant proportion of signal in the cytosol, which can obscure the mitochondrial signal if the cytosolic volume is sufficiently large.

If experimental questions demand a high temporal resolution—to follow changes in mitochondrial calcium concentration ($[Ca^{2+}]_m$), for example—then one is also obliged to compromise between optimizing temporal and spatial resolution. An alternative is to devise strategies that obviate the requirement to achieve spatial resolution in a way that one can still be confident that the signal is an accurate reflection of mitochondrial function without demanding the specific resolution of signals from individual mitochondria. Other technical issues that are exemplified by attempts to study mitochondria in cells are indicated where appropriate in this chapter (also see Chapters 2 and 5).

4. Measurements of Mitochondrial Potential

In isolated mitochondria, the potential has been measured for many years by following the distribution of lipophilic cations, such as tetraphenylphosphonium (TPP^+), using an electrode and a semipermeable membrane. The same principle has been employed in imaging studies using fluorescent lipophilic cations. Thus, the existence of a single, delocalized charge on the fluorescent compounds rhodamine-123 (Rh123), TMRE, TMRM, 5,5',6,6'-tetrachloro-1,1',3,3'-tetraethylbenzamidazolocarbocyanine (JC-1), $DiOC_6$, and DASPMI renders these dyes both cationic and membrane permeant. They cross the cell membrane easily and partition into compartments in response to their electrochemical potential gradients. Most of these indicators partition between the bathing saline and cytosol and then between the cytosolic and mitochondrial compartments in response to a series of potential differences.

4.1 Approaches to the Use of Indicators and Technical Requirements

For the rigorous measurement of mitochondrial potential, any of the dyes listed above (with the exception of JC-1) should be used at a concentration that is low but that is still consistent with a reasonable signal-to-noise ratio and that can be imaged with low light intensities. This is particularly necessary because all these indicators act as photosensitizing agents and are associated with photodynamic injury to cells, typically culminating in mitochondrial depolarization and ATP depletion. For TMRM or TMRE, in our hands the optimal concentration is approximately 15–30 nM. The problem is not only to avoid toxicity but also to avoid nonspecific binding and nonlinear fluorescence signals (see later) that become a greater problem as the intramitochondrial concentrations increase. The hope is that, at very low concentrations, one may have a linear relationship between dye concentration and fluorescence signal, and the dye concentration should simply reflect the Nernstian distribution in response to $\Delta\Psi_m$.

According to this principle, one would expect approximately a tenfold concentration of dye from saline to cytosol and then a concentration of some 400–600-fold from the cytosol into the mitochondria if $\Delta\Psi_m$ lies at about -150 mV or so. This means that if one is to resolve the signal in both cytosol and mitochondria at the same time, it is essential to digitize the signal to at least 12 bits (0–4095). Even then, this dynamic range is only just adequate to make accurate measurements, and thus ideally one needs digitization to 14 or 16 bits, which are generally not available on confocal systems.

The removal of contaminated out-of-focus signals is also necessary if measurements are to be made accurately. The images shown in Figure 6–3 emphasize the need for confocal resolution if mitochondrial structures are to be identified clearly and if quantitative measurements are to be made. Even with reasonably high resolution and good quality imaging on a cooled charge-coupled device camera, the loss of spatial resolution caused by out-of-focus signal makes it impossible to accurately measure discrete mitochondrial signals. These issues have been thoroughly discussed by Fink et al. (1998).

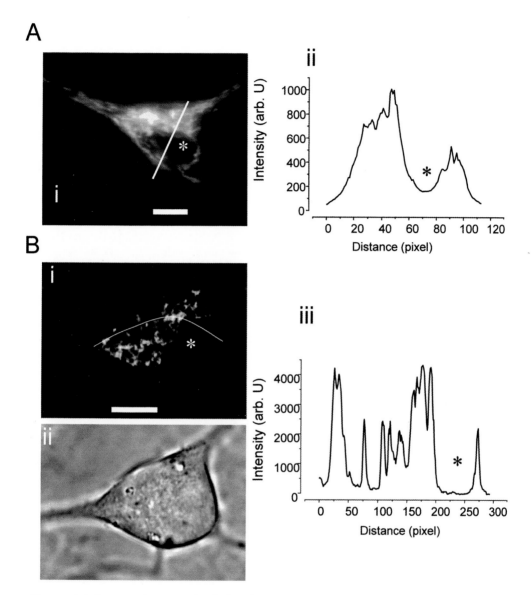

FIGURE 6–3 Images of neurons loaded with TMRE (30 nM) imaged on *A*, a cooled charge-coupled device (CCD) camera and *B*, a confocal microscope (Zeiss 510). In both cases, the cells were incubated in TMRE until the signal reached a steady state. The images were digitized in both cases to 12 bits. On the right (*Aii*) is shown a plot of the intensity profile along a line drawn as indicated across the cell. Note that while mitochondria can be clearly identified, the specific signal arising from mitochondrial and cytosolic structures (i.e., over the nucleus [asterisk]) cannot be separately resolved (image acquired using a Hamamatsu 4880 interline transfer charge-coupled device camera). This is no longer the case in confocal imaging, as shown in *Biii*. Note that now the outlines of the mitochondrial signal are much more distinct, and the specific measurements can be extracted for the signal arising from cytosol (or nucleus) and mitochondria. The signal over the cytosol was of the order of 8–16 (arb. units), while over individual mitochondria, the signal reached 3800, giving a predicted differential signal of about 400-fold. A bright-field image of the neuron is included (*Bii*) to assist orientation of the confocal image. Bars, 10 μm (objective lens, ×63 oil; numerical aperture, 1.4).

When the dyes are used in this way (at very low concentrations and at equilibrium), a fall in mitochondrial potential causes a redistribution of indicator between compartments—dye will leave mitochondria and move into the cytosol. In the short term, this can be resolved as a decrease in mitochondrial signal and an increase in signal over cytosol or nucleus (Fig. 6–4A,B). This may cause no change in signal in the overall cell (Fig. 6–4C). If depolarization is brief, the dye will simply be taken back up into mitochondria. If the mitochondrial depolarization is sustained, however, one would expect the gradual re-equilibration of the dye with movement of mitochondrial dye to cytosol and then from the cytosol to the bathing saline. Furthermore, if the plasma membrane is depolarized, cytosolic dye will be lost, the mitochondrial dye will re-equilibrate, and the signal may decrease, even though mitochondrial potential has not changed. It should be clear, then, that great care must be made in the interpretation of such signals. These problems have been discussed at some length by Nicholls and Ward (2000) and by Rottenberg and Wu (1998).

These considerations refer to attempts to follow changes in $\Delta\Psi_m$ with time in response to specific manipulations. A frequent demand, however, is that $\Delta\Psi_m$ be assessed after several hours exposure to some experimental manipulation. Under these conditions, the degree of loading with a very low concentration of dye should depend on $\Delta\Psi_m$, and thus careful quantitative confocal microscopy should allow a reasonable indication of a change in $\Delta\Psi_m$ as long as all conditions—dye concentration, laser power (or illumination intensity), detector sensitivity, and pinhole aperture (confocal slice thickness)—are all identical.

It should be clear that, while this approach is theoretically rigorous, it is also fraught with potential misinterpretations and errors. When we and others started this research in the late 1980s, imaging technology was not generally accessible, and we were obliged to use photomultiplier tubes to measure the average signal across a cell or small cluster of cells. In this case, if a dye redistributes from the mitochondria to the cytosol in response to a mitochondrial depolarization, it is quite possible to measure no change in signal at all. The signal could also increase or decrease depending on the mitochondrial volume fraction in the cell, which may vary considerably between cell types.

Therefore, several research groups adopted a strategy that may seem confusing: dyes were loaded into cells at relatively high concentrations (~10–20 μM) for short periods—usually 10–15 min—followed by washing. As the dyes accumulate into the mitochondria, they reach concentrations at which the signal shows a phenomenon called *autoquenching*. In this event energy is transferred by collisions between monomeric dye molecules, and the concentration of dye may promote formation of aggregates of dye molecules that may be nonfluorescent. Redistribution of the dye into the cytosol as the mitochondria depolarize relieves the quench, and the net fluorescence signal increases (Fig. 6–5). Thus, in this mode mitochondrial depolarization is associated with an increase in fluorescence. TMRM, TMRE, and rhodamine-123 have all been used in this mode by several groups. In our hands, these signals have always behaved exactly as predicted from chemiosmotic theory (see below and Fig. 6–7), and the signal provides a reliable way to follow changes in $\Delta\Psi_m$ with time.

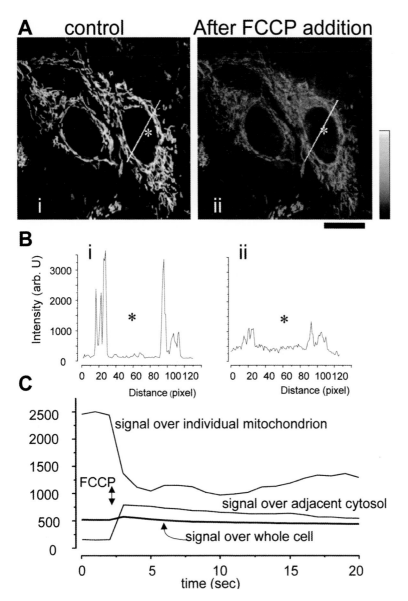

FIGURE 6–4 Imaging changes in mitochondrial membrane potential ($\Delta\Psi_m$) in the redistribution mode. *A*, Rat cortical astrocyte in culture in which TMRE was added to the bathing saline at a concentration of 15 nM and allowed to equilibrate for about 20 min before (*i*) and after (*ii*) application of 1 μM FCCP to collapse $\Delta\Psi_m$. The dye shows a redistribution throughout the cytosol. The display is "mapped" nonlinearly (see scale bar) in order to display both signals, so great is the apparent loss of signal from the mitochondria. The change in signal distribution is illustrated in *B*, which shows the intensity profile across a selected line of the image before (*i*) and after (*ii*) FCCP. Changes in signal over several mitochondria, over the nucleus, and over the whole cell are illustrated in *C*. Note that the signal decreases dramatically over the mitochondria and increases in the nucleus, but measurement of the average signal from the whole cell barely changed at all. Clearly, the signal measured from the whole cell would be uninterpretable. In this instance, images were obtained with a confocal system as described in Figure 6.3. Bars, 10 μm (Zeiss 510 CLSM; $\times$63 oil; numerical aperture, 1.4).

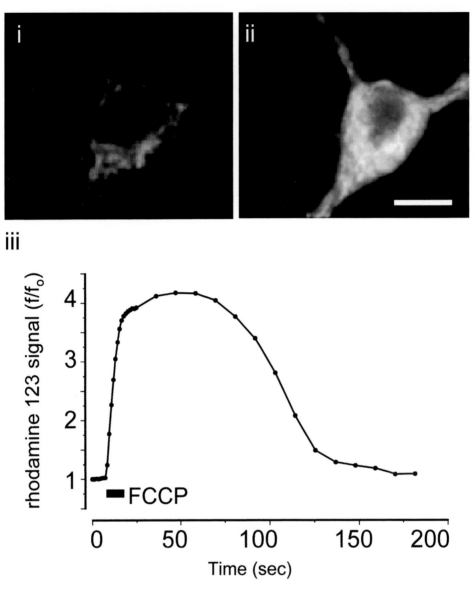

FIGURE 6–5 Imaging changes in mitochondrial membrane potential ($\Delta\Psi_m$) in the dequench mode. In these experiments, neurons were loaded with TMRE at 1.5 μM for 15 min and then washed. Under these conditions, application of FCCP (*ii*) causes a redistribution of dye throughout the cell, but this is now associated with an increase in intensity as the quench of dye fluorescence is relieved. The plot in *iii* shows the change in intensity as a function of time. Bars, 10 μm (Zeiss 510 CLSM, ×63 oil; numerical aperture, 1.4).

Another important point here is that measurements of the mean signal from a cell will give information about the average change in mitochondrial potential in that cell, and thus useful information can be obtained without demanding high-resolution imaging. Indeed, we have exploited this approach in low-power imaging of a field of cells in which we can bathe the cells with a

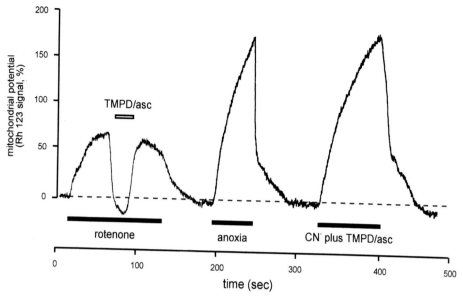

FIGURE 6–7 Manipulation of $\Delta\Psi_m$ using a sequence of inhibitors and electron donors. This record comes from a mouse sensory neuron in culture loaded with rhodamine-123. Rotenone was applied, as indicated, and TMPD with ascorbate (asc) added at the peak of the response. The TMPD donates electrons directly to complex IV, bypassing the block by rotenone at complex I. Anoxia and CN^- have similar effects on $\Delta\Psi_m$, as both inhibit respiration at complex IV. The response to CN^- was not affected at all by the presence of TMPD/asc, as it blocks respiration beyond the site of electron donation by the TMPD. TMPD/asc had almost no effect at all on the resting signal in these cells, but it can stimulate respiration and increase the potential. See text for further explanation.

given drug and follow changes in $\Delta\Psi_m$ in 20–30 neurons with time (e.g., see Vergun et al., 1999).

4.2 JC-1: A Ratiometric Indicator

JC-1 is often described as a "ratiometric indicator" of mitochondrial potential and has become quite fashionable, especially in fluorescence-activated cell sorter (FACS) analysis of cell populations. The promise of ratiometric measurements of the elusive $\Delta\Psi_m$ has an obvious appeal, as it holds promise of standardization and quantification of data. The principle of the indicator is that at high concentrations the dye forms complexes (J-complexes) that show a striking red shift in the fluorescence emission spectrum with excitation at approximately 490 nm, the peak emission of the monomer is in the green with a peak at about 539 nm, and the J-complex emission peaks at 597 nm in the red (Reers et al., 1991; Smiley et al., 1991). The concentration of dye into mitochondria should promote aggregation, and, thus, the greater the potential, the redder signal we should see.

Fluorescence of the monomer is, however, enhanced in a lipid environment (DiLisa et al., 1995), and therefore green fluorescence may reflect the accumulation of dye in membranes, while J-aggregates giving rise to red fluorescence appear in the aqueous phase. It is also not clear how specific and accurate the conditions are that cause J-aggregate formation; therefore, it is difficult to be sure that the aggregates arise only from mitochondria. Dual emission fluorimetry with continuous measurements at approximately 539 and 597 nm might be expected to provide a ratio measurement of $\Delta\Psi_m$. Indeed, we have found that an uncoupler tends to decrease the red signal and increase the green, although these typically change at different rates (Duchen et al., 1993). Experiences in different laboratories, however, seem to show a substantial variability. DiLisa et al. (1995) also suggested that the red fluorescence signal appears more sensitive to relatively small changes in mitochondrial potential compared with the green signal, which only changed significantly with larger excursions of $\Delta\Psi_m$. We have also noted effects of various reagents that do not seem consistent with predictions made from chemiosmotic theory. For example, oligomycin consistently causes changes in signal, suggesting a loss of mitochondrial potential and not an effect expected from the action of this drug. This effect was not seen with any other dye.

The red fluorescence from the "J"-complexes also appears sensitive to factors other than $\Delta\Psi_m$. A recent paper by Chinopoulos et al. (1999) showed that H_2O_2 has profound effects on the red fluorescence, independent of any evidence for a change in $\Delta\Psi_m$. Furthermore, the equilibrium between monomers, dimers, and polymers is not solely due to membrane potential, as JC-1 is pH sensitive and its absorption spectrum may be affected by the osmolarity of its environment (Reers et al., 1991).

Confocal imaging of JC-1 generates some strikingly beautiful images (Color Fig. 6–6), but clearly the interpretation is difficult. In some preparations mitochondria appear red, and in others they look green. We wondered if that meant that there are populations of mitochondria with different potentials. We have even seen single mitochondria that are mostly green, but have some red sections to them (see also Smiley et al., 1991). How can we best decide how to interpret these signals? Thus far, careful confocal imaging of the same cell types with very low concentrations of TMRE have failed to reveal the same differences in populations with JC-1 (M. Duchen, unpublished observations). Having said all this, others have clearly used the indicator with great effect (see Szalai et al., 1999). Nevertheless, given all the considerations outlined, we remain very cautious about the interpretation of JC-1 signals, especially in trying to follow changes in signal with time.

4.3 Validation of Dyes

Imaging of cells stained with most of the fluorophores listed above reveals obvious selective loading into mitochondria. Using JC-1, we have seen red fluorescent objects that are not mitochondria (including the nuclei of dead cells!). There is a tendency for reviewers to ask for validation using "tracker" dyes. This method can present problems, however, because the MitoTracker indicators typically load preferentially into mitochondria in exactly the same way as the potentiometric probes (such as TMRM), causing the indicators to co-localize. There are also concerns that the MitoTracker dyes themselves may impair mitochondrial function, and therefore if these dyes are used in functional studies, appropriate controls must be carried out in order to ensure that function has not been altered.

One must be careful to address several concerns regarding each dye and in any cell type:

1. Does the signal change reflect only changes in mitochondrial potential, or is there a substantial change with depolarization of the plasma membrane?
2. How quickly and reliably does the dye follow changes in $\Delta\Psi_m$ in response to a range of manipulations that should alter $\Delta\Psi_m$?
3. Is there any evidence of toxicity of the dye in terms of cell function, mitochondrial function, and so forth?

4.3.1 Plasma membrane potential

The most rigorous way to examine the effects of plasma membrane depolarization is to record changes in signal while the cell membrane potential is voltage clamped. Using this approach, we showed that a voltage step from -70 to $+60$ mV (the potential at which the Ca^{2+} current appears to reverse) caused no significant change in the rhodamine-123 signal, while a step to 0 mV caused a small but significant ($\sim$20%) increase in signal (Duchen, 1992). This latter response was shown to be due to the mitochondrial response to a rise in $[Ca^{2+}]_i$ and mitochondrial calcium uptake (which generates an inward current across the mitochondrial inner membrane and, hence, a transient depolarization). Whole-cell patch clamping as an approach to studying metabolic processes is always problematic, as $[Ca^{2+}]_i$ is usually buffered and ATP needs to be added to pipette solutions to avoid the washout of currents. An alternative is to use the amphotericin perforated patch technique (Nowicky and Duchen, 1998).

Plasma membrane depolarization is easily achieved using high potassium (50 mM will give a depolarization close to 0 mV), and this can be done with and without Ca^{2+} to establish the component of the response due simply to membrane depolarization. This is discussed in some detail by Rottenberg and Wu (1998), but in our hands, using rhodamine-123 or TMRE in the dequench mode or using TMRE at very low concentrations, we see no significant change in signal simply with plasma membrane depolarization, even if one might expect this result on theoretical grounds (see Nicholls and Ward, 2000). $DiOC_6(3)$, often used as a probe of mitochondrial potential, has also been used as a probe of plasma membrane potential, and great care must be taken when interpreting signals with this dye (see Rottenberg and Wu, 1998).

4.3.2 Manipulation of mitochondrial membrane potential

The typical published study employs (at most) a mitochondrial uncoupler to demonstrate the change in fluorescence signal with mitochondrial depolarization. While this is reasonable as a starting point, it hardly suffices to validate the use of the dye in full. One approach is to use a preparation of isolated mitochondria and to look in parallel at the changes in fluorescence signal and at a TPP^+ electrode trace. Once a potential is established, the TMRE or rhodamine-123 signal is quenched as the mitochondria accumulate the dye, and the fluorescence quench faithfully follows the change in TPP$^+$ electrical signal (Duchen, Leyssens and Crompton, unpublished data). In the following, we suggest a scheme to test the behavior of an indicator (illustrated in Fig. 6–7).

Application of rotenone inhibits respiration at complex I and usually causes a slow depolarization or none at all (varying in rate with cell type). After washout of the rotenone (which recovers only very slowly), addition of oligomycin may cause a small hyperpolarization of potential or no change at all (the hyperpolarization suggests some resting proton flux through the F_1F_0-ATPase, which causes a tonic depolarization). Application of rotenone should now cause a faster and larger depolarization, as the F_1F_0-ATPase cannot operate in reverse mode to maintain the potential. Now application of TMPD ($\sim$10 μg/ml) and ascorbate ($\sim$1 mM) should restore the potential, as this bypasses the inhibition by rotenone and delivers electrons directly to cytochrome c. A further addition of cyanide will depolarize the potential again, despite the TMPD/ascorbate, as CN blocks downstream of the TMPD site.

The procedure generates predictable and reliable responses using rhodamine-123 and TMRE/TMRM in the dequench mode. Using TMRE at a very low concentration and with high-resolution confocal microscopy, one can monitor the movement of dye between cytosol and mitochondria with this manipulation.

4.3.3 Some anomalies?

Rhodamine-123 has always appeared to be satisfactory as an indicator of $\Delta\Psi_m$ in all cell types that we have tried, except in cardiomyocytes. In these cells, the distribution of dye is clearly mitochondrial, but uncouplers failed to cause a change in signal (but see Griffiths et al. 1997). TMRE seems to behave in cardiomyocytes exactly as it does in all other cells. We do not yet know what the basis is for these differences.

4.4 Toxicity

4.4.1 Direct toxic effects

Most indicators have direct toxic effects on mitochondrial function that are usually ignored. Thus, carbocyanine dyes like DiOC$_6$(3) and JC-1 have long been known to inhibit complex I (NADPH dehydrogenase). At concentrations of 40–100 nM, concentrations often used in flow cytometric studies, DiOC$_6$ inhibits mitochondrial respiration by approximately 90%, equivalent to rotenone. Rhodamine-123 inhibits the F_1F_0-ATPase at high concentrations (Emaus et al., 1986).

4.4.2 Photodynamic toxicity

Photobleaching of fluorescent dyes, whereby light-induced oxidation of the fluorophore results in a gradual loss of signal, may occur if the intensity of illumination by the excitation light is too great. This is particularly relevant if dyes are loaded at low concentrations where low fluorescence signals may prompt an increase in excitation intensity. If this is unavoidable, limiting the duration of exposure by closing a shutter intermittently or by increasing the delay between images may reduce bleaching.

Light-induced oxidative damage can be equally problematic with mitochondrial probes loaded at high concentrations. Illumination of fluorophores may result in the production of reactive oxygen species, and the consequent oxidative stress can have wide-ranging effects, including induction of local calcium release, inhibition of the electron transport chain, and induction of the mitochondrial permeability transition. As with photobleaching, limiting the intensity of the excitation light by attenuating the excitation output or reducing the period of illumination will reduce the oxidative stress.

4.4.3 Other problems

Many of the lipophilic cations used for these measurements are substrates for the multidrug resistance transporter (MDR), and some have been used to try to quantify MDR activity experimentally. As cyclosporin A is an inhibitor of the MDR pump, the effects of cyclosporin A on fluorescence from these dyes must be treated with great caution, and other agents like verapamil, inhibitors of the MDR, should also be tested.

5. NADH and Flavoprotein Autofluorescence

A dissipation of mitochondrial potential can result either from an increase in "load" (an increase in ATP production or a pathophysiological increase in proton flux across the mitochondrial inner membrane) or from a lack of substrate, be it carbohydrate or oxygen. As introduced earlier, one can discriminate between the two by examining changes in autofluorescence. In this section, the source of autofluorescence, the methodology to measure it, and some applications are discussed.

It is important to understand these changes in autofluorescence not only for their own sake but also because these changes may contaminate signals from other fluorescence indicators, such as Indo-1, for which the signals may introduce changes in raw fluorescence signals in response to metabolic manipulations that have nothing to do with changes in $[Ca^{2+}]_i$.

5.1 Theoretical Basis

The endogenous compounds that act as hydrogen carriers, which ferry the protons and electrons from carbohydrates to the electron transport chain (see Fig. 6–1), have fluorescent properties. The term *autofluorescence* is used to distinguish this phenomenon from the fluorescence of indicators that are artificially introduced. This signal can be used as an indicator of changes in cellular metabolism, as its properties change when the carrier binds an electron.

Thus, the fluorescence of the pyridine nucleotide NADH is excited in the ultraviolet (UV; peak excitation at about 350 nm) and emits in the blue (with a peak at about 450 nm; see Chance et al., 1979). The oxidized form, NAD^+, is not fluorescent. Therefore, an *increase* in UV-induced blue fluorescence indicates an increase in the ratio of NADH to NAD^+ and a net shift in the pyridine nucleotide pool to the reduced state. It is important to emphasize that these changes in signal do not indicate net changes in the absolute size of the total pool, but rather a change in the balance of reduced to oxidized forms. NADPH is also fluorescent with very similar spectral properties. NADH and NADPH are present in both mitochondrial and cytosolic compartments: however, several properties tend to mitigate in favor of the mitochondrial signal—there is more of it, and the binding of NADH to membranes enhances the fluorescence while enzymatic binding tends to quench the cytosolic fraction (Chance and Baltscheffsky, 1958).

Flavoproteins ferry electrons using a flavin or FAD molecule. Flavoprotein fluorescence is excited in the blue (peak at about 450 nm) and emission is maximal in the green, with a peak at about 550 nm. In contrast to NADH, flavoprotein fluorescence decreases, but does not completely disappear when the carrier binds electrons. A *decrease* in flavoprotein autofluorescence reflects an increase in the ratio of reduced to oxidized flavoprotein, which is the inverse of the response of the pyridine nucleotides.

5.2 Methods

5.2.1 Pyridine nucleotides (NADH)

Because the transmission of microscope optics usually falls off rapidly below 350 nm, a good compromise between illumination intensity and excitation efficiency is obtained with a 350 ± 20 nm or 360 ± 5 nm bandpass filter and quartz optics, if available. Any light above 390 nm is likely to excite flavoprotein fluorescence, which will not only contaminate the NADH signal but also result in an underestimate of the NADH signal, because flavoprotein fluorescence changes inversely with the degree of reduction. The emission may be measured using a wide bandpass filter with a peak at 450 nm and a bandwidth from ±20 to 40 nm. The fluorescence tends to bleach, and excessive illumination can cause photodamage to the cell. Thus, as in every case, it is best to keep the illumination intensity to a minimum, consistent with a reasonable signal-to-noise ratio. In any fluorescence system, this is always maximized by increasing the efficiency of the light collection pathway. High numerical aperture lenses and sensitive, low noise detection systems should be used.

5.2.2 Flavoproteins

Flavin autofluorescence is best elicited with excitation at 450 ± 20 nm or so and is measured between 500 and 600 nm (we have used a bandpass filter at 550 ± 40 nm; or simply image with a longpass filter at >510 nm).

For both of these signals, the quality of signal tends to vary depending on the density of mitochondria within the cell. Thus, cardiomyocytes, in which the mitochondria may represent 40%–50% of the cell volume, and sensory neurons produce very large robust signals (see Fig. 6–8). In flat cells like astrocytes and fibroblasts, which have a low overall mitochondrial volume, it may be very hard to obtain acceptable signal-to-noise ratios to measure such signals, although good images of flavoprotein autofluorescence have been obtained with a good cooled CCD chip and relatively long exposures.

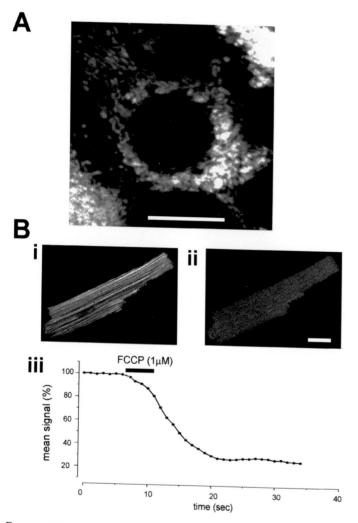

FIGURE 6–8 Imaging NADH autofluorescence. Images of NADH autofluorescence in cardiomyocytes (neonatal cultured cells in *A*; mature, freshly dissociated in *B*) acquired on the confocal LSM (Zeiss 510) with excitation at 351 nm and measured at 375–470 nm. The mitochondrial localization of the signal is evident when compared with equivalent images using TMRE. In *B*, the change in signal with FCCP is shown, with representative images (×40; numerical aperture, 1.3) from a sequence in *i* and *ii* and a plot of intensity with time in *iii*. Note the disappearance of identifiable mitochondrial localization of signal after application of the uncoupler. Bars, 10 μm (×63 oil; numerical aperture, 1.4).

5.3 Applications

The pyridine nucleotide and flavoprotein redox states depend on the net reactions that either oxidizes or reduces them (Figs. 6–1 and 6–2). For example, an increase in the activity of the TCA cycle through the Ca^{2+}-dependent upregulation of the enzymes (McCormack et al., 1990; Duchen, 1992) or through increased delivery of substrate (Fig. 6–2A) will increase the balance in favor of NADH. If respiration is inhibited by mitochondrial inhibitors like CN^- or by anoxia, the respiratory chain cannot oxidize the reduced forms that will accumulate to a new steady state. Again, NADH autofluorescence will rise (Fig. 6–2C), and flavoprotein autofluorescence will fall. In contrast, mitochondrial respiration responds to a maximal depolarization by an uncoupler by increasing respiratory rate (see above). This leads to maximal oxidation of NADH to NAD^+ and of $FADH_2$ to FAD, decreasing the autofluorescence from NADH and increasing that from FAD (Figs. 6–2B and 6–8B; and see above).

It is difficult to estimate the full range of autofluorescence, that is, when an electron carrier is completely oxidized or reduced. It is usually assumed that the bath application of CN^- stops all electron flux completely, leaving all NADH and flavoprotein in the reduced state. This assumption is valid as long as most NADH is contained in the mitochondria and relatively little cytosolic NADH is oxidized by lactate dehydrogenase. Application of uncouplers like FCCP will maximize O_2 consumption and, therefore, electron flux. This does not, however, necessarily mean that the NADH is completely oxidized, as the dehydrogenases of the TCA cycle will continue to supply reducing equivalents. Only when glycolysis and fatty acid oxidation are completely blocked can it be assumed that the NADH autofluorescence is minimal. Unfortunately, inhibition of glycolysis by iodoacetate or deoxyglucose is irreversible. It should be clear that an increase in NADH autofluorescence signal can be interpreted in different ways—either an increase in effective substrate supply or an inhibition of oxidation—and other experiments may be necessary to differentiate between these.

6. MEASURING MITOCHONDRIAL CALCIUM ($[Ca^{2+}]_{mt}$)

It has been clear for many years that isolated mitochondria have a massive capacity to accumulate calcium. What has been less clear is the functional significance of that pathway. Ca^{2+} is accumulated into mitochondria through the ruthenium red–sensitive uniporter in response to a large electrochemical potential gradient established by the mitochondrial membrane potential and a low intramitochondrial calcium concentration ($[Ca^{2+}]_{mt}$) maintained by the mitochondrial Na^+/Ca^{2+} exchange. Calcium plays a major role in the regulation of mitochondrial metabolism, as the major rate-limiting enzymes of the TCA cycle are all upregulated by calcium. Mitochondrial calcium uptake may also influence the cytosolic calcium concentration locally, controlling microdomains of calcium close to channels and influencing calcium signaling in subtle ways (Jouaville et al., 1995; Boitier et al., 1999). Furthermore, excessive mitochondrial calcium accumulation may be toxic and initiate pathways that lead to cell death.

It should be clear, then, that the measurement of changes in $[Ca^{2+}]_{mt}$ within cells is of considerable interest in both physiological and pathophysiological processes. Although such measurements can be made with relative ease in preparations of isolated mitochondria, the functional consequences of a change in $[Ca^{2+}]_{mt}$ can only be properly understood when the mitochondrion is studied in its native environment—the cell. This can be fraught with difficulties, however, and studies on the role and regulation of mitochondrial Ca^{2+} transport in cells were hampered for many years by the lack of available techniques to make such measurements.

6.1 Imaging $[Ca^{2+}]_{mt}$ with Fluorescent Indicators

When cells are loaded with standard acetoxymethylester (AM) derivatives of fluorescent indicators, the dyes inevitably tend to distribute throughout the cell, localizing into both organelle and cytosolic compartments. The degree of localization differs between cell types and between dyes and depends on dye concentration and loading temperature. In some cell types, predominant mitochondrial loading may be achieved with little effort. In many cells, standard loading techniques for dyes, such as Fura-2, may result in as much as 50% of dye within organelles. This can be exploited by exaggerating those aspects of the loading procedures that increase compartmentalization and encouraging dye loss from the cytosol. For example, Csordás et al. (1999) found that a mast cell line (RBL-2H3 cells) loaded with 5 μM Fura-2FF/AM at room temperature for 60 min yielded a predominantly mitochondrial signal. In those experiments, the residual cytosolic dye was removed as the cells were permeablized. The same result can be achieved using whole-cell patch-clamp recording to dialyse out the residual cytosolic dye.

In other instances, parameters of dye concentration and temperature need to be manipulated to a much greater degree in an attempt to get more selective loading into mitochondria. For example, Griffiths et al. (1997) found that isolated rat ventricular myocytes can be loaded with Indo-1 AM under conditions in which about half the dye is located within mitochondria (5 μM, 15 min at 30°C). If cells were then incubated for 1.5 h at 37°C, only the mitochondrial fluorescence remained. These protocols exploit the temperature sensitivity of plasmalemmal anion transporters to remove the cytosolic fraction of the indicator and clearly depend on the expression of the transporters by the cell type being studied. The extent to which protocols involving exposure to higher temperatures for the selective removal of cytosolic dye are applicable to a wider array of fluorescent Ca^{2+} indicators and cell types remains to be determined.

In an alternative approach to limit the contamination of mitochondrial signal with cytosolic dye, Miyata et al. (1991) loaded single cardiomyocytes with a high concentration (10 μM or greater) of Indo-1 and then exposed the cells to Mn^{2+}. Entry of Mn^{2+} into the cell quenches the cytosolic Indo-1 fluorescence so that the remaining signal is a reflection of mitochondrial dye. This approach cannot be readily applied to all cells, however. Miyata et al. (1991) found that in rat cardiomyocytes the cytosolic fluorescence was quenched rapidly upon superfusion with Mn^{2+}, and the fluorescence signal then remained stable for some time. Not all cells, however, readily permeate Mn^{2+}. Mn^{2+} is also sequestered by mitochondria via the mitochondrial Ca^{2+} uniporter, and in rabbit myocytes a rapid quenching of the mitochondrial fluorescent signal proves the Mn^{2+} quench method to be inappropriate (Chacon et al., 1996). Furthermore, intramitochondrial Mn^{2+} inhibits the efflux of Ca^{2+} from mitochondria. Mn^{2+} also depletes cellular energy supplies by interfering with oxidative phosphorylation at the level of the ATPase and complex I. The Mn^{2+}-quench method is now little used and perhaps should only be applied with caution and relevant controls to ensure that cell physiology is not otherwise unduly affected by this method.

6.2 Rhod-2 AM

The AM ester of Rhod-2 is currently the only cell permeant Ca^{2+} indicator that carries a delocalized positive charge, and so it is taken up preferentially into polarized mitochondria. Upon hydrolysis of ester moieties, the Rhod-2 free acid remains trapped inside the mitochondria where it reports increased $[Ca^{2+}]_{mt}$ as an increase in fluorescence intensity. In almost all cell types, the dye effectively partitions between cytosol and mitochondria so that a significant proportion of dye is left in the cytosolic compartment. At first glance of images of quiescent Rhod-2–loaded cells, this may not seem to be much of a problem. Images often show a beautiful mitochondrial localization that can be well resolved, and this suggests that changes in the mitochondrial compartment can be followed and monitored accurately. Stimulation of the cells to raise $[Ca^{2+}]_{cyt}$ and $[Ca^{2+}]_{mt}$ usually, however, causes a large increase in the fluorescence of residual Rhod-2 in the cytosolic compartment (see Boitier et al., 1999). We have found that this may completely obscure the mitochondrial Rhod-2 signal in hippocampal neurons during stimulation with glutamate (Keelan and Duchen, unpublished observations). A major caveat, then, is that it is essential when using this dye to be able to resolve clearly that the measured signal is coming from identifiable mitochondria or that appropriate measures have been taken (as above) to minimize residual cytosolic dye.

In an attempt to improve the mitochondrial localization of the dye, several groups have developed protocols that involve incubation at 37°C for several hours or overnight after initial standard loading of the Rhod-2 AM into the cells at room temperature for 30 min. During the incubation, plasma membrane transporters may eliminate the residual cytosolic dye without affecting the mitochondrial loading. Similar prolonged incubations apparently enhance the mitochondrial localization of Rhod-2 in myocytes, hepatocytes, and oligodendrocytes. This approach has clearly been used effectively by several groups. Indeed, Figure 6–9 shows images of hippocampal neurons loaded with Rhod-2 and then incubated at 36°C for 12 h to reveal a signal, which is clearly highly localized to mitochondria. Remarkably, however, we have found that, in our hands, the Rhod-2 signal seems to become significantly less responsive to changes in Ca^{2+} after a long incubation, and we have failed to see changes in signal even after massive changes in calcium induced by ionophore.

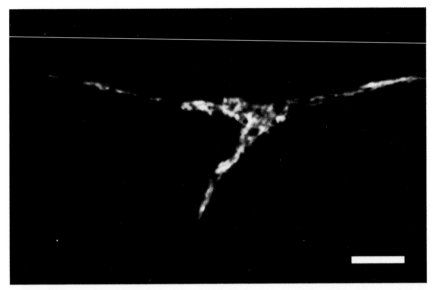

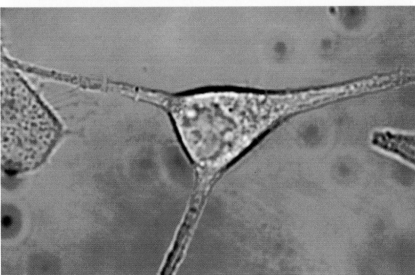

Figure 6–9 Rhod-2 localization in a rat hippocampal neuron in culture. The cells were loaded with 5 μM Rhod-2 AM for 30 min at room temperature and then incubated for 12 h at 36°C. The image acquired on the confocal LSM (Zeiss 510) shows selective mitochondrial loading and a minimal cytosolic signal (e.g., over the nucleus). The corresponding bright-field image (bottom) is included for orientation. Bars, 10 μm (×63 oil; numerical aperture, 1.4).

We have experienced a similar problem when using dihydro-rhod-2 AM, a nonfluorescent derivative that is readily made in the laboratory from rhod-2 AM (instructions provided by Molecular Probes). This is taken up into mitochondria in the same way as rhod-2 AM, and the cells are then washed and left in the incubator for some hours or over night. Oxidation of the dye within the mitochondria produces the fluorescent Rhod-2. We have tried this protocol and see beautiful and clear selective loading of dye into mitochondria. We found these signals completely unresponsive to manipulations that raise $[Ca^{2+}]_i$ massively, however, and have ceased using this approach. The lack of literature about using the dye this way suggests that others have also found it difficult to use, and it does raise questions about the performance of the dye.

Conventional imaging techniques (such as a CCD camera) can be readily applied to image Rhod-2 signals in flat, thin cells like fibroblasts or astrocytes in which mitochondria can be easily resolved (see Boitier et al., 1999). Other cell types, including "fatter" neurons, require the use of confocal microscopy to resolve the Rhod-2 fluorescence in different compartments unequivocally. Exactly the same constraints apply as described above in measuring mitochondrial potential except that the contaminating dye in the cytosol is now Ca^{2+} sensitive and may be exposed to larger dynamic changes than the mitochondrial signal. Confocal microscopy may permit improved resolution of the mitochondrial Rhod-2 and reduce contamination by cytosolic fluorescence, but even this approach is not without problems. Peng et al. (1998) have commented that measuring the fluorescence from a region of interest that includes large areas of cytoplasm outside the mitochondria will "dilute" the changes in $[Ca^{2+}]_{mt}$, for example, in response to N-methyl-D-aspartate, but, if the cytosol also contains a significant amount of dye, then the measured signal will not be diluted but rather contaminated or distorted.

One of the most powerful features of the confocal system is the ability to acquire simultaneous images at different wavelengths without needing filter changes or time delays. This allows the combination of two indicators to measure the cytosolic and mitochondrial signals, as shown in Color Figure 6–10. Neonatal cardiomyocytes in culture were loaded with both fluo-4 and Rhod-2. As shown, the signal from a given mitochondrion can be related to the change in signal from the restricted volume of cytosol surrounding it, and thus the local relationship between cytosolic calcium and mitochondrial uptake becomes accessible. Finally, subtraction of the prestimulus signal from that after stimulation reveals that the signal has changed only over the mitochondria.

6.3 Validation and Limitations

It should be evident now that when discussing the validations and limitations involved in measuring $[Ca^{2+}]_{mt}$, one cannot assume that dyes marketed as "cytosolic" or "mitochondrial" actually localize to these compartments when loaded as AM esters and that adequate controls need to be conducted to ensure that one knows the true origin of the fluorescence signal in a particular preparation. This is particularly important when using fluorometric or conventional imaging techniques in which mitochondria may not be adequately resolved. Fluorescence can be shown to originate primarily from mitochondria if the fluorescence signals can be altered by inhibition of mitochondrial Ca^{2+} uptake. This is best achieved as follows:

1. By inhibition of mitochondrial Ca^{2+} uptake and efflux by ruthenium red and clonazepam, respectively. Ruthenium red is a large cationic

molecule that should be impermeant. In fact, it has been reported to enter some cells (cardiomyocytes), but it generally requires specific methods of introduction into the cell, such as microinjection or lipofusion techniques. When microinjected, its effects on the plasma membrane are avoided. A newly marketed "cell permeant" derivative, RU360, is now available, although in our hands this compound altered physiological intracellular Ca^{2+} signaling in astrocytes (Jacobson and Duchen, unpublished observations).

2. By the collapse of the mitochondrial potential thus removing the driving force for calcium accumulation. This is readily achieved using a mitochondrial uncoupler, such as FCCP, to abolish $\Delta\Psi_m$. Caution is needed with this approach, however, to ensure that any observed changes in calcium signaling are not due to a secondary effect on cellular energy levels and the consequential inhibition of Ca^{2+} efflux or sequestration via the plasmalemmal Ca^{2+}-ATPase. Inclusion of low concentrations of oligomycin (e.g., 2.5 $\mu g/ml$) to inhibit any potential ATP hydrolysis via the F_1F_0-ATPase can help address this issue. Collapse of $\Delta\Psi_m$ may similarly be achieved using a respiratory inhibitor together with oligomycin, and this should also ensure that effects occurring when an uncoupler is used are not due to collapsing proton gradients in other compartments or to pH shifts.

Even with the judicious use of fluorescent dyes clearly localized to mitochondria, limitations apply as in all calcium imaging. For example, the K_d of Rhod-2 for Ca^{2+} is on the order of 570 nm, and it is perfectly possible that this will saturate and distort measurements of $[Ca^{2+}]_{mt}$ in some models. Currently, we do not have a low-affinity mitochondria-selective calcium indicator available.

6.4 Mitochondrial Directed Transfection of Recombinant Aequorin

It would seem negligent to discuss the measurement of intramitochondrial calcium without inclusion of the technique pioneered by Rizzuto and his colleagues in the early 1990s, whereby the Ca^{2+}-sensitive photoprotein aequorin targeted to the mitochondrial matrix is expressed in cells. Cells are transfected with a chimeric cDNA encoding aequorin, and a mitochondrial presequence from the mitochondrial enzyme cytochrome oxidase is encoded in order to target the fusion protein specifically to mitochondria (Rizzuto et al., 1993, 1994). This is an elegant technique ensuring specific measurement of Ca^{2+} changes within a discrete compartment. The signal generated by the photoprotein expressed this way is generally too weak to be useful for $[Ca^{2+}]_{mt}$ imaging at the single cell level using current technology. Also, the data are typically gathered from cells in suspension. The approaches discussed above are nicely complemented with the use of recombinant aequorin. Spatial information is inferred from the selective expression of the reporter protein in different compartments in the cells, and while single cell imaging using fluorescent dyes has limitations regarding accurate compartmentalization, the approach provides details of the spatiotemporal features of the signaling pathways.

Looking ahead, it seems likely that the specific targeting of protein reporters to specific compartments will be used to overcome many of the problems of specific probe localization to which we have drawn attention. Use of the brighter chameleon green fluorescent protein reporters is a burgeoning technology, which may solve many of these problems (Miyawaki et al., 1999).

We like thank the Wellcome Trust, the Medical Research Council, and the Royal Society for support. We also thank D. Lilian Patterson for her help in the preparation and maintenance of cells in culture for which we all depend in order to carry out our experiments.

REFERENCES

Boitier, E., R. Rea, and M. R. Duchen. Mitochondria exert a negative feedback on the propagation of intracellular Ca^{2+} waves in rat cortical astrocytes. *J. Cell Biol.* 145:795–808, 1999.

Chacon, E., H. Ohata, I. S. Harper, D. R. Trollinger, B. Herman, and J. J. Lemasters. Mitochondrial free calcium transients during excitation–contraction coupling in rabbit cardiac myocytes. *FEBS Lett.* 382:31–36, 1996.

Chance, B., and H. Baltscheffsky. Respiratory enzymes in oxidative phosphorylation: VII. Binding of intramitochondrial reduced pyridine nucleotide. *J. Biol. Chem.* 233:736–739, 1958.

Chance, B., B. Schoener, R. Oshino, F. Itshak, and Y. Nakase. Oxidation-reduction ratio studies of mitochondria in freeze-trapped samples. NADH and flavoprotein fluorescence signals. *J. Biol. Chem.* 254:4764–4771, 1979.

Chinopoulos, C., L. Tretter, and V. Adam-Vizi. Depolarization of in situ mitochondria due to hydrogen peroxide–induced oxidative stress in nerve terminals: Inhibition of alpha-ketoglutarate dehydrogenase. *J. Neurochem.* 73:220–228, 1999.

Csordás, G., A. P. Thomas, and G. Hajnóczky. Quasi-synaptic calcium signal transmission between endoplasmic reticulum and mitochondria. *EMBO J.* 18:96–108, 1999.

DiLisa, F., P. S. Blank, R. Colonna, G. Gambassi, H. S. Silverman, M. D. Stern, and R. G. Hansford. Mitochondrial-membrane potential in single living adult-rat cardiac myocytes exposed to anoxia or metabolic inhibition. *J. Physiol.* 486:1–13, 1995.

Duchen, M. Ca^{2+}-dependent changes in the mitochondrial energetics in single dissociated mouse sensory neurons. *Biochem. J.* 283:41–50, 1992.

Duchen, M., O. McGuinness, L. Brown, and M. Crompton. On the involvement of a cyclosporin A sensitive mitochondrial pore in myocardial reperfusion injury. *Cardiovasc. Res.* 27:1790–1794, 1993.

Emaus, R. K., R. Grunwald, and J. J. Lemasters. Rhodamine 123 as a probe of transmembrane potential in isolated rat-liver mitochondria: Spectral and metabolic properties. *Biochim. Biophys. Acta* 850:436–448, 1986.

Fink, C., F. Morgan, and L. M. Loew. Intracellular fluorescent probe concentrations by confocal microscopy. *Biophys. J.* 75:1648–1658, 1998.

Griffiths, E. J., M. D. Stern, and H. S. Silverman. Measurement of mitochondrial calcium in single living cardiomyocytes by selective removal of cytosolic indo 1. *Am. J. Physiol.* 273:C37–C44, 1997.

Jouaville, L. S., F. Ichas, E. L. Holmuhamedov, P. Camacho, and J. D. Lechleiter. Synchronization of calcium waves by mitochondrial substrates in *Xenopus laevis* oocytes. *Nature* 377:438–441, 1995.

McCormack, J. G., A. P. Halestrap, and R. M. Denton. Role of calcium ions in regulation of mammalian intramitochondrial metabolism. *Physiol. Rev.* 70:391–425, 1990.

Miyata, H., H. S. Silverman, S. J. Sollott, E. G. Lakatta, M. D. Stern, and R. G. Hansford. Measurement of mitochondrial free Ca^{2+} concentration in living single rat cardiac myocytes. *Am. J. Physiol.* 261:H1123–H1134, 1991.

Miyawaki, A., O. Griesbeck, R. Heim, and R. Y. Tsien. Dynamic and quantitative Ca^{2+} measurements using improved cameleons. *Proc. Natl. Acad. Sci. U.S.A.* 96:2135–2140, 1999.

Nicholls, D., and M. Ward. Mitochondrial membrane potential and neuronal glutamate excitotoxicity: Mortality and millivolts. *Trends Neurosci.* 23:166–174, 2000.

Nowicky, A., and M. Duchen. Changes in $[Ca^{2+}]_i$ and membrane currents during impaired mitochondrial metabolism in dissociated rat hippocampal neurons. *J. Physiol.* 507: 131–145, 1998.

Peng, T. I., M. J. Jou, S. S. Sheu, and J. T. Greenamyre. Visualization of NMDA receptor–induced mitochondrial calcium accumulation in striatal neurons. *Exp. Neurol.* 149:1–12, 1998.

Reers, M., T. W. Smith, and L. B. Chen. J-aggregate formation of a carbocyanine as a quantitative fluorescent indicator of membrane potential. *Biochemistry.* 30:4480–4486, 1991.

Rizzuto, R., C. Bastianutto, M. Brini, M. Murgia, and T. Pozzan. Mitochondrial Ca^{2+} homeostasis in intact cells. *J. Cell Biol.* 126:1183–1194, 1994.

Rizzuto, R., M. Brini, M. Murgia, and T. Pozzan. Microdomains with high Ca^{2+} close to IP_3-sensitive channels that are sensed by neighboring mitochondria. *Science* 262:744–747, 1993.

Rottenberg, H., and S. L. Wu. Quantitative assay by flow cytometry of the mitochondrial membrane. *Biochim. Biophys. Acta Mol. Cell Res.* 1404:393–404, 1998.

Smiley, S. T., M. Reers, C. Mottola-Hartshorn, M. Lin, A. Chen, T. W. Smith, G.D. J. Steele, and L. B. Chen. Intracellular heterogeneity in mitochondrial membrane potentials revealed by a J-aggregate–forming lipophilic cation JC-1. *Proc. Natl. Acad. Sci. U.S.A.* 88:3671–3675, 1991.

Szalai, G., R. Krishnamurthy, and G. Hajnoczky. Apoptosis driven by IP(3)-linked mitochondrial calcium signals. *EMBO J.* 18:6349–6361, 1999.

Vergun, O., J. Keelan, B. I. Khodorov, and M. R. Duchen. Glutamate-induced mitochondrial depolarization and perturbation of calcium homeostasis in cultured rat hippocampal neurons. *J. Physiol.* 519(Pt 2):451–466, 1999.

7

Diffusion Measurements
by Fluorescence Recovery
After Photobleaching

ALAN S. VERKMAN, LAKSHMANAN VETRIVEL, AND PETER HAGGIE

1. INTRODUCTION

Fluorescence recovery after photobleaching (FRAP) is a classic technique for measurement of the translational diffusion of fluorophores and fluorescently labeled macromolecules. In spot photobleaching, a brief intense light pulse irreversibly bleaches fluorophores in a defined volume of a fluorescent sample. With an attenuated probe beam, the diffusion of unbleached fluorophores into the bleached volume is measured as a quantitative index of fluorophore translational diffusion. A variety of optical configurations, detection strategies, and analysis methods have been used to quantify diffusive phenomena in FRAP measurements. In living cells, FRAP has been used widely for determination of solute/macromolecule diffusion in membranes and aqueous compartments, as well as for semiquantitative analysis of protein transport mechanisms. The introduction of green fluorescent protein (GFP) and other novel cellular labels has motivated a renewed interest in developing quantitative measurement and analysis methods to follow the diffusive and directed transport of defined targets in cells.

This chapter focuses on the instrumentation and analytic methods for FRAP measurements of diffusion in biological systems. Practical considerations in instrument design and data acquisition are described, as are issues regarding the selection of suitable fluorophores. Recent examples of diffusion measurements in cell membranes and aqueous compartments are given to demonstrate applications of FRAP in living cells. The concerns and challenges when making quantitative FRAP measurements are discussed, including issues involving reversible photobleaching, complex bleach geometry, and anomalous diffusive phenomena. Finally, the application of FRAP methods in modern biology is put in perspective with regard to important biological problems and advances in complementary technologies, such as fluorescence correlation spectroscopy and multiphoton spectroscopy (see Chapters 13).

2. Instrumentation and Technical Considerations

The instrumentation used to carry out FRAP measurements consists of a suitable light source to bleach and probe a defined sample volume, a fluorescence microscope, and a detector, all of which are computer controlled to accomplish precise timing of bleaching and detection events. An apparatus constructed in our laboratory is illustrated in Figure 7–1.

For spot photobleaching, a monochromatic beam from an Argon ion laser (4 W) is modulated by a high-contrast acousto-optic modulator and directed onto a fluorescent sample using an epifluorescence microscope with a dichroic mirror and objective lens. The excitation path also contains beam-focusing optics, a pinhole, a variable neutral density filter, and a fast electronic shutter. When closed, the shutter blocks excitation light before bleaching and between successive measurements when following very slow diffusive processes of seconds to minutes. For full-field illumination to visualize the sample, the zero-order beam is directed by a fiberoptic and lens/shutter system onto a mirror (as shown in Fig. 7–1). The emission path contains an interference filter, a pinhole, and a gated photomultiplier detector. The photomultiplier is gated off during the intense bleach pulse using a circuit that controls dynode voltage (Kao and Verkman, 1996). A computer coordinates the timing of acousto-optic modulator voltage, shutter signal, photomultiplier protection circuitry, and data acquisition.

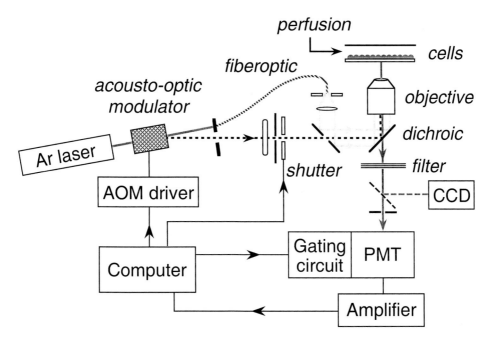

Figure 7–1 Spot photobleaching apparatus. A laser beam is modulated by an acousto-optic modulator and directed onto a sample using a dichroic mirror and objective lens. Emitted fluorescence is collected, filtered, and detected by a gated photomultiplier (PMT). The nondeflected zero-order beam is directed by a fiberoptic lens–mirror assembly for full-field illumination. See text for details. CCD, charge-coupled device; AOM, acousto-optic modulator.

The minimum bleach time is 1 μs, and processes down to approximately 20 μs can be followed. The apparatus is positioned on a large floating optical table in a temperature-controlled, air-filtered darkroom. For temperature-sensitive experiments, the microscope stage, cell perfusion chamber, and objective lens are temperature controlled.

In cell studies it is often desirable to record images of photobleaching events to complement the quantitative spot photobleaching measurements. We use a Nipkow-wheel confocal microscope and cooled charge-coupled device (CCD) camera detector to record fluorescence images from a full field after bleaching a defined spot (Dayel et al., 1999). With excitation and camera shutters, the sample is illuminated only during image acquisition to minimize photobleaching by the probe illumination. Similar semiquantitative imaging studies have been accomplished by scanning laser confocal microscopy (Lippincott-Schwartz et al., 1999; White and Stelzer, 1999); however, such systems have limited time resolution and sensitivity for studies of rapid diffusive processes.

Several other optical configurations have been used in FRAP measurements. Total internal reflection FRAP has been accomplished by directing the laser beam onto a prism in optical contact with the fluorescent sample (Swaminathan et al., 1996). Fluorescence recovery in the evanescent field is recorded using an objective lens and the detection optics described above. Spot photobleaching has been done recently by two-photon excitation using a focused Ti:sapphire femtosecond laser (Brown et al., 1999). The restricted bleach volume in the z-direction in two-photon excitation is a potential advantage when studying diffusion in three dimensions. In another configuration, two modulated interfering laser beams produce a spatial fringe pattern at the sample (Davoust et al., 1982). This approach was recently used to measure the diffusion of globular proteins in the cytoplasm of cultured muscle cells (Dupont et al., 2000).

3. CHROMOPHORES AND PHOTOPHYSICS

An ideal fluorescent probe for FRAP measurements should undergo irreversible photobleaching without reversible photophysical processes (e.g., triplet state relaxation, flicker) or photochemical reactions. The chromophore should be moderately sensitive to photobleaching in order to minimize bleach beam intensity and duration, but not so sensitive that significant bleaching occurs during the measurement of fluorescence recovery by the attenuated probe beam. Brightness, cellular toxicity, Stokes' shift, and excitation by available laser lines are additional factors to consider when selecting a chromophore.

A variety of fluorescent probes have been employed in photobleaching experiments. Fluorescein (and its derivatives, e.g., 2,7-bis[2-carboxyethyl]-5-[and 6-]-carboxyfluorescein [BCECF]) (Kao et al., 1993; Seksek et al., 1997; Lukacs et al., 2000), green fluorescent protein (GFP; Cole et al., 1996; Swaminathan et al., 1997; Partikian et al., 1998; Dayel et al., 1999), and rhodamine (Luby-Phelps et al., 1986) have been used extensively in the aqueous phase of cells, and lipophilic fluorophores, such as 1,1'-dihexadecyl-3,3,3',3'-tetramethylindocarbocynanine perchlorate ($DiIC_{16}$) and octadecyl rhodamine B chloride (R18), have been used to study diffusion in cell plasma membranes (Cowan et al., 1997; Niv et al., 1999). Less commonly used fluorophores include NBD, carbocyanines, and oxacyanines (Hochman et al., 1985; Jovin and Vaz, 1989). Our experience has mostly been with bleaching of fluorescein and GFP-labeled protein and solutes. Figure 7–2A shows spot photobleaching measurements of thin aqueous layers (5 μm thickness between cover glasses) of fluorescein, GFP, a 500 kDa fluorescein isothiocyanate (FITC)–labeled dextran, and a 6 kb FITC-labeled double-stranded circular DNA plasmid. In each case the brief bleach pulse produces an immediate drop in fluorescence, followed by a monophasic recovery of fluorescence to its initial level. The different recovery rates reflect the size-dependent diffusion coefficients.

Reversible photobleaching is the recovery of fluorescence by photochemical processes that are unrelated to fluorophore diffusion. It is important to identify and eliminate reversible photobleaching processes when possible because of their

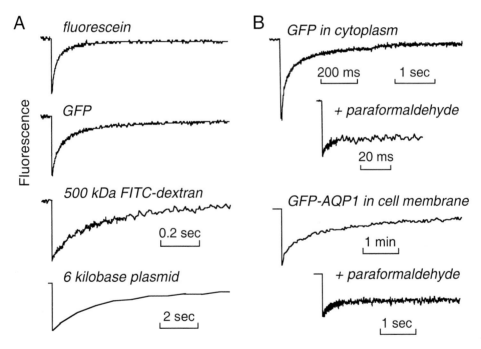

FIGURE 7–2 Irreversible and reversible photobleaching of common fluorophores. *A*, Irreversible photobleaching of aqueous solutions of fluorescein, purified recombinant green fluorescent protein (GFP)-S65T, 500 kDa fluorescein isothiocyanate (FITC)-dextran, and a 6 kb FITC-labeled circular double-stranded plasmid. Thin solution layers (~5 μm) were bleached using a ×20 objective to produce a spot of approximately 4 μm diameter. Fluorescence was bleached by 20%–30% using a brief 488 nm laser pulse. [Adapted from Lukacs et al., 2000; Dayel et al., 1999; Seksek et al., 1997.] *B, Top,* Reversible photobleaching of GFP expressed in the cytoplasm of Chinese hamster ovary cells and GFP-AQP1 expressed in the plasma membrane of LLC-PK1 cells. Cytoplasmic GFP was bleached by a brief 488 nm pulse using a ×40 objective. Where indicated, cells were fixed in 4% paraformaldehyde to immobilize GFP. *Bottom,* Membrane-associated GFP-AQP1 fluorescence was bleached using a ×100 objective. See text for explanations. [Adapted from Umenishi et al., 2000; Dayel et al., 1999.]

confounding effects on the interpretation of fluorescence recovery in terms of solute diffusional mobility. Regardless of the mechanism of the reversible process, there are several practical maneuvers that are useful in distinguishing irreversible from reversible fluorescence recovery (Periasamy et al., 1996; Swaminathan et al., 1997). Irreversible bleaching produces a fluorescence recovery whose rate depends strongly and predictably on spot diameter, whereas recovery by reversible photobleaching is independent of spot size. For reversible fluorescence recovery by triplet state relaxation (see below), changing bleach beam intensity and time also gives predictable effects on reversible versus irreversible photobleaching efficiencies. Subtle reversible photobleaching processes may be exposed by fluorophore immobilization to suppress diffusion-related recovery, as has been done for GFP in various cellular compartments using paraformaldehyde fixation (Partikian et al., 1998; Dayel et al., 1999).

Figure 7–2B (top) shows a FRAP measurement of GFP diffusion in cytoplasm using a ×40 objective to produce a 2 μm diameter spot. The efficient recovery indicates that GFP is mobile, and the recovery rate indicates that GFP diffusion in cytoplasm is slowed by three- to fourfold compared with that in water. Paraformaldehyde fixation abolishes the slow recovery, but reveals a rapid (half-time <5 ms) reversible recovery process whose recovery rate is independent of spot diameter. It is noted that the presence of reversible photobleaching does not preclude diffusion coefficient determination from analysis of the irreversible photobleaching component. As shown for GFP diffusion in cytoplasm, the recovery from reversible photobleaching is generally much faster than that from irreversible photobleaching or can be made so by increasing spot diameter. For some chromophores like fluorescein, triplet state quenchers such as oxygen can be used to accelerate or eliminate the reversible photobleaching process. This was done in total internal reflection-FRAP measurements of solute diffusion in membrane-adjacent cytoplasm (Swaminathan et al., 1996).

There are multiple photophysical processes that can produce reversible recovery. The best characterized is triplet state relaxation, in which light exposure populates a triplet state. A nonradiative transition from the triplet state to the singlet ground state produces apparent fluorescence recovery. Generally triplet-relaxation processes are rapid (<5 ms), sensitive to oxygen and triplet state quenchers, and are produced by relatively weak illumination. A very slow (>100 ms), oxygen-dependent reversible photobleaching process was reported for a surface-adsorbed FITC-labeled membrane protein (Stout and Axelrod, 1995). Although a definitive mechanism for the reversible recovery could not be established, it was concluded that a triplet state process was probably not involved. We have observed slow reversible recovery processes for GFP-labeled membrane proteins (Umenishi et al., 2000). Figure 7–2B (bottom) shows the slow membrane diffusion of a GFP-tagged AQP1 water channel protein in an epithelial cell. There is a hint of a fast recovery process. Paraformaldehyde fixation revealed a spot size–independent reversible recovery process of approximately 100 ms. The physical origin of this reversible recovery process is not clear, but it underscores the need to rigorously distinguish between diffusion-related and reversible fluorescence recovery processes in FRAP measurements.

4. Applications to Diffusion Measurements in Living Cells

The viscous properties of aqueous compartments in cells has been a topic of long-standing interest. Cellular viscosity has important implications for many cellular processes, such as metabolism, signaling, and replication. Our laboratory has used FRAP methods to measure the translational diffusion of fluorescent solutes and/or macromolecules in cytoplasm, nucleus, and various intracellular organelles. One parameter describing cytoplasmic rheology is fluid-phase viscosity, defined as the microviscosity sensed by a small solute in the absence of interactions with macromolecules and organelles. Measurements of the rotation of small fluorophores by time-resolved anisotropy (Fushimi and Verkman, 1991; Bicknese et al., 1993) and of intrinsically viscosity-sensitive fluorophores (Luby-Phelps et al., 1993) indicated that the fluid-phase viscosity of cytoplasm is not much greater than that of water. For transport of small solutes such as metabolites and nucleic acids, a more important parameter describing cytoplasmic viscosity is the translational diffusion coefficient.

Kao et al. (1993) measured the translational mobility of a small fluorescent probe in the cytoplasm of Swiss 3T3 fibroblasts utilizing spot photobleaching. Diffusion of a fluorescein-like molecule in cytoplasm was about four times slower than in water. Three independently acting factors that accounted quantitatively for the fourfold slowed diffusion were identified: slowed diffusion in fluid-phase cytoplasm, probe binding to intracellular components, and probe collisions with intracellular components. Probe collisions were determined to be the principal diffusive barrier that slow the translational diffusion of small solutes. Similar results were obtained for small solute diffusion in membrane-adjacent cytoplasm using total internal reflection–FRAP (Swaminathan et al., 1996).

In earlier studies (Luby-Phelps et al., 1986) spot photobleaching was used to measure the translational diffusion of larger solutes, including microinjected, fluorescently labeled dextrans and Ficolls. As dextran or Ficoll molecular size was increased, diffusion in cytoplasm progressively decreased relative to that in water, suggesting a cytoplasmic "sieving" mechanism that was proposed to involve the skeletal mesh. More recent data from our laboratory (Seksek et al., 1997; Lukacs et al., 2000) confirmed that very large solutes (>500 kDa) are impeded in their motion in cytoplasm; however, we did not observe sieving of smaller solutes.

It is noted that the measurements of dextran and Ficoll diffusion required direct cell microinjection, which is an invasive procedure with the potential for cell damage. Cytoplasmically expressed GFP was used as a noninvasive probe to measure the rotational and translational diffusion of a protein-sized molecule (Swaminathan et al., 1997; see also Fig. 7–2B). The three- to fourfold slowed diffusion of GFP in cytoplasm supports the view that protein-sized molecules are mildly impeded in their motion and that the primary barrier to GFP diffusion is collisional interactions between GFP and macromolecular solutes.

Recently we used FRAP to measure the mobility of FITC-labeled DNA fragments of oligonucleotide to plasmid size (20–6000 base pairs) microinjected directly into the cytoplasm and nucleus (Lukacs et al., 2000). Whereas DNA fragments of all sizes are nearly immobile in nucleus, small DNAs of a size below 200 bp were nearly as mobile in cytoplasm as comparably sized dextrans. The diffusional mobility of larger DNA fragments was considerably slowed, suggesting that the cytoplasm poses a significant barrier in gene delivery using nonviral vectors that enter cells by endocytosis followed by cytoplasmic diffusion and nuclear uptake.

Solute diffusion within the aqueous lumen of intracellular organelles is involved in many processes, such as metabolism in mitochondria and protein processing and recognition in endoplasmic reticulum. The mitochondrial matrix, the aqueous compartment enclosed by the inner mitochondrial membrane, is a major site of metabolic processes. The matrix is a particularly interesting compartment because of its presumably high density of enzymes and other proteins. Theoretical considerations have suggested that the diffusion of metabolite- and enzyme-sized solutes may be severely restricted in the mitochondrial matrix (Welch and Easterby, 1994). It has been proposed that biochemical events might occur by a "metabolite channeling" mechanism in which metabolites are passed from one enzyme to another in an organized complex without aqueous-phase diffusion.

The ability to target GFP to the mitochondrial matrix provided a unique opportunity to test the hypothesis that solute diffusion is greatly slowed in the matrix. The diffusion of GFP was measured in the mitochondrial matrix of fibroblast, liver, skeletal muscle, and epithelial cell lines (Partikian et al., 1998). Spot photobleaching of GFP with a $\times 100$ objective (0.8 μm spot diameter) gave a half-time for fluorescence recovery of 15–19 ms with greater than 90% of the GFP mobile (Fig. 7–3A, top curve). As predicted for aqueous-phase diffusion in a confined compartment, fluorescence recovery was slowed or abolished by increased laser spot size (second curve), paraformaldehyde fixation (third curve), or increased bleach time (fourth curve). The fluorescence recovery data were analyzed using a mathematical model of matrix diffusion.

As shown in Figure 7–3B, mitochondria were modeled as long, continuously open cylinders with a specified orientational distribution. Fluorescence recoveries were computed from analytical solutions to the diffusion equation. Predicted recovery curves for different diffusion coefficients (D) are shown in Figure 7–3C. The fitted values for D were 2–3 $\times 10^{-7}$ cm^2/s, only three- to fourfold less than that for GFP diffusion in water. Interestingly, little fluorescence recovery was found for bleaching of GFP in fusion with subunits of the fatty acid α-oxidation multienzyme complex (Fig. 7–3A, bottom curve) that are normally present in the matrix. The rapid and unrestricted diffusion of GFP in the mitochondrial matrix suggested that classic metabolite channeling may not be required. It was proposed that the clustering of matrix enzymes in membrane-associated complexes might serve to establish a relatively uncrowded aqueous space where solutes can freely diffuse, thus reducing metabolite transit times and pool sizes.

A similar photobleaching study was carried out to measure solute diffusion in the endoplasmic reticulum, which is the major compartment for the processing and quality control of newly synthesized proteins (Dayel et al., 1999). Green fluorescent protein was targeted to the aqueous endoplasmic reticulum lumen of Chinese hamster ovary cells by transient transfection with cDNA encoding GFP with C-terminus KDEL retention sequence and upstream preprolactin secretory

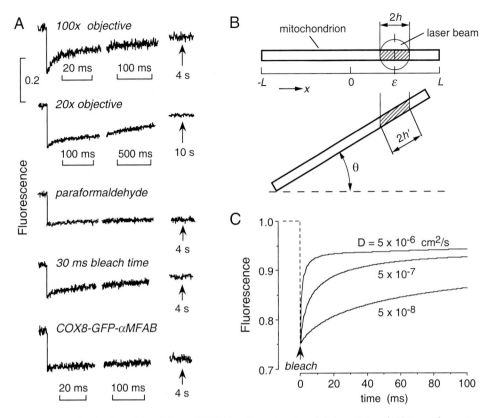

FIGURE 7–3 Spot photobleaching of GFP in the mitochondrial matrix of Chinese hamster ovary cells. *A*, Photobleaching with 40 μs bleach time using $\times$100 or $\times$20 objectives, top two curves; photobleaching with $\times$100 objective in cells after 6 h fixation in 4% paraformaldehyde, third curve; photobleaching as in the top curve with long (30 ms) bleach time, fourth curve; and photobleaching (as in the top curve) of a fusion protein of GFP with a matrix enzyme (α-subunit of the mitochondrial β-fatty acid oxidation pathway), fifth curve. *B*, Model of mitochondria as a long thin cylinder with unobstructed lumen oriented at an angle θ. See Partikian et al. (1998) for details. *C*, Model predictions for fluorescence recovery for different diffusion coefficients (D) for bleaching with a $\times$100 objective and short bleach time. [Adapted from Partikian et al., 1998.]

sequence. Figure 7–4A shows a time series of fluorescence images of the GFP-labeled endoplasmic reticulum after a large circular spot is bleached. The darkened zone produced by the bleach pulse was progressively filled in with unbleached GFP diffusing into the bleached zone by diffusion through the endoplasmic reticulum lumen. The bleach did not affect the fluorescence of an adjacent cell (Fig. 7–4, arrow). Spot photobleaching showed that nearly all GFP was mobile with a $t_{1/2}$ for fluorescence recovery of 143 ms ($\times$40 objective) and 88 ms ($\times$60 objective). Based on a mathematical model of photobleaching that accounts for endoplasmic reticulum geometry, a GFP diffusion coefficient of 0.5–1×10^{-7} cm^2/s was computed. The result was 9–18-fold less than that in water and three- to sixfold less than in cytoplasm. The data suggest that GFP is mobile in the endoplasmic reticulum lumen and that its translational diffusion is mildly hindered. Measurements utilizing various expressed GFP-fusion proteins should

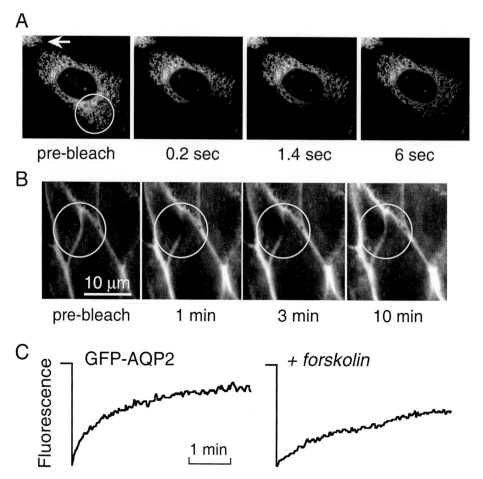

Figure 7–4 Photobleaching of green fluorescent protein (GFP) in the lumen of endoplasmic reticulum (ER) and in the cell plasma membrane. *A,* GFP was targeted to the aqueous lumen ER of Chinese hamster ovary cells. Serial micrographs show cell ER fluorescence before (pre-bleach) and at indicated times after bleaching (bleach time, 200 ms; 488 nm). The white circle denotes the bleach spot. Arrow indicates an adjacent unbleached cell. [Adapted from Dayel et al., 1999.] *B,* A GFP-AQP2 chimera was targeted to the plasma membrane of LLC-PK1 cells. A series of pre- and post-bleach images is shown as in *A. C,* Spot photobleaching of cells as in *B* using an ~0.8 μm diameter spot produced by a ×100 oil objective. Bleach time was 5 ms. Where indicated, cells were pre-incubated with the cAMP agonist forskolin for 15 min before the bleaching study. [Adapted from Umenishi et al., 2000.]

be useful in defining interactions between luminal contents and the endoplasmic reticulum wall, such as the binding of misfolded proteins to molecular chaperones.

These examples show that GFP diffusion is readily measurable in intracellular organellar compartments. However, because organelles often have a complex exterior geometry, as well as internal barriers, analytic methods are required to quantitatively interpret photobleaching experiments. The mitochondrial matrix is seen by electron microscopy as a cylindrical compartment with internal barriers called *cristae*. The endoplasmic reticulum is generally thought of as an interconnected reticular network of cylinders or plates in three dimensions. From a practical experimental perspective, a strategy is needed to deduce intrinsic solute diffusion from photobleaching recovery data. A more general issue is the elucidation of how organellar shape and barrier geometry influence particle diffusion in vivo. The analysis of solute diffusion in mitochondria described earlier utilized a model that assumes an open mitochondrial lumen without internal barriers. It is thus important to understand how cristae barriers affect the diffusional transport of metabolites along the mitochondrial axis.

A number of theoretical descriptions of particle diffusion in two-dimensional membranes have been reported for a variety of situations involving binding, crowding, and mobile and immobile obstacles (Saxton, 1994). We have utilized Monte-Carlo simulations in three dimensions to analyze the effects of barrier properties on diffusion in the aqueous matrix of mitochondria and the effects of reticular geometry on diffusion in endoplasmic reticulum. We have also used the simulations to relate experimental photobleaching data to intrinsic diffusion coefficients (Ölveczky and Verkman, 1998). Mitochondria were modeled as long closed cylinders containing fixed luminal obstructions of variable number and size, and the endoplasmic reticulum was modeled as a network of interconnected cylinders of variable diameter and density.

For mitochondria-like cylinders, significant slowing of diffusion required either large or wide single obstacles or multiple obstacles. In simulated spot photobleaching experiments a 25% decrease in apparent diffusive transport rate (defined by the time to 75% fluorescence recovery) was found for a single thin transverse obstacle occluding 93% of lumen area, a single 53%-occluding obstacle of width 16 lattice points (8% of cylinder length), or ten equally spaced 50%-obstacles alternately occluding opposite halves of the cylinder lumen. Recovery curve shapes with obstacles showed long tails indicating anomalous diffusion. Additional simulations indicated that significantly slowed particle diffusion can be produced by binding to barrier walls. For a reticulum-like network, particle diffusive transport was mildly reduced from that in unobstructed three-dimensional space. In simulated photobleaching experiments, apparent diffusive transport was decreased by 39%–60% in reticular structures where 90%–97% of space was occluded. These computations provided an approach to analyze photobleaching data in terms of microscopic diffusive properties and supported the paradigm that organellar barriers must be quite severe to seriously impede solute diffusion.

The examples discussed above involve solute diffusion in aqueous compartments. Diffusion in membranes is remarkably slower and may be quite complex because of membrane crowding with proteins, distinct lipid domains, and membrane–cytoskeletal interactions. As described for the analysis of solute diffusion in aqueous compartments, photobleaching recovery provides a quantitative method for the analysis of GFP-fusion protein diffusion in membranes. Photobleaching measurements in membranes are technically easier than in aqueous compartments because of the slower recovery rates, giving improved signal-to-noise ratio. In addition, rapid reversible photobleaching processes are generally less of a concern for the measurement of slow lateral diffusion in membranes.

Figure 7–4B shows serial fluorescence images of epithelial cells expressing a GFP-AQP1 fusion protein. AQP1 is a small integral membrane protein that functions as a molecular water channel (Verkman and Mitra, 2000). Fluorescence recovery into the large bleached zone occurred over several minutes. The fluorescence recovery was abolished by paraformaldehyde fixation and was strongly dependent on spot diameter, as expected for a diffusion-related process (Umenishi et al., 2000). A GFP-AQP1 diffusion coefficient of approximately 5×10^{-11} cm^2/s was determined.

Similar fluorescence recoveries were found for cells expressing GFP-AQP2. AQP2 is a cAMP-regulated water channel expressed in a kidney collecting duct. Antidiuretic hormone increases collecting duct water permeability by a membrane cycling mechanism. Figure 7–4C shows a spot photobleaching measurement using an 0.8 μm diameter spot produced by a $\times 100$ objective. Fluorescence recovery was nearly complete, indicating that GFP-AQP2 was fully mobile, and recovery occurred over the first minute. Interestingly, stimulation of cells by the cAMP agonist forskolin resulted in remarkable slowing of GFP-AQP2 mobility, suggesting interactions of AQP2 with skeletal or other proteins. The slowing of recovery was blocked by a specific mutation in AQP2 that prevented phosphorylation, as well as by actin filament disrupters. FRAP measurements provided a unique biophysical tool to investigate protein–protein interactions in living cells that are difficult to study with classic biochemical methods.

There is a large body of literature on the diffusion of membrane lipids and proteins (for review, see Jovin and Vaz, 1989; Lippincott-Schwartz et al., 1999). Photobleaching of GFP chimeras in the Golgi membrane was used to establish that membrane protein immobilization is not the mechanism of protein retention (Cole et al., 1996). Repetitive photobleaching of the GFP chimeras in one region of the Golgi membrane (a technique termed FLIP, fluorescence loss in photobleaching) revealed that the membranes of adjacent Golgi stacks are connected. In a similar study, the mobility of cytochrome P450 was measured (Szczesna-Skorupa et al., 1998). Cytochrome P450 is a protein that is restricted to the endoplasmic reticulum membrane despite containing no known retention/retrieval signals (such as the di-lysine motif KKXX). A high diffusion coefficient suggested that mechanisms other than immobilization were responsible for endoplasmic reticulum retention. Photobleaching in conjunction with the high-resolution imaging provides information about the distributions and dynamics of GFP-fusion proteins. Studies of the lamin B receptor–GFP fusion revealed a dynamic relationship between the nuclear envelope and the endoplasmic reticulum during the cell cycle (Ellenberg et al., 1997). Interphase cells contain a mobile fraction of lamin B in the endoplasmic reticulum membrane in addition to immobile lamin B in the nuclear envelope. During mitosis the nuclear envelope–associated lamin B redistributes into the endoplasmic reticulum membrane and become highly mobile. Similarly, modulation of the mobility of cadherin has been followed by photobleaching of GFP-fusion proteins during cell adhesion (Adams et al., 1998). Cadherin–GFP was initially mobile, but became immobilized and sequestered at the sites of cell contact.

At the plasma membrane, photobleaching of GFP chimeras has been used to investigate the mobility of transport proteins like the aquaporins (Umenishi et al., 2000) and proteins involved in signal transduction like Ras (Niv et al., 1999) and nitric oxide synthase (Sowa et al., 1999). The rapid diffusion coefficient of Ras ($\sim 2 \times 10^{-9}$ cm^2/s), only 1.6-fold slower than that of the membrane probe DiC$_{16}$, supported the contention that farnesyl moieties are involved in membrane attachment (Niv et al., 1999). Furthermore, the diffusion coefficient of Ras was increased by pharmacological agents that disrupt the interaction of Ras with the plasma membrane. FRAP experiments demonstrated that endothelial nitric oxide synthase (eNOS) had more restricted mobility in the plasma membrane than in the Golgi (Sowa et al., 1999). The qualitative and semiquantitative photobleaching of GFP-fusion proteins has been extremely informative in elucidating cellular processes such as in the analysis of protein trafficking mechanisms (Presley et al., 1997; Hirschberg et al., 1998; Nakata et al., 1998).

5. ANALYSIS METHODS

The determination of diffusion mechanisms from fluorescence recovery curves, F(t), requires quantitative analysis procedures. Spot photobleaching data from cell membrane and aqueous compartments have generally been analyzed in terms of a single diffusing component with or without an immobile fraction. For spot photobleaching in two dimensions (as in membranes), in which bleach and probe regions are identical, the diffusion equation has been solved analytically for a Gaussian beam profile and very short bleach time to compute F(t) from solute diffusion coefficient and spot diameter (Axelrod et al., 1976). The diffusion equation has been solved numerically for total internal reflection-FRAP measurements for arbitrary bleach times (Swaminathan et al., 1996). Direct application of the diffusion equation, however, has limited utility in real FRAP measurements where beam profiles are nonideal, bleach time may not be negligible compared with recovery time, sample geometry may be complex, and the assumption of simple diffusion of a single species may not be valid. It is generally impractical to solve the diffusion equation for these nonideal situations and for diffusion in three dimensions.

For analysis of photobleaching data in three dimensions, such as the aqueous phase of cell cytoplasm, we introduced a calibration procedure in which the half-time ($t_{1/2}$) for fluorescence recovery in cells is compared with the $t_{1/2}$ measured in thin layers of fluorophores dissolved in artificial solutions of known viscosity (Kao et al., 1993). These analytical and empirical methods are useful for the determination of single, time-independent diffusion coefficients, but are not easily adapted to complex diffusive phenomena such as anomalous diffusion or diffusion of multiple species with different diffusion coefficients. Empirical equations have been given to fit $t_{1/2}$ values from recovery curves, F(t) (Feder et al., 1996). A useful equation that generally provides a good empirical fit to FRAP data is $F(t) = [F_0 + F_1(t/t_{1/2})^\alpha]/[1 + (t/t_{1/2})^\alpha]$, where F_0 is fluorescence before bleaching, F_1 is fluorescence at infinite time, and α ($0 < \alpha < 1$) is an empirical parameter related to solute restriction/anomalous diffusion. The potential pitfalls in the assumption of simple diffusion and the significance of long tail kinetics in diffusive phenomena have been recognized, and recent work has begun to consider how to interpret photobleaching data in systems with complex or anomalous diffusion (Nagle, 1992; Feder et al., 1996).

Solute diffusion is described as "simple" or "nonanomalous" in a homogeneous medium like a liquid solvent, in which case solute transport is described by a single diffusion coefficient. There are, however, situations in which solute diffusion cannot be described in terms of a single diffusion coefficient, such as anomalous diffusion or diffusion of more than one species. The diffusion of a solute is defined to be anomalous if the mean squared displacement varies with time in a nonlinear manner. Here the diffusion coefficient is not constant, but is time and/or space dependent. A number of different physical mechanisms giving anomalous diffusion have been described (Bouchaud et al., 1991; Bhattacharya and Bagchi, 1997). Another example of nonsimple diffusion is the presence of two or more diffusing species, each of which is described by a single diffusion coefficient. Without prior knowledge of the presence of multiple diffusing species, as might be the case in cellular environments where heterogeneous binding can occur, photobleaching data can be wrongly interpreted as anomalous diffusion.

We introduced the idea that photobleaching recovery data, F(t), can be resolved in terms of a continuous distribution of diffusion coefficients, α(D) (Periasamy and Verkman, 1998). A regression method to recover α(D) from F(t) was developed based on the Maximum Entropy Method, which utilizes a "basis" recovery curve for simple diffusion obtained using a reference sample such as a thin uniform film of fluorescein in saline. The procedure recovered multiple narrow α(D) in simulations and experimental FRAP measurements in solutions containing fluorophore mixtures. Analysis of theoretical cases of anomalous diffusion defined characteristic α(D) patterns, including broad and asymmetric distributions. An independent method for identifying anomalous diffusive processes from F(t) data was developed in which an apparent time-dependent diffusion coefficient, D(t), was computed directly from F(t) and the reference recovery curve. The approach utilizes a reference F(t) curve for simple diffusion of a single species in order to deduce D(t) by a convolution procedure. As expected, D(t) was constant for simple diffusion of a single species and decreased and increased, respectively, for anomalous subdiffusion and superdiffusion. The determination of α(D) and D(t) from FRAP data provides a systematic approach to identify and quantify simple and anomalous diffusive phenomena. Further work on artificial and biological systems is required to assess the utility of the complementary methods to analyze photobleaching data.

6. PERSPECTIVE

Photobleaching measurements have provided fundamental information about the dynamics of biologically important solutes and macromolecules in cell membrane and aqueous compartments. Rapid advances in fluorophore targeting methods make possible the selective labeling of cellular components with a wide variety of chromophores. There remain many unresolved questions for which photobleaching and related methods, such as photoactivation, are applicable. The mechanisms and regulation of membrane protein mobility remain poorly understood. Issues about the role of molecular crowding and supermolecular enzyme organization in cell metabolism mandate dynamic measurements. Although much has been learned about the diffusional mobility of macromolecules in cytoplasm, nucleus, and organelles, the relationship between molecular diffusion and cell function remains incomplete.

The application of photobleaching methods to problems in cell biology is generally straightforward because bleaching and recovery events can be visualized and the resultant time-domain information is readily interpretable. The available instrumentation for photobleaching—lasers, beam modulators, and detectors—have adequate temporal and spatial resolution for most cellular applications. As discussed in this chapter, the challenge in photobleaching measurements is to confirm the diffusive nature of the fluorescence recovery and to deduce quantitative information about diffusive processes. Although beyond the scope of this chapter, other fluorescence-based technologies can provide complementary information about molecular diffusion. Measurements of fluorophore rotational diffusion by time-resolved anisotropy (Fushimi and Verkman, 1991) and polarization photobleaching recovery complement FRAP measurements of translational diffusion. Fluorescence correlation microscopy holds promise for cellular diffusion measurements, but issues related to background autofluorescence and complex multicomponent autocorrelation functions may limit its utility for quantitative cell studies.

This work was supported by NIH grants DK43840, DK35124, HL60288, DK16095, and HL59198 and by grant R613 from the National Cystic Fibrosis Foundation.

REFERENCES

Adams, C. L., Y. T. Chen, S. J. Smith, and W. J. Nelson. Mechanism of epithelial cell–cell compaction revealed by high-resolution tracking of E-cadherin-green fluorescent protein. *J. Cell Biol.* 142:1105–1119, 1998.

Alexrod, D., D. E. Koppel, J. Schlessinger, E. Elson, and W. W. Webb. Mobility measurements by analysis of fluorescence photobleaching recovery kinetics. *Biophys. J.* 16:1055–1069, 1976.

Bhattacharya, S., and B. Bagchi. Anomalous diffusion of small particles in dense liquids. *J. Chem. Phys.* 106:1757–1763, 1997.

Bicknese, S., N. Periasamy, S. B. Shohet, and A. S. Verkman. Cytoplasmic viscosity near the cell plasma membrane: Measurement by evanescent field frequency-domain microfluorimetry. *Biophys. J.* 65:1272–1282, 1993.

Bouchaud, J. P., A. Ott, D. Langevin, and W. Urbauch. Anomalous diffusion in elongated micelles and its Levy flight interpretation. *J. Phys. II France* 1:1465–1482, 1991.

Brown, E. B., E. S. Wu, W. Zipfel, and W. W. Webb. Measurement of molecular diffusion in solution by multiphoton fluorescence photobleaching recovery. *Biophys. J.* 77:2837–2849, 1999.

Cole, N. B., C. L. Smith, N. Sciaky, M. Terasaki, M. Edidin, and J. Lippincott-Schwartz. Diffusional mobility of Golgi proteins in membranes of living cells. *Science* 273:797–801, 1996.

Cowan, A. E., L. Nikhimovsky, D. G. Myles, and D. E. Koppel. Barriers to diffusion of plasma membrane proteins form early during guinea pig spermiogenesis. *Biophys. J.* 73:507–516, 1997.

Davoust, J., P. F. Devaux, and L. Leger. Fringe pattern photobleaching, a new method for the measurement of transport coefficients of biological macromolecules. *EMBO J.* 1:1233–1238, 1982.

Dayel, M., E. Hom, and A. S. Verkman. Diffusion of green fluorescent protein in the aqueous lumen of endoplasmic reticulum. *Biophys. J.* 2843–2851, 1999.

Dupont, A. M., F. Foucault, Vacher, M., P. F. Devaux, and S. Cribier. Translational diffusion of globular proteins in the cytoplasm of cultured muscle cells. *Biophys. J.* 78:901–907, 2000.

Ellenberg, J., E. D. Siggia, J. E. Moreira, C. L. Smith, J. F. Presley, H. J. Worman, and J. Lippincott-Schwartz. Nuclear membrane dynamics and reassembly in living cells: Targeting of an inner nuclear membrane protein in interphase and mitosis. *J. Cell Biol.* 138:1193–1206, 1997.

Feder, T. J., I. Brust-Mascher, J. P. Slattery, B. Baird, and W. W. Webb. Constrained diffusion or immobile fraction on cell surfaces: A new interpretation. *Biophys. J.* 70:2367–2373, 1996.

Fushimi, K., and A. S. Verkman. Low viscosity in the aqueous domain of cell cytoplasm measured by picosecond polarization microscopy. *J. Cell Biol.* 112:719–725, 1991.

Hirschberg, K., C. M. Millar, J. Ellenberg, J. F. Presley, E. D. Siggia, R. D. Phair, and J. Lippincott-Schwartz. Kinetic analysis of secretory protein traffic and characterization of Golgi to plasma membrane transport in living cells. *J. Cell Biol.* 143:1485–1503, 1998.

Hochman, J., S. Ferguson-Miller, and M. Schindler. Mobility in the mitochondrial electron transport chain. *Biochemistry* 24:2509–2516, 1985.

Jovin, T. M., and W. L. C. Vaz. Rotational and translational diffusion in membranes measured by fluorescence and phosphorescence methods. *Methods Enzymol.* 172:471–513, 1989.

Kao, H. P., J. R. Abney, and A. S. Verkman. Determinants of the translational diffusion of a small solute in cell cytoplasm. *J. Cell Biol.* 120:175–184, 1993.

Kao, H. P., and A. S. Verkman. Construction and performance of a photobleaching recovery apparatus with microsecond time resolution. *Biophys. Chem.* 59:203–210, 1996.

Lippincott-Schwartz, J., J. F. Presley, K. J. M. Zall, K. Hirschberg, C. D. Miller, and J. Ellenberg. Monitoring the dynamics and mobility of membrane proteins tagged with green fluorescent protein. *Methods Cell Biol.* 58:261–281, 1999.

Luby-Phelps, K., S. Mujundar, R. Mujundar, L. Ernst, W. Galbraith, and A. Waggoner. A novel fluorescence ratiometric method confirms the low solvent viscosity of the cytoplasm. *Biophys. J.* 65:236–242, 1993.

Luby-Phelps, K., D. L. Taylor, and F. Lanni. Probing the structure of the cytoplasm. *J. Cell Biol.* 102:2015–2022, 1986.

Lukacs, G. L., P. Haggie, O. Seksek, D. Lechardeur, N. Freedman, and A. S. Verkman. Size-dependent DNA mobility in cytoplasm and nucleus. *J. Biol. Chem.* 275:1625–1629, 2000.

Nagle, J. F. Long tail kinetics in biophysics? *Biophys. J.* 63:366–370, 1992.

Nakata, T., S. Terada, and N. Hirokawa. Visualization of the dynamics of synaptic vesicle and plasma membrane proteins in living axons. *J. Cell Biol.* 140:659–674, 1998.

Niv, H., O. Gutman, Y. I. Henis, and Y. Kloog. Membrane interactions of the constitutively active GFP-Ki-Ras 4B and their role in signaling. *J. Biol. Chem.* 274:1606–1613, 1999.

Ölveczky, B. P., and A. S. Verkman. Monte-Carlo analysis of obstructed diffusion in 3 dimensions: Application to molecular diffusion in organelles. *Biophys. J.* 74:2722–2730, 1998.

Partikian, A., B. Ölveczky, R. Swaminathan, Y. Li, and A. S. Verkman. Rapid diffusion of green fluorescent protein in the mitochondrial matrix. *J. Cell. Biol.* 140:821–829, 1998.

Periasamy, N., S. Bicknese, and A. S. Verkman. Reversible photobleaching of fluorescein conjugates in air-saturated viscous solutions: Molecular tryptophan as a triplet state quencher. *Photochem. Photobiol.* 63:265–271, 1996.

Periasamy, N., and A. S. Verkman. Analysis of fluorophore diffusion by continuous distributions of diffusion coefficients: Application to photobleaching measurements of simple and anomalous diffusion. *Biophys. J.* 75:557–567, 1998.

Presley, J. F., N. B. Cole, T. A. Schroer, K. Hirschberg, K. J. M. Zaal, and J. Lippincott-Schwartz. ER-to-Golgi transport visualized in living cells. *Nature* 389:81–84, 1997.

Saxton, M. J. Anomalous diffusion due to obstacles: A Monte Carlo study. *Biophys. J.* 66:394–401, 1994.

Seksek, O., J. Biwersi, and A. S. Verkman. Translational diffusion of macromolecule-size solutes in cytoplasm and nucleus. *J. Cell Biol.* 138:131–142, 1997.

Sowa, G., J. Liu, A. Papapetropoulos, M. Rex-Haffner, T. E. Hughes, and W. C. Sessa. Trafficking of endothelial nitric-oxide synthase in living cells. *J. Biol. Chem.* 274:22524–22531, 1999.

Stout, A. L., and D. Axelrod. Spontaneous recovery of fluorescence by photobleached surface-adsorbed proteins. *Photochem. Photobiol.* 62:239–244, 1995.

Swaminathan, R., C. P. Hoang, and A. S. Verkman. Photochemical properties of green fluorescent protein GFP-S65T in solution and transfected CHO cells: Analysis of cytoplasmic viscosity by GFP translational and rotational diffusion. *Biophys. J.* 72:1900–1907, 1997.

Swaminathan, R., N. Periasamy, S. Bicknese, and A. S. Verkman. Cytoplasmic viscosity near the cell plasma membrane: Translation of BCECF measured by total internal reflection-fluorescence photobleaching recovery. *Biophys. J.* 71:1140–1151, 1996.

Szczesna-Skorupa, E., C. D., Chen, S. Rodgers, and B. Kemper. Mobility of cytochrome P450 in the endoplasmic reticulum membrane. *Proc. Natl. Acad. Sci. U.S.A.* 95:14793–14798, 1998.

Umenishi, F., J. M. Verbavatz, and A. S. Verkman. cAMP regulated membrane diffusion of a green fluorescent protein–aquaporin-2 chimera. *Biophys. J.* 78:1024–1035, 2000.

Verkman, A. S., and A. K. Mitra. Structure and function of aquaporin water channels. *Am. J. Physiol.* 278:F13–F28, 2000.

Welch, G. R., and J. S. Easterby. Metabolic channeling versus free diffusion: Transition-time analysis. *TIBS* 19:193–197, 1994.

White, J., and E. Stelzer. Photobleaching GFP reveals protein dynamics inside live cells. *TICB* 9:61–65, 1999.

8

Processing Microscope-Acquired Images for Use in Multimedia, Print, and the World Wide Web

HAROLD L. NOAKES, JR. AND AMMASI PERIASAMY

1. INTRODUCTION

"One picture is equivalent to a thousand words." This Chinese proverb is as true in light microscopic imaging as elsewhere. Historically, scientists drew the images on a piece of paper that they saw while looking through the microscope, often with the assistance of a *camera lucida* attachment. Later, 35 mm still or movie cameras were used to record microscope images of cellular structures and their functional activities. Then, the video camera was coupled to the microscope, and the images were photographed from the monitor to present their microscopic images (Shotton, 1993; Aikens et al., 1998; Bright and Taylor, 1986; Herman, 1998). These technological advancements have revolutionized the methods of microscopic imaging and processing (see other chapters in this book).

Commercially available digital cameras and software packages are used to acquire and process microscopic images (Periasamy and Herman, 1994; Lockett et al., 1992; Axelrod et al., 1976). Image processing software helps scientists remove some of the optical noise introduced into the images by light scattering, uneven illumination, and autofluorescence from the cells and the medium (Inoué and Spring, 1997; Russ, 1999; Pawley, 1995; Meek and Elder, 1977; Tsien and Poenie, 1986). It is of great interest that microscopically acquired digitized and processed images are being applied to many analytical, explicative, and didactic purposes. These images can be displayed in print, video, and computer presentations, in the context of a Web page, or within a multimedia CD or DVD project.

2. COMMERCIAL SOFTWARE

In this chapter we describe the commercially available software programs that create digitized video microscope images in different formats. We also discuss related subjects such as file conversion and how to prepare images for the World Wide Web.

Many software tools are available to perform specific tasks in the preparation of images for multimedia purposes. To solve particular image manipulation problems, users and hobbyists have developed a variety of "shareware" and "freeware" tools. These programs are attractive because of their low price and specificity, but unattractive because of their inflexible narrowness of focus and lack of technical support. A step up from the free or almost free applications are the lower end commercial software packages designed for home use that are usually included with the purchase of a computer or bundled in an inexpensive "family suite" of all-purpose software. These low-end programs, however, are often buggy, produce files that adhere imperfectly to standards, and have native file formats that are unreadable by computers running other brands of graphics software. This is a very serious issue because much graphic work is done in collaboration with other users. It is important not to confuse purchase price with the real cost of software, which is the time it takes to acquire productive competence. Time spent learning to use poor software is time wasted.

Currently, Adobe, Macromedia, and Corel are producing the mainstream professional graphics software programs. These programs may initially cost more than the low-end packages, but they are often offered at substantial discounts for educational users. Their increased expense is more than compensated for by the strength of their image manipulation features and by the widespread acceptability of their native file formats to slide makers, pre-press bureaus, and scientific publications. This is also an issue when preparing files for high-end specialty software, such as that used in rasterizing poster image files for large-format printers. The process of learning these programs is also facilitated by the consistency of the interface among the complementary programs from each manufacturer and the widespread availability of third-party manuals and seminar tutorials. The loyal and enthusiastic users of these popular products, accessible locally and via on-line users newsgroups, are also helpful when problems occur. (To find the newsgroups, go to the main website of the maker [e.g., http://www.macromedia.com or http://www.adobe.com] and then to their support section. If you do not see the answer to your problem there, just post a detailed description and wait for the deluge of responses.)

3. IMAGE MANIPULATION SOFTWARE

3.1 PhotoShop

Adobe PhotoShop is the industry standard among the many bitmap image manipulation and editing programs available. Although it is somewhat more expensive than competing software, it is by far the most flexible and powerful. A benefit of using a standard application like PhotoShop is that scientific publications, graphics service bureaus, and colleagues worldwide will be able to accept and utilize files in the native PhotoShop file format without conversion and corruption difficulties. Users of PhotoShop will find use of Adobe Illustrator (the companion vector graphics package) very smooth, as these programs have quite similar user interfaces and feature cross-program file transportability. They are both available for PC or Mac platforms. PhotoShop is able to convert the Tiff images received from the microscope computer into GIF, JPEG, and PNG formats for use in various end applications.

3.2 Fireworks

Macromedia Fireworks is a professional quality, nimble, and very useful graphics manipulation and creation tool. It also has the advantage of being quite inexpensive for what it offers. Although Fireworks was created primarily as a Web graphics tool, it can be used on a regular basis to quickly optimize or convert graphics. It has a very useful export "wizard" that facilitates the optimization process. Fireworks, in keeping with its Web orientation, also make the creation of Web page special effects (pop-ups, rollovers, image mapping, and so forth) quite simple.

Fireworks is comfortable with the Adobe file formats and performs a unique combination of vector and bitmap editing in the graphics created with its drawing tools. This program is also available cross-platform, allowing Mac users and PC users to collaborate without conversion problems.

3.3 Adobe Illustrator and CorelDraw

Adobe Illustrator and CorelDraw are both quality "vector" drawing programs. That is, instead of each pixel on a line between two points in the drawing being enumerated in the stored file (like filled-in squares on a piece of graph paper), the line exists within the program as a mathematical formula describing the curve (vector) between the two coordinates. A filled area is represented by a series of coordinates, formulas, and a number for the fill color. This makes vector files much, much smaller than bitmap files and also allows vector files to be scaled without distortion or loss of clarity.

Bitmap images are collected from the microscope and then are processed in PhotoShop or Fireworks. They can then be imported into a vector-drawing program for the addition of vector lines, arrows, and explanatory text. This approach has the twin advantages of keeping the final file size down and maximizing the clarity and scalability of the text annotations.

Adobe Illustrator is a graphics industry standard and has a native file format of standards-compliant postscript. This makes Illustrator files very easy to import into other programs, acceptable to pre-press service bureaus and slide making companies, and comfortable to use in a cross-platform PC/Mac environment.

Adobe Illustrator and CorelDraw are our vector tools of choice. There are, of course, a number of other commercially available vector format graphics creation programs.

4. IMAGE CHARACTERISTICS

Digital images are characterized by the size of the pixel array (e.g., 512 × 512) of which they are composed and the color depth for each of those pixels.

4.1 Color Depth

Color depth is the total number of actual colors that will be used to create the seen image. The term *depth* comes from the imaginary three-dimensional cube representing the image. The X and Y dimensions are the size of the pixel array, and the Z dimension corresponds to the number of possible colors for each pixel. This can vary from the thousands or millions of colors available for print use to the 216 color cross-platform "Netscape Palette" used for the World Wide Web. The number 216 represents the colors, which are exactly the same in the 256 color Windows palette and the 256 color Apple palette. Those two "standard" palettes only partially coincide, so it is only by restricting color usage to those 216 colors that we can be sure that everyone on the Web will see the image the way it was intended.

A palette is the array of colors, chosen from the millions possible, that can be used within a given image. There are standard palettes, like the Windows Palette, the Macintosh Palette, the Pantone Palette, and the Netscape Palette. These have the benefits of predictability and uniformity. The user can also create an "optimized palette" for a particular image. If, for example, an image had a wide variety of subtle differences in a particular range of green and not much red, it would be advantageous to have the 256 colors weighted to show as many different greens as possible. This would be at the expense of the other parts of the spectrum, but this would have been done intentionally. This can be a useful tactic, but requires that a distant user of the file be given the palette information so that he or she knows what colors have mapped into the customized assortment. Custom palettes work well for printing to a personal printer, but can be problematic for Web use or when used in multimedia authoring environments like Macromedia Director, Macromedia Authorware, Adobe Acrobat, Microsoft PowerPoint, and so on.

An alternative to using custom "optimized" palettes is "dithering." Dithering is the optical creation of a particular color by placing two or more other colors from a basic palette side by side and letting them be mixed by the user's eye. The French Impressionist painter Georges Seurrat famously used this principle of optical mixing in his "pointalism." The use of this technique allows a standard palette to represent a much wider range of colors than would be otherwise possible, but has the downside of softening the resolution of the image. For example, an edge to a green shape, instead of being a single continuous curve of green pixels, might have to be a double line of mixed yellow and blue pixels. Changing the ratio of the blue to yellow in the mix can produce particular shades.

Color on a computer screen is significantly affected by the differences from monitor to monitor. These can be a result of peculiarities of a monitor brand, the color and brightness of the ambient light, and a user's settings for screen color depth, resolution, contrast, and brightness. The colors in a printed image can, and will, vary greatly from those seen on the computer screen. This is a result of highly idiosyncratic monitor settings and the transmissive, additive nature of monitor color versus the reflective, subtractive nature of printed colors. That is, we see blue on a page because all colors except blue have been absorbed and the blue light is reflected back to our eye. Unfortunately, no pigment is perfectly pure, and the bits of yellow or red in the blue pigment unavoidably skew the printed color slightly toward red or yellow. The color purity, which can be produced on a computer screen mathematically, can never be exactly reproduced in ink dyes and pigments. Use of standards like "Pantone" and controlled trials will give the user some control over the quality and accuracy of the output image. One could use a matrix of color CMYK (Cyan, Magenta, Yellow, Black) swatches to refer to when a specific color is needed in the output. There is a useful Adobe Acrobat (PDF) CMYK color chart on our website (http://www.cci.virginia.edu/book/cmykchart.pdf).

4.2. File Size

File size is always an issue. Oversized files will always adversely affect file storage, manipulation, and transfer. Thus, it is important not to wastefully select a resolution or color depth that exceeds that of the original image. It is equally important not to specify a resolution or color depth exceeding the capabilities of the output medium. For example, a VGA computer screen displays at about 72 dpi and displays 256 colors. If an image is being produced for Web use or projection display, this is all one can count on. Therefore, if a 200 dpi image is put on a Web page, the loading of the page will be slowed down significantly and nothing will be gained from the delay. If this is multiplied for several images on a page, the user will be long gone before the last picture finishes loading.

Table 8–1 displays typical file sizes for a 512 × 512 pixel TIFF microscope file at 256 colors. TIFF (TIF) files are uncompressed, and JPEG (JPG) files are compressed with a "lossy" compression. That is, areas are averaged and simplified to enable a higher degree of compression. When a JPEG file is being produced, there is a choice of quality. The higher the quality chosen, the less the compression, the "truer" the representation, and the larger the file. The quality level used is determined by the requirements of the output application and the nature of the image. Note in Table 8–1 that a 137 KB (Kilobyte) high-quality JPEG file becomes a 25 K medium-quality JPEG file. For Web use, the difference in quality between the two may not be noticeable, but the saving in time certainly will. Generally,

TABLE 8–1 File sizes depending on the chosen file compression method for storage

Pixel Array	TIFF	JPEG (high/medium)	PNG	GIF
		File Sizes (in KB) According to Method		
512 × 512	280	137/25	124*	104
256 × 256	88	53/15	109	99
128 × 128	76	6/3.3	10	8
64 × 64	28	2/1.3	3	2.5

*Use PNG-8 to save for Web use.

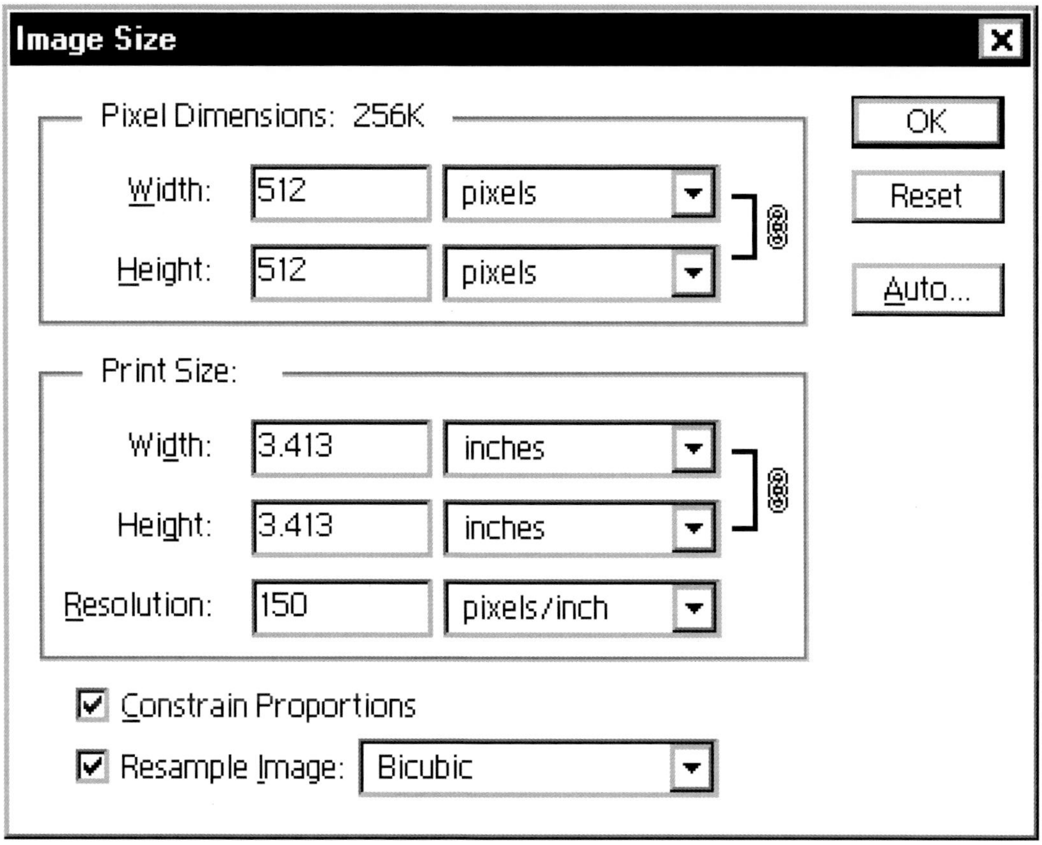

Figure 8–1 The PhotoShop IMAGE SIZE pull-down, used to set the size of the pixel array and the resolution for output.

photographic or continuous tone images work better with JPEG. If the original is at all fuzzy or indistinct, the loss in compression may be completely imperceptible. Line art, like cartoons or graphs, works better with the PNG or GIF compression formats.

Changing the image size in PhotoShop is much preferred to using the scaling capabilities of the output software to determine dimensions. By using the PhotoShop image size pull-down, one can set the size of the pixel array for output and the resolution for output (Fig. 8–1). The pixel array divided by the resolution gives the actual size of the image on paper or typical screen. In the example shown in Figure 8–1, 512 pixels/150 dpi = 3.413 inches. When the "scaling" capability of software is used (usually by dragging one of the corner bounding dots), the representation of the image changes, but the embedded file size remains the same.

4.3 Print

For print use, the size and resolution are chosen as appropriate to the file size, color depth, and image dimension limitations of the chosen output device.

If the image is being processed for commercial or professional publication, the editors of the publication should be contacted to determine preferred file format (JPEG, TIFF, Ai, and so forth), image size, color depth (2 bit, 8 bit, 16 bit), and resolution in dots per inch (dpi). If the image is going to be printed on a personal printer, one has to determine pragmatically the best-looking output versus a file size compatible with transfer and storage capabilities.

5. IMAGE PREPARATION

5.1 Still Image Optimization in PhotoShop

First, one should start up PhotoShop and open the file. To do this, double click on the Photoshop "eye" icon to open Photoshop. Select FILE, OPEN (filename of image); or, in another window, open the folder containing the TIFF files to be processed and drag the file icon for the desired file into the Photoshop window. The images will instantly appear. Immediately save (with SAVE AS) the image under a new name (usually a variant of the original) so that the original file will not be corrupted. Many compression methods involve irreversible, although inconspicuous, changes to the image, so an unaltered "raw" image should always be kept if it might be used in another size or context. For example, if the original file was named "PollenRaw.tif," (http://www.cci.virginia.edu/book/PollenRaw.tif) the altered file could be named "Pollen-a.tif." Putting the character changes at the end of the file name will keep all of the variants of the same image contiguous in the directory listing. This is a great convenience and time saver when different versions of the same image are being made for different purposes in several Web pages or documents.

Now, any desired changes to the brightness and contrast of the image can be made, and any of the many PhotoShop filters can be applied. PhotoShop filters are ready-made special effects that can be applied to all or part of one's image. Many filters are included with PhotoShop, and many more can be purchased as after-market add-ons. The truly motivated person can design and program his or her own special purpose filters, as well. If, when the image has been opened in PhotoShop, the FILTER option on the top menu is "grayed out" and inactive, the image will need to be converted to RGB before a filter can be applied. This is done by selecting IMAGE, then MODE from the drop-down menu, then RGB. Now the filters will be available. There are so many filters available that it would be most useful to take a sample image and try them all.

For example, the Pollen image shown in this chapter could be rendered as three-dimensionally solid. This may be done by selecting FILTER, STYLIZE, or EMBOSS. One can dramatically change the appearance of an image by changing the ANGLE, HEIGHT, and AMOUNT parameters in the menu box. As changes are made, the image on the screen will instantly be altered to reflect the choice. If the alteration is unwanted, just hit CANCEL and everything will return to the prior state.

To apply a filter to a portion of an image, first click on SELECT, DESELECT to clear any previous selections. Now click on the area selection tool (the dashed square on the upper left corner of the tool palette), and then click on one corner of the desired area; while holding down the mouse button, drag a rectangle until it encompasses the area of interest. If a filter is applied now, it will only alter the selected area. This can be useful for simultaneously showing the "raw" image and the enhanced image in the same graphic.

The PhotoShop filters have numerical settings on their adjustable parameters. By taking notes as a filter is applied, one can be sure it will be applied in the exact same manner to all of the other images in that set. It is important to save the image in the "before" state so that other changes can be made to the original image later on, which would be unavailable with the altered image. If the purpose of microscopy is considered as an ongoing effort to make the obscure apparent, then digital processing, such as the use of filters, should be embraced. Obviously, if several images were being presented for critical comparison, one would need to make the exact same changes to all of the images in the series to avoid inadvertent misrepresentation of the data.

To adjust the brightness or contrast in an image, select IMAGE, ADJUST, LEVELS (Fig. 8–2), and then change the settings by sliding one of the triangles below the histogram left or right. In Figure 8–3, the left triangle (representing the darkest point) is slid over to the left end of the histogram. One can also click on the left eyedropper, then click-select the darkest area of the image, and then click on the right eyedropper and click-select the brightest area of the image. All other values in the image will then be apportioned between those two arbitrary extremes. To choose a computer-selected range, click on AUTO. Remember that it is possible to click UNDO to go back to the way things were.

Before declaring an image finished, it is always a good idea to do a test print and determine how the contrast and brightness on the computer screen compare with the printed output. It is helpful to make some notes as this is done so that other images in the same set will be processed to the same specifications, and scientific and aesthetic consistency will be preserved.

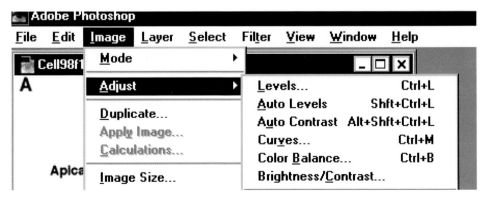

FIGURE 8–2 The PhotoShop IMAGE, ADJUST menu, used to alter the brightness, contrast or color balance of an image.

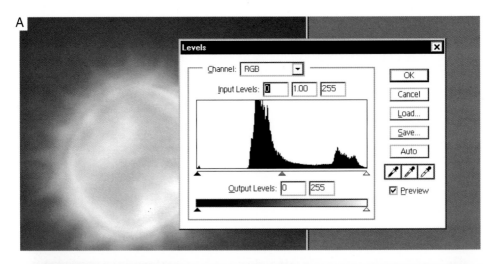

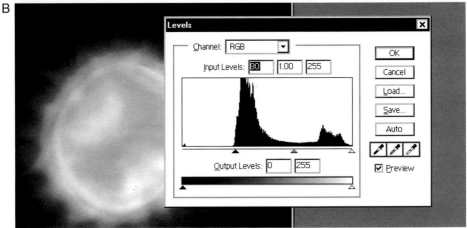

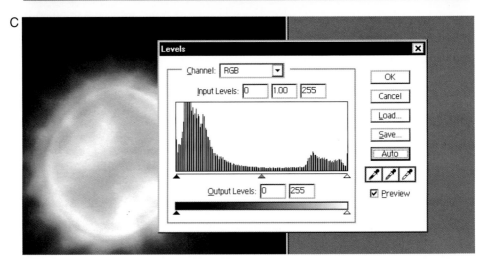

FIGURE 8–3 (*A*) Initial presentation of "raw" image. Note the uneventful left side of the histogram. (*B*) The left edge of active area of histogram is selected as "black" by sliding the left triangle indicator to the right. (*C*) In the automatic LEVELS mode, PhotoShop AUTO selects "black" and adjusts rest of image across the 256 output levels.

On some occasions a reduced size version of an image may be wanted at a particular location on a page. This can occur when submitting an image for publication in a book or dissertation. Simply selecting IMAGE, IMAGE SIZE, and then typing in a percentage of reduction can do this. Copy the image to the clipboard by clicking SELECT, SELECT ALL and then EDIT, COPY. Open a new file and select the preferred page size.

For example, if the printer has an unprintable margin area of one half inch, select a PhotoShop page size of 7 and one half inches wide and 10 inches high. Then paste the image from the clipboard into the new file. The image can be dragged around on the screen to the desired location on the finished page. For precision, click VIEW and then SHOW RULERS. Click on the left ruler and drag a guideline to the left offset, and then drag the guideline down from the top ruler can set the top location. Now the image can be dragged to the guidelines. Save the new image (under a different name) and it will be ready for publication.

The books in the Adobe "Classroom in a Book" series come highly recommended. They are simple, straightforward, and inexpensive. They contain the training material that is used by Adobe personnel for their in-house training. These books are widely available through bookstores or on-line.

5.2 Fireworks

Fireworks, an inexpensive, Web-oriented, graphics program is also very useful for quickly processing an image for different end uses. Fireworks is produced by the Macromedia Company and is well supported by on-line user groups, the Macromedia website, and third-party manuals available in any computer stores. A free fully functional 30-day free demonstration version of Fireworks can be downloaded from Macromedia (http://www.macromedia.com).

6. PRESENTATIONS

PowerPoint is, by far, the most popular and widely used electronic presentation software. It is used for the preparation of slides, computer projection, overhead transparencies, custom graphics, and even poster preparation.

To create a new slide or slide set, start PowerPoint. Select BLANK presentation and click OK, and then select the desired slide layout type. The empty first slide will be open on the screen for editing. Once the slide set has been defined and the basic page design set, the figures can be imported and annotated. At this point, it is worth double-checking FILE, PAGE SETUP, SLIDES SIZED FOR, to make sure the size and orientation are correct. Note that letter-size paper and transparencies have a different height-to-width ratio than do 35 mm slides. Failure to pay attention to this (bitter experience!) will result in severely cropped slides.

To import a picture click on INSERT, PICTURE, FROM FILE. Change the path as necessary to find the directory where the processed images have been stored. Highlight the appropriate image file name (there will be a preview); then click INSERT and the image will appear in the first slide. The image can be positioned in the slide by clicking on it and dragging it to the desired area. If the size of the image needs to be changed, it can be resized in PhotoShop and reinserted into the slide. Although the size of the image could be adjusted by clicking on the image and dragging a corner in or out, this will ultimately put more processing overhead on the computer that displays the show. Because scaling does not reduce the file size of the embedded image, this will unnecessarily slow the loading of the presentation.

Text is added by selecting INSERT, TEXT BOX, and click-dragging the text box to the size, shape, and location of choice. Click within the box and type in the text. It is a good idea to leave an inch or so of free space around the edges of the slide to allow for mounting variations. Text in slides and transparencies should be larger and bolder than would be used for a comparable print application or it will be difficult to read. By clicking INSERT, NEW SLIDE, a new slide will be created after the previous one. To change the presentation order of the slides, click on VIEW, SLIDE SORTER, and drag the slides into the desired sequence.

When using a commercial slide service it is a very good idea to make up a series of test slides with different fonts, line widths, typical images, and color combinations and have them made into 35 mm slides. It will also help later to include annotation on the slides as to CMYK color specifications for the colors used, fonts, and line widths. Then, take the same set of sample slides and print them out on transparencies and on quality paper with a personal color printer. These real-world examples will be a valuable reference and will serve as disaster prevention when an important and time-critical project is being prepared. For example, thin red lines on a dark blue background look very cool on a computer screen but very bad on a slide or transparency. Twelve point Times Roman looks just fine on paper, but is very hard to read on a transparency, especially if the ink density is at all diminished.

7. PROGRAMS THAT CAN BE USED FOR MOVIE CREATION

Some presentations require a series of related pictures in a rapid sequence to give the illusion of traveling through a cell or having a cell appear to rotate before the viewer. This can be accomplished by creating an animated GIF with an image series. It can also be done by importing the images into a digital video-editing program like Adobe Premiere or by using a multimedia-authoring environment like Macromedia Director and then saving the edited animation as a Microsoft AVI movie or as a QuickTime Movie. Both of these formats are viewable on almost all computers through the use of utilities that are either included with the operating system or available via free download from the Web.

Animated GIFs, part of the GIF 89a standard, are composed of a series of still images that are grouped within an individual file and flipped into view, one after another, to give the illusion of motion. On the positive side, these files are very useful as they are easy to produce and to insert into a Web page. The HTML that produces the Web page calls animated GIFs in exactly the same way it would call a static image file (<img src = filename.gif>). The mechanism to do the flipping is built into the GIF 89a standard, and any reasonably current browser or image viewer can produce the animation on call. When using an image that is composed of relatively few pixels per frame and not very many frames, this is a good approach. With larger images or many frames, however, significant download time may be required before the animation can begin. Unless a "lossy" compression is applied to it (with the corresponding decrease in image quality), the size of the animated GIF is approximately equal to the sum of the files that comprise it.

Macromedia Flash, Macromedia Fireworks, and Adobe PhotoShop (via the ImageReady program, which comes included with PhotoShop version 5.5) allow the relatively simple creation of animated GIF files. Again, although there are free GIF editing programs download-available on the Web, they do not have the processing and annotating capabilities that come with the professional-grade software used for other image manipulation tasks.

8. Making Animated GIF Files with Macromedia Flash

To make an animated GIF using Macromedia Flash, first open Flash and start a new file by clicking FILE, NEW. Like celluloid movies, animated GIF movies are composed of a series of frames that are passed before the viewer.

When a new file is opened, the first frame is shown on the screen. Set the frame size of the movie by clicking MODIFY, MOVIE and set the DIMENSIONS. Note that, because of margins, borders, and menus, the reliably clear space on a Web page is only about 550 pixel wide by 440 pixel high. If there is going to be annotation, menus, or other material on the page, they need to be considered in this space, as well. Finally, the larger the GIF, the larger the file, and the longer it will take to load. If the viewers are highly motivated academics, they may be willing to wait, but the curious amateur may not, so a different version available by hot links may be wanted, with a suitable warning by the link to the large version.

To place an image in the first frame of the movie, click FILE, IMPORT, and select the image from the directory lists on the left side of the window. Then, click ADD and IMPORT, and the image will appear on the first frame. Text, arrows, and other graphics can be added to the frame by using the standard tools accessed through WINDOW, TOOLBAR.

Add a new empty frame by pressing F7 or by clicking INSERT, BLANK KEYFRAME, and then import the image for the second frame. If an image needs to appear on more than one frame, just insert a frame (F5) after the current frame, and the image will be carried over to the new frame. When the whole image set has been assembled, select FILE, EXPORT MOVIE. Then, in the window that appears, select FORMAT, ANIMATED GIF, and name the movie. Note that the movie could be produced in a number of other formats at this point, including QuickTime and AVI (video for windows). This flexibility of standard file formats in the production is very useful if one plans to incorporate the new movie into a multimedia CD project or on a Web page.

9. MAKING ANIMATED GIF FILES WITH MACROMEDIA FIREWORKS

The process of making an animated GIF with Macromedia Fireworks is quite similar to that of making one with Macromedia Flash. Macromedia, like most major software manufacturers, strives to help the user leverage his or her hard-won proficiency by maintaining an interface consistency across their product lines.

Open Fireworks, select FILE, NEW and enter the dimensions of the new GIF, which will appear on the screen as the empty first frame of the animated GIF movie. Use FILE, IMPORT to select the first image to be placed. To add more frames, WINDOW, FRAMES will show the frames submenu. On the right hand side of that window is a little right-pointing triangle. Clicking on that will give the option of adding or deleting frames. Click on ADD FRAMES, and select the number of desired frames (more can be added later, if needed). Note that the duration of each slide can be adjusted later during EXPORT, so extra duplicate frames do not need to be added to prolong the stay of a particular image on screen.

On the FRAMES list, select the frame to edit, and, using FILE, IMPORT, place the desired graphic on that frame. The usual editing tools can also be used to add annotation, and so forth, to the frame. After the editing of the movie is done, it can be saved as a Fireworks file (in case further edits or additions become necessary), and then exported as an animated GIF file using FILE, EXPORT WIZARD, SELECT AN EXPORT FORMAT, CONTINUE, ANIMATED GIF, CONTINUE.

In the EXPORT PREVIEW screen, one can preview the movie using the VCR-looking controls at the bottom left of the screen. To extend the duration of any of the frames, click on the ANIMATION tab at the top, which will show a list of the frames in the movie with their durations on screen in 100ths of a second. Above the frame list is a small white box with a number in it. Highlight the frame whose duration is to be changed, and edit that number as desired.

Click the "Play" button on the bottom left of the screen to preview the altered movie. The movie can just play through once, or it can loop a specified number of times or take the default, which is forever. To produce the animated GIF, click on EXPORT (bottom right of the screen), and choose a filename and a location for the new movie.

10. MAKING ANIMATED GIF FILES WITH ADOBE IMAGEREADY (BUNDLED WITH PHOTOSHOP 5.5)

Start up ImageReady and open a new file (FILE, NEW), give it a name, and set the dimensions. Making animations in ImageReady is a two-step process. First, the image set is assembled on layers in ImageReady (or in PhotoShop and imported into ImageReady), and then the layers are converted, one at a time, into frames of the animation. If the LAYERS window is not showing, click WINDOW, SHOW

LAYERS. To put an image in the first layer, click FILE, PLACE, and select the position on the layer for the image. The image can be automatically centered or placed at a user-selected horizontal and vertical offset. Then CHOOSE the first image file.

Repeat this process (layers will be created automatically) until all of the images for the animation have been placed on layers. Click on the "eye" icon on the left side of each of the layers, which will make all of the layers invisible. Now, one at a time, make layers visible, and make the contents of each layer into the contents of a frame of the movie.

To begin assembling the move, click on WINDOWS, SHOW ANIMATION, and a new window will appear, which is the animation project management tool. On the left side of this window is an empty square with the number 1 in the upper left corner. This empty square is the first frame of the movie. On the bottom right corner of this square is a 0, which is the time duration of the frame on the screen when the movie is played. By clicking on the 0, time choices appear. Make the visible image for the first frame of the animation by turning on the "eye" icon in the appropriate layer. When this image appears on the editing window, it will also appear in frame number 1 in the animation project window.

Place a new frame in the animation by clicking on the little icon on the bottom of the animation window that looks like two little overlapping boxes. It is just to the left of the trashcan icon. Now, turn off the visibility of the layer used for the first frame and turn on the visibility for the layer, which contains the image for the second frame. Preview the animation by clicking the "play" button on the bottom of the animation window. The time duration can be changed for any frame that needs it, and the order of the frames can be rearranged by clicking on a frame and, while holding down the left mouse button, dragging that frame into a new spot in the sequence. When the animation is complete, save it using FILE, SAVE OPTIMIZED AS.

11. Conclusion

The software programs described in this chapter are all highly capable and sophisticated graphics manipulation tools, and this brief introduction is intended only to help users get started on a few basic tasks. Both Macromedia and Adobe have great websites filled with many examples and helpful tutorials. Computer stores (Staples, Office Depot, Circuit City, and so forth) and some websites (e.g., Amazon.com) also have many books with tutorials and instruction for their use. When choosing a book, make sure that too much of the book is not taken up explaining the most basic material. The "Classroom in a Book" series from Adobe is very good, as are the books in the "Nutshell" series from the O'Reilly Publishing Company (http://www.oreilly.com).

This work was supported by grants from the W. M. Keck Foundation, the Academic Enhancement Program (AEP), and the National Science Foundation (NSF) Center for Biological Timing.

REFERENCES

Aikens, R., D. Agard, and J. Sedat. Solid state imagers for optical microscopy. *Methods Cell. Biol.* 29: 291–313, 1988.

Axelrod, D., D. E. Koppel, J. Schlessinger, E. Elson, and W. W. Webb. Mobility measurements by analysis of fluorescence photobleaching recovery kinetics. *Biophys. J.* 16: 1055–1069, 1976.

Becker, P. L. Quantitative fluorescence measurements. In: *Fluorescence Imaging Spectroscopy and Microscopy, Chemical Analysis,* Vol. 137, edited by X. F. Wang and B. Herman. New York: John Wiley & Sons, 1996, pp. 1–29.

Bookman, R. Temporal response characterization of video cameras. In: *Optical Microscopy for Biology,* edited by B. Herman and K. Jacobson. New York: Wiley-Liss, 1990, pp. 235–250.

Bright, G. R., and D. L. Taylor. Imaging at low light level in fluorescence microscopy. In: *Applications of Fluorescence in the Biomedical Sciences,* edited by D. L. Taylor, F. Lanni, A. S. Waggoner, R. F. Murphy, and R. R. Birge. New York: Alan R. Liss, 1986, pp. 257–288.

Herman, B. *Fluorescence Microscopy,* 2nd Ed., New York: Springer-Verlag, 1998.

Inoué, S., and K. R. Spring. *Video Microscopy: The Fundamentals.* New York: Plenum, 1997.

Lockett, S. J., K. Jacobson, and B. Herman. Application of 3D digital deconvolution to optically sectioned images for improving the automatic analysis of fluorescent labeled tumor specimens. SPIE-Biomed. *Image Process. Three Dimen. Microsc.* 1660:130–139, 1992.

Meek, G. A., and H. J. Elder. Analytical and Quantitative Methods in Microscopy. Cambridge: Cambridge Univ., 1977.

Pawley, J. *Handbook of Biological Confocal Microscopy.* 2nd Ed. New York: Plenum Press, 1995.

Periasamy, A., and B. Herman. Computerized fluorescence microscopic vision in the biomedical sciences. *J. Comput. Assist. Microsc.* 6:1–26, 1994.

Russ. J. C. *The Image Processing Handbook,* 3rd Ed. Boca Raton: CRC Press, 1999.

Shotton, D. *Electronic Light Microscopy.* New York: Wiley-Liss, 1993.

Tsien, R. Y., and M. Poenie. Fluorescence ratio imaging: A new window into intracellular ionic signaling. *Trends Biochem. Sci.* 11:450–455, 1986.

II

MULTIPHOTON EXCITATION
FLUORESCENCE MICROSCOPY

INTRODUCTION

Two-photon absorption was theoretically predicted by Göppert-Mayer in 1931 and was experimentally observed for the first time in 1961 with a ruby laser used as the light source. Now, multiphoton (two- or three-photon) excitation microscopy is used more commonly by scientists as a result of the contributions made by Watt Webb's group (Denk, Strickler, and Webb, *Science* 248:73–76, 1990) from Cornell University. More importantly, the commercialization of multiphoton imaging systems by Bio-Rad has increased awareness and application of this technology in biomedical imaging. It should also be recognized that laser companies played a key roll in introducing a tunable (700–1100 nm) high-speed femtosecond infrared-pulsed laser system (Ti: Sapphire) for multiphoton imaging. Moreover, the future will bring easier-to-use equipment and increased sensitivity, which will allow greater flexibility in the simultaneous imaging of multiple fluorophores while images are collected over time and at greater depths inside tissue.

To describe the multiphoton process in brief, an infrared femtosecond pulsed laser is required to create multiphoton absorption in a biological sample. Multiphoton excitation occurs at only a single, diffraction-limited spot where the photon flux is great enough to allow absorption of more than one photon. For one-photon excitation, a wavelength of 480 nm CW (continuous wave) is selected for the FITC molecule, but for two-photon excitation a single 960 nm wavelength from a Ti: Sapphire pulsed laser is used. But for most of the fluorophore the two-photon absorption cross-section is blue shifted (for example: FITC could be excited using 780 or 920 nm). The multiphoton excitation microscopic images have better signal-to-noise ratios than confocal images because of a considerably less amount of light scattering. In addition, autofluorescence and photobleaching are minimized because there is no absorption throughout the specimen due to illumination (for more details, see Chapters 9–14).

The basic principles of multiphoton absorption, absorption cross-section calculation, and demonstration of deep tissue imaging are described in Chapter 9. Chapter 9 also includes instructions on building a multiphoton system using a wide-field fluorescence microscope. Chapter 10 describes the conversion of an existing confocal system to a multiphoton system and compares the one-photon and multiphoton microscopy systems. Chapter 11 addresses many aspects of multiphoton imaging including depth penetration for different wavelengths, Monte Carlo simulation demonstrating the excitation and emission, point spread functions at different depths, and the decrease in fluorescence signal with increase in excitation pulse width.

Chapter 12 focuses on the use of multiphoton microscopy in developmental biology. Additionally, Chapter 12 reviews the implementation of the multiphoton system to monitor the dynamic aspects of development of frog, sea urchin, and mammalian embryos and the optimization of the fluorescent probes for labeling embryonic cells. Also described is the use of the two-photon system for photo uncaging for sea urchin embryogenesis. Chapter 13 describes an in situ measurement of the diffusion mobility of biologically relevant molecules using multiphoton fluorescence photobleaching recovery and fluorescence correlation spectroscopy (FCS) methods. The theory behind these two methods details about their instrumentation and data collection, and comparison of this methodology with conventional fluorescence recovery after photobleaching (FRAP) techniques (Chapter 7) are provided. Chapter 14 focuses primarily on the interaction between excitation radiation and single living cells in laser microscopes. In particular, the influence of continuous wave ultraviolet (UV) and near infrared (NIR) microbeams, as well as that of femtosecond and picosecond NIR laser pulses, on cellular metabolism, ultrastructure, and viability is described.

9

Basic Principles of Multiphoton Excitation Microscopy

Peter T. C. So, Ki H. Kim, Christof Buehler, Barry R. Masters, Lily Hsu, and Chen-Yuan Dong

1. Introduction

Multiphoton microscopy is one of the fastest growing areas in biomedical imaging. The potential of multiphoton excitation was first theorized by Maria Göppert-Mayer in 1931. Generating three-dimensionally resolved microscopic images based on nonlinear optical excitation was postulated in the 1970s (Gannaway and Sheppard, 1978; Wilson and Sheppard, 1984). The definitive experiment was done by Denk, Webb, and co-workers (1990), who accomplished two-photon, three-dimensional (3D) imaging of biological specimens. Furthermore, they demonstrated 3D localized uncaging and photobleaching using two-photon excitation to trigger a photochemical reaction in a subfemtoliter volume.

In this chapter we focus on the basic physical principles and image formation theories of multiphoton microscopy. We also discuss the properties of two-photon fluorophores and their excitation cross sections as well as the construction of two-photon microscopes by modifying standard fluorescence microscopes. Finally, we describe the application of two-photon microscopy to deep-tissue imaging.

2. Basic Physical Principles Underlying Multiphoton Microscopy

2.1 The Quantum Theory of Two-Photon Excitation

The two-photon excitation process was predicted by Göppert-Mayer (1931). The basic physics of this phenomenon has been reviewed in the literature (Baym, 1973; Callis, 1997). One-photon fluorescence excitation of a fluorophore results from the absorption of a single photon. One-photon excitation is a linear process in which the fluorescence generated is proportional to the excitation power. In contrast, two-photon excitation is a nonlinear process in which two lower energy, near-infrared photons are simultaneously absorbed. If their combined energies are

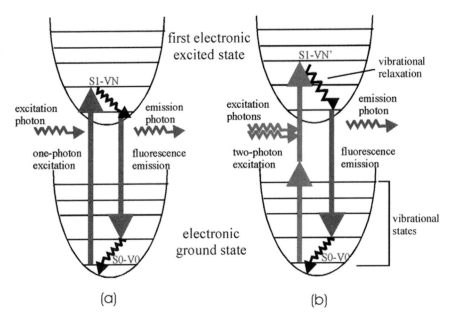

FIGURE 9–1 Jablonski diagram illustrating one-photon (a) and two-photon (b) excitation and de-excitation pathways.

sufficient to allow an electronic transition, the fluorophore can still reach an excited electronic state. A Jablonski diagram illustrating the differences between one- and two-photon absorption is shown in Figure 9–1. Single-photon excitation involves the direct transition between the ground state and the excited state. In two-photon excitation, the first photon excites the molecule to a virtual intermediate state, and the molecule is eventually brought to the final excited state by the absorption of a second photon.

Fluorescence excitation can be understood as a quantum mechanical interaction between a fluorophore and light. This interaction can be quantified by a time-dependent Schrödinger equation. The lowest order interaction term of the Hamiltonian corresponding to an electric dipole interaction is $\vec{E}_\gamma \cdot \vec{r}$, where $\vec{E}_\gamma$ is the electric field vector of the photons and $\vec{r}$ is the position operator. Using perturbative expansion, the first-order solution corresponds to the one-photon excitation, and the multiphoton transitions are represented by higher order solutions. For two-photon excitation between molecular initial state $|i\rangle$ and the final state $|f\rangle$, we can calculate the transition probability:

$$P \sim \left| \sum_m \frac{\langle f|\vec{E}_\gamma \cdot \vec{r}|m\rangle\langle m|\vec{E}_\gamma \cdot \vec{r}|i\rangle}{\varepsilon_\gamma - \varepsilon_{mi}} \right|^2 \tag{1}$$

where ε_γ is the photonic energy associated with the electric field vector $\vec{E}_\gamma$ the summation is over all the possible intermediate states m, and ε_{mi} is the energy difference between the intermediate state m and the ground state. The dipole operator has odd parity (i.e., absorbing one-photon changes the parity of the state), and the one-photon transition moment $\langle f | \vec{E}_\gamma \cdot \vec{r} | i \rangle$ requires the initial and final states having opposite parity. On the other hand, the transition of two states with the same parity can be coupled by the two-photon moment $\langle f | \vec{E}_\gamma \cdot \vec{r} | m \rangle \langle m | \vec{E}_\gamma \cdot \vec{r} | i \rangle$ (Baym, 1973; Callis, 1997).

2.2 The Optical Basis for 3D Resolution in Two-Photon Microscopy

For microscopy, the two-photon effect has an important consequence of limiting the excitation region to within a subfemtoliter volume. The 3D confinement of the two-photon excitation volume can be understood based on optical diffraction theory. Using excitation light with wavelength, λ, the intensity distribution at the focal region of an objective with numerical aperture NA = $\sin(\alpha)$ is described by

$$I(u, v) = \left| 2 \int_0^1 J_0(v\rho) e^{\frac{i}{2}u\rho^2} \rho \, d\rho \right|^2 \tag{2}$$

where J_0 is the zero-order Bessel function, $u = 4k \sin^2(\alpha/2)z$, and $v = k \sin(\alpha)r$ are the respective dimensionless axial and radial coordinates normalized to the wave number $k = 2\pi/\lambda$ (Sheppard and Gu, 1990; Gu and Sheppard, 1995).

Quantum theory dictates that the fluorescence intensity has a linear dependence on the excitation photon flux for one-photon excitation and has a quadratic dependence for two-photon excitation. Furthermore, to excite a given fluorophore species, two-photon excitation uses infrared light that has approximately twice the wavelength of the light used in one-photon excitation. The fluorescence distribution at the focal region inside a uniform fluorescence specimen has a functional form of $I(u, v)$ for the one-photon case and $I^2(u/2, v/2)$ for two-photon excitation. These fluorescence distributions are called the *point spread functions* of the microscope.

The two-photon point spread function (PSF) is axially confined as compared with the one-photon PSF. The fluorescence signals originating from each z-plane for one- and two-photon excitation are proportional to

$$I_z^{1P}(u) = \int I(u, v) v \, dv \tag{3}$$

$$I_z^{2P}(u) = \int I^2(u/2, v/2) v \, dv \tag{4}$$

The 3D discrimination of two-photon microscopy can be understood by examining the numerical behavior of these functions (Fig. 9–2). In general, two-photon microscopy has a radial resolution comparable with one-photon conventional microscopes, but has a very narrow depth of focus that allows 3D resolved imaging.

2.3 Two-Photon Excitation Cross Sections

Because two-photon absorption is a second-order process involving the almost simultaneous interaction of two photons with one fluorophore, this process has a

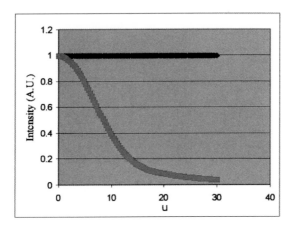

FIGURE 9–2 Total fluorescence generated at each z-plane is plotted against the distance from the focal plane. Distances are measured in reduced optical unit, u, as defined in the text. In the one-photon example (red), equal fluorescence intensity is observed in all planes and there is no depth discrimination. In the two-photon example (blue), the integrated intensity decreases rapidly away from the focal plane.

small cross section, δ, on the order of 10^{-50} cm^{4}s (defined as 1 GM; Göppert-Mayer, 1931). The instantaneous intensity of the fluorescence signal generated, $F(t)$ (photon/s/ molecule), is proportional to the incident photon flux, $I(t)$ (photon/s/m^2), as

$$F^{2P}(t) \approx \delta I(t)^2 \tag{5}$$

$$F^{2P}(t) \approx \delta P(t)^2 \left(\frac{(NA)^2}{2\hbar c\lambda}\right)^2 \tag{6}$$

where $P(t)$ is the laser instantaneous power, λ is the wavelength of light, NA is the numerical aperture of the focusing objective lens, and $\hbar$ and c are Planck's constant and the speed of light, respectively. The excitation probability is enhanced by increasing the NA of the focusing lens corresponding to spatially confining the excitation power to a smaller focal volume (Denk et al., 1990).

The time-averaged two-photon fluorescence intensity per molecule, F^{2P}, is defined as

$$F^{2P} = \frac{1}{T}\int_0^T F^{2P}(t)dt = \delta\left(\frac{(NA)^2}{2\hbar c\lambda}\right)^2 \frac{1}{T}\int_0^T P(t)^2 dt \tag{7}$$

T is an arbitrary time interval for continuous wave (CW) lasers. For pulsed lasers with f_P as the pulse repetition rate, $T = 1/f_P$. For CW lasers, where $P(t) = P_0$, the time-average fluorescence signal (Eq. 7) can be simplified to

$$F_{cw}^{2P} = \delta P_0^2 \left(\frac{(NA)^2}{2\hbar c\lambda}\right)^2 \tag{8}$$

For pulsed lasers with pulse width, τ, repetition rate, f_P, and averaged power, P_0, one can make the simplest approximation on the pulse profile as

$$P(t) = \frac{P_0}{f_P \tau} \quad \text{for} \quad 0 < t < \tau$$

$$P(t) = 0 \quad \text{for} \quad \tau < t < \frac{1}{f_P} \tag{9}$$

In this case,

$$F_P^{2P} = \delta \frac{P_0^2}{f_P^2 \tau^2} \left(\frac{(NA)^2}{2\hbar c\lambda}\right)^2 \frac{1}{T} \int_0^\tau dt = \delta \frac{P_0^2}{f_P \tau} \left(\frac{(NA)^2}{2\hbar c\lambda}\right)^2 \tag{10}$$

A comparison of Equations 8 and 10 shows that, for CW and pulsed lasers to have equivalent excitation efficiencies, the average power of the CW laser has to be higher by a factor of $1/(\sqrt{\tau f_P})$. For a typical pulsed laser source with a repetition rate of 100 MHz and a pulse width of 100 fs, this factor is approximately 300.

It is also important to consider the excitation probability of a fluorophore illuminated by a single laser pulse:

$$\mathrm{Pr}_P^{2P} = \delta \frac{P_0^2}{f_P^2 \tau} \left(\frac{(NA)^2}{2\hbar c\lambda}\right)^2 \tag{11}$$

The excitation process reaches saturation level when Pr_P^{2P} approaches unity. Excitation saturation is undesirable because two-photon PSF broadens in this situation.

The use of Equations 10 and 11 allows one to choose laser parameters that maximize excitation efficiency without saturation. As an example, the common fluorophore fluorescein has a two-photon excitation cross section, δ, of 38 GM at 780 nm. We can estimate the excitation probability per laser pulse using typical instrumentation parameters, $f_P = 80$ MHz, $\tau = 150$ fs, NA = 1.2, and $\lambda = 780$ nm as a function of laser average power:

$$\mathrm{Pr}_P^{2P} = 38 \times 10^{-50} \, cm^4 s \frac{P_0^2}{(80 \times 10^6 \, Hz)^2 \cdot 150 \times 10^{-15} s}$$

$$\times \left(\frac{(1.2)^2}{2 \cdot 1.05 \times 10^{-34} \, Js \cdot 3 \times 10^{10} \frac{cm}{s} \cdot 780 \times 10^{-7} \, cm}\right)^2$$

$$\mathrm{Pr}_P^{2P} \approx (3400 \cdot W^{-2}) \cdot P_0^2 \tag{12}$$

The fractions of fluorescein molecules excited with an average power of 1, 10, and 20 mW are 3.4×10^{-3}, 3.4×10^{-1}, and 6.8×10^{-1}, respectively. Saturation occurs at an average excitation power of about 10 mW. We can also calculate the rate of photon emission per fluorescein molecule under 10 mW excitation assuming no photobleaching:

$$F_P = \mathrm{Pr}_P^{2P} \cdot f_P = 2.7 \times 10^7 \frac{photons}{s} \tag{13}$$

3. TWO-PHOTON CHROMOPHORES

Most conventional fluorophores can be used in two-photon microscopes. As a guideline, fluorophores can be excited in two-photon mode at twice their one-photon absorption wavelength. It should be recognized, however, that this "twice wavelength" rule is only a rough approximation because one- and two-photon absorption processes have different quantum mechanical selection rules. Currently, the two-photon excitation spectra of many fluorophores have been measured. In general, a fluorophore's two-photon excitation spectrum, scaled to half the wavelength, can be very different from its one-photon counterpart (Xu et al., 1995; Albota et al., 1998b).

3.1 Extrinsic and Endogenous Two-Photon Fluorophores

Accurate spectral characterization of fluorophore cross sections under nonlinear excitation is challenging, requiring a well-characterized excitation pulse train profile and detection geometry. Xu et al. (1995) and Albota et al. (1998b) have measured the absorption spectra of a wide variety of extrinsic and intrinsic fluorophores. Their work has covered common fluorophores, such as fluorescein and rhodamine, cellular structural labels, such as DAPI and DiI, and ion probes, such as Indo-1 and Fura-2. Equally important, the excitation spectrum has been determined for green fluorescence protein, which is a key molecular marker for gene expression (Chalfie et al., 1994; Niswender et al., 1995; Potter et al., 1996).

In addition to extrinsic fluorophores, two-photon–induced fluorescence from tryptophan and tyrosine in proteins has been studied (Lakowicz and Gryczynski, 1992; Kierdaszuk et al., 1996). Multiphoton imaging also provides a novel method to monitor the secretion of a neurotransmitter, serotonin, in vivo (Shear et al., 1997). The fluorescence spectra of endogenous fluorophores, reduced pyridine nucleotides, reduced nicotinamide adenine dinucleotide (NADH) and reduced NADH phosphate (NADPH), and flavoproteins have also been measured. NAD(P)H levels are closely tied to cellular metabolic rates (Masters and Chance, 1999) and have allowed the in vivo study of the redox states of cornea (Piston et al., 1995) and skin (Masters et al., 1997).

3.2 Recent Efforts in Two-Photon Probe Development

The development of novel fluorophores with optimal two-photon absorption properties is essential. Because one- and two-photon excitations have different selection rules, one-photon probes are not necessarily optimized for two-photon absorption. Molecules with phenomenally high two-photon absorption cross sections of over 1000 GM have been synthesized (Albota et al., 1998a). Because the excitation of endogenous fluorophores is a major photodamage mechanism for cells and tissues, using a more efficient extrinsic fluorophore will allow two-photon imaging at reduced excitation power and will enhance specimen viability. Furthermore, with the availability of fluorophores with high two-photon cross sections, lower peak power and more economical lasers can be used.

Using two-photon excitation to trigger 3D localized chemical reactions is an exciting new microscopy technique. The ability to create active signaling molecules inside a subfemtoliter volume in cells and tissues allows the study of intracellular signal propagation. Localized uncaging of fluorescein has been demonstrated by Denk and co-workers (1990). The uncaging of neurotransmitters has been used to map cellular receptor distribution (Denk, 1994; Pettit et al., 1997). The uncaging studies of some other cellular messengers, such as calcium, have been less successful due to the lack of an efficient two-photon photolabile cage. The two-photon excitation power that is needed to uncage calcium in cells is sufficiently high to trigger cellular damage. This situation is improving with the development of new cage groups targeted for two-photon applications (Furuta et al., 1999).

4. DESIGNING A MULTIPHOTON MICROSCOPE BY MODIFYING CONVENTIONAL FLUORESCENCE MICROSCOPES

Although two-photon microscopes are now commercially available (see Chapter 10), it is not difficult to modify an existing standard fluorescence microscope for two-photon work inexpensively, with high sensitivity and excellent image quality (Fig. 9–3).

4.1 Light Source Selection

Two-photon microscopes have been developed based on femtosecond, picosecond, and CW laser sources. Femtosecond titanium-sapphire (Ti:Sapphire) systems are the most common light sources used for two-photon imaging. The typical pulse train of Ti:Sapphire lasers consists of 100 fs pulses at a repetition rate of 80 MHz. These systems are capable of delivering peak powers of over 100 kW. The wide tuning range of a Ti:Sapphire laser (700–1000 nm) also allows selective excitation of a wide variety of fluorophores. Cr:LiSAF and pulse-compressed Nd:YLF lasers are other femtosecond light sources that have been utilized (Wokosin et al., 1996).

Picosecond laser systems, such as mode-locked Nd:YAG (~100 ps), picosecond Ti:Sapphire lasers, and pulsed dye lasers (~1 ps) can also be used for two-photon excitation. Because these systems have wider pulses, a higher average power is required to achieve equal excitation efficiency as a femtosecond laser. In addition, two-photon excitation using a CW laser has been successful with CW ArKr lasers and Nd:YAG lasers (Hell et al., 1998). As expected, even higher average power lasers are required. The significant reduction in system cost is the main advantage of using CW lasers.

The choice of laser sources with wavelengths in the range of 700–1100 nm is dictated by two factors. First, the most commonly used fluorophores in microscopy have excitation wavelength ranges of 350–600 nm. Under the "twice wavelength" rule, these fluorophores are excited by near-infrared laser sources. Second, the absorption coefficients of most biological specimens, cells and tissues,

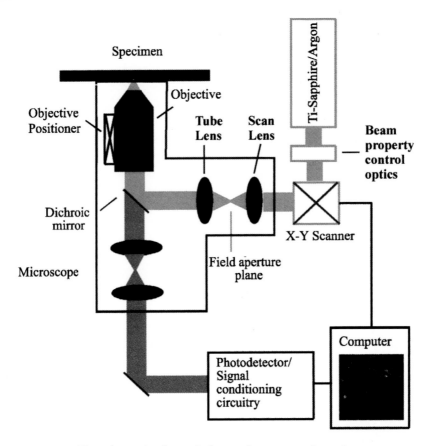

FIGURE 9–3 The schematic of a typical two-photon scanning microscope.

are minimized in this wavelength range (Color Fig. 9–4). Throughout this wavelength range, heating due to water absorption of the excitation light is mostly negligible at a typical power level of about 10 mW. In the presence of strong infrared absorbers such as melanin, however, thermal damage of biological specimens can easily occur.

4.2 Excitation Beam Path Modifications

The power, polarization, and pulse width of the laser light are controlled by external optical elements. The excitation light enters a fluorescence microscope via its epiluminescence light path directed by a galvanometer-driven x–y scanner. The epiluminescence light path is modified such that a scan lens can be mounted at a position where the x–y scanner is at its eye-point and the field aperture plane is at its focal point. A tube lens is inserted to re-collimate the excitation light for standard infinity-corrected microscope optics. It should also be noted that the scan lens and the tube lens function together as a beam expander that overfills the back aperture of the objective lens. Overfilling ensures that the full aperture of the objective is used and that the light is focused into a diffraction limited spot.

A dichroic mirror reflects the excitation light to the objective. The dichroic mirrors are shortpass filters that reflect in the near infrared and transmit in the blue-green region of the spectrum. Typically, high numerical aperture objectives are used to maximize excitation efficiency. The x–y galvanometer-driven scanners allow raster scanning of the focal point in the object space. An objective positioner, which can be piezo- or servo-motor driven, translates the focal point axially to allow 3D imaging. A major challenge in adapting an existing fluorescence microscope for two-photon scanning is the development of interface hardware and software to coordinate scanner motion, signal acquisition, and digital data archival.

4.3 Detection Beam Path Modifications

A high-sensitivity detection system is critical. The fluorescence emission is collected by the imaging objective and transmitted through the dichroic mirror along the emission path. Because high excitation intensity is used, an additional barrier filter is needed to further attenuate the scattered excitation light. The fluorescence signal can be further enhanced by placing an additional reflector in the condenser assembly (J. White and D. Wokosin, University of Wisconsin, personal communication). This arrangement effectively increases the collection solid angle and the fluorescence detection efficiency. Unlike confocal microscopes, emission pinholes and descanning optics are not necessary to achieve axial depth discrimination for a two-photon microscope.

Two-photon excitation localizes to the focal volume based on the nonlinear nature of the excitation process, and there is little off-focal plane fluorescence. The addition of a pinhole can enhance resolution, but at a cost of signal loss (Centonze and White, 1998; Gauderon et al., 1999). Another consideration in detection light path design is the implementation of a de-scan lens. For an optical sensor with a uniformly sensitive detection area, de-scanning is typically not necessary. For nonuniform detectors such as some photomultipliers, however, it may be desirable to implement de-scanning optics to prevent signal variations during scanning.

4.4 Detector Choices

Photodetectors that have been used in two-photon microscope systems include photomultiplier tubes (PMTs), avalanche photodiodes, and charge-coupled device (CCD) cameras. Due to their low cost and robust nature, PMTs are the most commonly used detectors. In addition, they have good sensitivity in the blue-green spectral region with a typical quantum efficiency of 20%–40%. Their efficiency can, however, drop down to below 1% for wavelengths above 600 nm. The large size of the photocathode area of PMTs is also an advantage, which allows efficient collection of light without de-scanning optics. In the analog mode, PMTs have a typical dynamic range on the order of 10^8. In the single-photon–counting mode, PMTs can handle a maximum count rate of about 1 MHz.

For higher sensitivity detection, single-photon–counting avalanche photodiodes are excellent. They have a quantum efficiency of about 70%–80% throughout the visible range. The drawbacks of these detectors are their high cost and failure rate. Another concern is the avalanche photodiodes' small active photo-cathode area (less than 1 mm), which requires special de-scanning optics (Farrer et al., 1999).

Change-coupled device cameras are also commonly used detectors, especially for video rate "multifocal multiphoton" imaging (Straub and Hell, 1998). Although CCD cameras can be easily integrated into a scanning microscope, they are relatively expensive, lack single-photon sensitivity, and are more sensitive to PSF degradation due to the scattering of the emitted light.

5. DEEP-TISSUE IMAGING APPLICATIONS OF MULTIPHOTON MICROSCOPY

Two-photon microscopy has found application in many areas of biology and medicine. The study of thick and highly scattering tissues is clearly one area in which two-photon microscopy will have a significant impact. Much progress has been made in neurobiology and embryology, and the application of two-photon microscopy in other tissue systems is also underway. The use of this technique for clinical tissue diagnosis is another important area where promising data are emerging.

5.1 Advantages of Two-Photon Deep-Tissue Imaging

A recent comparison study has convincingly demonstrated that two-photon microscopy is a superior method with which to image thick specimens (Centonze and White, 1998). The successful application of two-photon microscopy for deep-tissue fluorescent imaging is a result of four factors: (1) Because two-photon imaging uses infrared radiation that has reduced absorption and scattering coefficients in tissues, it allows deeper tissue penetration. (2) Because two-photon excitation occurs only in a subfemtoliter-sized focal volume, it greatly reduces tissue photodamage. (3) Typically, two-photon light sources allow efficient excitation of many common tissue fluorophores, such as NAD(P)H, flavoproteins, collagen, or elastin, that have absorption spectra in the near-ultraviolet range. (4) Because two-photon microscopes need no emission pinhole aperture to achieve 3D imaging, they are insensitive to emission PSF degradation due to specimen scattering.

5.2 Two-Photon Tissue Physiology Study

The potential of using two-photon microscopy to study tissue physiology has long been recognized (Denk et al., 1990). Two-photon tissue imaging has been used extensively in the study of the cornea structure of rabbit eyes (Piston et al., 1995; Buehler et al., 1999), the human and mouse dermal and subcutaneous structures (Color Fig. 9–5) (Masters et al., 1997, 1998; So et al., 1998), and the metabolic processes of pancreatic islets (Bennett et al., 1996; Piston et al., 1999). Two-photon microscopy has a particularly high impact in neurobiology and embryology. For neurobiology, two-photon microscopy has allowed minimal invasive imaging of living neuronal structures in brain slices and in animals (Denk et al., 1994, 1995; Yuste et al., 1999; Svoboda et al., 1999; Helmchen et al., 1999; Shi et al., 1999;

Mainen et al., 1999; Maletic-Savatic et al., 1999; Engert and Bonhoeffer, 1999; Kleinfeld et al., 1998). In embryology studies, two-photon imaging allows the study of developing embryos without damage (Jones et al., 1998; Mohler and White, 1998; Mohler et al., 1998; Squirrell et al., 1999).

Seminal work by Squirrel et al. (1999) demonstrates that mammalian embryos can be monitored continuously by two-photon microscopy over a time span of hours without compromising their viability. The embryos were subsequently implanted, and live animals were born from the irradiated embryos. Two-photon activation of caged dye was also used to map cell lineages in developing sea urchin embryos (Summers et al., 1996). Two-photon microscopic imaging will certainly be found useful in many more applications for noninvasive studies of tissue physiology.

5.3 Two-Photon Optical Biopsy and Photodynamic Therapy

Two-photon deep-tissue imaging may also find application in clinical diagnosis. The traditional biopsy procedure involves the removal, fixation, and microscopic imaging of tissues. Although the standard histological biopsy procedure is the gold standard, two-photon microscopy imaging has the potential to augment this important technique. After tissue has been extracted, histology analysis is highly accurate. There are, however, significant uncertainties that reside in choosing the correct site for excisional biopsy. Given the invasive nature of this procedure, biopsy sampling has to be sparse. Sparse sampling entails the risk of missing malignant tissues. The development of an optical biopsy technique that is noninvasive, but has cellular resolution may serve to guide the traditional biopsy.

Instead of random biopsy, two-photon imaging may be used to identify questionable tissue sites that should be excised for further examination. Another potential clinical use of two-photon imaging involves the identification of surgical margins, particularly in the removal of tumors. The accurate determination of a tumor margin will reduce the chance of tumor recurrence and will minimize the amount of tissue removal.

Today, two-photon microscopy is being successfully applied to obtain 3D tissues images in living animals and humans. Two-photon microscopy has successfully imaged skin structures of human volunteers down to a depth of 150 μm (Masters et al., 1997; Kim et al., 1998). One can resolve four distinct structural layers in the epidermis and the dermis (Color Fig. 9–5). The stratum corneum, consisting of cornified cells, is the surface protective layer. Epidermal keratinocytes form a stratified cellular layer in the epidermis. At the epidermodermal junction, the germinative basal cells are clearly visible. In the dermal layer, collagen/elastin fibers are the most prominent structure observed.

In addition to diagnosis, two-photon excitation may also find application in treatments based on photodynamic action. Photodynamic therapy is used to destroy targeted tissues, such as a tumor. The malignant tissues are targeted by preferential loading with a photosensitizer. Tissues loaded with photosensitizers are subsequently exposed to laser illumination. The photosensitizer enhances light absorption and causes localized tissue damage based on the phototoxic effect of the chemical. Although photosensitizers are designed to localize in tumors, non-negligible uptake in normal tissues is common and may cause peripheral damage of healthy tissue. The ability of two-photon excitation to trigger a 3D localized photochemical reaction allows selective exposure of the tumor region and sparing of the surrounding tissue. There are a number of promising studies in this area (Bhawalkar et al., 1997; Bodaness et al., 1986; Fisher et al., 1997).

6. CONCLUSION

Two-photon microscopy is an exciting technique for many areas of biology and medicine where 3D resolved imaging is needed. The basic physical principles behind two-photon excitation and 3D imaging are not difficult to understand. The conversion of a standard fluorescence microscope for two-photon imaging is relatively straightforward. The major hurdle remains the high cost of the pulsed laser sources required for efficient two-photon excitation. Nevertheless, two-photon imaging has found many important biomedical applications.

P. T. C. So acknowledges kind support from the National Institutes of Health, grant R29GM56486-01; the American Cancer Society, grant RPG-98-058-01-CCE; and the Unilever Corporation. C. Y. Dong acknowledges fellowship support from the National Institutes of Health, grant 5F32CA75736-02.

REFERENCES

Albota, M., D. Beljonne, J.-L. Bredas, J. E. Ehrlich, J.-Y. Fu, A. A. Heikal, S. E. Hess, T. Kogej, M. D. Levin, S. R. Marder, D. McCord-Maughon, J. W. Perry, H. Rockel, M. Rumi, G. Subramaniam, W. Webb, X.-L. Wu, and C. Xu. Design of organic molecules with large two-photon absorption cross sections. *Science* 281:1653–1656, 1998a.

Albota, M. A., C. Xu, and W. W. Webb. Two-photon fluorescence excitation cross sections of biomolecular probes from 690 to 960 nm. *Appl. Opt.* 37:7352–7356, 1998b.

Baym, G. *Lectures on Quantum Mechanics.* Menlo Park, CA: Benjamin/Cummins Publishing Co., 1973.

Bennett, B. D., T. L. Jetton, G. Ying, M. A. Magnuson, and D. W. Piston. Quantitative subcellular imaging of glucose metabolism within intact pancreatic islets. *J. Biol. Chem.* 271:3647–3651, 1996.

Bhawalkar, J. D., N. D. Kumar, C. F. Zhao, and P. N. Prasad. Two-photon photodynamic therapy. *J. Clin. Laser Med. Surg.* 15:201–204, 1997.

Bodaness, R. S., D. F. Heller, J. Krasinski, and D. S. King. The two-photon laser-induced fluorescence of the tumor-localizing photosensitizer hematoporphyrin derivative. Resonance-enhanced 750 nm two-photon excitation into the near-UV Soret band. *J. Biol. Chem.* 261:12098–12101, 1986.

Buehler, C., K. H. Kim, C. Y. Dong, B. R. Masters, and P. T. C. So. Innovations in two-photon deep tissue microscopy. *IEEE Eng. Med. Biol. Mag.* 18:23–30, 1999.

Callis, P. R. The theory of two-photon induced fluorescence anisotropy. In: *Nonlinear and Two-Photon-Induced Fluorescence,* edited by J. Lakowicz. New York: Plenum Press, 1997, pp. 1–42.

Centonze, V. E., and J. G. White. Multiphoton excitation provides optical sections from deeper within scattering specimens than confocal imaging. *Biophys. J.* 75:2015–2024, 1998.

Chalfie, M., Y. Tu, G. Euskirchen, W. W. Ward, and D. C. Prasher. Green fluorescent protein as a marker for gene expression. *Science* 263:802–805, 1994.

Denk, W. Two-photon scanning photochemical microscopy: Mapping ligand-gated ion channel distributions. *Proc. Natl. Acad. Sci. U.S.A.* 91:6629–6633, 1994.

Denk, W., K. R. Delaney, A. Gelperin, D. Kleinfeld, B. W. Strowbridge, D. W. Tank, and R. Yuste. Anatomical and functional imaging of neurons using 2-photon laser scanning microscopy. *J. Neurosci. Methods* 54:151–162, 1994.

Denk, W., J. H. Strickler, and W. W. Webb. Two-photon laser scanning fluorescence microscopy. *Science* 248:73–76, 1990.

Denk, W., M. Sugimori, and R. Llinas. Two types of calcium response limited to single spines in cerebellar Purkinje cells. *Proc. Natl. Acad. Sci. U.S.A.* 92:8279–8282, 1995.

Engert, F., and T. Bonhoeffer. Dendritic spine changes associated with hippocampal long-term synaptic plasticity. *Nature* 399:66–70, 1999.

Farrer, R. A., M. J. R. Previte, C. E. Olson, L. A. Peyser, J. T. Fourkas, and P. T. C. So. Single-molecule detection with a two-photon fluorescence microscope with fast-scanning capabilities and polarization sensitivity. *Opt. Lett.* 24:1832–1834, 1999.

Fisher, W. G., W. P. Partridge, Jr., C. Dees, and E. A. Wachter. Simultaneous two-photon activation of type-I photodynamic therapy agents. *Photochem. Photobiol.* 66:141–155, 1997.

Furuta, T., S. S. Wang, J. L. Dantzker, T. M. Dore, W. J. Bybee, E. M. Callaway, W. Denk, and R. Y. Tsien. Brominated 7-hydroxycoumarin-4-ylmethyls: Photolabile protecting groups with biologically useful cross-sections for two photon photolysis. *Proc. Natl. Acad. Sci. U.S.A.* 96:1193–1200, 1999.

Gannaway, J. N., and C. J. R. Sheppard. Second harmonic imaging in the scanning optical microscope. *Opt. Quantum Electronics* 10:435–439, 1978.

Gauderon, R., R. B. Lukins, and C. J. R. Sheppard. Effects of a confocal pinhole in two-photon microscopy. *Microsc. Res. Tech.* 47:210–214, 1999.

Göppert-Mayer, M. Uber Elementarakte mit zwei Quantensprungen. *Ann. Phys. (Leipzig).* 5:273–294, 1931.

Gu, M., and C. J. R. Sheppard. 1995. Comparison of three-dimensional imaging properties between two-photon and single-photon fluorescence microscopy. *J. Microsc.* 177:128–137, 1995.

Hell, S. W., M. Booth, and S. Wilms. Two-photon near- and far-field fluorescence microscopy with continuous-wave excitation. *Opt. Lett.* 23:1238–1240, 1998.

Helmchen, F., K. Svoboda, W. Denk, and D. W. Tank. In vivo dendritic calcium dynamics in deep-layer cortical pyramidal neurons. *Nat. Neurosci.* 2:989–996, 1999.

Jones, K. T., C. Soeller, and M. B. Cannell. The passage of Ca^{2+} and fluorescent markers between the sperm and egg after fusion in the mouse. *Development* 125:4627–4635, 1998.

Kierdaszuk, B., H. Malak, I. Gryczynski, P. Callis, and J. R. Lakowicz. Fluorescence of reduced nicotinamides using one- and two-photon excitation. *Biophys. Chem.* 62:1–13, 1996.

Kim, K. H., P. T. C. So, I. E. Kochevar, B. R. Masters, and E. Gratton. Two-photon fluroescence and confocal reflected light imaging of thick tissue structures. *SPIE Proc.* 3260:46–57, 1998.

Kleinfeld, D., P. P. Mitra, F. Helmchen, and W. Denk. Fluctuations and stimulus-induced changes in blood flow observed in individual capillaries in layers 2 through 4 of rat neocortex. *Proc. Natl. Acad. Sci. U.S.A.* 95:15741–15746, 1998.

Lakowicz, J. R., and I. Gryczynski. Tryptophan fluorescence intensity and anisotropy decays of human serum albumin resulting from one-photon and two-photon excitation. *Biophys. Chem.* 45:1–6, 1992.

Lakowicz, J. R., B. Kierdaszuk, P. Callis, H. Malak, and I. Gryczynski. 1995. Fluorescence anisotropy of tyrosine using one- and two-photon excitation. *Biophys. Chem.* 56:263–271, 1995.

Mainen, Z. F., R. Malinow, and K. Svoboda. Synaptic calcium transients in single spines indicate that NMDA receptors are not saturated. *Nature* 399:151–155, 1999.

Maletic-Savatic, M., R. Malinow, and K. Svoboda. Rapid dendritic morphogenesis in CA1 hippocampal dendrites induced by synaptic activity. *Science* 283:1923–1927, 1999.

Masters, B. R., and B. Chance. Redox confocal imaging: Intrinsic fluorescent probes of cellular metabolism. In: *Fluorescent and Luminescent Probes for Biological Activity,* edited by W. T. Mason. London: Academic Press, 1999, pp. 361–374.

Masters, B. R., P. T. So, and E. Gratton. Multiphoton excitation fluorescence microscopy and spectroscopy of in vivo human skin. *Biophys. J.* 72:2405–2412, 1997.

Masters, B. R., P. T. C. So, and E. Gratton. Optical biopsy of in vivo human skin: Multiphoton excitation microscopy. *Laser Med. Sci.* 13:196–203, 1998.

Mohler, W. A., J. S. Simske, E. M. Williams-Masson, J. D. Hardin, and J. G. White. Dynamics and ultrastructure of developmental cell fusions in the *Caenorhabditis elegans* hypodermis. *Curr. Biol.* 8:1087–1090, 1998.

Mohler, W. A., and J. G. White. Stereo-4-D reconstruction and animation from living fluorescent specimens. *Biotechniques* 24:1006–1010, 1012, 1998.

Niswender, K. D., S. M. Blackman, L. Rohde, M. A. Magnuson, and D. W. Piston. Quantitative imaging of green fluorescent protein in cultured cells: Comparison of microscopic techniques, use in fusion proteins and detection limits. *J. Microsc.* 180:109–116, 1995.

Pettit, D. L., S. S. Wang, K. R. Gee, and G. J. Augustine. Chemical two-photon uncaging: A novel approach to mapping glutamate receptors. *Neuron* 19:465–471, 1997.

Piston, D. W., S. M. Knobel, C. Postic, K. D. Shelton, and M. A. Magnuson. Adenovirus-mediated knockout of a conditional glucokinase gene in isolated pancreatic islets reveals an essential role for proximal metabolic coupling events in glucose-stimulated insulin secretion. *J. Biol. Chem.* 274:1000–1004, 1999.

Piston, D. W., B. R. Masters, and W. W. Webb. Three-dimensionally resolved NAD(P)H cellular metabolic redox imaging of the in situ cornea with two-photon excitation laser scanning microscopy. *J. Microsc.* 178:20–27, 1995.

Potter, S. M., C. M. Wang, P. A. Garrity, and S. E. Fraser. Intravital imaging of green fluorescent protein using two-photon laser-scanning microscopy. *Gene* 173:25–31, 1996.

Shear, J.B ., C. Xu, and W. W. Webb. Multiphoton-excited visible emission by serotonin solutions. *Photochem. Photobiol.* 65:931–936, 1997.

Sheppard, C. J. R., and M. Gu. Image formation in two-photon fluorescence microscope. *Optik* 86:104–106, 1990.

Shi, S. H., Y. Hayashi, R. S. Petralia, S. H. Zaman, R. J. Wenthold, K. Svoboda, and R. Malinow. Rapid spine delivery and redistribution of AMPA receptors after synaptic NMDA receptor activation. *Science* 284:1811–1816, 1999.

So, P. T. C., H. Kim, and I. E. Kochevar. Two-photon deep tissue ex vivo imaging of mouse dermal and subcutaneous structures. *Opt. Exp.* 3:339–350, 1998.

Squirrell, J. M., D. L. Wokosin, J. G. White, and B. D. Bavister. Long-term two-photon fluorescence imaging of mammalian embryos without compromising viability. *Nat. Biotechnol.* 17:763–767, 1999.

Straub, M., and S. W. Hell. Multifocal multiphoton microscopy: A fast and efficient tool for 3-D fluorescence imaging. *Bioimaging* 6:177–184, 1998.

Summers, R. G., D. W. Piston, K. M. Harris, and J. B. Morrill. The orientation of first cleavage in the sea urchin embryo, *Lytechinus variegatus,* does not specify the axes of bilateral symmetry. *Dev. Biol.* 175:177–183, 1996.

Svoboda, K., W. Denk, D. Kleinfeld, and D. W. Tank. In vivo dendritic calcium dynamics in neocortical pyramidal neurons. *Nature* 385:161–165, 1997.

Svoboda, K., F. Helmchen, W. Denk, and D. W. Tank. Spread of dendritic excitation in layer 2/3 pyramidal neurons in rat barrel cortex in vivo. *Nat. Neurosci.* 2:65–73, 1999.

Wilson, T., and C. J. R. Sheppard. *Theory and Practice of Scanning Optical Microscopy.* New York: Academic Press, 1984.

Wokosin, D. L., V. E. Centonze, J. White, D. Armstrong, G. Robertson, and A. I. Ferguson. All-solid-state ultrafast lasers facilitate multiphoton excitation fluorescence imaging. *IEEE J. Selected Top. Quantum Electronics* 2:1051–1065, 1996.

Xu, C., J. Guild, W. W. Webb, and W. Denk. Determination of absolute two-photon excitation cross sections by in situ second-order autocorrelation. *Opt. Lett.* 20:2372–2374, 1995.

Yuste, R., A. Majewska, S. S. Cash, and W. Denk. Mechanisms of calcium influx into hippocampal spines: Heterogeneity among spines, coincidence detection by NMDA receptors, and optical quantal analysis. *J. Neurosci.* 19:1976–1987, 1999.

10

Building a Two-Photon Microscope Using a Laser Scanning Confocal Architecture

Alberto Diaspro

1. Introduction

Although conceived over 20 years ago (Gannaway and Sheppard, 1978; Sheppard and Kompfner, 1978) and developed in its modern form 10 years ago (Denk et al., 1990), two-photon excitation (TPE) fluorescence microscopy can be considered a comparatively young technique in far-field fluorescence optical microscopy. This technique has advantages over both widefield and confocal laser scanning microscopy (Wilson and Sheppard, 1984; Wilson, 1990; Pawley, 1995; Webb, 1996; see also Chapter 5) for the study of the three-dimensional (3D) structure and dynamic properties of biological systems (Denk et al., 1995; Hell, 1996; Denk, 1996; Centonze and White, 1998; Diaspro, 1999a,b).

The development of mode-locked lasers that provide moderate average power at high repetition rates and ultrafast pulses (Fisher et al., 1997), matched with the increased dissemination of confocal laser scanning microscopes (CLSMs), has created an increased demand for the development of TPE architectures. Moreover, in accordance with this trend, new "ad hoc" organic molecules, endowed with large two-photon absorption cross sections, have been developed (Xu and Webb, 1996; Albota et al, 1998; Furuta et al., 1999).

In my opinion, the simplest and fastest method for implementing two- or multiphoton excitation microscopy is to modify a commercial CLSM. Less inviting alternatives are to modify a common optical microscope or to buy a commercial TPE microscope, but these choices have some drawbacks. In fact, referring to the former alternative solution, one has to consider that a commercial CLSM comes with optimized solutions for acquisition that should be implemented in a common microscope, including control operations, scanning, and digitalization (Pawley, 1995; Webb, 1996). Moreover, confocal imaging is still useful and constitutes an important reference for TPE imaging. On the other hand, commercial TPE microscopes are expensive, not as flexible and easy to use, and generally not mature. An important issue to consider is that it is difficult at this time to compare commercial instruments, especially with respect to relevant arguments like sample damage (Schönle and Hell, 1998; König et al., 1997, 1999; Koester et al., 1999; see also Chapter 14) and multiphoton capabilities (Xu et al., 1996; Hell et al., 1996; Wokosin et al., 1996). In sum, it is my opinion that it is too early to invest a large amount of money in a turnkey TPE microscope. Table 10–1 lists a few main features related to some of the emerging commercial TPE imaging systems.

There are currently two popular approaches to realize TPE imaging architectures based on a CLSM (Denk et al., 1995; So et al., 1996; Potter et al., 1996; König et al., 1996; Soeller and Cannell, 1996; Periasamy et al., 1999; Soeller and Cannell, 1999; Diaspro and Robello, 1999). One approach uses the very same optical pathway and mechanism employed in CLSMs. Pinholes are removed or set to their maximum aperture, and the emission signal is captured using the galvanometric scanning mirrors. This is called the *de-scanned mode.* This approach has a better spatial resolution, and it is also allows confocal pinholes. Unfortunately, in some practical experimental situations the low efficiency of the TPE fluorescence process may rule out such a solution. When pinhole insertion is possible, however, the major advantage is attained in terms of axial resolution that can be ameliorated by 40% (Gu and Sheppard, 1993; Nakamura, 1993; Gauderon et al., 1999).

Another approach is called the *non-de-scanned mode.* In the non-de-scanned mode, the signal-to-noise ratio is dramatically increased. The signal is collected using dichroic mirrors on the emission path or external detectors without passing through the galvanometric scanning mirrors on the way back. Unfortunately, this approach introduces some problems related to external detector positioning and environmental light noise. This means that it is not as immediate as the de-scanned solution. In the next paragraphs I will refer only to the de-scanned method, assuming that all one wants to do is to make some modifications to a CLSM and to couple it with an appropriate laser source.

2. A TWO-PHOTON EXCITATION MICROSCOPE: GENERAL CONSIDERATIONS

The main components of a de-scanned TPE microscope are shown in Figure 10–1. The components for TPE microscopy are almost identical to those for the CLSM. The two main differences are the type of laser light source and the absence of a pinhole. Moreover, an increased number of options for detection are possible in TPE microscopy. As an extra feature there is the ability of the TPE microscope to perform time-resolved fluorescence measurements using the mode-locked laser (So et al., 1996; see also Chapter 9). Currently, the mode-locked Ti:Sapphire laser (Spence et al., 1991) is preferred because of the small amplitude fluctuations produced that could become critical in a TPE application where the fluorescence intensity depends on the square of the excitation intensity. Moreover, the trend commercially is to produce compact and easy to use mode-locked lasers; this reduces the cost and problems related to maintenance, the complexity of the operations related to laser functioning, and the room needed for the system. The important features of

TABLE 10-1 Commercially available multiphoton excitation microscopy systems

Model	Company	Dimesion	Pulse Width Regime	Wavelength Range (nm)	Average Power (mW)	Laser Coupling	Acquisition	Other Features
MRC 1024 MP	Bio-Rad	Normal/ large	fs	690–1000	Not reported	Direct box	De-scanned/ non-de-scanned	Simul. confocal
Radiance 2000 MP	Bio-Rad	Compact/ normal	fs	690–1000	Not reported	Direct box	De-scanned/ non-de-scanned	Faster scanning (>750 Hz); simul. confocal
RTS2000 MP	Bio-Rad	Large	fs	690–1000	Not reported	Direct box	De-scanned/ non-de-scanned?	130 frames/s video rate
TCS SP2	Leica	Normal/ large	ps	720–900	Not reported (120 max at the sample)	Fiber	De-scanned/ non-de-scanned (in preparation)	Spectral capability
LSM 510 NLO	Zeiss	Compact/ normal	fs	700–900	50	Direct box/fiber	De-scanned	Simul. confocal

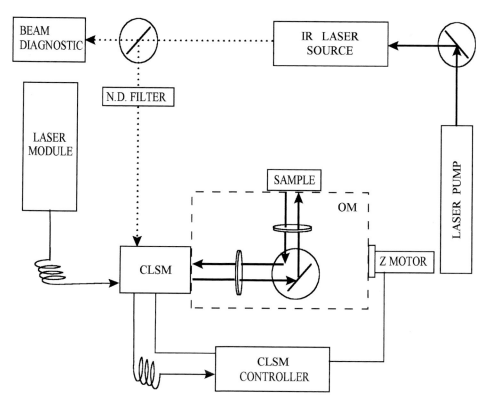

Figure 10–1 Layout of a general TPE microscope in which the ability to use the microscope as a confocal laser scanning microscope (CLSM) is retained (laser module is for conventional excitation) in order to be able to operate in one-photon confocal and two-photon imaging modes. This creates a complete system for three-dimensional fluorescence imaging and performing easy comparisons, when needed, of the two imaging modes. Light from the laser sources enters into the CLSM scanning head; from there it is delivered to the sample passing through an optical microscope (OM) stage endowed with the proper objectives and a z-axis translator. The fluorescence emitted from the sample travels back into the CLSM scanning head to be driven to acquisition units (CLSM controller). A beam diagnostic stage is useful for controlling laser functioning and beam characteristics. IR, infrared; N. D., neutral density.

infrared pulsed lasers are high average power, 80 MHz high repetition rate, and 100 fs short pulse width (Gratton and van de Ven, 1995).

All recommended laser sources for multiphoton excitation in the femtosecond pulse width regime have a number of common characteristics. Ti:Sapphire lasers cover a wide range of fluorophores commonly used in conventional fluorescence excitation. Table 10–2 lists the main features of the most utilized lasers in the field. The lasers listed in Table 10–2 can also operate in the picosecond regime and can allow optical fiber delivery, even with the drawbacks discussed in this chapter. The use of short pulses and small duty cycles is mandatory to allow image acquisition in a reasonable amount of time while using power levels that are biologically tolerable (Denk et al., 1995; Schönle and Hell, 1998; König et al., 1997, 1999; Koester et al., 1999). To take advantage of any available fluorescent molecule, a suitable laser source should also be widely tunable. The average power

TABLE 10–2 Detailed information about laser systems used for multiphoton microscopic imaging

Company/ Model	Tuning Range	Wavelength (nm)	Pulse Width	Average Power (mW) for 5 W Pump	Dimension	Pump
Spectra Physics, Tsunami	Wide	680–1050	25 fs, <100–130 fs up to 100 ps	100–800	Normal/ large	Solid state 5–10 W compact
Coherent, Mira	Wide	680–1000	<100 fs, 5–10 ps	100–700	Large	Solid state 5–10 W compact
Spectra Physics, Mai Tai	Limited	750–850 780–920	<100 fs, 25 fs	750	Compact	Solid state 5 W integrated
Coherent, Vitesse	Limited	700–760 760–860	100 fs	<1000	Compact	Solid state 5 W integrated

outputs are commonly excessive (200 mW to 1 W), which allows them to serve different instruments at the same time.

The laser beam can be directly coupled to the scanning head, but enclosures and boxes should be used to allow safe operations. Some research groups are working with optical fiber coupling, and different commercial systems are available for femtosecond and picosecond operating laser sources. Some problems and drawbacks, however, still remain. Due to high-order nonlinearities in the fiber, not much power can be put through the fiber (60–100 mW). Moreover, because the fiber coupler needs certain devices for maintaining pulse width and beam shape, it is not totally turnkey demanding for further control and alignment procedure, especially when wavelength changes are required. It is my opinion that, due to the availability of very compact appropriate laser sources (i.e., the newborn Mai-Tai by Spectra Physics has dimensions of 59 × 35 × 12 cm, allowing 100 fs pulses, up to 1 W average power, and a 100 nm tuning range), air coupling in a protected path is a very good solution. All that matters is that the full aperture of the objective is filled, and this can easily be checked by removing the objective and looking at the size of the laser spot. Practicing this operation can be useful for simply verifying alignment of the long-range tunability of the laser source.

In principle, any confocal scanning unit can be converted to a TPE microscope. In practice, one needs the simplest confocal unit, optically speaking. The best solution is a confocal scanning head with a reduced optical path that requires easy manual operations for controlling dichroic positions, pinhole aperture, and excitation power. The different excitation wavelengths require the replacement of a dichroic beam splitter, which is usually inserted in any confocal unit and which separates fluorescence from excitation. In terms of filters, changes also have to be made in the output module where fluorescence is collected. In fact, the excitation of light needs to be suppressed by 6–10 orders of magnitude. Tests can be performed by recording images from large fluorescent spheres at different power levels to check for quadratic power behavior. If it is less than two, some reflections are affecting the acquisition. Toward the complete elimination of such contributions, an optical fiber collection can help. Moreover, if optical fiber collection is performed and photomultiplier tubes are kept outside the scanning head, one has the extra advantages of working without enclosure and of controlling beam displacements within the head, if needed. Currently, several producers of fluorescence optical filters provide them with high infrared rejection. Another common solution is to use a BG39 filter in front of the acquisition channels.

The scanning head is provided with one of the most precious devices for TPE, which is the scanning system itself. Usually, scanning is performed with galvanometer-scanning mirrors. These mirrors should be metal coated (silver) and should have thermal resistance in the way they are mounted on the effective scanning device. Moreover, they should be maintained as clean as possible because in a de-scanned scheme they also deliver low-level fluorescence. Dust on any surface impinged by the laser source is generally burned, deteriorating its optical characteristics. The following section contains a practical example and should clarify the situations described above.

3. A Two-Photon Excitation Microscope: An Example of Realization

This section discusses a TPE microscope made through minor modifications of a commercial CLSM in which the ability to operate as a standard CLSM has been preserved (Diaspro and Robello, 1999; Diaspro et al., 1999c). A panoramic view of the experimental set-up is given in Figure 10–2.

The core of the design is a mode-locked Ti:sapphire infrared pulsed laser (Tsunami 3960, Spectra Physics Inc.). Mountain View, CA) pumped by a high-power (5 W at 532 nm) solid state laser (Millennia V, Spectra Physics Inc.). The Ti:sapphire laser output can be tuned anywhere from 680 to 830 nm, with a bandwidth from 4 to 12 nm, allowing excitation of a variety of fluorescent molecules normally excited by visible and ultraviolet radiation (Xu et al., 1996; Xu and Webb, 1996). The restriction of the tunable range is given by the set of mirrors installed for our purposes. Table 10–3 lists measured values for pulse width and average power as function of the operating wavelength. Measurements have been performed using an RE201 model ultrafast laser spectrum analyzer (Ist-Rees, Godalming, Surrey, UK) and an AN2/10A-P model thermopile detector power meter (Ophir, Jerusalem, Israel) that constitute the beam diagnostics module of the system. A model 409-08 scanning autocorrelator (Spectra Physics) has been also used for precise pulse width evaluation, but it is not comprised within routine beam diagnostics.

A dichroic mirror, optimized for high-power ultrashort infrared pulses (CVI, Albuquerque, NM, USA), is used to bring the Tsunami beam to the scanning

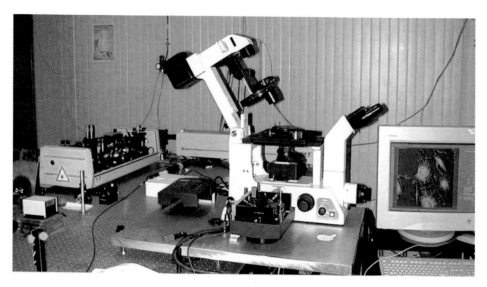

FIGURE 10–2 Photograph of a TPE microscope operating at the Laboratory of Physics of Biosystems (LFB, Department of Physics, University of Genoa, Italy) within the strategic framework of a national project of the National Institute of the Physics of Matter (INFM). This microscope, realized in 1999, is part of a multipurpose architecture that includes lifetime imaging and fluorescence correlation spectroscopy modules (Diaspro and Robello, 1999). On the left is the visible opening of the Tsunami cavity. The microscope and the PCM2000 scanning head mounted on its lateral port are visible in the center, while on the right is the video unit. An oscilloscope is used for monitoring beam conditions (left), and in the center is the Tsunami control unit integrated with a Lok-to-clock module that provides a feedback signal related to repetition frequency (Model 3930, Spectra Physics). On the right is the Tsunami source with the pump in the back and a dichroic mirror for laser delivering to the scanning head (not visible on the right).

TABLE 10–3 Measured values for pulse width and average power as a function of the operating wavelength

Wavelength (nm)	Average Power (mW)	Pulse Width (fs)
690	170	88
700	300	76
710	450	70
720	570	70
730	660	63
740	710	79
750	740	72
760	780	72
770	780	76
780	800	72
790	795	69
800	790	69
810	690	69
820	535	94

head. Before entering into the scanning head, the average beam power is brought to values ranging from 2 to 50 mW using a neutral density rotating wheel (Melles Griot, Irvine, CA, USA). For an average power of 20 mW at the entrance of the scanning head, the average power before the microscope objective is about 9–13 mW and at the sample is estimated between 3 and 5 mW (Pawley, 1995).

Due to the complexity of the procedure, we do not perform routine measurements of the pulse width at the sample, assuming that at the focal volume a 1.5–2 times broadening occurs using a high numerical aperture objective and a reduced amount of optics within the optical path (Soeller and Cannell, 1996; Hanninen and Hell, 1994; Muller et al., 1995). In our case, for a measured laser pulse width of about 80 fs at the Tsunami output window, the estimate at the sample is about 150 fs in the more favorable situation. We continuously display the pulse condition by means of an oscilloscope connected to the output of the spectrum analyzer. The pulse condition can also be tested using a simple reflective grating. In this case the reflected image on a screen will appear sharp for continuous emission and blurred for pulsed emission. This is due to the fact that the output of the pulsed laser beam is spectrally broadened. This spectrum is what you see on the screen of the above-mentioned oscilloscope. For a transform-limited $Sech^2$ pulse the relationship between pulse width (dT) and frequency width (df) is as follows: $dT \cdot df = 0.315$. Unfortunately, often the pulse is not transform limited so that this product can exceed 0.315. This can, however, be a useful check. The laser beam can be aligned using the conventional laser source of the scanning head by marking some reference positions inside the scanning head itself.

The scanning and acquisition system for microscopic imaging is based on a commercial single-pinhole scanning head Nikon PCM2000 (Nikon Instruments, Florence, Italy) mounted on the lateral port of a common inverted microscope, Nikon Eclipse TE300. The Nikon PCM2000 has a simple and compact light path that makes for an easy conversion to a two-photon microscope (Diaspro et al., 1999a). The optical resolution performances of this microscope when operated in conventional confocal mode, and using a ×100/1.3 numerical aperture oil immersion objective, have been reported in detail and are 178 ± 21 nm laterally and 509 ± 49 nm axially (Diaspro et al., 1999b). The scanning head operates in the "open pinhole" condition; that is, a wide-field de-scanned detection scheme is used (Diaspro et al., 1999b). Figure 10–3 illustrates the simple but effective optical path of the PCM2000 scanning head, and Figure 10–4 shows a detailed picture of the device itself.

The first dichroic mirror has been substituted in the original scanning head to allow excitation from 680 to 1050 nm (Chroma Inc., Brattleboro, VT). The substituted dichroic mirror reflects very efficiently (>95%) from 680 to 1050 nm. The 50% cut-off is around 640 nm. The mirror transmits the best (>90%) from 410 to 620 nm. The neutral density filter at the open-pinhole location has been removed. The galvanometer mirrors are metal coated (silver) on fused silica and exhibit a high damage threshold. The minimum pixel residence time is $3\mu s$, and it is related to the mechanical response of the scanners. A series of custom-made emission filters that block infrared radiation (>650 nm) to an optical density of 6–7

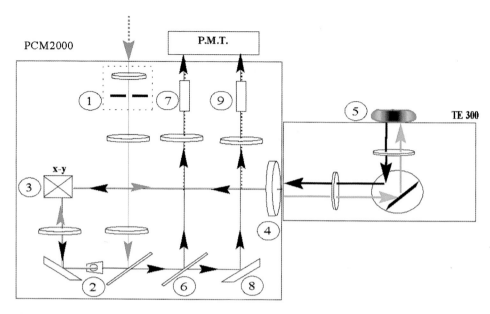

FIGURE 10–3 Optical scheme of the Nikon PCM2000 confocal scanning head. The excitation beam (red) enters into the scanning head through a coupling lens *(1)* in order to reach the sample on the x–y–z stage *(5)*. The beam passes through the pinhole holder *(2)* kept in the open position, the galvanometric mirrors. *(3)*, and the scaning lens *(4)* that optically couple the beam to the microscope objective. Fluorescence (green) coming from the sample *(5)* is directed to the acquisition channels by two selectable mirrors *(6, 8)* via optical fiber *(7, 9)* coupling.

within 50 mW of beam power incident on the filters themselves have been utilized, namely, E650P, HQ 460/50, HQ 535/50, HQ 485/30, and HQ 405/30 (www.chroma.com). The E650P filter has been initially tested to check its blocking performances with respect to the infrared/near-infrared reflections coming from stray rays within the scanning head or from the sample and constitutes the base for the other HQ filters.

One-photon and two-photon modes can be easily accomplished by switching from the single-mode optical fiber (one-photon), coupled to a module containing conventional laser sources (Ar-Ion, He-Ne green), to the optical path in air, delivering the Tsunami laser beam (two-photon). To minimize architectural changes of the PCM2000 scanning head, one should use a lens having a numerical aperture of about 0.11, the numerical aperture of the optical fiber used for conventional excitation laser delivering (see Fig. 10–4).

A high-throughput optical fiber delivers the emitted fluorescence from the scanning head to the PCM2000 control unit where photomultiplier tubes (R928, Hamamatsu, Bridgewater, NJ, USA) are physically plugged. This solution is particularly useful for three main reasons: *(1)* electrical noise is reduced, *(2)* background light noise is reduced, and *(3)* it is possible to directly verify optical conditions keeping the scanning head without enclosure.

Axial scanning for confocal and TPE 3D imaging is actuated by means of two different positioning devices depending on the experimental circumstances and

FIGURE 10–4 Photograph of the opened PCM2000 scanning head. The main components described in the text and schematically shown in Figure 10.3 are labeled.

axial accuracy needed, namely, a belt-driven system using a DC motor (RFZ-A, Nikon, Tokyo, Japan) and a single-objective piezo nano-positioner (PIFOC P-721-17, Physik Instrumente, Waldbronn, Germany). The piezoelectric axial positioner allows an axial resolution of 10 nm within a motion range of 1000 nm at 100 nm steps and utilizes a linear variable differential transformer (LVDT) integrated feedback sensor.

Acquisition and visualization are completely computer controlled by a dedicated software, EZ2000 (Coord, Apeldoorn, The Netherlands; http://www.coord.nl). The main available controls are related to PMT voltage, pixel dwell time, frame dimensions (1024 × 1024 maximum), field of scan (from 1 to 140 μm using a ×100 objective). It has to be remembered that decreasing the size of the field of scan increases the radiation exposure time when the resulting pixel dimension is smaller than one half the dimension of the diffraction limited spot (i.e., <200 nm).

4. Two-Photon Excitation Microscope Behavior

Point spread function (PSF) measurements are referred to a planachromatic Nikon (×100, 1.4 numerical aperture immersion oil objective) with enhanced transmission in the infrared region. Blue fluorescent carboxylate-modified microspheres of 0.1 μm diameter (F-8797, Molecular Probes, Eugene, OR) were used. A drop of the diluted bead suspension was spread between two coverslips of nominal thickness (0.17 mm). These microspheres constitute a very good compromise between the utilization of subresolution point scatters and acceptable fluorescence emission. An object plane field of 18 × 18 μm was imaged in a 512 × 512 frame at a pixel dwell time of 17 μs.

Axial scanning was performed, and 21 optical consecutive and parallel slices were collected at steps of 100 nm. The x–y scan step was 35 nm. The scanning head pinhole was set to the open position. The 3D data sets of several specimens were analyzed. The measured full width at half maximum (FWHM) lateral and axial resolutions were 210 ± 40 nm and 700 ± 50 nm, respectively. Intensity profiles, along with the x–y–z directions of experimental data and theoretical expectations, are reported in Figure 10–5A. To be sure of operating in the TPE regime, the quadratic behavior of the fluorescence intensity versus excitation power has been demonstrated. Figure 10–5B shows the quadratic behavior obtained from a solution of fluorescein. Moreover, during any fluorescence acquisition a simple and effective test for the TPE condition can be performed by delivering continuous radiation instead of pulsed radiation. This can be accomplished by interrupting the pumping at the Ti:Sapphire laser for awhile and switching off the pulse control. When the pump is activated, if there are not too many vibrations, it is possible to get a continuous beam that is not appropriate for TPE even if it is endowed with the same average power as the pulsed one. Restoring the pulse at any moment during the scanning fluorescence becomes visible, confirming the TPE imaging condition.

Another interesting issue is given by the demonstration of the localized photobleaching compared with the single-photon confocal case. Large fluorescent microspheres (i.e., 22 μm diameter) can be used in a 3D acquisition session. Figure 10–6 shows 3D selective photobleaching in both one-photon confocal (left column) and two-photon (right column) imaging modes. Photobleaching was produced by zooming in on a 1.96 μm^2 squared area in the center plane of the microsphere. Then 3D imaging was performed using the confocal and TPE modes. The images shown are intended only for qualitative comparison; the selective bleaching can be useful in determining the power commonly used for this operation.

In Figure 10–6, in the case of confocal imaging, it is possible to see a double-conelike image (Wilson and Sheppard, 1984; Periasamy et al., 1999; Bianco and Diaspro, 1989; Gu and Sheppard, 1995) produced within the bead, while in the TPE case the dark volume is extremely localized (Denk et al., 1995; Periasami et al., 1999; Nakamura, 1993; Patterson and Piston, 2000). This demonstrates the main difference between the two 3D imaging methods: Even though out-of-focus fluorescence is not detected, it is generated. Table 10–4 lists some of the differences between confocal and two-photon imaging. A more detailed analysis can be found

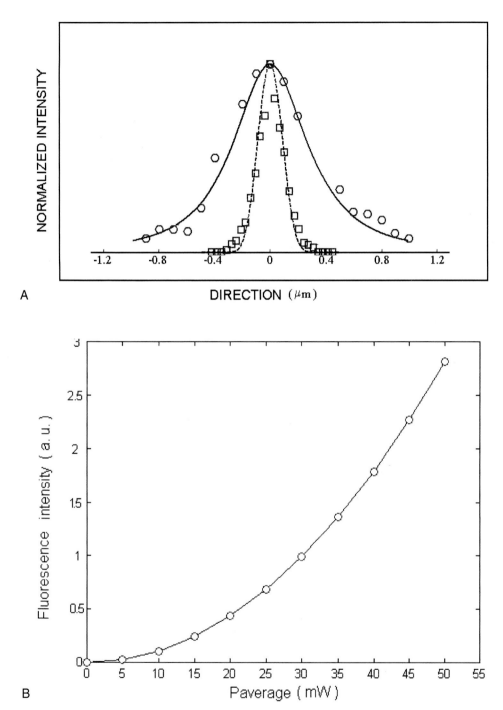

Figure 10–5 *A,* The radial and axial intensity profiles of TPE experimental and theoretical (solid line) point spread function (see text for details). *B,* The quadratic dependence on the excitation intensity, which forms the basis for the application of TPE for 3D microscopy, obtained with a solution of fluorescein (Merck, Darmstadit, Germany) in water. All intensity data, corrected only for background and measured in the central subarea of the scanned region, fell within a logarithmic best-fit straight line with a slope of 2. 03 ± 0.18.

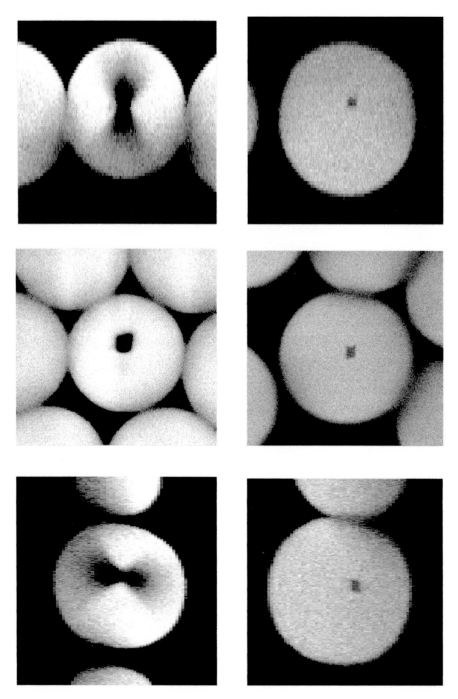

Figure 10–6 Three-dimensional projections (x–y center, y–z right, x–z left) of localized photobleaching within a large fluorescent sphere made in single-photon confocal (left column) and TPE (right column) mode. In confocal mode, a 488 nm laser beam has been used for both photobleaching and imaging, keeping the pinhole at its smallest size (Diaspro et al., 1999b). In TPE, that 720 nm excitation wavelength has kept a 10 mW average power at the entrance of the scanning head. In both cases, a 3 μs dwell time has been used for bleaching and for imaging, and a 0.5 μm axial step has been performed producing 128 optical slices. See text for further details.

TABLE 10–4 Comparison of two-photon excitation (TPE) and confocal imaging systems*

	TPE	*Confocal*
Excitation source	Femto- or picosecond pulsed-infrared (IR) laser, 80–100 MHz repetition rate, tunable 680–1050 nm	Continuous wave vis/ultraviolet laser (365, 488, 514, 543, 568, 633, 647 nm)
Excitation/emission separation	Wide	Close
Detectors	Photomultiplier tube (typical)	Photomultiplier tubes (typical)
Volume selectivity	Intrinsic (fraction of femtoliter)	Pinhole not required
Image formation	Scanning (or rotating disks or microlens array)	Scanning (or rotating disks or microlens array)
Deep imaging	>400 μm (problems related to pulse shape modifications and scattering)	Yes, but limited with respect to TPE (problems related to shorter wavelength scattering)
Spatial resolution	Less than confocal because of the focusing of IR radiation, compensated by the higher signal-to-noise ratio; better resolution if using pinhole at reduction of the collected intensity expenses	Diffraction limited, depending on pinhole size
Real-time imaging	Possible	Possible
Signal-to-noise ratio	High (especially in non-de-scanned mode)	Good
Fluorophores	All available for conventional excitation plus newly designed specifically for TPE	Selected fluorophores, depending on laser line used
Photobleaching	Only in the focus volume defined through resolution parameters. Less photobleaching compared with confocal	Within all of the double cone of excitation defined by the lens characteristics
Contrast mechanisms	Fluorescence, second harmonic generation	Fluorescence, reflection, transmission
Other features	Lifetime, single molecule detection, spectroscopy	Lifetime, spectroscopy
Commercially available	Yes (but not mature)	Yes (very affordable)

* See also Periasamy et al. (1999).

in earlier literature (Nakamura, 1993; Pawley, 1995; Gu and Sheppard, 1995; Periasamy et al., 1999).

In a further comparison of confocal and two-photon imaging, Figure 10–7 shows a TPE image of *Saccharomyces cerevisiae* cells marked with DAPI. This image conjugates the autofluorescence of the sample with the DAPI-DNA fluorescence coming from the nucleus. In this case the effect of the excitation radiation

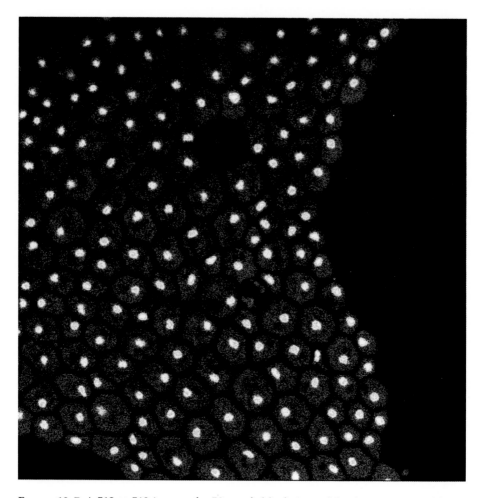

FIGURE 10–7 A 512 × 512 image of a 50 μm field of view of *Saccharomyces cerevisiae* yeast cells exhibiting autofluorescence and DAPI-DNA emission. The specimen has been excited using 720 nm radiation.

being used (i.e., 720 nm) is the same that is obtainable with ultraviolet excitation. Another example is given in color Figure 10–8, where a common and very useful test specimen (F-147780, Molecular Probes) is used to show the effect of two-photon excitation on bovine pulmonary artery endothelial cells marked with MitoTracker red, Bodipy FL phallacidin, and DAPI specific for mitochondria, F-actin, and DNA, respectively. In the case of confocal imaging, due to the absence of ultraviolet excitation, the nucleus could not be imaged, while the two other fluorescent molecules were excited using two different laser lines. In the case of two-photon imaging, the three different fluorescent molecules have been excited with the very same excitation beam (operating at 720 nm), allowing a better co-localization. The two-photon image has been realized using false colors, and one acquisition channel has been used without emission separation (the purpose of this image is related only to the demonstration of the simultaneous excitation of the three different fluorophores).

5. CONCLUSION

Two-photon excitation microscopy broke into the arena bringing its intrinsic 3D resolution, the absence of background fluorescence, and the possibility of exciting ultraviolet excitable fluorescent molecules at infrared wavelengths, increasing sample penetration. Moreover, TPE microscopy can be considered an essential imaging system for thick samples and live cell imaging.

At present, the main disadvantages of TPE fluorescence microscopy come from the cost of femtosecond laser sources and the difficulty in predicting or measuring two-photon absorption spectra of the fluorescent molecules. Because the idea of a TPE microscope is surprisingly straightforward, this chapter also discussed the conversion of a CLSM to a two-photon (multiphoton) microscope. A TPE microscope requires many of the same elements as a CLSM, but results in optical simplifications because of the high localization of the physical event.

The successful construction of a TPE microscope from simple modifications of a compact, new-generation, low-cost CLSM, with the ability to switch between TPE and confocal modes in seconds by changing the active light source, is reported here. It is shown that all one needs to do is to change a dichroic mirror inside the confocal unit, get rid of the pinhole, change the filter sets in the detection unit, and drive the beam of an appropriate laser source to the confocal unit. Once this technology becomes less expensive and simpler, there will no longer be confocal or light microscopes that are not also a two-photon microscope.

I would like to acknowledge my wife and my daughter for patience and love. Moreover, I am indebted to Massimo Fazio, Cesare Fucilli, and Marco Raimondo for their graphical efforts. Enrico Gratton, Weyming Yu, David Piston, Joseph Lakowicz, Peter So, and Hans Gerritsen helped me in defining the realized architecture. This work is in memory of Mario Arace, who purchased my first oscilloscope.

REFERENCES

Albota, M. ,D. Beljonne, J. L. Bredas, J. E. Ehrlich, J. Y. Fu, A. A. Heikal, S. E. Hess, T. Kogej, M. D. Levin, S. R. Marder, D. McCord-Maughon, J. W. Perry, H. Rockel, M. Rumi, G. Subramaniam, W. W. Webb, X. L. Wu, and C. Xu. Design of organic molecules with large two-photon absorption cross sections. *Science* 281:1653–1656, 1998.

Bianco, B., and A. Diaspro. Analysis of the three dimensional cell imaging obtained with optical microscopy techniques based on defocusing. *Cell Biophys.* 15:189–200, 1989.

Centonze, V. E., and J. G. White. Multiphoton excitation provides optical sections from deeper within scattering specimens than confocal imaging. *Biophys. J.* 75:2015–2024, 1998.

Denk, W. Two-photon excitation in functional biological imaging. *J. Biomed. Opt.* 1:296–304, 1996.

Denk, W., J. H. Strickler, and W. W. Webb. Two-photon laser scanning fluorescence microscopy. *Science* 248:73–76, 1990.

Denk, W., D. W. Piston, and W. W. Webb. Two-photon molecular excitation in laser scanning microscopy. In: *The Handbook of Biological Confocal Microscopy,* edited by J. Pawley. New York: Plenum Press, 1995, pp. 445–458.

Diaspro A. (guest editor). Two-photon excitation microscopy. *IEEE Eng. Med. Biol. Mag.* 18(5):16–99, 1999a.

Diaspro, A. (guest editor). Two-photon microscopy. *Microsc. Res. Tech.* 47:163–212, 1999b.

Diaspro, A., S. Annunziata, M. Raimondo, P. Ramoino, and M. Robello. A single-pinhole CLSM for 3-D imaging of biostructures. *IEEE Eng. Med. Biol.* 18(4):106–110, 1999a.

Diaspro, A., S. Annunziata, M. Raimondo, and M. Robello. Three-dimensional optical behavior of a confocal microscope with single illumination and detection pinhole through imaging of subresolution beads. *Microsc. Res. Tech.* 45:130–131, 1999b.

Diaspro, A., M. Corosu, P. Ramoino, and M. Robello. Adapting a compact confocal microscope system to a two-photon excitation fluorescence imaging architecture. *Microsc. Res. Tech.* 47:196–205, 1999c.

Diaspro, A., and M. Robello. Two-photon excitation of fluorescence in three-dimensional microscopy. *Eur. J. Histochem.* 43(3):70–79, 1999.

Fisher, W. G., E. A. Watcher, M. Armas, and C. Seaton. Titanium: Sapphire laser as an excitation source in two-photon spectroscopy. *Appl. Spectrosc.* 51(2):218–226, 1997.

Furuta, T., S. S. H. Wang, J. L. Dantzker, T. M. Dore, W. J. Bybee, E. M. Callaway, W. Denk, and R. Y. Tsien. Brominated 7-hydroxycoumarin-4-ylmethyls: Photolabile protecting groups with biologically useful cross-sections for two photon photolysis. *Proc. Natl. Acad. Sci. U.S.A.* 96:1193–1200, 1999.

Gannaway, J. N., and C. J. R. Sheppard. Second harmonic imaging in the scanning optical microscope. *Opt. Quant. Electronics.* 10:435–439, 1978.

Gauderon, R., P. B. Lukins, and C. J. R. Sheppard. Effect of a confocal pinhole in two-photon microscopy. *Microsc. Res. Tech.* 47:210–214, 1999.

Gratton, E., and M. J. van de Ven. Laser sources for confocal microscopy. In: *The Handbook of Biological Confocal Microscopy,* edited by J. Pawley. New York: Plenum Press, 1995, pp. 69–97.

Gu, M., and C. J. R. Sheppard. Effects of a finite sized pinhole on 3D image formation in confocal two-photon fluorescence microscopy. *J. Mod. Opt.* 40:2009–2024, 1993.

Gu, M., and C. J. R. Sheppard. Comparison of three-dimensional imaging properties between two-photon and single-photon fluorescence microscopy. *J. Microsc.* 177:128–137, 1995.

Hanninen, P. E., and S. W. Hell. Femtosecond pulse broadening in the focal region of a two-photon fluorescence microscope. *Bioimaging* 2:117–121, 1994.

Hell, S. W. (guest editor). Non linear optical microscopy [special issue]. *Bioimaging* 4(3):121–224, 1996.

Hell, S. W., K. Bahlmann, M. Schrader, A. Soini, H. Malak, I. Gryczynski, and J. R. Lakowicz. Three-photon excitation in fluorescence microscopy. *J. Biomed. Opt.* 1:71–74, 1996.

Koester, H. J., D. Baur, R. Uhl, and S. W. Hell. Ca^{2+} fluorescence imaging with pico- and femtosecond two-photon excitation: Signal and photodamage. *Biophys. J.* 77:2226–2236, 1999.

König, K., T. W. Becker, P. Fischer, I. Riemann, and K. J. Halbhuber. Pulse-length dependence of cellular response to intense near-infrared laser pulses in multiphoton microscopes. *Opt. Lett.* 24:113–115, 1999.

König, K., P. T. C. So, W. W. Mantulin, and E. Gratton. Cellular response to near-infrared femtosecond laser pulses in two-photon microscopes. *Opt. Lett.* 22:135–136, 1997.

Muller, M., J. Squier, and G. J. Brakenhoff. Measurements of femtosecond pulses in the focal point of a high numerical aperture lens by two-photon absorption. *Opt. Lett.* 20:1038–1040, 1995.

Nakamura, O. Three-dimensional imaging characteristics of laser scan fluorescence microscopy: Two-photon excitation vs. single-photon excitation. *Optik* 93:39–42, 1993.

Patterson, G. H., and D. W. Piston. Photobleaching in two-photon excitation microscopy. *Biophys. J.* 78:2159–2162, 2000.

Pawley, J. B. *Handbook of Biological Confocal microscopy.* 2nd Ed. New York: Plenum, 1995.

Periasamy, A., P. Skoglund, C. Noakes, and R. Keller. An evaluation of two-photon excitation versus confocal and digital deconvolution fluorescence microscopy imaging in *Xenopus* morphogenesis. *Microsc. Res. Tech.* 47:172–181, 1999.

Potter, S. M., C. M. Wwang, P. A. Garrity, and S. E. Fraser. Intravital imaging of green fluorescent protein using 2-photon laser-scanning microscopy. *Gene* 173:25–31, 1996.

Schönle, A., and S. W. Hell. Heating by absorption in the focus of an objective lens. *Opt. Lett.* 23:325, 1998.

Sheppard, C. J. R., and R. Kompfner. Resonant scanning optical microscope. *Appl. Opt.* 17:2879–2885, 1978.

So, P. T. C., K. M. Berland, T. French, C. Y. Dong, and E. Gratton. Two-photon fluorescence microscopy: Time resolved and intensity imaging. In: *Fluorescence Imaging Spectroscopy and Microscopy, Chemical Analysis Series*, Vol. 137, edited by X. F. Wang and B. Herman. New York: J. Wiley & Sons, 1996, pp. 351–373.

Soeller, C., and M. B. Cannell. Construction of a two-photon microscope and optimization of illumination pulse duration. *Pflugers Arch.* 432:555–561, 1996.

Soeller, C., and M. B. Cannell. Two-photon microscopy: Imaging in scattering samples and three-dimensionally resolved flash photolysis. *Microsc. Res. Tech.* 47:182–195, 1999.

Webb, R. H. Confocal optical microscopy. *Rep. Prog. Phys.* 59:427–471, 1996.

Wilson, T. *Confocal Microscopy.* London: Academic Press, 1990.

Wilson, T., and C. J. R. Sheppard. *Theory and Practice of Scanning Optical Microscopy.* London: Academic Press, 1984.

Wokosin, D. W., V. E. Centonze, S. Crittenden, and J. G. White. Three-photon excitation fluorescence imaging of biological specimens using an all-solid-state laser. *Bioimaging* 4:208–214, 1996.

Xu, C., and W. W. Webb. Measurement of two-photon excitation cross sections of molecular fluorophores with data from 690 to 1050 nm. *J. Opt. Soc. Am.* 13:481–491, 1996.

Xu, C., W. Zipfel, J. B. Shear, R. M. Williams, and W. W. Webb. Multiphoton fluorescence excitation: New spectral windows for biological nonlinear microscopy. *Proc. Natl. Acad. Sci. U.S.A.* 93:10763–10768, 1996.

11

Two-Photon Microscopy in Highly Scattering Tissue

Vincent P. Wallace, Andrew K. Dunn, Mariah L. Coleno, and Bruce J. Tromberg

1. Introduction

Molecular excitation by two-photon absorption and the subsequent fluorescence have proved to be a useful tool for imaging biological systems using laser-scanning microscopy (Denk et al., 1990). In two-photon fluorescence microscopy (TPM), near-infrared (NIR) light is used to excite transitions of twice the energy of a single photon. One of the main advantages of TPM over conventional fluorescence imaging is that the NIR excitation light penetrates more deeply into tissue than the corresponding one-photon excitation wavelength. This is because factors governing light propagation (i.e., tissue absorption and scattering) are substantially reduced in this spectral region. As a result, TPM offers the possibility of probing relatively thick tissue with submicron resolution.

Recent intravital TPM studies that take advantage of this principle include video-rate imaging of skin (So et al., 1999; see also Chapter 9) and imaging of blood flow in exposed rat neocortex (Kleinfeld et al., 1998). In general, the investigators were able to image structure and function at depths between 300 and 600 μm.

In previous studies of TPM at depth, Centonze and White (1998) found that the point spread function (PSF) in a phantom does not deteriorate with depth and the maximum imaging depth was at least twice that for the same samples using confocal fluorescence microscopy. This increased penetration is only applicable for the NIR excitation light. The resultant fluorescence, typically in the blue-green region of the spectrum, is still limited by high absorption and scattering in biological tissue. More fluorescent photons can, however, be detected in TPM versus conventional confocal microscopy for two reasons: (1) Scattered fluorescent photons do not have to be filtered out with a pinhole, and (2) the fluorescence does not need to be de-scanned before detection (Denk et al., 1995). Only the numerical aperture (NA) of the objective, the throughput of the microscope, and the efficiency of the detectors limit the sensitivity to emitted photons.

Other techniques, such as confocal reflectance microscopy and optical coherence tomography, provide the ability to obtain depth-resolved images. While it is understood that scattering is the primary effect that limits the imaging depth of these techniques, including TPM, the detailed factors are not completely understood. Scattering plays a role in TPM by decreasing the number of excitation photons reaching the focus area and by decreasing the number of fluorescent photons collected by the objective lens. Due to their flexibility, Monte Carlo models have been used to characterize different imaging modalities in highly scattering media (Dunn et al., 1996; Gan et al., 1998). Here we use a Monte Carlo model to investigate several factors that degrade TPM images, including source pulse width (τ), objective NA, and tissue optical properties (i.e., absorption [μ_a], scattering [μ_s], and anisotropy [g] for excitation and emission photons).

Insights gained from these simulations enable us to assess the complex relationship between image quality and tissue depth. A number of important statistical parameters, such as the PSF, image contrast, and detectability, can be obtained. With a realistic light propagation model, questions that are difficult to address experimentally, such as the importance of multiple scattering, detector configuration, excitation pulse width, and microscope objective design can be formulated and answered by simulation studies. In conjunction with the Monte Carlo simulations, TPM images of fluorescent spheres embedded in turbid gels are used to validate the model and to assess instrument performance. Finally, we apply this approach to estimate imaging depth of a biological tissue with known optical properties and to extract an estimate of the optical properties of tumor spheroids.

2. OPTICAL PROPERTIES OF TISSUES

The optical properties of tissues are dictated by their structural and chemical composition. When light is transmitted into tissue it is either scattered or absorbed. Anderson and Parrish (1981) used a simple model to estimate how the depth of penetration (to 1/e of the incident light intensity) in skin varies with wavelength; the data are plotted in Figure 11–1 and clearly show how penetration depth increases with wavelength.

2.1 Absorption Theory

Biological tissues contain many absorbing compounds, the principal NIR absorbers being water and hemoglobin. Molecules absorb light when the incident photon energy is equal to the energy difference between two allowed electronic states. Under these conditions, the photon promotes the transition of an electron from a lower (ground) to a higher (excited) energy state, thus making absorption highly wavelength dependent. De-excitation occurs when energy is lost either by emission of a less energetic photon or by vibrations/rotations within the same molecule (i.e., heat).

The result of absorption is the termination of photon propagation. In a homogeneous absorbing medium, the Beer-Lambert extinction law applies:

$$I = I_0 e^{-\mu_a d} \tag{1}$$

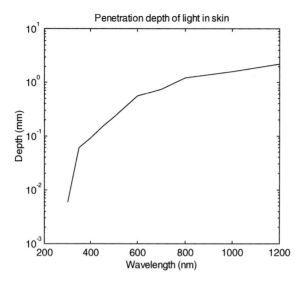

FIGURE 11–1 The optical penetration depth in skin at which the intensity of light has been reduced by 1/e (37%) of the incident intensity (Anderson and Parrish, 1981).

where I is the intensity at some depth (d) beyond the depth in the medium where the intensity is (I_0) and μ_a is a constant known as the absorption coefficient.

2.2 Scattering Theory

Scattering effects dominate light propagation in the visible and NIR spectral regions. Scattering arises from variations in the refractive index of the constituent parts of the tissue, which, for purposes of discussing their optical scattering characteristics, can be roughly divided into small and large particles. Small particles, such as organelles, mitochondria, and large proteins, have diameters significantly less than the wavelength of the light (Hielscher et al., 1997). These particles scatter light almost uniformly in all directions and are known as *Rayleigh scatterers*. Large particles with diameters much greater than the wavelength of the light, such as cells and collagen bundles, scatter light in a highly anisotropic manner with most of the light scattered in the forward direction. These are known as *Mie scatterers*. Because biological tissue is a mixture of small and large scatterers, to a first approximation, scattering is inversely proportional to wavelength.

The ability of a particle to scatter light is expressed in terms of an effective area presented to the incident light called the *total scattering cross section, σ_s*, which is the integral over all possible angles of the differential scattering cross section. The scattering coefficient μ_s, for a scattering medium consisting of one type of particle with number density ρ (number of particles per unit volume), is given by:

$$\mu_s = \rho\sigma_s \tag{2}$$

The scattering coefficient is a measure of the likelihood per distance traveled that a photon will be scattered as it passes through the scattering medium and is usually expressed in units of mm^{-1}. The average scattering length, that is, the mean distance the photon will travel before being scattered, is $1/\mu_s$ (mm).

The parameter that is used to define the directionality of scattering is called the *mean cosine of the scattering angle* or the *anisotropy factor* (g). For isotropic scattering, $g = 0$; when the scattering is totally in the forward direction, $g = 1$; and when it is solely backward, $g = -1$. The parameter g can be used in conjunction with the scattering coefficient, μ_s' to give a parameter known as the *transport scattering coefficient*, μ_s'. This concept is used to express the distance over which multiple anisotropic scattering events have the same effect as a single isotropic event:

$$\mu_s' = \mu_s(1 - g) \tag{3}$$

Combining the absorption and scattering coefficients, we get the total attenuation coefficient; for a homogeneous isotropic scattering medium,

$$\mu_t = \mu_s + \mu_a \tag{4}$$

For an anisotropic scattering medium, the analogous expression, known as the *transport coefficient*, is

$$\mu_t' = \mu_s' + \mu_a \tag{5}$$

In the red/NIR spectral regions used for TPM excitation, tissue scattering is generally much more important than absorption, with the scattering coefficient being as much as 100 times larger than the absorption coefficient. The net effect of absorption and scattering on two-photon imaging in tissues is to attenuate the excitation light reaching the focus and diminish the intensity of fluorescence escaping from the tissue.

3. MONTE CARLO

The term *Monte Carlo* refers to a method in which numerical evaluations are based on random sampling from appropriate probability distributions. The photons are treated like classic particles, that is, the polarization and wave nature are neglected. For photon propagation in tissue this allows us to inject photons one by one and follow their history until they are absorbed or scattered out of the region of interest. Events are scored to make an estimate of quantities like absorption and diffuse reflectance by summing over many photons. The number of photons used is dictated by the desired accuracy, but is typically on the order of 10^6. Hence, this technique is normally computationally intensive. However, it has many advantages, because

1. It can be applied to complex geometries, inhomogeneities can be included, and boundary conditions can easily be incorporated.
2. The required parameters for a Monte Carlo simulation are the scattering and absorption coefficients and the anisotropy.

These parameters determine the probability distributions for the mean free path, the likelihood of absorption, and the photon scattering angles. Each step between photon positions is variable and equals $\ln(\alpha)/(\mu_a + \mu_s)$, where α is a random number and μ_a and μ_s are the absorption and scattering coefficients, respectively. The weight of the photon packet is decreased from an initial value of 1 as it moves through the tissue to a value of a^n after n steps, where a is the albedo defined by $a = \mu_s/(\mu_a + \mu_s)$. When the photon strikes the surface of the tissue, a fraction of the photon weight escapes, and the remaining weight is internally reflected and continues to propagate. Eventually, the photon weight drops below a threshold level or escapes from the sample. At this point the simulation for that photon packet is terminated.

4. MONTE CARLO SIMULATION OF TWO-PHOTON IMAGING IN SCATTERING SAMPLES

The Monte Carlo model used to investigate the influence of optical properties on TPM imaging in turbid samples has been described elsewhere (Dunn et al., 2000). Briefly, the Monte Carlo model used for this work was split into two halves: excitation of two-photon fluorescence and collection of the emitted fluorescent photons.

In the first part of the calculation, excitation photons were launched into the tissue in a focused beam geometry as illustrated in Figure 11–2a. The optical

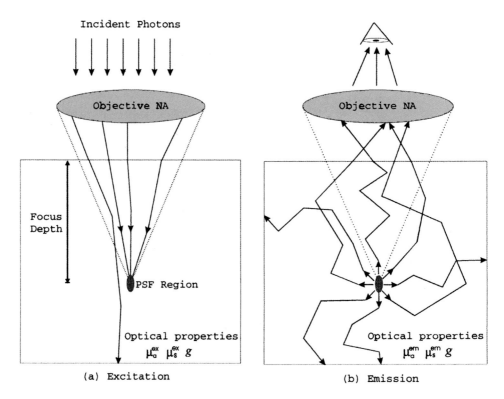

FIGURE 11–2 Geometry used in the Monte Carlo simulation demonstrating the excitation (*a*) and the emission (*b*) parts of the model. NA, numerical aperture; PSF, point spread function.

properties of the sample were those at the source wavelength. Photons were uniformly distributed at the objective lens, and the initial direction of each photon was determined from the focal depth and the starting point on the objective lens. The initial directions were varied so that, in the absence of tissue scattering, the intensity profile at the focus would approximate a diffraction-limited spot. As a photon propagated through the medium the coordinates $\mathbf{r} = (x, y, z)$ of all interaction points were recorded and stored. Each photon was propagated until it was either absorbed or reflected out of the medium or until its path length exceeded a given value, which was usually ten times the focal depth.

Once the first part of the simulation was run and the entire photon trajectories for all photons were stored, we could determine the spatial distribution, $F_{ex}(\mathbf{r})$, of the generated two-photon fluorescence sources within the medium. This is related to the instantaneous intensity squared using

$$F_{ex}(\mathbf{r}) = \frac{1}{2} \phi \sigma C(\mathbf{r}) \int_{-\infty}^{\infty} I_{ex}^2(\mathbf{r}, t) dt \tag{6}$$

where ϕ is the fluorescence quantum efficiency, σ is the two-photon absorption cross section, and $C(\mathbf{r})$ is the spatially dependent fluorophore concentration (Xu and Webb, 1996).

The second half of the Monte Carlo simulation was used to determine the fraction of the generated two-photon fluorescence collected by the detector. Fluorescent photons were launched from within the sample with a spatial distribution given by $F_{ex}(\mathbf{r})$ and isotropic initial directions; this is represented in Figure 11–2b. The optical properties of the sample were those at the emission wavelength. For those photons exiting the top surface of the sample, a geometrical ray trace was performed to determine whether the photon was collected by the objective lens. This procedure is similar to that used in confocal Monte Carlo models (Dunn et al., 1996).

Once the fraction of fluorescent photons reaching the detector, $F_{em}(\mathbf{r})$, was determined from the second half of the simulation, the total number of photons reaching the detector due to a single laser pulse focused at a depth, z, could be determined from the product of the generation and collection of fluorescence given by

$$S(z) = \eta \int F_{ex}(\mathbf{r}) F_{em}(\mathbf{r}) d\mathbf{r} \tag{7}$$

where $F_{em}(\mathbf{r})$ describes the fraction of fluorescent photons generated at $\mathbf{r}$ that reach the surface within the acceptance angle of the objective lens, and η describes the collection efficiency of the optics and filters and the detector quantum efficiency. Using the two-part simulation, the generation and collection of two-photon fluorescence at increasing focal depths was studied for each set of optical properties. The resulting detector signal as a function of focal depth, $S(z)$, was fit to a function of the form

$$S(z) = S_0 \exp\left[-(b_{ex}\mu_t^{ex} + b_{em}\mu_t^{em})z\right] \tag{8}$$

TABLE 11–1 Parameters and their values used in the
Monte Carlo Simulation

Parameter	Values
Absorption, μ_a	0.01 mm^{-1}
Scattering, μ_s	5, 10, 15 mm^{-1}
Anisotropy, g	0.6–0.98
Pulse width, τ	10–500 fs
Numerical aperture, N.A.	0.2–1.3
Focus depth, z	0–250 μm

where μ_t^{ex} and μ_t^{em} are the total attenuation coefficients at the excitation (*ex*) and emission (*em*) wavelengths, respectively, and b_{ex} and b_{em} are parameters determined by fitting the computed decay of the generation and collection of two-photon fluorescence. The term S_0 in Equation 8 is the amount of fluorescence generated in a nonscattering medium for the same NA, fluorophore concentration, and quantum efficiency. The variable parameters used in the simulation and their ranges of values are given in Table 11–1.

5. TWO-PHOTON IMAGING OF PHANTOM

The simulated results were compared with measurements of the depth-dependent signal decay of a point emission source in a phantom with similar scattering properties as tissue. The phantom consisted of 0.1 μm diameter fluorescent (Bodipy) spheres (Fluorospheres, F-8803, Molecular Probes) suspended in an agarose matrix. Intralipid (20% solids, Baxter) was added to the agarose at a final concentration of 2% to provide scattering comparable with that in tissues. The optical properties of the phantom at the excitation (780 nm) and emission (520 nm) wavelengths (Table 11–2) were verified by measuring the reduced scattering coefficient, $\mu_s' = \mu_s (1 - g)$, and the absorption coefficient, μ_a, with a frequency domain photon migration instrument (Fishkin et al., 1997).

The two-photon microscope system used in this study has been previously described (So et al., 1998; Dunn et al., 2000). The beam scanning and image acquisition are computer controlled through a data acquisition board developed at the Laboratory for Fluorescence Dynamics at the University of Illinois, Urbana—Champaign.

TABLE 11–2 Optical properties of the 2% intralipid agarose gel used in the measurements and simulations at the excitation and emission wavelengths

	Excitation Wavelength (780 nm)	Emission Wavelength (515 nm)
μ_s (mm^{-1})	6	16
μ_a (mm^{-1})	4×10^{-4}	4×10^{-4}
g	0.65	0.8

Values taken from van Staveren, 1996.

6. RESULTS

6.1 Simulation

6.1.1 Effect of μ_s

Figure 11–3 illustrates the effect of μ_s on the generation and collection of two-photon fluorescence as a function of focus depth in the sample. In these simulations, the scattering coefficients of the medium were varied at 5, 10, and 15 mm^{-1}, and the anisotropy and absorption coefficients were held fixed at $g = 0.9$ and $\mu_a = 0.01$ mm^{-1}. The objective NA was 1.3, and the pulse width and repetition rate were 100 fs and 76 MHz, respectively. The excitation data at each set of optical properties were fit to an exponential decay of the form $\exp(-b_{ex}\mu_t^{ex}z)$, where μ_t is the total attenuation coefficient at the excitation wavelength, and z is the focal depth. The parameter b_{ex} was used as a measure of the rate of decay of the excitation with depth. For the excitation data shown in Figure 11–3a, the fit coefficients b_{ex} were 2.63, 2.69, and 2.53 for scattering coefficients of 5, 10, and 15 mm^{-1}, respectively.

The collection efficiency was computed for the same three scattering coefficients and the results are shown in Figure 11–3b as a function of depth. These data were also fit to an exponential decay of the form $\exp(-b_{em}\mu_t^{em}z)$, and the fit

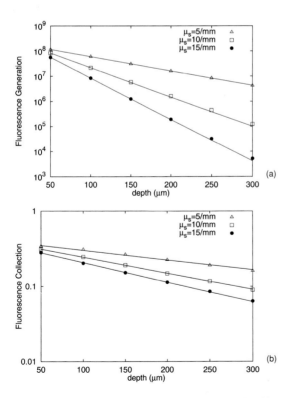

(a)

(b)

FIGURE 11–3 Fluorescence generation (*a*) and collection (*b*) in arbitrary units versus depth at three different scattering coefficients (5, 10, and 15 mm^{-1}). The solid lines indicate the least-squares linear fits to the data.

coefficients, b_{em}, were 0.59, 0.49, and 0.40 for scattering coefficients of 5, 10, and 15 mm^{-1}, respectively. Comparison of the generation and collection efficiency decay rates shows that the generation of two-photon fluorescence falls off with depth considerably faster than the signal collection.

An example of the radial PSF computed with the Monte Carlo simulation is shown in the image given in Color Figure 11–4a. The plot in Color Figure 11.4b demonstrates the radial distribution of the fluorescence intensity at the focal plane. This plot shows that as the focus progresses deeper into the medium, the intensity of the generated fluorescence decreases (which agrees with the data in Figure 11–3). Interestingly, although absolute intensity decays with depth, the full width half maximum of the intensity profile (i.e., the radial PSF) remains constant with depth. This result is confirmed later from measurements using the experimental model.

6.1.2 Effect of g

While typical values of the scattering anisotropy, g, for tissue lies in the range 0.8 to 0.95 (Cheong, 1990), in cells g can be significantly higher (Dunn and Richards-Kortum, 1997). The effect of g on the generation of two-photon fluorescence is shown in Figure 11–5a. The generated fluorescence intensity is plotted as a function of anisotropy at a focal depth of 200 μm, an NA of 1.3, and a pulse width of 100 fs. The fluorescence intensity remains relatively constant for all values of $g < 0.95$, after which it rises sharply. The effect of g can be clearly seen in the two images in Figure 11–5b, c. The fluorescence region in Figure 11–5c ($g = 0.98$) is larger and brighter than that in Figure 11–5b ($g = 0.80$). The large increase in the fluorescence generation at high values of g is due to the fact that, for $g > 0.95$, photons can undergo a scattering event and still reach the focal region. When $g < 0.9$, however, the angular deviation of the photon trajectory is large enough that the photon scatters away from the focal region.

6.1.3 Effect of numerical aperture

In addition to the optical properties of the sample, the objective NA can play an important role in determining the amount of two-photon fluorescence generated and collected. Typically a high NA (1.0–1.4) objective is used to obtain the smallest possible spot size and therefore generate the largest amount of fluorescence. To study the role that the NA plays in two-photon imaging in highly scattering samples at depth, simulations were run at focal depths of 100, 200, and 300 μm for NAs ranging between 0.2 and 1.3. The optical properties of the sample were $\mu_s = 10$ mm^{-1}, $\mu_a = 0.005$ mm^{-1}, and $g = 0.9$.

The total detected two-photon fluorescence is plotted in Figure 11–6a as a function of NA. At a focus depth of 100 μm, the detected fluorescence increases as the NA increases. At depths of 200 and 300 μm, however, the detected fluorescence peaks at an NA of about 0.6, suggesting that the optimal TPM NA varies with depth.

6.1.4 Effect of pulse length

The effect of increasing pulse length (τ) on the generation of fluorescence is illustrated in Figure 11–6b. From the simulation, the fluorescence generation at a depth of 150 μm and a scattering coefficient of 10 mm^{-1} is plotted as a function

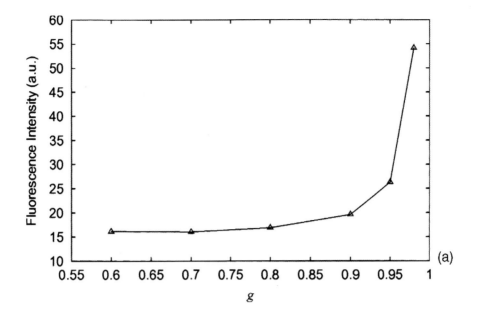

0.80 0.98

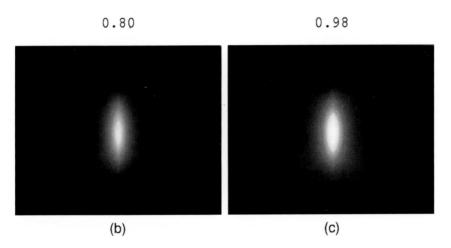

(b) (c)

FIGURE 11–5 Effect of anisotropy, *g*, on the generation of two-photon fluorescence at a focus depth of 200 μm (*a*) and computer-generated plots of the point spread function at a depth of 200 μm for *g* values of 0.80 (*b*) and 0.98 (*c*).

of pulse length. As expected, the amount of fluorescence decreases with increasing pulse length. This is apparent from the following relationship:

$$I \propto \frac{\langle P \rangle^2}{\tau F} \tag{9}$$

where *I* is the two-photon fluorescence intensity, P_{av} is the average power, τ is the pulse width, and *F* is the repetition rate. For a fixed average power and repetition rate, the fluorescence intensity will increase as the pulse width is shortened.

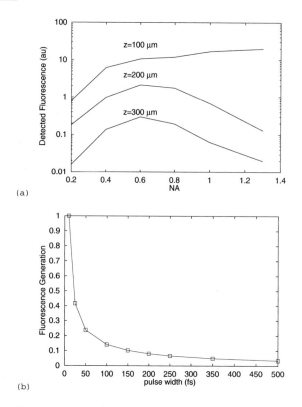

(a)

(b)

FIGURE 11–6 *a*, The total detected fluorescence intensity in arbitrary units as a function of objective numerical aperture (NA) for focal depths of 100, 200, and 300 μm. *b*, Fluorescence generation decreases with increasing pulse length.

At the moment, commercial ultrafast laser systems typically have pulse widths of around 100 fs.

The only way to probe deeply into scattering samples is to increase the average power as the depth of focus in the sample is increased, although compressing pulses using grating or prism pairs (Fork et al., 1984) can achieve shorter pulses. This approach, however, is primarily used to compensate for pulse broadening due to group velocity dispersion that occurs as the pulse travels through optical elements on its way to the sample. There are ultrafast laser systems that can produce pulses of less than 10 fs (Morgner et al., 1999); however, these systems are difficult to operate, and they intrinsically have a wide spectral bandwidth (~400 nm). Because this bandwidth is substantially broader than the absorption bands of typical fluorophores, excitation efficiency goes down (Denk et al., 1995) and background signals can increase.

6.2 Phantom Measurements

As an example an image 100 μm into the phantom is shown in Figure 11–7a. Individual 0.1 μm fluorescent spheres are visible. An expanded view of a single sphere is shown in Figure 11–7b. A Gaussian fit to the pixel intensities shown in Figure 11–7c illustrates the procedure that was followed to measure the signal characteristics as a function of depth. The intensity at a given focal depth was determined from the average amplitude of the Gaussian fits to approximately

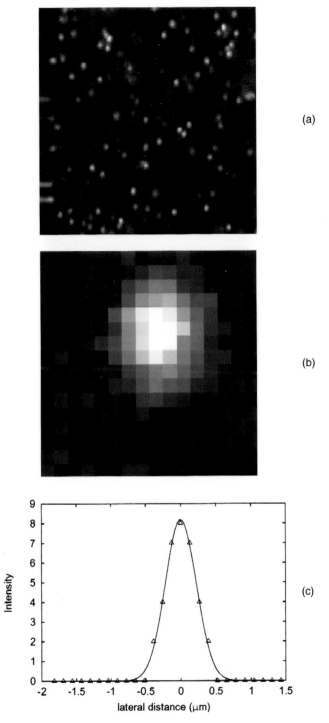

(a)

(b)

(c)

FIGURE 11–7 Measurement of the lateral point spread function of the system. *a*, Image of the tissue phantom at a depth of approximately 100 μm. *b*, Image of an individual sphere from the image in *a*. *c*, Gaussian fit to the intensity (arbitrary units) of a single sphere. Images were obtained with a Zeiss C-Apochromat ×63, 1.2 NA, water immersion objective, with an average power of approximately 10 mW.

25 spheres in each image plane. The width of the lateral PSF was estimated from the 1/e reduction of the Gaussian profile.

In Figure 11–8a, the width of the PSF is plotted against depth. The measured width of the 0.1 μm spheres shows no significant change to a depth of 250 μm, roughly the working distance of the objective, confirming the result from the Monte Carlo simulation. The theoretical width of the lateral PSF in a nonscattering medium for this set of parameters is approximately 0.35 μm (Sheppard and Gu, 1990), which is close to the measured value.

In Figure 11–8b, the natural logarithm of intensity is plotted against depth. The depth-dependent intensity decay was fit to an exponential of the form $\exp(-az)$. This form is slightly different from that used for the simulated data because the generation and collection processes cannot be separated in the experiment. Therefore, the fit parameter a, in the measured data is related to the simulated data by $a = b_{ex}\mu_t^{ex} + b_{em}\mu_t^{em}$. The value of a determined from the fit shown in Figure 11–8a was 16 mm^{-1}, which is close to the value predicted by the model using $b_{ex} = 2.5$ and $b_{em} = 0.5$.

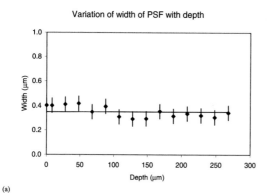

(a)

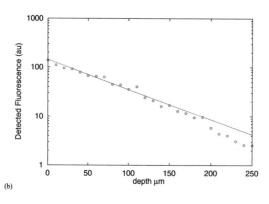

(b)

FIGURE 11–8 a, Measured two-photon fluorescence as a function of focal depth for 0.1 μm fluorescent spheres embedded in a 2% intralipid phantom. b, Measured width of 0.1 μm fluorescent spheres embedded in 2% intralipid as a function of focal depth. The width at each depth is the average width of approximately 25 spheres as measured by Gaussian fits to the individual spheres. PSF, point spread function.

7. Discussion

TPM signal degradation has been decomposed into two components in our model: attenuation of excitation photons and attenuation of fluorescence. The product of the two components yields a quantity proportional to the measured signal intensity. Figure 11–3 illustrates these two components at different levels of scattering and shows that each is well described by an exponential attenuation.

The slope of the exponential decay for the collection efficiency of the fluorescent photons ranges between 0.4 and 0.59. Because this value is less than unity (i.e., $b_{em} = 1$ for single scattering behavior), there is a clear indication that multiple scattered photons are collected by the objective lens and subsequently detected. The excitation of two-photon fluorescence, however, decays more rapidly with a relatively large decay coefficient, b_{ex}, of about 2.5. Because $b_{ex} = 2$ would be expected for nonscattering attenuation in a two-photon process, this result suggests that ballistic photons are the primary source of excitation. The calculated value of 2.5 being larger than 2 suggests that the average photon path length is slightly greater than the linear distance between the sample surface and the focal plane. This observation is consistent with our understanding of the enhanced scattering probability for high incidence angle photons emerging from high NA objectives.

In TPM, objectives with the highest available NA are typically used because this results in the smallest spot size and the greatest instantaneous excitation intensity. In highly scattering media, a tradeoff exists between the smallest spot size and the increased attenuation of excitation photons at high NAs. This is due to the fact that photons propagating along the outer cone of light at high NA have an incidence angle as high as 60°. Although this effect is not seen in transparent media, in turbid samples these photons must travel a longer distance to reach the focus. Because of the higher probability of long path length photons undergoing scattering events, fewer photons may be available for excitation.

As illustrated in Figure 11–6a, the impact of numerical aperture on the detected fluorescence changes with depth. At a depth of 100 μm (approximately one mean free path), increasing the NA improves signal collection. In contrast, at depths of 200 and 300 μm, the optimal NA is between 0.6 and 0.8. Although the high NAs still produce a smaller spot size than the lower NAs at these depths, the increased probability of scattering for high-angle paths leads to their selective attenuation and a decrease in the overall detected signal.

In a scattering sample, a lower NA is advantageous for deep two-photon excitation because it results in an increase in the number of fluorescent photons generated. However, as shown in Figure 11–6a, because multiply scattered emission light is collected, a high NA is beneficial at any depth for optimal collection of fluorescent light. This suggests that the optimal TPM for deep-tissue imaging would employ a lower NA for excitation and a higher NA to collect the emission, a feat that could be accomplished by underfilling the microscope objective in the back focal plane.

Figure 11–5a demonstrates that the anisotropy factor, g, can significantly alter the generation of two-photon fluorescence at values above about 0.95.

At lower values of g (<0.95), the generation of fluorescence is dominated by photons reaching the focus unscattered. Those photons that are scattered encounter an angular deviation sufficiently large to prevent them from reaching the focus and therefore do not contribute to the generation of fluorescence. At higher values of g, however, the angular deviation imparted to a photon upon scattering is sufficiently small to permit the photon to still reach the focal volume and contribute to the excitation. Therefore, when tissues like densely packed epithelial cell layers are imaged, photons may encounter multiple small-angle scattering and still contribute to the generation of two-photon fluorescence.

The measured signal decay in the tissue phantom (Fig. 11–8b) demonstrates that the detected fluorescence signal declines exponentially with depth as predicted by the model. By fitting Equation 8 to the measured data, a total decay constant is calculated to be 16 mm^{-1}. This value is comparable to model simulations for 2% intralipid, which range between 19 and 26 mm^{-1} assuming a 10% uncertainty in the optical properties of the intralipid. This measured decay coefficient, a, is actually a composite function of the optical properties at the excitation and emission wavelengths $a = b_{ex}\mu_t^{ex} + b_{em}\mu_t^{em}$, and our calculated values show good agreement with previously reported measurements of the optical properties of 2% Intralipid (van Staveren et al., 1991).

The lateral resolution is unaffected by scattering up to a depth of 250 μm, as illustrated in Figure 11–8a, where the measured width of the lateral point spread remains relatively constant. This indicates that the limiting factor in imaging in turbid samples is a lack of signal rather than a loss of resolution, which is in agreement with previous studies (Smithpeter et al., 1998).

8. Two-Photon Imaging in Tissues

8.1 How Deep?

As tissue engineering and biomechanics research progress, it becomes increasingly necessary to observe changes in tissue structure and function in situ over time without irreversibly damaging or sectioning the tissue. Two-photon imaging will play an important role in studying tissue growth and regeneration after chemical or physical stimuli. The question that always comes up is how deep can we image? As described by our model, knowledge of tissue optical properties can help provide an answer to this question.

For example, the maximum depth for two-photon excited autofluorescence imaging of cells in a highly scattering tissue, such as cartilage, can be estimated using the measured optical properties of cartilage at the excitation (780 nm) and emission (440 nm) wavelengths as listed in Table 11–3 (Youn et al., 2000). With the expression for the two-photon fluorescence signal decay given by Equation 8, predictions on the maximum imaging depth can be made for a given set of experimental parameters and tissue optical properties based on an estimation of the

TABLE 11–3 Optical properties of cartilage used in the estimation of imaging depth

Wavelength (nm)	μ_s' (mm^{-1})	μ_a (mm^{-1})
780	0.41	0.016
440	1.52	0.034

The absorption and reduced scattering coefficients were taken from reference Youn *et al.,* 2000 and *g* was assumed to be 0.97 (Madsen, 2000a).

signal-to-noise ratio. The signal-to-noise ratio (*SNR*) in photon counting mode can be written as (Hamamatsu Photonics, 1994)

$$SNR = \frac{N_s \sqrt{t}}{[N_s + 2(N_d + N_b)]^{1/2}} \tag{10}$$

where N_s is the number of fluorescence photons/second that can be computed from Equation 10, N_b and N_d are the number of background and dark counts/second, and t is the integration time.

The objective NA was assumed to be 1.3, and the laser characteristics were those of a typical mode-locked titanium:sapphire source (76 MHz repetition rate and 100 fs pulse width) with an average power of 10 mW at the sample. The detector was assumed to have a quantum efficiency of 0.2 with a dark count, $N_d = 20$ counts/second, and a background level of 100 counts/second (Hamamatsu Photonics, 1994). The image is 256 × 256 pixels with an acquisition time of 2 s. Based on these values, and by using Equations 8 and 10, the signal level and signal-to-noise ratio can be estimated for a complete set of experimental parameters and tissue characteristics. The term S_0 in Equation 8 represents the signal intensity in the absence of scatterers and can be calculated as described by Xu and Webb (1996). The maximum imaging depth can then be determined through a calculation of the depth at which the signal-to-noise ratio falls to a value of 3. Assuming the reduced form of nicotinamide adenine dinucleotide (NADH) or NADH phosphate (NADPH) is the primary fluorophore in cells at a protein bound concentration of 30–100 μm (König et al., 1995) and a two-photon absorption cross section estimated to be about 0.005 GM (Xu et al., 1996), the maximum imaging depth is predicted to be 140–180 μm.

In Color Figure 11–9a we show three images at different depths within rabbit septal cartilage tissue that demonstrate that this prediction is in good agreement with the experiment. The left image in Color Figure 11–9a, obtained at a depth of 50 μm below the surface, clearly shows chondrocyte cells with nonfluorescent nuclear shadows and bright cytosolic zones characteristic of mitochondrial NAD(P)H. The middle image in Color Figure 11–9a was obtained at 150 μm. Here at a depth of nearly seven times the scattering mean free path, cellular structure is clearly visible. The signal-to-noise ratio, however, has fallen to less than half of that in the image at 50 μm. Finally, in the right image in Color Figure 11–9a, at 200 μm, cells are no longer visible. The specimen was 1 mm thick. Therefore, using this procedure, the predicted imaging depth and signal-to-noise ratio can be studied for a complete set of experimental conditions, including laser source characteristics, tissue optical properties, and fluorophore properties.

8.2 Attenuation Coefficient in Tumor Spheroids

Multicellular tumor spheroids are aggregates of tumor cells grown in culture and have been widely used in cancer research as an experimental tool for novel therapies. They share many characteristics of tumors, and their optical properties are of interest for diagnostic purposes and for investigation of therapeutic methods based on photodynamic therapy (Madsen et al., 2000b). An estimate for the attenuation coefficient can be made by studying the fluorescence intensity decay with depth as given by Equation 8.

Three glioblastoma tumor spheroids, 0.5 mm in diameter, were incubated with 5-aminolevulinic acid for 6 h. The ALA is converted to protoporphyrin IX (PpIX), a potent photosensitizer that has an absorption peak at 400 nm. For this study, two excitation wavelengths of 800 and 840 nm were used. The fluorescence emissions from NADH (autofluorescence), centered at 440 nm, and PpIX at 635 nm, were detected in two separate detection channels using appropriate bandpass filters. Images were obtained using a ×10, 0.3 NA objective at 10 μm intervals to a depth of about 150 μm. Three example images of a whole spheroid, composed of approximately 2×10^5 cells, are shown in Color Figure 11–9b; from left to right, the depths are 25, 50, and 75 μm. As the image plane cuts through the sphere, many more cells are passed through when the center rather than a side is imaged.

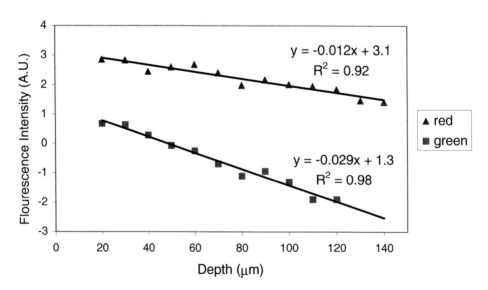

FIGURE 11–10 The log of fluorescence emission intensity (measured in arbitrary units) is plotted versus the depth into the spheroid (in μm) for an excitation wavelength of 800 nm. The plot clearly shows how the green fluorescence is attenuated more rapidly with depth than the red fluorescence.

TABLE 11–4 Scattering coefficients extracted from the measured decay of fluorescence in tumor spheroids

Wavelength (nm)	440	635	800	840
μ_s (mm^{-1})	24	7.1	5.6	5.3

This clearly shows how the detected fluorescence is attenuated by scattering and absorption and how it is a function of tissue thickness.

To determine the decay of the cellular fluorescence intensity with depth for each spheroid, the intensities of the central region were measured. In Figure 11–10, the natural logarithm of fluorescence emission intensity (measured in arbitrary units) is plotted on the y-axis and the depth into the spheroid is plotted on the x-axis for an excitation wavelength of 800 nm. This plot clearly shows that the green fluorescence is attenuated more rapidly with depth than the red fluorescence. Four fit parameters can be determined for each combination of excitation and emission wavelengths: 800/440, 800/635, 840/440, and 840/635.

For the given experimental parameters, the model described in Equation 8 predicts $b_{ex} = 2.2$ and $b_{em} = 0.45$. To obtain scattering values (μ_s) for glioblastoma spheroids at each wavelength, we assume for dense cell suspensions that $\mu_s(\lambda) \propto \lambda^{-1}$ (for $\lambda > 500$ nm), $g = 0.98$ (Mourant et al., 1998), $\mu_a \sim 0.4$ mm^{-1} at 635 nm, and $\mu_a \sim 0.01$ for $\lambda > 700$ nm (Hargrave et al., 1996). Using least-squares fits to the measured intensity decays, optical parameters at 635, 800, and 840 nm are obtained. Once the scattering coefficients are known for 635, 800, and 840 nm, a value for the scattering coefficient at 440 nm can be calculated algebraically.

Table 11–4 summarizes the results of the scattering coefficient calculations for all excitation and emission wavelengths. The much larger value of μ_s at 440 nm is due to the fact that shorter wavelengths are more sensitive to smaller particle Rayleigh scattering, which has a strong wavelength dependence of $1/\lambda^4$. In contrast, at longer wavelengths, large particles dictate the scattering behavior. These coefficients are in good agreement with values obtained from cell suspensions using elastic scattering techniques and a diffusion approximation (Mourant et al., 1998).

9. CONCLUSIONS

Our results show that TPM can be used to form high-resolution images at depths predicted with prior knowledge of the tissue's optical properties. The overall signal intensity is a function of the experimental parameters, including excitation source pulse width, objective NA, tissue optical properties, and the photophysical characteristics of the fluorophore. Although the two-photon fluorescence signal intensity decreases due to scattering-induced losses in the ballistic photons, image resolution is essentially unaffected by tissue optical properties. As a result, transform-limited resolution can be obtained at depths of two to seven times the scattering mean free path.

With a Monte Carlo model and measurements from tissue phantoms, the characteristics of the two-photon fluorescence signal with depth in a highly scattering medium have been investigated. The model and measurements have shown that the two-photon excited fluorescence signal in a highly scattering medium depends on the properties of the medium and the instrument. Fluorescence was found to decay exponentially with a slope determined by the scattering and absorption coefficients of the medium at the excitation and emission wavelengths. The scattering anisotropy influences the generation of two-photon fluorescence only at values greater than 0.95, where multiply scattered photons can still contribute to the signal. The optimal objective NA was found to vary with focal depth, illustrating the tradeoff between higher instantaneous excitation intensity at high NAs and the increased probability of scattering for photons entering turbid samples at high incidence angles.

The model predictions for the decay of the two-photon signal with depth were comparable with the measured decay in a scattering phantom. Based on the predicted form of the two-photon signal decay with depth and a minimum acceptable signal-to-noise ratio, the maximum imaging depth can be predicted for a complete set of experimental parameters. For two-photon imaging of cartilage tissue, the maximum imaging depth is predicted to be approximately 150 μm, assuming a fluorophore with characteristics similar to NAD(P)H. By measuring the slope of the exponential decay of fluorescence with depth, estimates of tissue optical properties can also be made.

This work was supported by the National Institutes of Health under grants RR-01192 and GM-50958; the Office of Naval Research (ONR N00014-91-C-0134), and the Department of Energy (DOE DE-FG03-91ER61227). Assistance from Peter So, Enrico Gratton, and the Laboratory for Fluorescence Dynamics in the construction of the two-photon microscope is gratefully acknowledged. The authors would also like to thank Chung-ho Sun for preparation of glioblastoma spheroids and Brian Wong for supplying the cartilage.

REFERENCES

Anderson, R. R., and J. A. Parrish. The optics of human skin. *J. Invest. Dermatol.* 77:13–19, 1981.

Centonze, V., and J. White. Multiphoton excitation provides optical sections from deeper within scattering specimens than confocal imaging. *Biophys. J.* 75:2015–2024, 1998.

Cheong W. A review of the optical properties of biological tissues. *IEEE J. Quantum Electronics* 26:2166–2185, 1990.

Denk, W., J. Strickler, and W. Webb. Two-photon laser scanning fluorescence microscopy. *Science* 248:78–76, 1990.

Denk, W., D. Piston, and W, Webb. Two-photon molecular excitation in laser-scanning microscopy. In: *Handbook of Biological Confocal Microscopy,* edited by J. B. Pawley. New York: Plenum Press, 1995, pp. 445–458.

Dunn, A., and R. Richards-Kortum. Three-dimensional computation of light scattering from cells. *IEEE J. Spec. Top. Quantum Electronics* 2:898–905, 1997.

Dunn, A., C. Smithpeter, R. Richards-Kortum, and A.J. Welch. Sources of contrast in confocal reflectance imaging. *Appl. Opt.* 35:3441–3446, 1996.

Dunn, A. K., V. P. Wallace, M. Coleno, M. W. Berns, and B. J. Tromberg. Influence of optical properties on two-photon fluorescence imaging in turbid samples. *Appl. Opt.* 39:1194–1201, 2000.

Fishkin, J., O. Coquoz, E. Anderson, M. Brenner, and B. Tromberg. Frequency-domain photon migration measurements of normal and malignant tissue optical properties in a human subject. *Appl. Opt.* 36:10–20, 1997.

Fork, R. L., O. E. Martinez, and J. P. Gordon. Nagative dispersion using prism pairs. *Opt. Lett.* 9:150, 1984.

Gan, X., S. Schilders, and M. Gu. Image formation in turbid media under a microscope. *JOSA A* 15:2052–2058, 1998.

Hamamatsu Photonics. *Photomultiplier Tube.* Hamatsu Photonics, Bridgewater, NJ, USA, 1994.

Hargrave, P., P. W. Nicholson, D. T. Delpy, and M. Firbank. Optical properties of multicellular tumour spheroids. *Phys. Med. Biol.* 41:1067–1072, 1996.

Hielscher, A. H., et al. Influence of particle size and concentration on the diffuse backscattering of polarized light from tissue phantoms and biological cell suspensions. *Appl. Opt.* 36:125–135, 1997.

Kleinfeld, D., P. Mitra, F. Helmchen, and W. Denk. Fluctuations and stimulus-induced changes in blood flow observed in individual capillaries in layers 2 through 4 of rat neocortex. *Proc. Natl. Acad. Sci. U.S.A.* 95:15741–15746, 1998.

König, K., L. Yagang, G. J. Sonek, M. W. Berns, and B. J. Tromberg. Autofluorescence spectroscopy of optically trapped cells. *Photochem. Photobiol.* 62:830–835, 1995.

Madsen, S. J., C. Sun, B. J. Tromberg, V. P. Wallace, and H. Hirschberg. Photodynamic therapy of human glioma spheroids using 5-aminolevulinic acid. *Photochem. Photobiol.* 72:128–134, 2000b.

Morgner, U., F. X. Kärtner, S. H. Cho, Y. Chen, H. A. Haus, J. G. Fujimoto, and E. P. Ippen. Sub-two-cycle pulses form a kerr-lens mode-locked Ti:sapphire laser. *Opt. Lett.* 24:411–413, 1999.

Mourant, J. R., J. P. Freyer, A. H. Hielscher, A. A. Eick, D. Shen, and T. M. Johnson. Mechanisms of light scattering from biological cells relevant to noninvasive optical-tissue diagnostics. *Appl. Opt.* 37:3586–3593, 1998.

Sheppard, C., and M. Gu. Image formation in two-photon fluorescence microscopy. *Optik* 86:104–106, 1990.

Smithpeter, C., A. Dunn, A. Welch, and R. Richards-Kortum. Penetration depth limits of in vivo confocal reflectance imaging. *Appl. Opt.* 37:2749–2754, 1998.

So, P. T. C, C. Buehler, K. Kim, and I. Kochevar. Tissue imaging using two-photon video rate microscopy. *SPIE BiOS* 99:3604–3607, 1999.

So, P. T. C, H. Kim, and I. E. Kochevar. Two-photon deep tissue ex-vivo imaging of mouse dermal and subcutaneous structures. *Opt. Express* 3:339–350, 1998.

van Staveren, H., C. Moes, J. van Marle, S. Prahl, and M. van Gemert. Light scattering in Intralipid-10% in the wavelength range of 400–1100 nm. *Appl. Opt.* 30:4507–4514, 1991.

Xu, C., and W. Webb. Measurement of two-photon excitation cross-sections of molecular fluorophores with data from 690 to 1050 nm. *JOSA B* 13:481–491, 1996.

Xu, C., W. Zipfel, J. Shear, R. Williams, and W. Webb. Multiphoton fluorescence excitation: New spectral windows for biological nonlinear microscopy. *Proc. Natl. Acad. Sci. U.S.A.* 93:10763–10768, 1996.

Youn, J., S. A. Telenkov, E. Kim, N. C. Bhavaraju, B. J. F. Wong, J. W. Valvan, and T. E. Milner. Optical and thermal properties of nasal septal cartilage. *Laser Surg. Med.* 27:119–128, 2000.

12

Multiphoton Laser Scanning Microscopy and Dynamic Imaging in Embryos

Mary E. Dickinson and Scott E. Fraser

1. Introduction

For hundreds of years developmental biologists have yearned to peer inside the embryo as it develops, to appreciate the beauty and intricacies of morphogenesis from a front row seat. As microscopes have been improved and technology has flourished, our vantage point is improving. Advances in lasers, computer hardware and software, optical engineering, microscopy, and digital imaging have greatly changed not only how observations of living systems can be made, but what can be observed. Alongside the development of advanced technology, there has been an explosion of knowledge in the area of molecular signaling mechanisms.

Fluorescence microscopy has become an indispensable way not only to observe how cells move and behave during development at a cellular level but also as a means to understand what signals may be controlling these movements on a molecular level. One of the latest advances in fluorescence imaging technology, multiphoton laser scanning microscopy (MPLSM), is fast becoming the imaging method of choice to study vital developmental processes. In this chapter, we discuss some of the practical considerations for using MPLSM for dynamic imaging in embryos and highlight some of the advances that have been made using this imaging method.

2. Practical Considerations of MPLSM for Vital Imaging

Many practical considerations, such as embryo preparation or viability on the microscope stage, are not unique to MPLSM and have been reviewed elsewhere (see Further Readings below and other chapters in this volume). Several factors particular to this imaging modality should, however, be considered. In this section, we highlight the benefits and limitations of MPLSM imaging for studying developmental events and provide some insight into how to choose the best fluorescent probes for particular applications.

2.1 The Benefits of MPLSM for Imaging Live Embryos

Using fluorescence microscopy to study dynamic processes in embryos can be challenging. Embryos of many species are large turbid structures that scatter light, limiting the ability to produce clear images at a cellular level. Embryos must also be kept healthy during imaging, which can be compromised by long-term exposure to intense visible-wavelength light and repeated excitation of fluorescent probes.

For vital imaging of embryos, MPLSM offers substantial benefits over confocal microscopy. Because multi-photon excitation occurs only at the focal plane, less background signal is produced and fewer toxic byproducts are created in the sample. For two-photon excitation, the excitation rate of a fluorochrome depends on the square of the light intensity (or on the third power for three-photon excitation) and thus falls off very quickly, on the order of the fourth power of the distance from the focal plane. Single infrared (IR) photons do not induce fluorescence and are generally not absorbed by endogenous molecules; thus, multiphoton excitation minimizes toxic effects by minimizing needless out-of-focus fluorescence and by preventing the destruction of endogenous molecules that are harmed by the absorption of ultraviolet (UV) and visible photons.

Excitation in a defined voxel also benefits the quality of the image because there is no out-of-focus fluorescence to contribute to the background or to be rejected with a pinhole. In addition, MPLSM allows for more efficient detection of the fluorescent signal. Because emission photons are not filtered through a pinhole and do not have to be reflected back through the scan mechanism, all of the emission photons can be collected to make the image, and non-de-scanned detection systems can be used to enhance the sensitivity of the collected signal (Wokosin et al., 1998; Tan et al., 1999). Furthermore, because IR light can penetrate deeper into the specimen than can visible light, MPLSM enables more information to be gathered deeper along the z-axis. Thus, MPLSM allows clearer and more complete data sets to be obtained from healthier samples than do other methods.

2.2 Misconceptions About MPLSM

Several misconceptions circulate about MPLSM that should be discussed. First, MPLSM does not enhance optical resolution. Resolution is wavelength dependent, and, in fact, less overall resolution might be expected from MPLSM, although this is not usually noticeable (for discussion, see Gu and Sheppard, 1995; Denk, 2000). By reducing background fluorescence and improving the signal-to-noise ratio, however, smaller objects that can be obscured by background fluorescence may appear brighter or more defined.

Second, while MPLSM reduces overall bleaching of the sample by limiting excitation to a small voxel, bleaching within the excited area is not reduced. In fact, in some cases accelerated rates of bleaching have been observed using two-photon excitation compared with single-photon excitation (Patterson and Piston, 2000; M. Dickinson, unpublished observation), a phenomenon that is not clearly understood.

Third, while MPLSM can often enhance the viability of embryos, IR light can sometimes be very damaging. For instance, IR light focused on pigment cells generates enough local heating to cause these cells to explode (Scott Fraser, Fraser Laboratory, unpublished observations). Because many embryos are pigmented, care should be taken to use albino embryos or to block the formation of pigment by adding propylthiouracil or another such agent before attempting to image pigment-containing cells.

2.3 Choosing Fluorescent Probes for MPLSM

2.3.1 Evaluating commonly used fluorophores

There has been substantial progress in the development of specific fluorescent probes for vital imaging due to the dramatic technical advances in fluorescence applications. In addition to fluorescence imaging techniques, other applications, such as micro- and nano-fabrication, digital data storage, high-throughput screening, fluorescence-activated cell sorting (FACS), and clinical genetic diagnostic procedures, have fueled the intense development of new fluorochromes. This progress has greatly impacted biological imaging. Presently, there are thousands of different fluorochrome derivatives and conjugates that can be used to study a variety of vital events. There are probes available to study membrane dynamics, ion concentration fluctuations, localization of organelles, cell death, cell proliferation, protein–protein interactions, signal transduction, protein and gene expression, cell lineage analysis, and changes in cellular metabolism (see Haugland [1996] and Chapters 5–7 and 15–17).

Although there has been an intense development of fluorescent dyes, many of these fluorochromes have been optimized for single-photon excitation, and scientists are just beginning to concentrate on developing optimized fluorochromes for multiphoton imaging applications. Furthermore, it is difficult to predict whether a particular dye will be useful for multiphoton imaging, so many of us have relied on trial and error while a greater understanding of the physical principles of multiphoton excitation is evolving. That being said, there are some important parameters that one can use to help predict which dyes may be best for a particular application.

Similar to the way one would choose a fluorescent probe for conventional fluorescence microscopy, the best fluorochromes for vital multiphoton imaging are the ones that efficiently absorb light, reliably emit fluorescence, and are photostable. Values such as the two-photon cross section and the fluorescence quantum yield are important ones to consider when choosing a fluorochrome.

The two-photon cross section is a measure of how strongly a fluorochrome absorbs photons at a given wavelength and is expressed in units of 10^{-50} cm^4 s/photon or 1 GM (Goeppert-Mayer unit, after the scientist who first predicted two-photon absorption, Maria Goeppert-Mayer). A comprehensive list of cross sections measured in organic solvents can be found in Xu (2000) or at www.microcosm.com, a website with several tutorials on advanced imaging. Currently available dyes have two-photon cross-section values ranging from 40 to 200 GM (see Xu, 2000).

Although these values can be used as an indication of how well a dye will perform in biological imaging using MPLSM, a one-to-one correlation between the cross section and dye performance cannot always be drawn. Measurements of two-photon cross sections are usually performed with the dye dissolved in organic solvents, and it is unclear how the cellular environment affects the performance of a particular fluorochrome. Also, these values are determined at an optimal pulse width that cannot always be achieved at the level of the specimen. Although the cross section is a valuable first approximation of how a dye will perform, trial and error provides the truest test for a particular application.

The two-photon cross section not only provides information about how well a particular molecule is excited by pulsed IR light but also indicates the two-photon absorption peak, a value that has been surprisingly difficult to predict. While it may seem logical that the two-photon absorption peak should be twice the one-photon peak, this has not proved to be the case. An emerging rule of thumb is that the peak is usually shorter (more blue shifted) than twice the one-photon absorbance peak (Xu, 2000). One should also keep in mind, however, that the brightest signal may not be obtained at the predicted excitation peak, but at the point where the excitation peak and the power peak of the laser overlap. For tunable Ti:Sapphire lasers, there is a peak output at or near 800 nm, with less overall power produced at lower and higher wavelengths. As a result, it is possible to excite a given dye at a nonoptimal wavelength by using more power. More power, however, may also mean more heat or bleaching, so one should be careful to optimize the signal-to-power ratio.

In addition to how well a molecule absorbs light, the best molecules to choose are the ones that also efficiently release light as the molecule relaxes back to the ground state. The fluorescence quantum yield is a measure of the proportion of emission photons that are shed per excitation event. In general, this value is the same for one- or two-photon excitation, and values for common fluorochromes can be found in Haugland (1996).

As mentioned above, photostability is an important factor when choosing a dye. Photobleaching is minimal in out-of-focus regions using MPLSM, but bleaching still occurs at the point of focus. In fact, some dyes that are relatively stable with one-photon excitation have more rapid bleaching rates with MPLSM (Patterson and Piston, 2000; M. Dickinson, unpublished observation). This phenomenon is not fully understood, so the bleaching rate of dyes of interest should be tested for each application. When rapid bleaching occurs, we have found that ProLong (Molecular Probes) helps stabilize the signal in fixed samples.

2.3.2 Using autofluorescence

In addition to dyes produced in a test tube by a chemist, some of the most popular dyes for MPLSM are ones produced by nature. Many studies have taken advantage of autofluorescence, inherent in unstained tissues. For instance, scientists have used two- or three-photon excitation to examine levels of nicotinamide adenine dinucleotide (and its reduced form) and serotonin in live cells (Piston et al., 1995; Bennett et al., 1996; Maiti et al., 1997; Williams et al., 1999).

2.3.3 Fluorescent proteins

The most profound impact on MPLSM, and on vital imaging in general, has been the discovery and cloning of a fluorescent protein naturally produced by jellyfish called *green fluorescent protein* (GFP) (Prasher et al., 1992; Chalfie et al., 1994). Fluorescent proteins, which are now available in many color variants, are versatile, stable, and uniform labels that have enabled researchers to study a variety of dynamic events in live cells (for review, see Tsien, 1998). While traditional dyes are introduced into cells via injection or by soaking the sample in dye, fluorescent proteins are introduced into cells as a DNA sequence that is transcribed and translated into a fluorescent protein. Fluorescent proteins are an ideal tool for studying the dynamics of gene expression in live cells and can be fused with other proteins to study a wide range of processes, such as protein–protein interactions, the distribution of cellular organelles, cell mitosis, membrane dynamics, and calcium signaling (for review, see Tsien, 1998; see also Chapters 15–17).

Fluorescent proteins can be efficiently detected using MPLSM, and they have become very popular labels. Through genetic engineering, these proteins have been mutated to develop different color variants that are optimal for one-photon excitation (see Tsien, 1998). Mutations that enhance single-photon excitation of GFP also improve the two-photon excitation (see Xu, 2000). Furthermore, certain color variants, such as EYFP, appear less fluorescent than others using two-photon excitation (R. Lansford and S. Fraser, unpublished observation), suggesting that particular mutations may adversely affect the two-photon cross section. Thus, as more is learned about the relationship between protein structure and the factors that contribute to enhanced multiphoton excitation, it may be possible to optimize fluorescent proteins further for MPLSM.

3. IMAGING DYNAMIC ASPECTS OF DEVELOPMENT

Perhaps the best way to illustrate the impact that the new technology is making on developmental biology is to recapitulate some specific experiments that have used MPLSM to study developmental processes. In this section, we review both published work and experiments in progress, highlighting the methods used, and offer references for further information. This is not meant to be a comprehensive review of the field, but rather an introduction to a few representative studies that emphasize the diversity of approaches.

3.1 A Comparison of Fluorescence Microscopy Techniques for Imaging Frog Embryos

The frog, *Xenopus laevis*, is considered one of the classic systems for studying developmental events. *Xenopus* embryos are large and easy to manipulate, and many interesting aspects of cell morphogenesis have been studied in these embryos. Fluorescence imaging is not, however, often used to study these organisms. Many of the current studies could be enhanced using time-lapse fluorescent imaging, but *Xenopus* embryos are very opaque as because the cells contain yolk granules that absorb and scatter light, making fluorescence microscopy difficult.

To determine the best way to image fluorescent cells in the frog embryo, Periasamy et al. (1999) compared digital deconvolution microscopy (DDM), confocal microscopy, and MPLSM. They labeled cells by injecting RNA encoding a GFP-GAP-43 fusion protein. This fusion protein targets GFP to the cell membrane, highlighting the boundaries between the cells. For these experiments, the authors compared the images they obtained from their own custom-configured confocal, DDM, and MPLSM units (for further details, see Periasamy et al., 1999). They found that the best images of live cells were obtained by using MPLSM.

Periasamy et al. (1999) used a tunable Ti:Sapphire laser to determine the optimum excitation wavelength and power level. An excitation wavelength of 870 nm with a power level at the specimen of 22 mW produced the best MPLSM images of GFP-GAP-43 expression. Higher and lower average power levels resulted in decreased signal-to-noise ratios, and less detail could be recognized. The authors also indicate that even further improvements of the signal-to-noise ratio can be made by combining deconvolution and MPLSM. In addition, they observed less signal deterioration because images were acquired from deeper sections with MPLSM. This is a critical difference, as studies using confocal microscopy (Wallingford et al., 2000) have been limited to the cell layer closest to the objective lens, requiring tissues to be excised and mounted between coverslip fragments in order to obtain information from deeper cells. Although the details of this comparison may depend on the systems that were used, the study represents one of the only published comparisons between all three of these popular techniques for imaging embryos and makes the point that detailed fluorescence imaging is possible in frog embryos despite the apparent challenges.

3.2 Imaging Cleavage-Stage Mammalian Embryos with MPLSM

Because mammalian development takes place within the female reproductive system, much of what we know about early development has come from static images of embryos at different developmental stages. Thus, in many cases, developmental events have been deduced through rough reconstructions, pieced together from data collected from different specimens fixed at different time points. In vitro growth conditions have been established for preimplantation embryonic stages (Bavister, 1995); however, there are limitations to studying the development of these embryos using standard light microscopy because many mammalian species are extremely light sensitive at early cleavage stages (Bavister, 1995).

Squirrell et al. (1999) recently took advantage of MPLSM to acquire images of mitochondria during the early development of hamster embryos. The methods used are detailed by the authors and are also reviewed by Mohler and Squirrell (2000). Briefly, embryos are stained with MitoTracker-Rosamine (Molecular Probes), rinsed, cultured in a drop of HECM-9 culture medium, and covered with mineral oil. Embryos are imaged in micro-drop cultures arranged on specially prepared imaging dishes. To control the temperature and gas composition of the environment, mixed gas is pre-warmed and humidified and then passed into the imaging chamber. To compensate for cooling on the stage, gas is heated to three degrees above the desired culture temperature of 37°C (for details of the imaging system, see Mohler and Squirrell, 2000).

Using this system, Squirrell et al. (1999) have been able to image mitochondrial dynamics in developing embryos for 24 hours. For these studies, they used a diode-pumped, solid-state, doubled neodymium-doped:yttrium lithium fluoride-based (Nd:YLF) laser (emitting 1047 nm, 175 fs pulses) and a custom configured multi-photon microscope (see Squirrell et al. [1999] and references therein for details about the microscope configuration).

Embryos imaged every 15 or every 2.5 min developed from two-cell to morulae and blastocyst stages in the same proportions as nonimaged controls. Moreover, embryos imaged in this way can undergo normal implantation and fetal development when transferred into pseudopregnant females. In fact, on at least one occasion, the embryos gave rise to a healthy hamster pup. This is in stark contrast to embryos imaged using visible wavelengths (514, 532, and 548 nm). None of the labeled embryos imaged using confocal microscopy developed to the morula or blastocyst stage when images were taken at 15 minute intervals for 8 hours. In fact, only half of the embryos progressed beyond the two-cell stage.

Interestingly, omission of the fluorochrome did not improve embryo development. Unstained embryos imaged in an identical way using 514, 532, or 548 nm laser light also failed to develop to morula or blastocyst stages, and only half of the embryos underwent cell division at all. Thus, toxic effects unrelated to fluorochrome excitation, and perhaps due to photo-induced peroxide production, are observed in these embryos with confocal microscopy. This can be avoided by using infrared light. These studies have provided a valuable example of the benefits of MPLSM for enhancing cell viability during imaging.

3.3 Membrane Dynamics of Epithelial Structures in Caenorhabditis Elegans Embryos

Worm embryos are particularly well suited for time-lapse microscopy. These embryos are transparent and develop quickly at room temperature, so a great deal of information can be obtained during short time-lapse sessions. In addition, simple but elegant culture chambers have been developed for optimal imaging results (see Mohler and Squirrell, 2000). Caenorhabditis elegans is also a well-studied genetic system. Many mutants are available for studying a large number of developmental processes, and transgenic worms can be generated easily, making it possible to introduce fluorescent proteins and fluorescent protein fusions to study particular cellular events. Thus, this system offers the opportunity not only to study how cells behave during development but also to link the data to molecular genetic mechanisms.

Mohler et al. (1998), in a microscopic tour de force, have used confocal microscopy, MPLSM, and transmission electron microscopy (TEM) to study cell fusion in the hypodermis of worm embryos. To study membrane dynamics in embryos, the authors used FM 4-64 (Molecular Probes), a membrane label with an excitation peak of 543 nm, and MPLSM to obtain clear three-dimensional images of membrane boundaries over time as cell fusions took place. Two-photon images of FM 4-64 were collected using a custom-configured MPLSM system with an Nd:YLF laser emitting 1047 nm pulses (see discussion of materials and methods by Mohler et al. [1998] or Squirrell et al. [1999]).

The study revealed that cell fusions quickly begin with the loss of apical adherens junctions. To study adherens junctions during fusion, transgenic worms were created that express a GFP-MH27 fusion protein, which directs GFP to the apical zonulae adherens. Confocal microscope analysis of both FM 4-64 and GFP-MH27 showed that the initial site of cell fusion is localized at or near this junction. Fusion then propagates as a wave along the length of the cell. Furthermore, TEM was used for ultrastructural analysis of the cell fusion borders by fixing the embryos at precise times during time-lapse analysis. The images showed lipid vesicles aligned at the fusion border, suggesting a mechanism for membrane breakdown. All three lines of evidence support the idea that cell fusion in the *C. elegans* embryo is mechanistically similar to virus-cell fusion.

In a more recent paper by this group (Raich et al., 1999), the authors used MPLSM to investigate the role of cadherins and catenins in the sealing of the epithelial sheets in the epidermis of the *C. elegans* embryo during ventral enclosure. As in the earlier study, the researchers used a GFP fusion protein to mark the apical zonulae adherens. Time-lapse analysis with MPLSM was then used to analyze GFP expression at the adherens junctions in mutant worms lacking maternal and zygotic cadherin, β-catenin, and α-catenin. These studies show that, as the epidermis closes, four cells at the leading edge extend filopodia toward the midline. These filopodia adhere to one another via cadherin–catenin-mediated interactions that result in the rapid formation of adherens junctions between cells. The authors have referred to this event as "filopodial priming," and it is specific to the cells at the leading edge. In contrast, cells that are subsequently involved in the fusion process, but that do not extend filopodia, can fuse in a cadherin–catenin-independent manner. These data may provide insight into other processes that utilize rapid cell sealing, such as wound healing.

In both of the studies described, the investigators successfully combined robust imaging protocols with the strong genetics of the worm. These studies involve considerable three-dimensional reconstruction of optical sections (see Thomas et al. [1996] and Mohler and white [1998] for more information on image manipulation) in order to study dynamic events in four dimensions. In addition, the researchers combined three different forms of microscopy and extensive mutant analysis to extend their observations from cellular to subcellular levels. The ability to integrate the analysis of cell dynamics and ultrastructure during embryogenesis, with a means to understand the molecular components that are involved in these processes, illustrates the great strength of the MPLSM system.

3.4 Cell Lineage Analysis of Sea Urchin Embryogenesis Using Two-Photon Uncaging of Vital Dyes

Although this chapter is focused on using MPLSM for image analysis, some studies that have taken advantage of two-photon excitation to perform localized photochemistry should be acknowledged. Caged fluorescein dextran (D-3310, Molecular Probes) can be activated by a brief pulse of ultraviolet (UV) light to produce a fluorescent lineage tracer that can easily be imaged with a confocal microscope. Fluorescent dextrans have been used in numerous cell lineage studies because, once the labeled dextrans are introduced into cells, they remain intrinsic and can only be passed on to daughter cells via cell division. Although these molecules are useful lineage tracers, they must be microinjected into cells, which is often difficult to do after early cleavage stages. The ability to fill an entire embryo with invisible dextran that can be selectively activated in certain cells of interest is a tremendous advantage. Two-photon uncaging of dextrans is a great improvement over UV photorelease. Ultraviolet light is known to produce many toxic effects in cells, and few laboratories have microscopes equipped with UV lasers that can be used to precisely focus UV light. Furthermore, conventional UV lasers do not offer a means to expose only a single cell or a small area of cells. Because two-photon excitation is limited to a very defined voxel, this method offers a safer, more controllable source for photo-uncaging.

Two-photon uncaging has been used in two studies to analyze different aspects of sea urchin embryogenesis. Sea urchin embryos and larvae are small and transparent, develop at room temperature, and can be sustained in simple salt solutions; thus, they are ideal for imaging experiments. In the first study, Summers et al. (1996) compared lineage analysis using uncaged dextran and the carbocyanine dye DiI and showed that photo-uncaging could determine the relationship between early cleavage planes and the establishment of the dorsoventral and bilateral axes. To follow the lineage of one cell of a two-cell embryo, caged fluorescein dextran was injected into the single cell zygote, and then fluorescence was activated in a single blastomere using two-photon microscopy. The caged dextran quickly became uniformly distributed throughout the single cell and was inherited by daughter cells. Then two-photon excitation was focused precisely on a part of a single blastomere, ensuring that only a single cell was labeled. For these experiments, a custom-configured two-photon microscope with a Coherent Mira tunable Ti:Sapphire laser tuned to 700 nm was used to uncage the fluorescent lineage tracer. The fluorescence from the uncaged dye was then imaged with a Zeiss 410 confocal microscope. Using this technique, investigators confirmed that the first cleavage plane is variable in its orientation with regard to the embryonic axes that later become fixed in the gastrula and pluteus larva. Moreover, two-photon photo-uncaging of fluorescein dextran resulted in greater embryo viability and reproducibility than injection of DiI at the two-cell stage.

In a second study by this group, Piston et al. (1998) used two-photon uncaging to study the dynamics of archenteron formation in sea urchin embryos. In the sea urchin embryo gastrulation occurs in two phases, primary and secondary invagination of the archenteron. The involution of ventral (vegetal) cells into the blastocoel cavity occurs during primary invagination, whereas secondary invagination involves the extension of the archenteron, which will eventually contact the animal pole (or top) of the embryo. The archenteron is thought to extend through cell proliferation and through the convergence and extension of cells within the epithelium. It had previously been assumed that gastrulation concludes after secondary invagination in some species of sea urchin (particularly *Lytechinus variagatus*) and that the cells within the archenteron are directly transformed into the larval esophagus, stomach, and intestine. To learn whether vegetal cells around the blastopore opening continue to contribute to the archenteron after secondary invagination concludes, Piston et al. (1998) used the localized uncaging of fluorescein dextran to determine the fate of cells in and around the archenteron following secondary invagination.

Fertilized sea urchin embryos were injected at the single-cell stage with caged fluorescein dextran that was inherited by all cells in the gastrula stage embryo. To activate fluorescence in a small set of cells within the developing archenteron, a custom electronic scanning mirror (for details, see Piston et al., 1998) was used, in conjunction with the system described above, to control the pattern of labeling in sea urchin gastrulae. Cells in particular regions of the archenteron and cells adjacent to the archenteron were studied over time by imaging the lineage tracer to determine the contribution of these cells to the larval gastrointestinal tract. Interestingly, cells that will contribute to the larval gut continue to be recruited into the archenteron after secondary invagination. These results extend previous fate maps, and the continued recruitment of cells into the archenteron may suggest an additional mechanism for archenteron elongation. The contribution of cells around the blastopore was variable, indicating that positional signaling, and not inherited determinants, control the contribution of cells to endodermal organs in the sea urchin larva.

3.5 Using Lipids to Introduce Green Fluorescent Protein into Embryonic Somites

One focus of our research is the development of mesodermal structures called *somites*. Somites are blocks of mesoderm cells that transiently form on either side of the neural tube. These cells ultimately give rise to axial skeletal structures, muscle cells, and cells in the dermal layer of the skin. During development, these cells undergo interesting epithelial–mesenchymal transitions and are involved in cell–cell signaling with neighboring cells in and around the neural tube (see Pourquiè, 2000; Stern and Vasiliauskas, 2000; Dockter, 2000). We have begun to investigate whether somite cells contact other cells during development via filopodial extensions. There is some evidence from invertebrate embryos that such extensions are involved in cell signaling during development (Ramirez-Weber and Kornberg, 1999).

To image fine membrane extensions, we first tried labeling chick presomitic mesoderm cells with small injections of DiI. Although many cells were labeled, DiI precipitates out of solution near the injection site, obscuring the signal from nearby labeled cells (data not shown). In addition, DiI and other carbocyanine dyes have relatively poor two-photon cross sections, making them less well-suited for MPLSM, which may be necessary for detecting these thin extensions beneath the surface of the embryo. Thus, we were interested in introducing GFP-expressing plasmids into somite cell to ensure even labeling of the somite cells and to be able to use MPLSM for vital imaging.

While GFP expression constructs can be introduced into chick embryos by viral infection or electroporation of plasmid DNA, both methods have drawbacks. Neither method makes it easy to produce localized expression; furthermore, cells infected by retroviral constructs do not have enough GFP signal for imaging until at least 24 hours after infection (see Okada et al. 1999; R. Lansford, personal communication). For these reasons, we sought to identify alternative ways to introduce GFP-expressing plasmids into mesodermal cells. With the help of Ken Longmuir and colleagues at the University of California, Irvine, we used a novel multicomponent lipid transfection system to introduce GFP expression constructs into early somite cells (Longmuir et al., 2000). This system has proved to be highly effective for plasmid transfection into tissue culture cells into chick embryo cells. This method provides an easy and quick way to introduce plasmid DNA because the lipid complexes can be concentrated to more than 1 mg plasmid DNA per milliliter for localized nanoliter delivery via injection. GFP expression is seen in these cells as quickly as 6 hours after injection.

For our experiments, a small bolus of lipid solution can be injected into the presomitic mesoderm, a group of loosely associated mesenchymal cells, which will aggregate to form the epithelial ball of cells known as a *somite*. For our initial experiments we used a plasmid transfection vector that contains a CMV enhancer and a chick β-actin promoter driving EGFP (Okada et al., 1999). We analyzed cells 20 hours after injection of the plasmid using 860 nm pulsed light from a Coherent Mira Ti:Sapphire laser delivered to a Zeiss 510 NLO system via an optical fiber (Wolleschensky et al., 1998). We observed that somite cells extend many filopodial extensions, even after the formation of an epithelial structure (Color Fig. 12–1). In addition, optical sections through the somite reveal that fine extensions reach from epithelial cells in one somite to cells in the neighboring somite. We are currently studying how these extensions are modulated during development and if they have distinct points of contact with other cells.

3.6 Optimized Fluorescent Probes for Labeling Embryonic Cells: Work in Progress

Interestingly, many of the common fluorochromes that are currently used have relatively poor two-photon cross sections (see Xu, 2000). Although many currently available fluorochromes have been used successfully for MPLSM, optimized fluorochromes will greatly enhance the sensitivity of this approach. Thus, it is important to define chemical structures that show enhanced multi-photon excitation in order to design better probes for biological imaging, as well as for other multi-photon applications. Albota et al. (1998) recently reported the design and synthesis of a number of stilbene derivatives that have greater two-photon absorption cross sections than any of the fluorochromes that are currently in use. While many conventional dyes have measured cross sections of 40–200 GM, the fluorochromes described by Albota et al. (1998) measure between 900 and 1200 GM. The results show that the two-photon cross section can be improved by enhancing the extent of intramolecular charge transfer or the distance over which charge is transferred upon excitation. Furthermore, the addition of a particular donor or acceptor group not only enhances the extent of charge transfer but also can be used to tune the excitation peak of the fluorochrome. This study was the first to describe the systematic optimization of fluorochromes for two-photon excitation and suggests that better fluorescent probes for MPLSM may be within our reach. In collaboration with the Marder and Perry groups at the Optical Materials and Technology (OMAT) center at the University of Arizona, we have begun to analyze how these dyes can be used for imaging applications in vivo and have begun to develop derivatives of these dyes that can be used for a variety of biological imaging applications.

As a first assessment of whether these dyes can be used to label cells in vivo, we took advantage of the lipophilic nature of the dyes to label cell membranes. By injecting stilbene dyes dissolved in dimethylformamide (DMF) or dimethylsulfoxide (DMSO) into the neural tube of the chick embryo, we were able to efficiently label the epithelial cells of the neural tube, as well as migrating neural crest cells (Color Fig. 12–2). These embryos were imaged with a custom-configured system (for a description, see Potter, 2000) with 840 nm pulsed excitation from a Coherent Mira Ti:Sapphire tunable laser. In these experiments, cells continued to survive up to 48 hours postinjection, and no obvious signs of cell death were present. This protocol is identical to the one we use to label neural crest cells with the carbocyanine dye DiI.

DiI has been used to study the migration of these cells with confocal microscopy and time-lapse video analysis (Sechrist et al., 1993; Kulesa and Fraser, 1998, 2000). Although DiI-labeled cells have revealed important details about the early migration patterns of emerging neural crest cells, they cannot be used to image neural crest cells as they continue to migrate into the interior of the embryo. Although MPLSM could be used to penetrate deep into the tissue, carbocyanine dyes (such as DiI and DiO) have relatively small two-photon cross sections and neural crest cells dilute the dye as they divide. Thus, signal-to-noise ratios are not adequate for detailed imaging studies. Currently, we are testing whether using MPLSM and optimized two-photon dyes will expand our ability to image these cells in vivo.

To test whether these dyes show enhanced two-photon–induced fluorescent signals in biological media as compared with commonly used dyes, we developed an immunostaining assay to compare the brightness of stilbene dyes with Rhodamine B. To do this, we tagged a particular donor-II-donor dye with biotin to take advantage of biotin–streptavidin affinity. In this assay, we incubate quail embryos with QCPN, an antibody specific for quail nuclei (Developmental Studies Hybridoma Bank, University of Iowa). A biotinylated secondary antibody recognizes the primary antibody, and then either Rhodamine B–labeled streptavidin or unlabeled streptavidin is bound to the biotinylated secondary antibody. Biotinylated stilbene dye is reacted with the unlabeled streptavidin in an attempt to saturate the biotin-binding glycoprotein with dye. Results from a typical experiment are shown in Color Figure 12–3.

Images were collected using a custom-configured system as described by Potter (2000) that uses 740 nm pulsed excitation light from a Coherent Mira Ti:Sapphire laser. By averaging the signal of many nuclei from these experiments, we determined that the biotinylated bis-donor stilbene dye, which has a two-photon cross section of 900 GM, is approximately 10 times brighter than Rhodamine B, which has a two-photon cross section of 300 GM (M. Dickinson et al., unpublished data). A tenfold greater improvement in signal intensity was somewhat unexpected given that the cross section was only threefold better. This enhancement may relate to the fact that this dye shows very little self-quenching at high concentrations (M. Dickinson and Y. Gong, unpublished observations). Studies using this prototype stilbene dye have shown very promising results, but not all derivatives are soluble in aqueous solutions. Therefore, one focus of our research is to produce dyes that are more hydrophilic and thus more compatible with biological imaging applications.

These studies have provided some important first insights into how modified stilbene dyes behave in biological media and suggest that the dyes can indeed be used to improve imaging sensitivity. We are currently developing more dye derivatives and testing their efficacy in a variety of imaging applications. In addition, Kim et al. (2000) have reported that other chemical modifications can improve the two-photon cross section of fluorochromes. Thus, development of better dyes for two-photon applications is an important focus, and biological imaging will benefit greatly from the influence of these chemical studies.

4. CONCLUSION

MPLSM is an exciting revolution in vital imaging. With all of the advantages we have discussed here, it is clear why this mode of imaging is becoming more and more popular. The truth is that the revolution has just begun. Already in the last few years, there have been significant improvements. Both the lasers and the microscopes have become more advanced, and the number of available two-photon fluorescent probes has increased. Also, there is greater availability of financial support from funding agencies, such as the National Institutes of Health and the National Science Foundation, so more users can afford to apply this technology.

Perhaps the greatest impact of this technology is that multi-photon microscope systems are becoming more user-friendly. There has been a considerable amount of effort put toward marketing laser systems that are simple to maintain and tune. Ti:Sapphire lasers, such as the Coherent Vitesse XT and the Spectra Physics Mai Tai, are now available in a limited number of fixed wavelengths or with a limited wavelength range, making it effortless for the user to select different incident wavelengths. In addition, potential users no longer have to build systems from scratch. All three major microscope manufacturers in this field (Bio-Rad, Leica, and Zeiss) are developing systems that easily bring this technology to the specimen with minimal training of the user.

Because multi-photon imaging does not require that the light from the specimen be directed back through the scan optics, the design of these systems can be very simple. They can also offer more flexibility for the inclusion of non-de-scanned detectors and more ways to filter emission wavelengths. The future is headed toward easier-to-use, more sensitive systems that will allow greater flexibility in imaging multiple fluorochromes simultaneously while collecting images over time and at greater depths along the z-axis. Moreover, these systems will be completely software controlled, making it possible for the user to monitor imaging from the luxury of a computer screen next to the microscope or over the Internet from anywhere in the world. With the help of technology, front row tickets for viewing live events during development are now becoming available for everyone.

We give special thanks to Drs. Andres Collazo and Steve Potter for their helpful comments and discussions on this chapter.

REFERENCES

Albota, M., D. Beljonne, J. L. Bredas, J. E. Ehrlich, J. Y. Fu, A. A Heikal, S. E. Hess, T. Kogej, M. D. Levin, S. R. Marder, D. McCord-Maughon, J. W. Perry, H. Rockel, M. Rumi, C. Subramaniam, W. W. Webb, X. L. Wu, and C. Xu. Design of organic molecules with large two-photon absorption cross sections. *Science* 281:1653–1656, 1998.

Bavister, B. Culture of preimplantation embryos: Facts and artifacts. *Hum. Reprod. Update* 1:91–148, 1995.

Bennett, B. D., T. L. Jetton, G. Ying, M. A. Magnuson, and D. W. Piston. Quantitative subcellular imaging of glucose metabolism within intact pancreatic islets. *J. Biol. Chem.* 271: 3647–3651, 1996.

Chalfie, M., Y. Tu, G. Euskirchen, W. W Ward, and D. C. Prasher. Green fluorescent protein as a marker for gene expression. *Science* 263:802–805, 1994.

Denk, W. Principles of multiphoton-excitation fluorescence microscopy. In: *Imaging Neurons: A Laboratory Manual,* edited by R. Yuste, F. Lanni, and A. Konnerth. Cold Spring Harbor, NY: Cold Spring Harbor Laboratory Press, 2000.

Dockter, J. L. Sclerotome induction and differentiation. In: *Current Topics in Developmental Biology,* Vol. 48: *Somitogenesis* (Part 2), edited by C. P. Ordahl. San Diego: Academic Press, 2000.

Gu, M., and C. J. R. Sheppard. Comparison of 3-dimensional imaging properties between 2-photon and single-photon fluorescence microscopy. *J. Microsc.* 177:128–137, 1995.

Haugland, R. *Handbook of Fluorescent Probes and Research Chemicals.* Eugene, OR: Molecular Probes, 1996.

Kim, O. K., K. S. Lee, H. Y. Woo, K. S. Kim, G. S. He, J. Swiatkiewicz, and P. N. Prasad. New class of two-photon–absorbing chromophores based on dithienothiophene. *Chem. Mater.* 12:284, 2000.

Kulesa, P. M., and S. E. Fraser. Neural crest cell dynamics revealed by time-lapse video microscopy of whole embryo chick explant cultures. *Dev. Biol.* 204:327–344, 1998.

Kulesa, P. M., and S. E. Fraser. In ovo time-lapse analysis of chick hindbrain neural crest cell migration shows cell interactions during migration to the branchial arches. *Development* 127:1161–1172, 2000.

Longmuir, K. J., E. A. Murphy, A. J. Waring, M. E. Dickinson, and S. M. Haynes. A nonviral liposomal complex designed to overcome the multiple barriers to gene transfer. *Mol. Ther.* 5:S243, 2000.

Maiti, S., J. B. Shear, R. M. Williams, W. R. Zipfel, and W. W. Webb. Measuring serotonin distribution in live cell with three-photon excitation. *Science* 275:530–532, 1997.

Mohler W. A., J. S. Simske, E. M. Williams-Masson, J. D. Hardin, and J. G. White. Dynamics and ultrastructure of developmental cell fusions in the *Caenorhabditis elegans* hypodermis. *Curr. Biol.* 8:1087–1090, 1998.

Mohler, W. A., and J. M. Squirrell. Multiphoton imaging of embryonic development. In: *Imaging Neurons: A Laboratory Manual,* edited by R. Yuste, F. Lanni, and A. Konnerth. Cold Spring Harbor, NY: Cold Spring Harbor Laboratory Press, 2000.

Mohler, W. A., and J. G. White. Stereo 4-D reconstruction and animation from living fluorescent specimens. *Biotechnology* 24:1006–1012, 1998.

Okada, A., R. Lansford, J. M. Weimann, S. E. Fraser, and S. K. McConnell. Imaging cells in the developing nervous system with retrovirus expressing modified green fluorescent protein. *Exp. Neurol.* 156:394–406, 1999.

Patterson, G. H., and D. W. Piston. Photobleaching in two-photon excitation microscopy. *Biophys. J.* 78:2159–2162, 2000.

Periasamy, A., P. Skoglund, C. Noakes, and R. Keller. An evaluation of two-photon excitation versus confocal and digital deconvolution fluorescence microscopy imaging in *Xenopus* morphogenesis. *Microsc. Res. Tech.* 47:172–81, 1999.

Piston, D. W., B. R. Masters, and W. W. Webb. Three-dimensionally resolved NAD(P)H cellular metabolic redox imaging of the in situ cornea with two-photon excitation laser scanning microscopy. *J. Microsc.* 178:20–27, 1995.

Piston, D. W., R. G. Summers, S. M. Knobel, and J. B. Morrill. Characterization of involution during sea urchin gastrulation using two-photon excited photorelease and confocal microscopy. *Microsc. Microanal.* 4:404–414, 1998.

Potter, S. M. Two-photon microscopy for 4D imaging of living neurons. In: *Imaging Neurons: A Laboratory Manual,* edited by R. Yuste, F. Lanni, and A. Konnerth. Cold Spring Harbor, NY: Cold Spring Harbor Laboratory Press, 2000.

Pourquiè, O. Segmentation of the paraxial mesoderm and vertebrate somitogenesis. In: *Current Topics in Developmental Biology,* Vol. 47: *Somitogenesis* (Part 1), edited by C. P. Ordahl. San Diego: Academic Press, 2000.

Prasher, D. C., V. K. Eckenrode, W. W. Ward, F. G. Prendergast, and M. J. Cormier. Primary structure of the auquorea victoria green fluorescent protein. *Gene* 111:229–233, 1992.

Raich, W. B., C. Agbunag, and J. Hardin. Rapid epithelial-sheet sealing in the *Caenorhabditis elegans* embryo requires cadherin-dependent filopodial priming. *Curr. Biol.* 9:1139–1146, 1999.

Ramirez-Weber, F. A., and T. B. Kornberg. Cytonemes: Cellular processes that project to the principal signaling center in *Drosophila* imaginal discs. *Cell* 97:599–607, 1999.

Sechrist, J., G. Serbedzija, T. Scherson, S. E. Fraser, and M. Bronner-Fraser. Segmental migration of the hindbrain neural crest does not arise from segmental generation. *Development* 118:691–703, 1993.

Squirrell, J. M., D. L. Wokosin, J. G. White, and B. D. Bavister. Long-term two-photon fluorescence imaging of mammalian embryos, without compromising viability. *Nat. Biotechnol.* 17:763–767, 1999.

Stern, C. D., and D. Vasiliauskas. Segmentation: A view from the border. In: *Current Topics in Developmental Biology,* Vol. 47: *Somitogenesis* (Part 1), edited by C. P. Ordahl. San Diego: Academic Press, 2000.

Summers, R. G., D. W. Piston, K. M. Harris, and J. B. Morrill. The orientation of first cleavage in the sea urchin embryo, *Lytechinus variegatus,* does not specify the axes of bilateral symmetry. *Dev. Biol.* 175:177–183, 1996.

Tan, Y. P., I. Llano, et al. Fast scanning and efficient photodetection in a simple two-photon microscope. *J. Neurosci. Methods* 92:123–135, 1999.

Thomas, C., P. DeVries, J. Hardin, and J. White. Four-dimensional imaging: Computer visualization of 3D movements in living specimens. *Science* 273:603–607, 1996.

Tsien, R. The green fluorescent protein. *Annu. Rev. Biochem.* 67:509–544, 1998.

Wallingford, J. B., B. A. Rowning, K. M. Vogeli, U. Rothbacher, S. E. Fraser, and R. M. Harland. Disheveled controls cell polarity during *Xenopus* gastrulation. *Nature* 405:81–85, 2000.

Williams, R. M., J. B. Shear, W. R. Zipfel, S. Maiti, and W. W. Webb. Mucosal mast cell secretion processes imaged using three-photon microscopy of 5-hydroxytryptamine autofluorescence. *Biophys. J.* 76:1835–1846, 1999.

Wokosin, D. L., B. Amos, et al. Detection sensitivity enhancements for fluorescence imaging with multi-photon excitation microscopy. *IEEE Embs.* 20:1707–1714, 1998.

Wolleschensky, R., T. Feurer, R. Sauerbrey, and I. Simon. Characterization and optimization of a laser-scanning microscope in the femtosecond regime. *Appl. Phys. B-Lasers Opt.* 67:87–94, 1998.

Xu, C. Two-photon cross sections of indicators. In: *Imaging Neurons: A Laboratory Manual,* edited by R. Yuste, F. Lanni, and A. Konnerth. Cold Spring Harbor, NY: Cold Spring Harbor Laboratory Press, 2000.

Selected Readings

MPLSM-users@egroups. com. A well organized user-list for discussion about topics related to multi-photon microscopy.

Paddock, S., editor. *Confocal Microscopy: Methods and Protocols.* Totowa, NJ: Humana Press, 1999.

Pawley, J., editor. *Handbook of Biological Confocal Microscopy.* New York: Plenum Press, 1995.

Yuste, R., F. Lanni, and A. Konnerth, editors. *Imaging Neurons: A Laboratory Manual.* Cold Spring Harbor, NY: Cold Spring Harbor Laboratory Press, 2000.

13

In vivo Diffusion Measurements Using Multiphoton Excitation Fluorescence Photobleaching Recovery and Fluorescence Correlation Spectroscopy

Warren R. Zipfel and Watt W. Webb

1. Introduction

In situ measurements of the diffusional mobility of biologically relevant molecules are critical to understanding the rate and manner in which a biological process can occur. Diffusion measurements can be made using fluorescence photobleaching recovery (FPR) methods and/or fluorescence correlation spectroscopy (FCS), depending primarily on the concentration of diffusing species.

In conventional FPR, also known as fluorescence recovery after photobleaching (FRAP), (see Chapter 7), a fraction of the fluorescently labeled molecules are rapidly photobleached using single-photon excitation by a high-intensity laser pulse of a duration much less than the diffusion time of the molecules. The bleached region is confined to a small area on a membrane (two-dimensional diffusion measurement) or other known volume (e.g., a tiny actinic cylinder formed inside a cell created by illumination with a low numerical aperture (NA) objective).

The time course of fluorescence recovery after the pulse is monitored using low intensity, and the resulting curve is fit to extract a diffusion coefficient (Axelrod et al., 1976). For an accurate determination of the diffusion coefficient (D), the profile of the illumination (I[x,y,z]) must be known. For a spot on a membrane, I(x,y) can be adequately approximated as a radial Gaussian function (Axelrod et al., 1976). In three dimensions, if the illumination point spread function I(x,y,z) is known, the distribution of bleached and unbleached molecules immediately after the pulse is theoretically calculable, and the diffusion coefficient can then be accurately determined. In practice, however, this is only true for a cylinder of illumination produced using a low NA lens in a restricted volume (e.g., Seksek et al., 1997). In general, the bleaching volume of a high NA lens in three dimensions cannot be approximated in a simple enough form to derive an analytic fitting function for the determination of three-dimensional (3D) diffusion coefficients from FPR curves obtained using single-photon excitation (Brown et al., 1999).

Photobleaching methods are useful when the concentrations of labeled diffusing species are high enough such that measurable fluorescence remains after the bleaching pulse (so that the initial portion of the recovery curve can be accurately recorded). With conventional photomultiplier tubes (PMTs) and photon-counting techniques, FPR can be utilized at concentrations as low as approximately 250 nM under certain conditions. With the recent availability of GaAsP PMTs, we have been able to collect multiphoton FPR data at a usable signal-to-noise (S/N) ratio at concentrations of about 100 nM.

For nM concentrations or lower however, FCS may be a better alternative. First invented nearly 30 years ago (Madge et al., 1972), FCS obtains information on the diffusional mobility of a fluorescent species by autocorrelation analysis of fluorescence intensity fluctuations. The technique is analogous to dynamic light-scattering methods used to determine the macromolecular size in pure solutions, but it is more sensitive and can be used in complex mixtures as long as only the species of interest is fluorescent. If the number of molecules in the illuminated volume is small enough, fluorescence fluctuations due to diffusion of molecules in and out of the volume become measurable and the diffusion coefficient determinable. Single-photon excited FCS using confocal collection optics, first demonstrated by Koppel et al. (1976), produces a small enough focal volume to make FCS practical. Coupled with high quantum efficiency photodetectors, such as avalanche photodiodes, high-quality correlation curves can be obtained in vitro at femtomolar concentrations (Eigen and Riger, 1994; Maiti et al., 1997a).

More recently, both of these methods have been carried out using two-photon excitation (Brown et al., 1999; Berland et al., 1995; Mertz et al., 1995) in which two low-energy (infrared) photons are absorbed, simultaneously causing the emission of a single higher energy (bluer) fluorescence photon. Although the term *multiphoton excitation* (MPE) is used in this chapter, unless otherwise noted, it refers to a two-photon excitation event. Three-photon excitation is also possible (Maiti et al., 1996) and can be used for multiphoton FPR (MPFPR).

2. OVERVIEW OF MPFPR AND MPFCS

2.1 Multiphoton Excitation Focal Volume

The advantages that MPE brings to FPR and FCS stem from the natural 3D localization of the excitation volume inherent in nonlinear excitation; that is, the excitation depends on I(x,y,z) squared. The multiphoton focal volume is both mathematically and spatially well defined, and this property simplifies the accurate determination of the diffusion coefficient. Figure 13–1 demonstrates the 3D localization of excitation obtained with MPE compared with single-photon excitation.

The dimensions of the localized region of excitation created by MPE are needed to obtain diffusion coefficients from MPFPR and multiphoton FCS (MPFCS), and these values can be calculated knowing the NA of the objective lens and the excitation wavelength. The axial and lateral profiles of the square of I(x,y,z) are shown in Figure 13–1 (solid circles), calculated by using the Integral method of

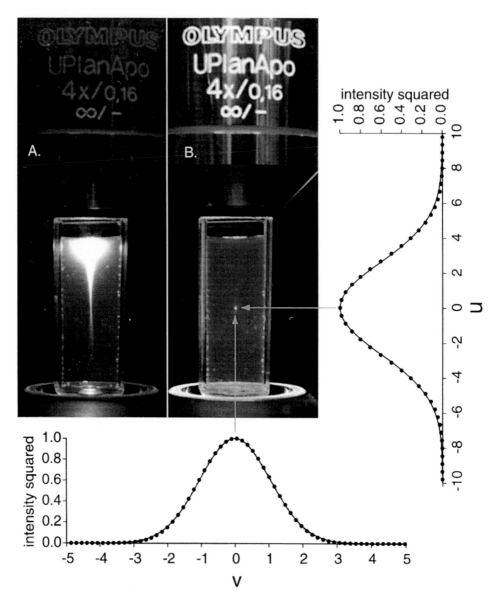

FIGURE 13–1 Localization of the multiphoton excitation. *A*, Single-photon excitation. The total number of excitations is equal in any x–y plane of the single-photon illumination point spread function; the decreasing fluorescence shown in *A* is due to absorption of photons along the z-axis. Calculation of the bleaching profile is extremely difficult. *B*, Multiphoton excitation (two photon in this case) is localized in all three dimensions, and the intensity squared profile is well fit by Gaussian functions. Side and bottom plots: Circles indicate calculated intensity squared in optical units ($v = [2\pi/\lambda (x^2 + y^2)^{1/2} n \sin(\alpha)]$ and $u = [2\pi/\lambda z n \sin^2(\alpha)]$) calculated according to Richards and Wolf (1959). α is arcsin(NA/n), where n is the index of refraction of the immersion medium. The solid line indicates the best-fit Gaussian function.

Richards and Wolf (1959). Best-fit Gaussian functions of the central lobe in the axial and lateral directions (lines) are also plotted. The use of a 3D Gaussian function to describe the dimensions of the MPE focal volume allows for derivation of simple analytic expressions that accurately describe MPFPR and MPFCS experiments. The lateral $1/e^2$ radius is well described by

$$NA < \sim 0.7: \omega_{xy} = \frac{0.32\lambda}{\sqrt{m}NA} \quad NA > 0.7: \omega_{xy} = \frac{0.325\lambda}{\sqrt{m}NA^{0.91}} \tag{1}$$

where m is the excitation order (e.g., $m = 2$ for two photon) and ω_{xy} is the average of the x- and y-axes for polarized light. The axial $1/e^2$ radius is given by

$$\omega_z = \frac{0.266\lambda}{n_i\sqrt{m}\sin^2(\alpha/2)} = \frac{0.532\lambda}{\sqrt{m}}\left[\frac{1}{n_i - \sqrt{n_i^2 - NA^2}}\right] \tag{2}$$

where α is the arcsin (NA/n_i) and n_i is the index of refraction of the immersion media (and ideally the sample). Both forms in Equation 3 are equivalent. All equations in this section assume diffraction-limited optics, that is, the objective lens entrance pupil (back aperture) is overfilled. In practice this requires that the $1/e$ diameter of the beam be at least equal to the entrance pupil diameter.

The size of the focal volume can also be readily calculated by integration over all space:

$$\int_x\int_y\int_z e^{-\left[\frac{2x^2}{\omega_{xy}} + \frac{2y^2}{\omega_{xy}} + \frac{2z^2}{\omega_z}\right]} dx\, dy\, dz = \left(\frac{\pi}{2}\right)^{3/2} \omega_{xy}^2 \omega_z \tag{3}$$

The above volume approximation is 0.68 of the value calculated by numerical integration of the two-photon illumination spread function (calculated according to Richards and Wolf, 1959), so Equation 3 should be further multiplied by 1.47 for the best approximation of the effective excitation volume.

2.2 Multiphoton Fluorescence Photobleaching Recovery

Figure 13–2 illustrates the basic principle of MPFPR. The fluorescence of a labeled probe is monitored at low intensity for a short period to establish F_o. The monitoring intensity must be low enough so that no measurable bleaching occurs. A brief, high-intensity pulse is given, bleaching a fraction of the molecules in the focal volume instantaneously, relative to the rate of diffusion in and out of the volume element. In practice this usually means a submillisecond pulse when 3D diffusion measurements are being carried out unless the labeled molecules are extremely high molecular weight and/or in a viscous or highly interacting environment. Immediately after the bleaching pulse, the fluorescence recovery is monitored at the same intensity used to measure F_o.

For the case of a single freely diffusing species in three dimensions, MPFPR data is fit to the equation given by Brown et al. (1999), which is a generalized

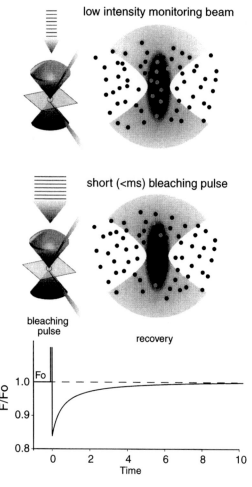

FIGURE 13–2 Multiphoton florescence photobleaching recovery (MPFPR). A short photobleaching pulse is used to create a concentration gradient in the fluorescent species, and the recovery kinetics of fluorescence is measured to extract a diffusion coefficient. F/F_o is the fluorescence after the bleaching pulse divided by the mean pre-bleach fluorescence.

form for the *mth* photon excitation of fluorescence and the *bth* photon photobleaching (two or higher). For two-photon excitation and two-photon bleaching, the equation becomes

$$\frac{F}{F_o} = \sum_{n=0}^{\infty} \frac{(-\beta)^n}{n!} \frac{1}{\left(1 + n\left(\frac{b}{m} + \frac{4bDt}{\omega_{xy}^2}\right)\right)\sqrt{1 + n\left(\frac{b}{m} + \frac{4bDt}{\omega_z^2}\right)}}$$

$$= \sum_{n=0}^{\infty} \frac{(-\beta)^n}{n!} \frac{1}{\left(1 + n\left(1 + \frac{8Dt}{\omega_{xy}^2}\right)\right)\sqrt{1 + n\left(1 + \frac{8Dt}{\omega_z^2}\right)}} \qquad (4)$$

where the bleaching parameter (β) and the diffusion coefficient (D) are determined using the Marquardt-Levenburg algorithm. Summation to $n = 6$ in the nonlinear fit is sufficient for convergence.

In vivo diffusion is most always more complex than simple Fickian (or normal) diffusion due to the presence of binding and barriers in the cell or tissue. It is usually necessary to fit in vivo MPFPR data to either multiple-component models (Periasamy and Verkman, 1998), which assume two or more diffusion coefficients, or to an anomalous diffusion model, in which $8Dt$ is replaced by Γt^α (Bouchard and Georges, 1990). In anomalous diffusion the mean squared distance is no longer linearly proportional to time, but to time raised to a power α. In the absence of directed or facilitated transport, α is usually less than or equal to 1.0 (anomalous subdiffusion) in cells. Γ is the transport coefficient with dimensions of length^2 time$^{-\alpha}$. A time-dependent effective diffusion coefficient can be calculated as $D_{eff}(t) = \Gamma t^{\alpha - 1}$.

Data can be fit to the two above models by expanding the fitting function (Eq. 4) to include additional weighted diffusion coefficient terms or by substituting Γt^α for Dt (Feder et al., 1996; Brown et al., 1999). An immobile fraction can also be added to Equation 4 for a model assuming normal diffusion with a fraction of fluorophores that are tightly bound.

2.3 Multiphoton Fluorescence Correlation Spectroscopy

Figure 13–3 illustrates MPFCS. The correlation function $G(\tau)$ is usually directly obtained with a digital autocorrelator that performs a real-time correlation on the detector pulses. τ is the correlation lag time. In practice a modified autocorrelation is used to improve the numerical performance and statistical accuracy at long lag times. The most common form is the following symmetrical normalization scheme (Schatzel et al., 1988):

$$G(j) = \frac{N \sum_{i=j}^{N} n_i n_{i-j}}{\sum_{i=0}^{N} n_i \sum_{i=j}^{N} n_{i-j}} - 1 \tag{5}$$

where there are N time bins of n_i photons in each bin and τ equals j times the bin size (time).

Other sources of fluorescence fluctuation, such as triplet state formation (Widengren et al., 1995), other reversible dark states (Haupts et al., 1998), or chemical kinetics (Magde et al., 1974), will also have characteristic times that will be reflected in the correlation curve. Models that account for these processes can be found in the above references and in Schwille et al. (1999), which provides a detailed overview of both single-photon and multiphoton FCS in cells. The simplest case, a single diffusion coefficient $G(\tau)$ in the absence of other processes, is fit to the following form:

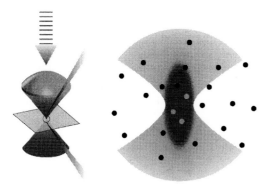

Spontaneous fluorescence fluctuations:

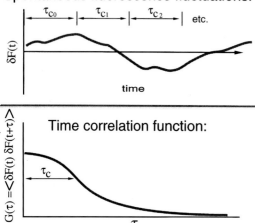

Time correlation function:

FIGURE **13–3** Multiphoton florescence correlation spectroscopy (MPFCS). Fluorescence fluctuations at low fluorophore concentrations due to diffusion are used to obtain a characteristic time (τ_c) from which a diffusion coefficient can be obtained. (For more details, see text.)

$$G(\tau) = G(0) \cdot \frac{1}{\left[1 = \frac{4D\tau}{\omega_{xy}^2}\right]} \cdot \frac{1}{\sqrt{1 + \frac{4D\tau}{\omega_z^2}}} \tag{6}$$

where $G(0)$ is inversely proportional to N, the average number of molecules in the FCS observation volume; and ω_{xy} and ω_z are the $1/e^2$ widths of the point spread function squared (Eqs. 1 and 2).

The characteristic diffusion time, τ_D, which is often used in the FCS literature, is given by $\tau_D = \omega_{xy}^2/4D$. Because the two-photon excitation volume is well defined, we prefer to fit for D directly using known values for the $1/e^2$ radii, reducing the ambiguity in the fit. FCS has the unique property of returning not only a diffusion time, but the concentration of probe as well. $G(0) = 1/N$ where N is the number of molecules in the observation volume. The FCS observation volume differs from the illumination (or excitation) volume discussed in section 2 by the factor γ (Thompson, 1991; Mertz et al., 1995), where $\gamma = 0.26$ for the diffraction-limited two-photon case (Xu and Webb, 1997). Therefore, $C = \gamma /(G(0)N)_e$ if the calculations presented in section 2 are used to estimate the excitation volume. $G(0)$ may also contain a contribution from noncorrelatable signal (e.g., shot noise) and can be corrected as discussed below. For in vivo MPFCS measurements, multiple diffusing species models can be tested by fitting to a weighted sum of two or more terms of Equation 6. Raising τ to a fitting parameter α in Equation 2 produces an anomalous diffusion FCS model.

3. INSTRUMENTATION REQUIREMENTS

3.1 Excitation Source

Multiphoton excitation requires a pulsed laser in order to achieve nonlinear excitation at low average powers. For multiphoton microscopy (MPM), mode-locked lasers that produce pulses of approximately 100 fs (10^{-13} s) duration at 80 MHz are typically used, although picosecond lasers (10^{-12} s) have also been applied. The probability of a two-photon absorption event is small, and, without the bunching of photons into extremely short pulses, MPE would not be practical in biological systems. For example, it would require 1.5 W of continuous wave laser power to achieve the same number of excitations as 5 mW of 100 fs pulses.

The most common type of laser for MPM, and therefore for MPFPR and MPFCS, is a mode-locked Ti:Sapphire laser, which can produce 400 to more than 1000 mW from about 700 to 1000 nm, depending on the laser optics and pump laser. The two most common sources for mode-locked femtosecond laser systems are Spectra-Physics Lasers (Mountain View, CA) and Coherent Inc. (Sunnyvale, CA). The total loss through the scanning system can be 80% or higher, including objective lens overfilling and transmission losses in the infrared (especially at >900 nm). For this reason, relatively high laser power is required both for imaging in thick specimens (a routine use for MPM) and for ensuring adequate power for photobleaching during the short pulse duration required for quantitative MPFPR measurements. In most cases, it typically requires that a minimum of 10 mW at the sample be available during the bleaching pulse to achieve enough photobleaching in a time much less than the diffusion time.

3.2 MPFPR-Specific Instrumentation

3.2.1 Pockels cell

The pulse train required for MPFPR is best produced using an electro-optic device known as a Pockels cell. Acousto-optic devices (AOMs) are generally too dispersive for use with a femtosecond pulses, and mechanical shutters are too slow for 3D diffusion measurements. A Pockels cell is an optical modulator capable of adjusting the laser intensity between approximately 0 and 95% of the maximum, with rise times in the microsecond range or faster. The least expensive type of Pockels cell, which is sufficient to create square pulses of duration around 10 μs, is sufficient for MPFPR. (More expensive models are capable of picking single pulses from a mode-locked laser running at 80 MHz.)

Pockels cells utilizing potassium dideuterium phosphate (KD*P) crystals have minimal dispersion in the infrared region compared with other crystal types (e.g., ADP or LBO). Piezoelectric resonances in the KD*P can, however, cause a pronounced ringing in the intensity after a short pulse is generated, causing fluctuations in the monitoring intensity immediately after the pulse. To our knowledge, only KD*P-based modulators from one source (ConOptics Inc, Danbury, CT) are free of this problem.

3.2.2 Detectors

In vivo MPFPR is most optimally implemented on a multiphoton microscope so that the specimen can be imaged and the location of the diffusion measurement recorded. The scanner is parked at the desired location in the cell or tissue, and the measurement is carried out. The same detectors used for multiphoton imaging can be used for MPFPR. Multiphoton microscope detectors are usually PMTs run in analog mode. For MPFPR, however, it is advantageous to use photon-counting mode, which requires a discriminator to convert the analog PMT output to TTL pulses for counting.

Gating of the PMT during the bleaching pulse is usually not required because the bleaching pulse duration is short. A simple test for artifacts arising from the absence of PMT gating is to measure a dye such as rhodamine in solution (rhodamine is difficult to bleach, and relatively high intensity is required to bleach during the 10 or 20 μs pulse). If the PMT response becomes nonlinear immediately after the bleaching pulse due to bright fluorescence during the pulse, the recovery will deviate from a single component fit. Test solutions should always be used to check the functioning of an MPFPR system in general because free diffusion of a low-molecular-weight dye in solution is one of the more difficult measurements to carry out.

3.2.3 Driving pulse generation and data acquisition

The required electronics for Pockels cell control and data acquisition can be built in the laboratory or acquired as a collection of generic instrumentation put together to perform the measurements (e.g., a pulse generator and multi-channel scalar), or a complete MPFPR system consisting of a Pockels cell, Pockels cell driver, and acquisition hardware and software can be purchased. Required specifications include the following:

1. Temporal control of the laser intensity via the Pockels cell with accuracy in the microsecond range is needed. For diffusion measurements of small molecules in three dimensions, a square pulse as short as 10 μs is sometimes required.
2. The equipment must be able to monitor fluorescence recovery by photon counting with flexible bin sizes. This is required because recovery times can vary from less then a millisecond to seconds. Repeat traces should also be able to be averaged and displayed in real time during acquisition.

3. A final useful option in an in vivo MPFPR system is the ability to use patterned illumination and data acquisition during fluorescence recovery measurements to minimize photodamage and bleaching. It is not always necessary to acquire continuous data, and the equipment should allow for continuous acquisition early in the recovery (to ~60% recovery) followed by short periods of acquisition in between which the laser intensity can be brought to minimum intensity.

3.3 MPFCS-Specific Instrumentation

3.3.1 Detection

As with MPFPR, MPFCS is best carried out on a scanning multiphoton microscope so that the specimen can be imaged with the same illumination and excitation optics as used for the point FCS measurements. In this way the position within a cell or tissue will be precisely known. For FCS, photon counting is required, and high quantum efficiency detectors should be used. The conventional bi-alkali or multi-alkali PMTs used in a multiphoton microscope have quantum efficiencies ranging from about 10% to 28% in the visible range, which are usable for MPFCS but are not optimal (see Chapter 3).

Single-photon FCS with confocal detection is usually carried out with an Avalanche Photodiode (APD), which have quantum efficiencies of approximately 70% at their peak, but extremely small active areas (180 μm diameter for the EG&G SPCM). For single-photon FCS, focusing optics are required to achieve confocal detection and the active area of the APD is on the order of the confocal spot size, so it does not present any special problem. With MPFCS, however, the complexity of the collection optics can be reduced because there is no need for confocal detection. The small active area of an APD often makes alignment more difficult than necessary. In addition, the high sensitivity of an APD in the infrared requires the use of multiple blocking filters to reject stray red light, which reduces the overall collection efficiency, as well.

An alternative detector for MPFCS is a GaAsP PMT (Hamamatsu H7421-40), which has a 5 × 5 mm photocathode (active area), about 42% quantum efficiency at 550 nm, and low quantum efficiency in the infrared. Figure 13–4 shows the quantum efficiencies of these two devices. Although there is a significant difference between the quantum efficiencies in the 500–650 nm range, in practice we

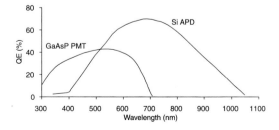

FIGURE 13–4 Quantum efficiencies (QE) comparison between an Si Avalanche Photodiode (APD) and a GaAsP photo-multiplier tube (PMT).

find that this is often not realized. The detection system's effective quantum efficiency will often be reduced to below what is obtainable with a GaAsP PMT. For example, in the green region an APD has approximately 48% quantum efficiency compared with 42% for a GaAsP PMT. A typical shortpass filter that blocks the excitation and passes the green (e.g., 550DF150), however, has 80% transmittance in the 500 nm region. If one additional filter needs to be added to an MPFCSsystem using an APD to reject the infrared excitation, the effective quantum efficiency drops to about 30% compared with 33% for a system using a GaAsP PMT. Additional losses can also occur due to extra optics needed to achieve a tight focus on to the small active area of the APD. In terms of dark noise, both detectors are equivalent (typically <200 cps), and the GaAsP PMT is currently the best choice for a MPFCS detector.

3.3.2 Other considerations for an optimal MPFCS system for in vivo work

First, the same detector used for MPFCS should also be capable of being used for scanned image acquisition, as well. If the exact location of the FCS data acquisition within the cell needs to be known (as is usually the case), the system must be capable of acquiring an image of the specimen loaded at low concentrations of labeled probe. Because photon counting detectors (APDs and GaAsP PMTs) output TTL pulses, provisions should be made to interface their output to the analog inputs of the scanner.

Second, the scanner should be stable so that, when parked, it will not introduce artifacts into the data due to residual motion of the galvanometers. The simple test for stability is to park the beam and acquire an MPFCS curve of a solution of a fluorescent dye. At low intensity, it should fit well into a single-component model for normal diffusion and return a reasonable diffusion coefficient based on its molecular weight. (Rhodamine-Green is a common standard that should yield $D = {\sim}2.8 \times 10^{-6}$ cm^2 s^{-1}.) A more stringent test involves FCS measurements of fluorescently labeled beads stuck on a surface. There should be no correlation function because there is not diffusion of fluorophores in and out of the excitation volume. Other sources of nonrandom fluctuations, however, such as stage vibration, can produce a correlation, and this test is not always totally conclusive.

3.3.3 Acquisition

The autocorrelation curve is usually obtained using digital corrector cards that were originally designed for light-scattering experiments. These cards either continuously count the number of photons in a short time period (typically 200 ns or less) or measure the time between incoming photon pulses. In either case, the lag time (τ) in the autocorrelation calculation is usually incremented in a logarithmic fashion because τ must cover several orders of magnitude (nanoseconds to a few seconds). Sources of digital autocorrelator cards include ALV (Langen, Germany), Correlator.com (Brigdewater, NJ), and ISS, Inc. (Champaign, IL). It is also possible to use other types acquisition cards (e.g., generic high-speed counter cards) with laboratory-written software to perform the correlation.

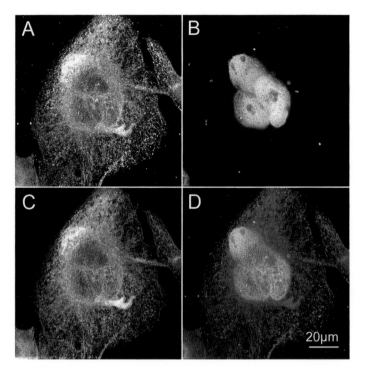

COLOR FIGURE 2–1 Triple labeling of Hela cervical carcinoma cells transfected with *following antibodies* to overexpress the protein ISG60. *A,* Alexa 488-linked secondary antibody staining (488/530 ± 30 nm excitation/emission). *B,* Propidium iodide (568/590 ± 30 nm ex./em.). *C,* Cy5-linked anti-alpha-tubulin antibody <(647/LP680 nm ex./em.). *D,* RGB color overlay produced from placing *A, B,* and *C* into green, red, and blue color planes, respectively. This clearly demonstrates co-localization (in cyan) of the two antibodies. LP, long pass.

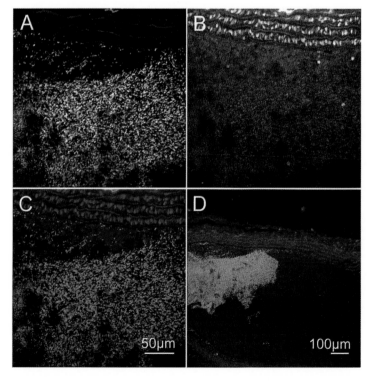

COLOR FIGURE 2–4 Cell tracking using CMFDA. Rat platelets were isolated and loaded with CMFDA (1 μM, 30 min at 37°C) and then injected back into the circulation of a second rat, which was subsequently subjected to crush injury of the carotid artery. The carotid was rapidly removed, fixed in formaldehyde, and counterstained with 1 μM propidium iodide. Dual-channel imaging facilitates analysis of platelet adhesion and coagulation in the occlusive thrombus (CMFDA channel, *A*), the vessel wall (propidium iodide channel, *B*) and hence establishes the location of the thrombus within the vessel (*C* and *D*). *A–C* are optical sections, and *D* is a projection of a partial series used to calculate volume. [Unpublished work by I. Harper, K. Heel, and S. Jackson, Department of Medicine, Monash University.]

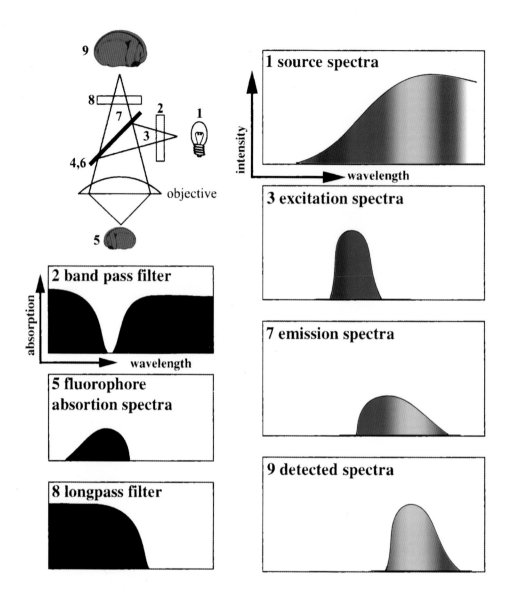

COLOR FIGURE 4–2 Spectral properties of light along the optical path of an epifluorescent microscope. Schematic of the optical train of an epifluorescence microscope. *1, 3, 7, 9,* Hypothetical spectra at different points along the optical train from the source to the detector. *2, 8,* Hypothetical absorption curves of the excitation and emission filters. *5,* Hypothetical absorption spectra of the fluorophore for excitation wavelengths. *4, 6,* The effects of the dichroic beam splitter are omitted for simplicity, but it could be thought of as a perfect longpass filter that reflects the excitation spectra from the light source toward the sample and passes the emitted spectra toward the detector (*9*).

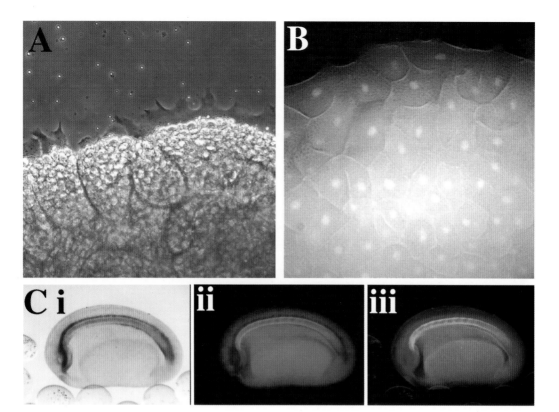

Color Figure 4–4 Practical imaging examples. *A*, An example of a phase contrast image showing a migrating sheet of embryonic mesoderm from the frog *Xenopus laevis.* Phase contrast highlights protrusions extending from the leading edge of the sheet but does not allow visualization of boundaries between cells. This image was taken with the ORCA camera and Metamorph software. *B*, Example of an epifluorescence image. With the same camera and software used for the phase contrast mage, the epifluorescence image is taken of the same type of sample as in *A*, but cells have been previously loaded with rhodamine-dextran-amine (Molecular Probes). Contrast from the fluorophore highlights nuclei as well as cell boundaries. Note that the protrusions visible in the phase contrast image are no longer seen. *C*, Example of stereoscope-collected images using visible light, epifluorescence, and a combined image. *i*, A gene expression pattern of sonic hedgehog in a *Xenopus* embryo at mid-neurulation. *ii*, Protein localization of the tor70 epitope marking notochord in the same embryo. *iii*, Merged image where the in situ pattern of panel *ii*, has been rendered as a pseudofluorescent image in the red channel and combined with the green channel of the fluorescent image of panel *i*, using Adobe Photoshop (version 5.5).

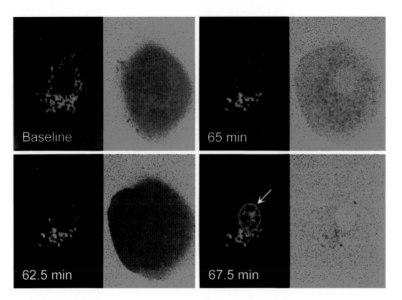

COLOR FIGURE 5–2 Calcein entry and propidium iodide nuclear labeling at onset of cell death in a cultured rat hepatocyte. A rhodamine-dextran–labeled hepatocyte was incubated with 300 μM calcein and 3 μM propidium iodide. In the baseline images, green calcein surrounds the hepatocyte, whereas the only red fluorescence is that of rhodamine-dextran within lysosomes. After chemical hypoxia with 2.5 mM KCN and 0.5 mM iodoacetate for 62.5 min, cell swelling was marked and lysosomes began to disintegrate, but both calcein and propidium iodide remained excluded. After 65 min, calcein was just beginning to enter the hepatocyte. By 67.5 min, calcein had equilibrated between the intracellular and extracellular compartments, and nuclear staining with red-fluorescing propidium iodide was intense (arrow). [Adapted from Zahrebelski et al., 1995.]

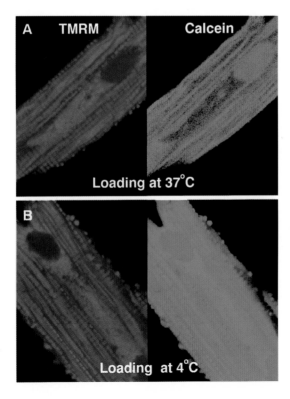

COLOR FIGURE 5–7 Temperature dependence of the ester-loading of calcein. Myocytes were loaded with calcein AM for 1 hour at 37°C (A) or for 2 hours at 4°C (B) and then with tetramethylrhodamine methyl ester (TMRM). The red fluorescence of TMRM (left) and the green fluorescence of calcein (right) were imaged simultaneously. Note that after warm loading, green calcein fluorescence was cytosolic and did not co-localize with TMRM-labeled mitochondria. After cold loading, calcein was present in both the cytosol and the mitochondria with a slight mitochondrial predominance. [Adapted from Ohata et al., 1998.]

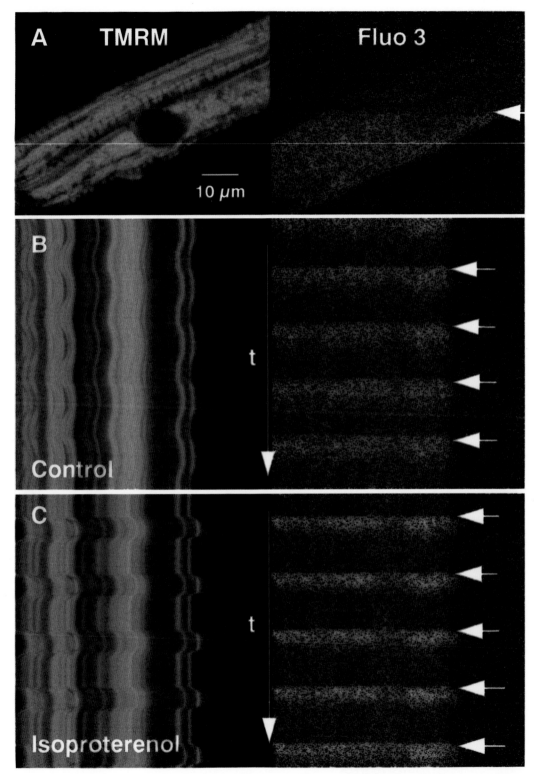

COLOR FIGURE 5–8 Mitochondrial and cytosolic Ca^{2+} transients during field stimulation. A myocyte was loaded with Fluo-3 and tetramethyl rhodamine methyl ester (TMRM) at 4°C. x–y images of red TMRM (*A*, left) and green Fluo-3 (*A*, right) fluorescence and the effect of a single field stimulation are shown. Subsequently, a 10 s line-scan image was collected during continuous stimulation at 0.5 Hz (*B*). The white line in *A* indicates the location of the line scan. Isoproterenol (1 μM) was then added, and another line scan was collected (*C*). [Adapted from Ohata et al., 1998.]

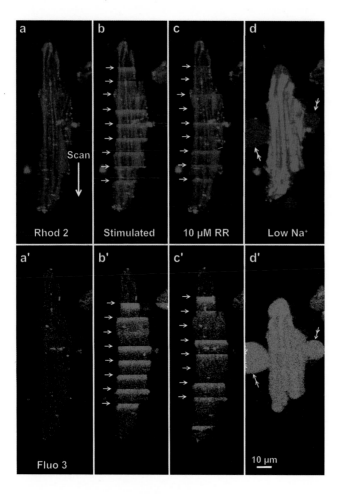

COLOR FIGURE 5–10 Visualization of cytosolic and mitochondrial Ca²⁺ transients in myocytes co-loaded with Rhod-2 and Fluo-3. Myocytes were labeled with Rhod-2 AM by the cold loading/warm incubation protocol. After warm incubation, the myocytes were then loaded with Fluo-3 AM at 37°C. Red Rhod-2 fluorescence excited by 568 nm light (upper panels) and green Fluo-3 fluorescence excited by 488 nm light (lower panels) were collected in sequential 16 sec scans. In *a*, before stimulation, Rhod-2 fluorescence showed a mitochondrial pattern, whereas in *a'* Fluo-3 fluorescence showed predominantly a diffuse pattern, except for small spots of lysosomal fluorescence. In *b* and *b'*, both Rhod-2 and Fluo-3 fluorescence showed horizontal banding when confocal images were collected during electrical stimulation, which indicated mitochondrial and cytosolic Ca²⁺ transients, respectively (arrows). In *c*, Rhod-2 fluorescence transients were suppressed after addition of 10 μM ruthenium red (RR), but in *c'* the Fluo-3 fluorescence transients remained. Electrical stimulation was close to threshold, and some electrical stimulations failed to induce Ca²⁺ transients. Subsequently, K⁺-substituted, low Na⁺ KRH was added to increase intracellular Ca²⁺ by reverse Na⁺/Ca²⁺ exchange (*d* and *d'*). In *d*, Rhod-2 fluorescence increased inside mitochondria, but not in cell surface blebs (*d*, double arrows), whereas in *d'* Fluo-3 fluorescence increased in both the cytosol and blebs (*d'*, double arrows). [Adapted from Trollinger et al., 2000.]

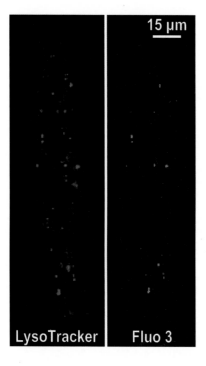

COLOR FIGURE 5–11 Lysosomal localization of Fluo-3 after ester loading. Myocytes were loaded with Fluo-3 AM by the cold loading/warm incubation protocol and then incubated with LysoTracker Red to label lysosomes. LysoTracker Red (left) and Fluo-3 (right) co-localized in heterogeneous punctate structures that represent lysosomes and associated acidic endosomes. [Adapted from Trollinger et al., 2000.]

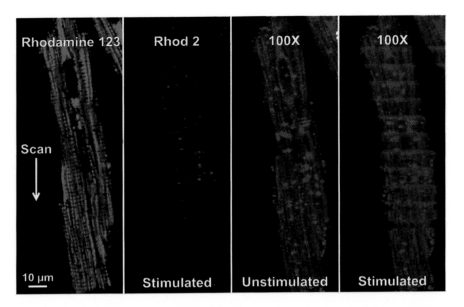

COLOR FIGURE 5–12 Discrimination of lysosomal versus mitochondrial loading of Ca^{2+}-indicating fluorophores. A cardiac myocyte was loaded with Rhod-2 by the cold ester-loading/warm incubation protocol and then with rhodamine-123. Green Rhodamine-123 fluorescence identifies mitochondria within the myocyte (*A*). A red fluorescence image of Rhod-2 was then collected during electrical stimulation at 1 Hz (*B*). Laser excitation was attenuated so that none of the pixels was saturated (gray level <255). This Rhod-2 image showed heterogeneous punctate fluorescence that did not correspond to the rhodamine-123–labeled mitochondria. Moreover, no fluorescence transients occurred during the electrical stimulation. The myocyte was imaged again at 100× more excitation energy, but without electrical stimulation (*C*). At 100-fold greater excitation energy, fluorescence arising from lysosomes was now saturated, and a mitochondrial pattern of fluorescence emerged as a background to the original lysosomal pattern. An identical image was then collected, except during electrical stimulation at 1 Hz (*D*). The background mitochondrial fluorescence then showed horizontal banding that indicated Ca^{2+} transients. [Adapted from Trollinger et al., 2000.]

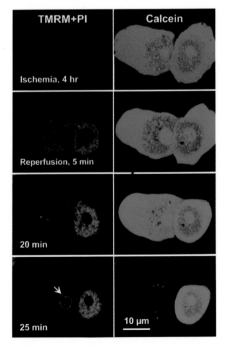

COLOR FIGURE 5–13 Onset of the mitochondrial permeability transition after ischemia/ reperfusion. Rat hepatocytes were loaded with tetramethyl rhodamine methyl ester (TMRM) and calcein, and red (TMRM, left) and green (calcein, right) fluorescence images were collected by confocal microscopy at the end of 4 hours of anoxia at pH 6.2 (simulated ischemia) and after 5, 20, and 25 min of reoxygenation at pH 7.4 (simulated reperfusion). Virtually all mitochondrial TMRM fluorescence was lost after 4 h of ischemia, indicating mitochondrial depolarization. Dark mitochondrial voids in the calcein fluorescence remained, indicating that mitochondrial permeabilization had not occurred. Within 5 min after reperfusion, TMRM began to enter the mitochondria of both hepatocytes. After 20 min, the hepatocyte to the left lost TMRM labeling, indicating mitochondrial depolarization. Simultaneously, the voids of calcein fluorescence were lost, indicating onset of the mitochondrial permeability transition. The hepatocyte subsequently lost viability after 25 min, as shown by loss of calcein fluorescence and nuclear uptake of propidium iodide (PI; arrow). In the hepatocyte to the right, TMRM continued to accumulate, and the dark voids in the calcein fluorescence persisted. [Adapted from Qian et al., 1997.]

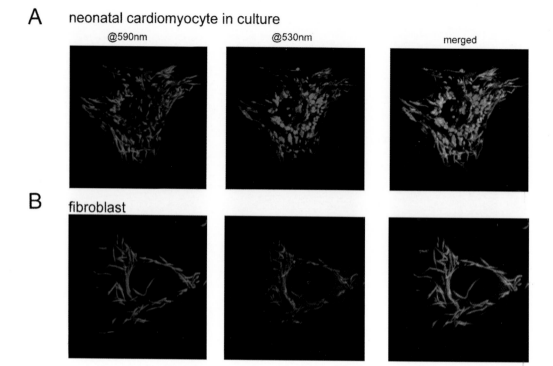

A neonatal cardiomyocyte in culture

@590nm @530nm merged

B fibroblast

COLOR FIGURE 6–6 Confocal images of JC-1 signals in *A*, a neonatal cardiomyocyte in culture and *B*, a fibroblast in the same culture. Cells were loaded with JC-1 under standard conditions (10 μM for 10 min followed by washing), and then images were acquired with a confocal system (Zeiss 510) with excitation at 488 nm and emission measured on two channels at 505–530 nm and at >585 nm displayed as green (*i*) and red (*ii*), respectively. The panels in *iii* show the overlay of the two signals. See text for further discussion. Bars, 10 μm (63× oil; numerical aperture, 1.4).

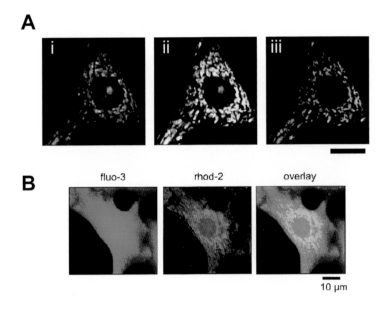

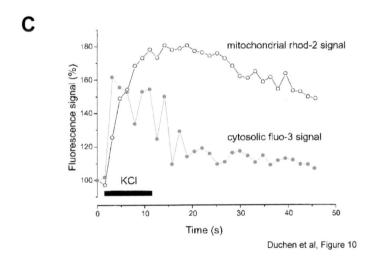

Duchen et al, Figure 10

COLOR FIGURE 6–10 Rhod-2 and Fluo-3. *A,* Images of a neonatal cardiomyocyte in culture loaded with Rhod-2 before (*i*) and after (*ii*) stimulation to raise $[Ca^{2+}]_c$ (depolarization with 50 mM KCl). The mitochondrial localization of the change in signal is emphasized by examining the pixels in which the signal has changed (*iii*) by subtracting the first image from the second, showing clearly that only the mitochondrial Rhod-2 signal has changed significantly. *B,* Neonatal cardiocytes loaded with Fluo-4 and Rhod-2 to measure changes in cytosolic and mitochondrial $[Ca^{2+}]$, respectively. The overlay of the two channels is also shown. *C,* Membrane depolarization with 50 mM KCl raised $[Ca^{2+}]_{cyt}$ (closed circles) and mitochondrial $[Ca^{2+}]$ (open circles), and the time course and relative amplitudes of the signals originating from each dye plotted from an area immediately over two mitochondria were clearly different. These images were acquired with a confocal system, excited at 488 nm and measured on two channels with one filter set at 505–550 nm (Fluo-4) and the other at >560 nm (Rhod-2). Bars, 10 μm (63× oil; numerical aperture, 1.4).

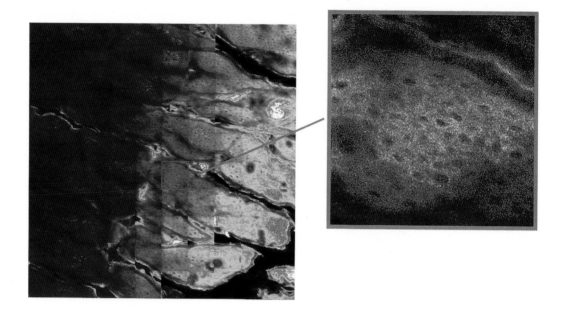

COLOR FIGURE 9–4 Two-photon images of ex vivo human skin acquired at a depth of 30 μm. The field size of the left picture is 1 × 1 mm. The resolution of the image is 0.3 μm. A subsection of the image is presented in higher resolution to the right. The ability to acquire a large tissue area with high resolution is critical for tissue physiological analysis.

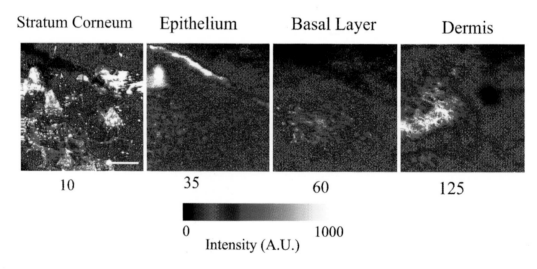

COLOR FIGURE 9–5 A montage of two-photon in vivo images acquired at different depths. In these images, four distinct structural layers are clearly observed: the strata corneum, the epidermis, the epidermodermal junction, and the dermis.

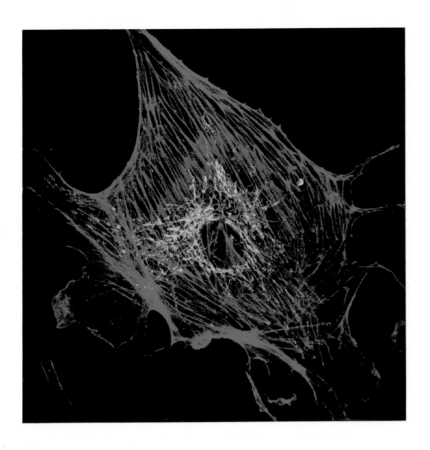

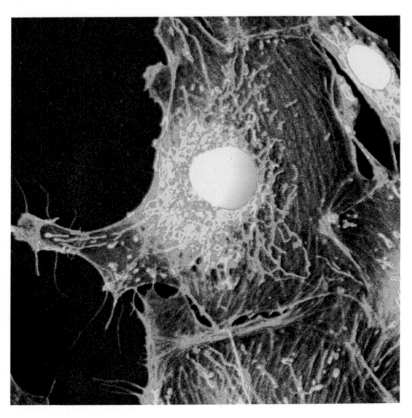

COLOR FIGURE 10–8
Bovine pulmonary artery endothelial cells (F-147780, Molecular Probes) marked with MitoTracker red, Bodipy FL phallacidin, and DAPI specific for mitochondria, F-actin, and DNA, respectively. Confocal imaging (*A*) was performed at 488 and 543 nm; TPE imaging (*B*) was performed at 720 nm.

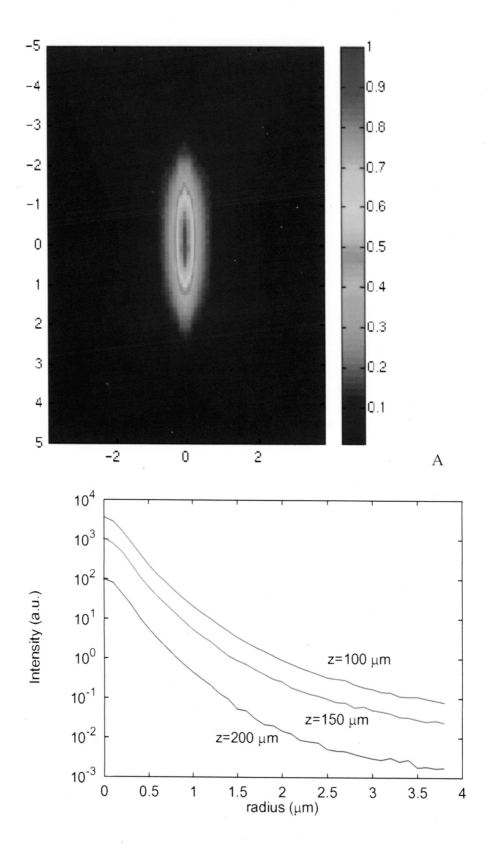

COLOR FIGURE 11–4 Example of computer-generated point spread function (*A*) and the associated radial curves at depths of 100, 150, and 200 µm (*B*).

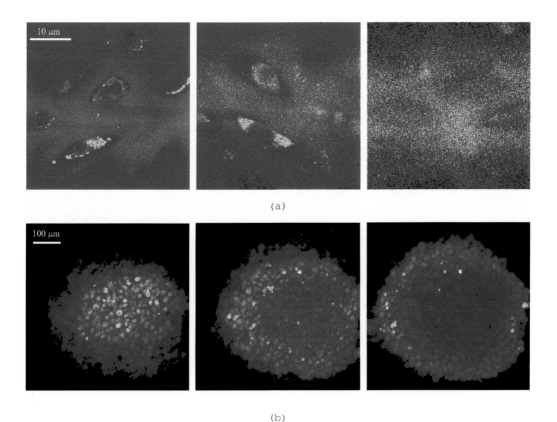

(a)

(b)

COLOR FIGURE 11–9 *a,* Three images at different depths within rabbit septal cartilage. The image on the left was obtained at a depth of 50 μm below the surface and clearly shows chondrocyte cells. The middle image was obtained at 150 μm and the right image at 200 μm, and cells are no longer visible. All images were obtained with a Zeiss C-Apochromat 63×, 1.2 NA, water immersion objective, with an average power of approximately 10 mW. *b,* Three example images of a whole spheroid; from left to right, the depths are 25, 50, and 75 μm. As the image plane cuts though the sphere, many more cells are passed through when the center rather than a side is imaged. All images were obtained with a Zeiss Plan-Neofluar 10×, 0.3 NA dry objective, with an average power of 15 mW.

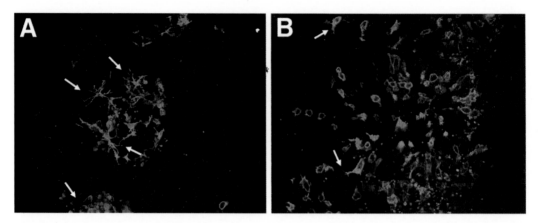

COLOR FIGURE 12–1 Optical sections through a somite expressing EGFP. *A*, Cells in the dorsal most part of the somite extending filopodia (arrows). *B*, Cells 30 μm below the section presented in *A* showing that cells from one somite are contacting adjacent somites via filopodial extensions (arrows). Two-photon images were collected with 860 nm excitation and a shortpass 680 nm emission filter.

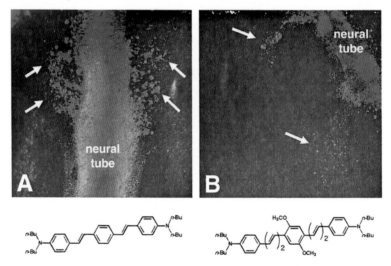

COLOR FIGURE 12–2 Neural crest (arrows) and neural tube cells in 2-day-old chick embryos (*A, B*) labeled with the stilbene dyes shown directly below the photomicrographs. Images were taken with a two-photon microscope using 840 nm excitation light and a shortpass 680 nm emission filter.

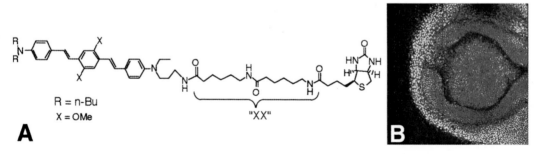

COLOR FIGURE 12–3 Biotinylated bis-donor stilbene dye conjugates for immunocytochemistry. *A*, Structure of the biotinylated bis-donor stilbene used for immunocytochemistry. *B*, Example of immunocytochemistry results. QCPN primary antibody bound to the nuclei in and around the eye in a 4-day-old quail embryo is revealed by stilbene dye fluorescence using 740 nm excitation and a shortpass 680 nm emission filter.

Color Figure 15–3 Monitoring Bcl-2 and Bax interactions in single living cells using FRET microscopy. BHK cells co-expressing GFP–Bax and BFP-Bcl-2 were obtained following transient co-transfection with GFP-Bax and BFP-Bcl-2. The left panel shows a cell that is co-expressing both proteins when viewed with a filter set for the acceptor only (Ex=480/20; Dm=500 nm; Em=515/30). The middle panel shows the same cell when viewed with a filter set to detect fluorescence emission from the donor only (Ex=390/20; Dm=420 nm; Em=450/40). The right panel shows the cell when viewed with a filter set to detect FRET (Ex=390/20; Dm=420 nm; Em=515/30).

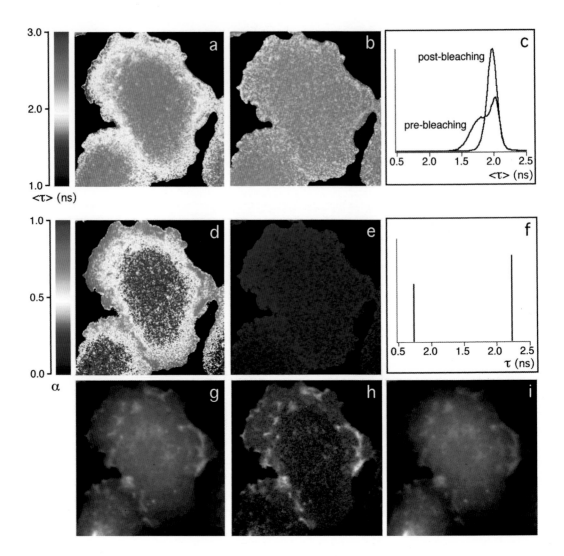

COLOR FIGURE 16–7 Imaging of phosphorylation of the receptor tyrosine kinase ErbB-1 in MCF7 cells after 5 minutes of stimulation with epidermal growth factor (EGF). Phosphorylation is detected by FRET between a fused green fluorescent protein (GFP) and an indocyanine dye (Cy3) bound to an antibody against phosphotyrosines. *a,* GFP fluorescence lifetime maps calculated from the average of phase and modulation lifetimes. *b,* GFP fluorescence lifetime maps after photobleaching of the acceptor. *c,* Histograms of the lifetime values from *a* and *b*. *d,* Fractional population image of the phosphorylated receptor (with $\tau = 0.76$ ns) obtained by global analysis. The fractional population image of the unphosphorylated receptor (with $\tau = 2.26$ ns) is not shown. *e,* Fractional population image after acceptor photobleaching. *f,* Histograms of the lifetime values from the global analysis. *g,* Steady-state fluorescence. *h,* Fluorescence image of the species with 0.76 ns lifetime, which is proportional to the concentration of phosphorylated receptor. *i,* Fluorescence image of the species with 2.26 ns lifetime, which is proportional to the concentration of unphosphorylated receptor.

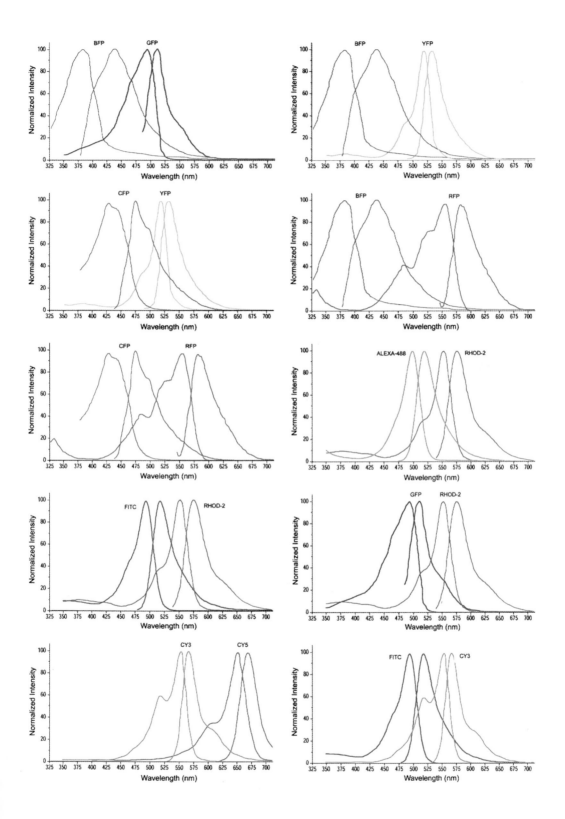

COLOR FIGURE 17–1 Illustrations of different fluorophore pair spectra for FRET imaging.

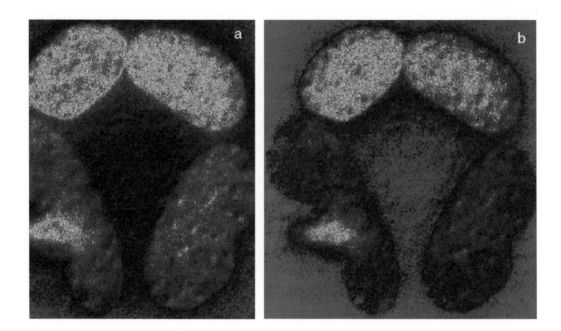

Color Figure 17–2 DV-FRET bleedthrough-corrected images for BFP–GFP–Pit-1 protein. As is described in the text, the donor and the acceptor images were acquired and then processed for bleedthrough correction. To obtain the FRET signal, the donor and the acceptor images are ratioed (acceptor/donor) before (*a*) and after (*b*) bleedthrough correction.

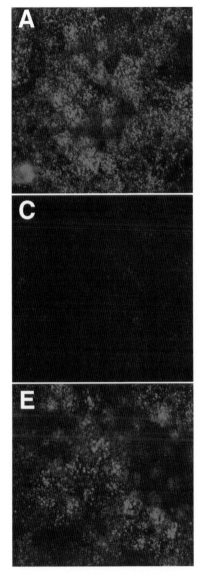

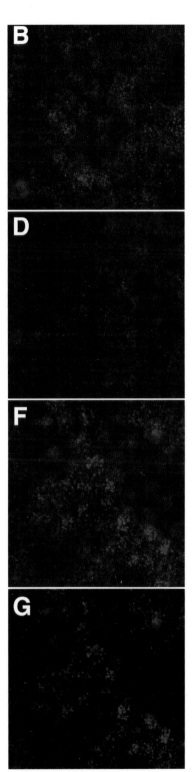

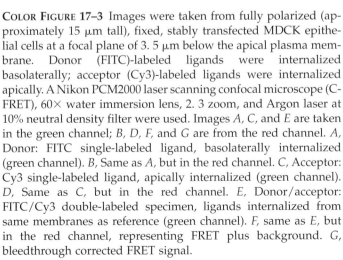

COLOR FIGURE 17–3 Images were taken from fully polarized (approximately 15 μm tall), fixed, stably transfected MDCK epithelial cells at a focal plane of 3. 5 μm below the apical plasma membrane. Donor (FITC)-labeled ligands were internalized basolaterally; acceptor (Cy3)-labeled ligands were internalized apically. A Nikon PCM2000 laser scanning confocal microscope (C-FRET), 60× water immersion lens, 2. 3 zoom, and Argon laser at 10% neutral density filter were used. Images *A, C,* and *E* are taken in the green channel; *B, D, F,* and *G* are from the red channel. *A,* Donor: FITC single-labeled ligand, basolaterally internalized (green channel). *B,* Same as *A,* but in the red channel. *C,* Acceptor: Cy3 single-labeled ligand, apically internalized (green channel). *D,* Same as *C,* but in the red channel. *E,* Donor/acceptor: FITC/Cy3 double-labeled specimen, ligands internalized from same membranes as reference (green channel). *F,* same as *E,* but in the red channel, representing FRET plus background. *G,* bleedthrough corrected FRET signal.

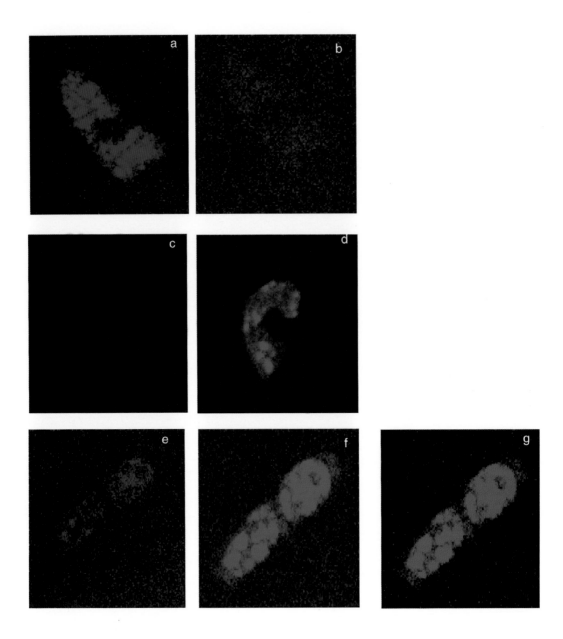

Color Figure 17–4 2P-FRET images of pituitary GHFT1-5 cells expressing BFP– and RFP–C/EBPα. As is described in the text, 740 nm was tuned from the Ti:Sapphire laser (see Fig. 17–7) to excite the donor-alone BFP–C/EBPα-expressed cells, and the images were collected in both the channels (donor and acceptor), as shown in *a* and *b*. The same wavelength was used to excite the acceptor-alone RFP–C/EBPα-expressed cells, and the images were collected in both the channels, as shown in *c* and *d*. The energy transfer images are collected for doubly expressed cells, BFP– and RFP–C/EBPα, as shown in *e* and *f*. The bleedthrough correction is implemented using the signal from *b* and *d*, and the resultant image is shown in *g*. These results demonstrate that the two-photon excitation considerably reduces the photobleaching of the BFP molecules, and the C/EBPα proteins are clearly shown in *g*.

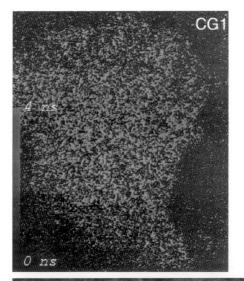

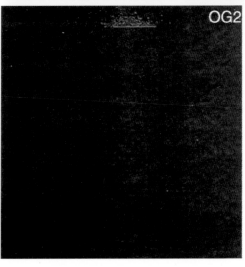

COLOR FIGURE 17–6 Comparison of single- and double-component decay of BHK21 cells labeled with GFP. The images were acquired at different times during the decay for contiguous and overlap gating after femtosecond laser pulse excitation (not shown). These acquired images were processed using decay equations for single- and double-exponential decays. The lifetime image of overlapped gating (OG1) provides good S/N ratio compared with contiguous gating image (CG1). Because the GFP is not targeted to any proteins, the second component image showed no signal (OG2).

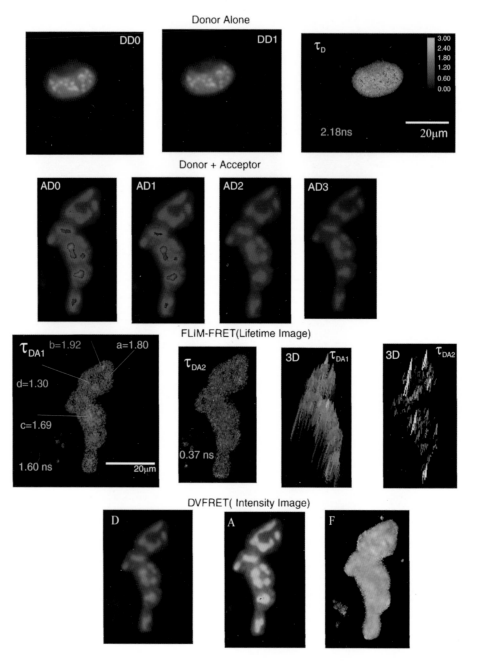

COLOR FIGURE 17–8 FLIM–FRET imaging of GHFT1-5 cells expressing CFP– and YFP–C/ EBPα. Time-resolved donor images were acquired in the presence (AD₀, AD₁, AD₂, AD₃,) and absence (DD₀, DD₁) of the acceptor, and then processed pixel-by-pixel using decay equations (Sharman et al., 1999). In the absence of the acceptor, one obtains donor (CFP–C/EBPα) two-dimensional distribution of a single component lifetime image (τ_D = 2. 18 ns), and there is no second component. But, in the presence of the acceptor (CFP– and YFP-C/EBPα), the processed images provide two-dimensional distributions of double component lifetime images (τ_{DA1} = 1. 6 ns and τ_{DA2} = 0. 37 ns), indicating the occurrence of protein dimerization ($\tau_{DA1} < \tau_D$) and distance distribution (see τ_{DA1}, and 3D images of τ_{DA1} and τ_{DA2}). On the other hand, the intensity FRET image (A/D = F) using DVFRET technique were acquired and processed for the same cell (same camera for FLIM–FRET and DVFRET). As shown, the intensity FRET image (F) does not provide any detailed information regarding the energy transfer process like lifetime FRET images (see τ_{DA1}).

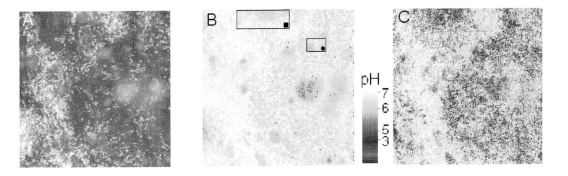

Color Figure 18–6 Typical intensity (*A*) and lifetime (*B, C*) images of the biofilm. The 100 × 100 μm² images were recorded 5 μm below the biofilm surface. *B,* was recorded before and *C,* 20 min after the addition of 14 mM of sucrose. Information from the blocked areas in *B* is shown in Figure 18.7.

First Harmonic Amplitude   One-photon Image

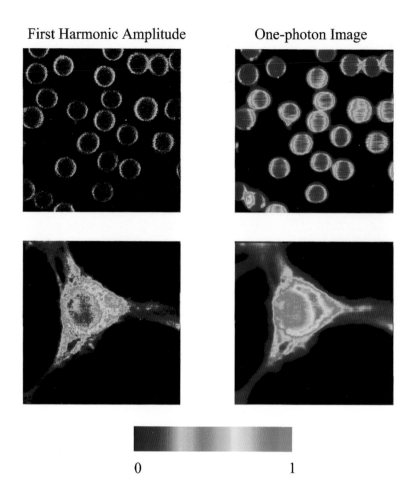

0   1

CoLor FiguRe 19–2 Pump-probe images (at 76.2 MHz excitation) of erythrocytes and rhodamine-DHPE–labeled mouse fibroblasts. [Adapted from Dong et al., 1995.]

First Harmonic Amplitude   Phase

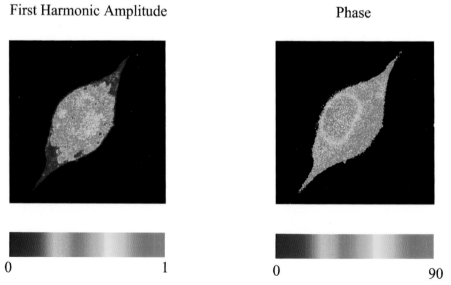

0   1   0   90

CoLor FiguRe 19–3 First harmonic amplitude and phase image of a mouse fibroblast double labeled with rhodamine-DHPE and the nucleic acid stain ethidium bromide. As determined from the phase data, representative lifetimes of the cytoplasmic region and nuclear regions are 2.0 ± 0.5 and 6.6 ± 4.8 ns, respectively. [Adapted from Dong et al., 1995.]

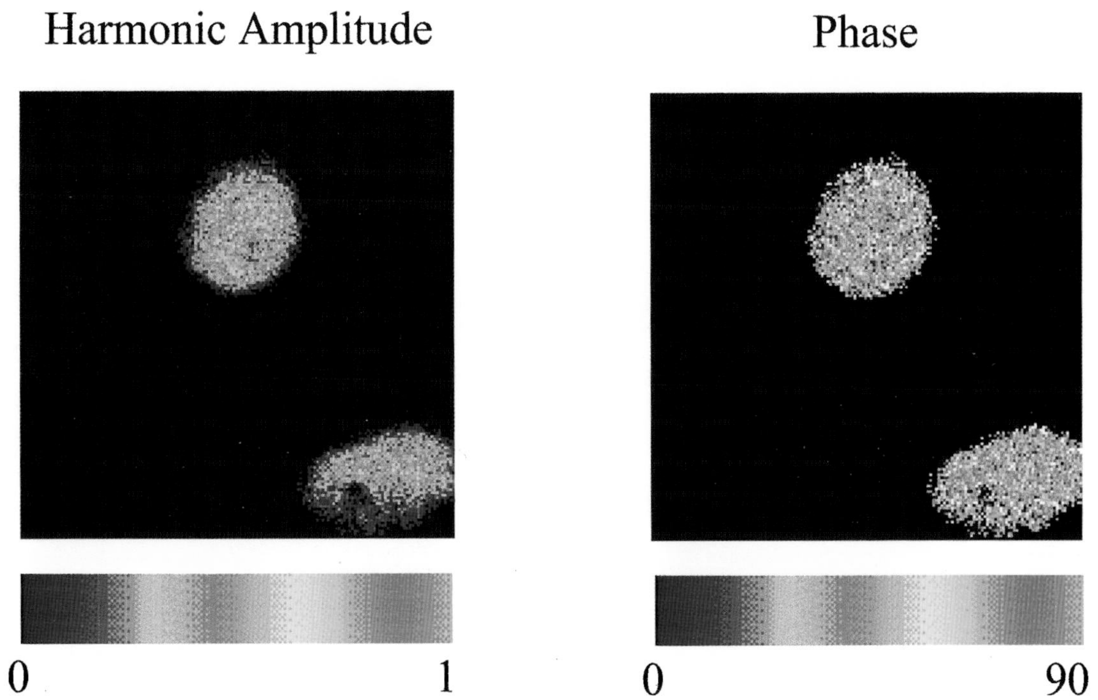

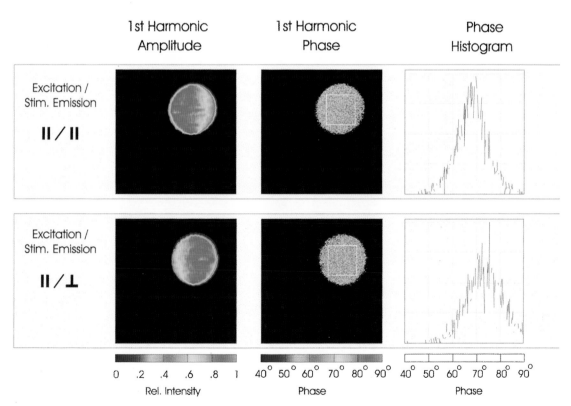

Harmonic Amplitude

0   1

Phase

0   90

COLOR FIGURE 19–4 A pump-probe image (80 MHz) of mouse STO cells labeled with nucleic stain TOTO-3 obtained using intensity-modulated laser diodes. The phase lifetime of the nuclear region is about 2.82 ns.

	1st Harmonic Amplitude	1st Harmonic Phase	Phase Histogram
Excitation / Stim. Emission **‖ / ‖**			
Excitation / Stim. Emission **‖ / ⊥**			

0   .2   .4   .6   .8   1
Rel. Intensity

40° 50° 60° 70° 80° 90°
Phase

40° 50° 60° 70° 80° 90°
Phase

COLOR FIGURE 19–5 Time-resolved, pump-probe polarization images at 76.2 MHz of an orange 15 μm fluorescent latex sphere. There is a differential phase of 5.8° between the parallel and perpendicular components of fluorescence. [Adapted from Buehler et al., 2000.]

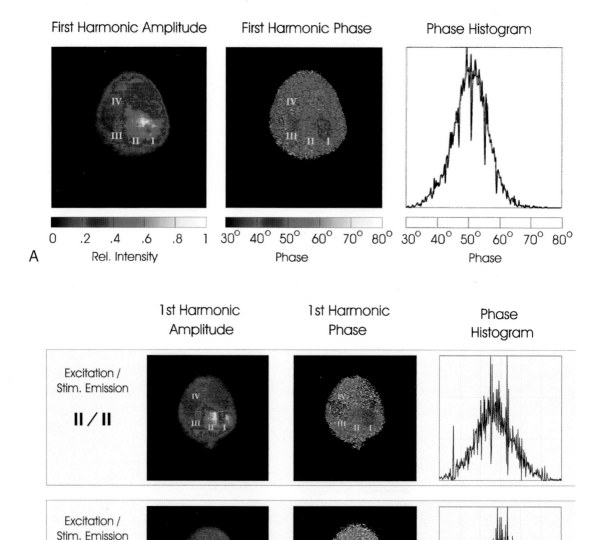

First Harmonic Amplitude

First Harmonic Phase

Phase Histogram

0   .2   .4   .6   .8   1
Rel. Intensity

30° 40° 50° 60° 70° 80°
Phase

30° 40° 50° 60° 70° 80°
Phase

A

1st Harmonic
Amplitude

1st Harmonic
Phase

Phase
Histogram

Excitation /
Stim. Emission

II / II

Excitation /
Stim. Emission

II / ⊥

0   .2   .4   .6   .8   1
Rel. Intensity

20° 30° 40° 50° 60° 70° 80°
Phase

20° 30° 40° 50° 60° 70° 80°
Phase

B

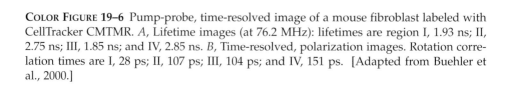

COLOR FIGURE 19–6 Pump-probe, time-resolved image of a mouse fibroblast labeled with CellTracker CMTMR. *A*, Lifetime images (at 76.2 MHz): lifetimes are region I, 1.93 ns; II, 2.75 ns; III, 1.85 ns; and IV, 2.85 ns. *B*, Time-resolved, polarization images. Rotation correlation times are I, 28 ps; II, 107 ps; III, 104 ps; and IV, 151 ps. [Adapted from Buehler et al., 2000.]

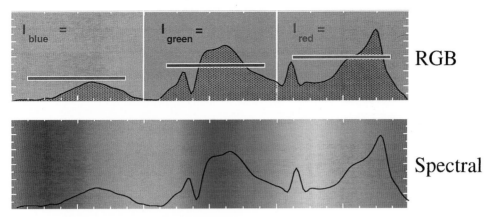

RGB

Spectral

Intensity

COLOR FIGURE 20–1 Comparison of three-color and spectral imaging. Graphic illustration of the fundamental difference between multicolor imaging, exemplified here by its most prevalent representative, the three-color red-green-blue (RGB) scheme (top), and by full spectral imaging (bottom). If the sample to be analyzed has delicate spectral details (black line in both panels), these get overlooked by the "pooling" of intensity data into a single average value I for each color "bin" (red, green, blue). An arbitrarily high number of spectral slices (represented by the rainbow color pattern) allows wavelength-based discrimination (bottom).

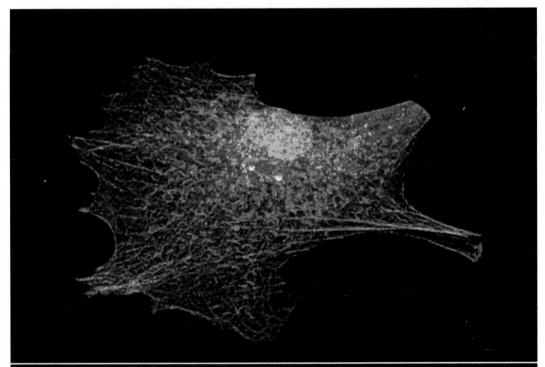

Pseudocolor: Actin - red; Endosomes - magenta; Golgi - light blue; Mitochondria - dark blue; Microtubules - green; Nucleus - yellow

COLOR FIGURE 20–2 Six-color simultaneous labeling of structures in a cell. Cellular components in a mouse 3T3 fibroblast were labeled as indicated, and their fluorescence was imaged (in fast sequence) in wavelength channels defined by use of excitation/ dichroic/emission filter cubes in a multimode microscopy workstation. The overlaid six-color image was created after re-registration of the images by assigning pseudocolors to the fluorescent signals collected in the chosen channels. The resulting color overlap (including some image degradation) is a good representation of the unavoidable spectral overlap between the chosen wavelength intervals.

Oxygen Tension **Oxygen Saturation**

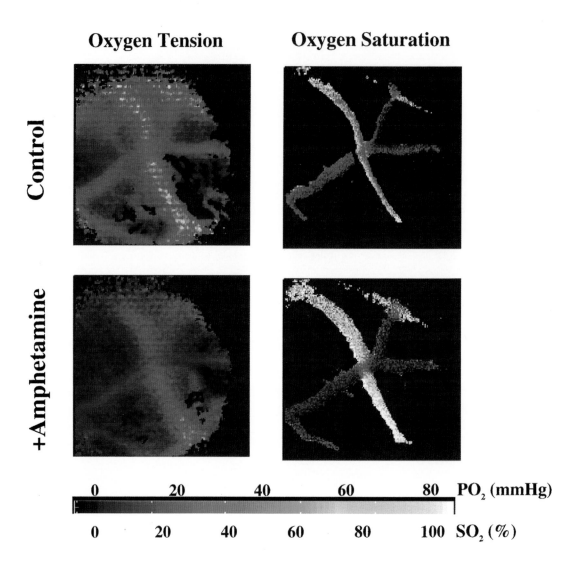

COLOR FIGURE 20–3 Unusual effect of amphetamine stimulation on brain oxygenation. In a mouse cranial window, mapping of oxygen tension (left panels) and hemoglobin oxygen saturation (right) was performed with an AOTF microscope we developed. Top panels show controls, and bottom panels show maps after injection of 10 mg/kg amphetamine sulfate. For the relatively rare event shown, amphetamine decreased oxygen tension while increasing hemoglobin oxygen saturation. More typical (but still somewhat surprising) responses and more details are given in Shonat et al. (1998).

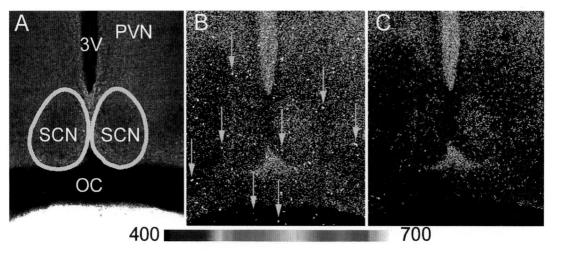

Color Figure 23–3 Bioluminescence from a *fos::luc* brain slice. *A,* The bright-field image shows the approximate boundary of the suprachiasmatic nucleus (SCN) in yellow, optic chiasm (OC), and the third ventricle (3V) in this 300 μm-thick brain slice maintained for 33 hours with 1 mM luciferin in culture medium and 10% newborn calf serum. Culture and imaging were at 37°C. PVN, paraventricular nucleus. *B,* Corresponding raw bioluminescence image from a 1 h exposure and 2 × 2 binning using an LN-cooled CCD camera (Roper Scientific CH260). Small white spots are attributed to cosmic ray events, some of which are indicated with yellow arrows. *C,* The processed image made by taking the minimum from two 1 h exposures to remove cosmic ray noise and plotted to show only pixels with 400–900 ADU. Both SCN show transgene expression along with particularly high expression in the ependymal cell layer surrounding the ventricle, the PVN, and an area at the ventral edge of the SCN overlapping the optic chiasm. Expression in the lateral hypothalamus is contiguous with expression in the SCN. A 10×, 0. 25 numerical aperture Olympus objective lens was used.

(a)

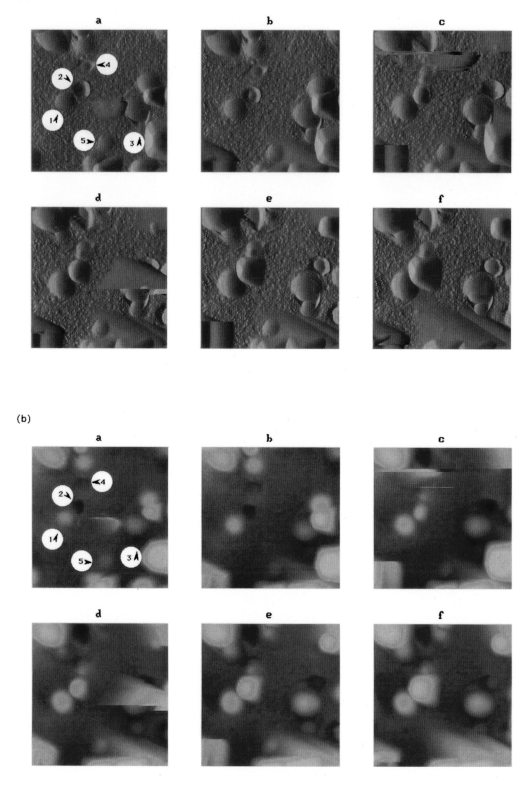

(b)

COLOR FIGURE 24–2 *A*, Deflection mode real-time images of yeast cells' growth and budding. The time lapse between each subsequent image is 6 min. Notice the removal of the growing cell in frame c (indicated by an arrowhead in a) when it reaches a critical height in frame d. *B*, The corresponding height mode images of those in *A*.

(a)

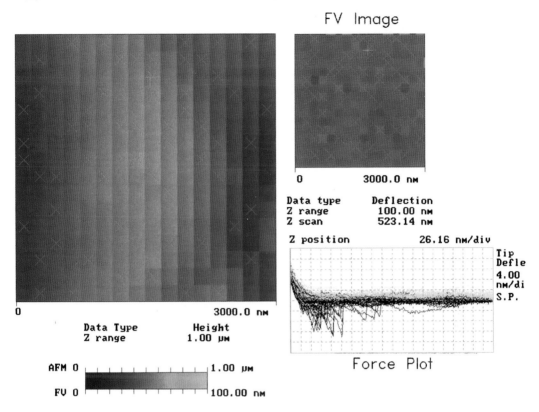

FV Image

Data type   Deflection
Z range   100.00 nm
Z scan   523.14 nm

Z position   26.16 nm/div

Tip
Defle
4.00
nm/di
S.P.

Force Plot

0   3000.0 nm

Data Type   Height
Z range   1.00 µm

AFM 0 ‖   ‖ 1.00 µm
FV 0   ‖   ‖ 100.00 nm

(b)

Extending
⟶ Retracting

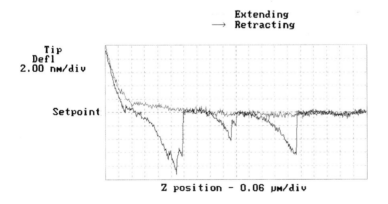

Tip
Defl
2.00 nm/div

Setpoint

Z position - 0.06 µm/div

Force Calibration Plot

COLOR FIGURE 24–5 *a*, Typical force volume data frame showing different types of data that can be collected at the same time with this mode. Upper left is a height image, upper right is a force volume image, and lower right is a force curve display window. *b*, Representative force curve showing the stretching phenomenon of mannoprotein polymer on the cell surface by the applied pulling force. The parameters used in this measurement were as follows: Z scan size, 1. 186 µm; set point, 4. 5 V; Z scan rate, 0. 5008 Hz; Z range, 20 nm.

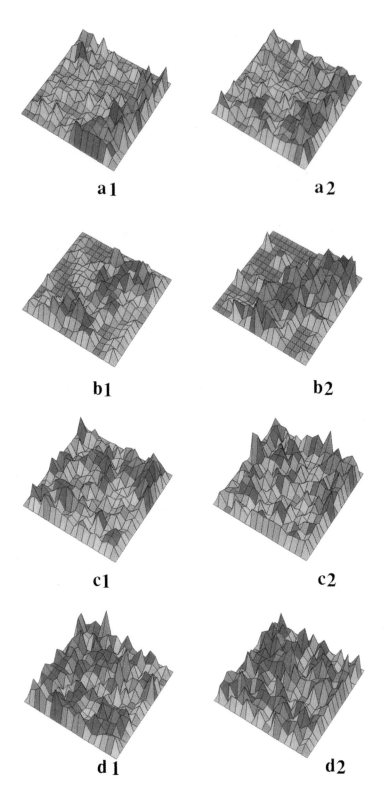

COLOR FIGURE 24–6 A collection of adhesion force distributions shown by three-dimensional surface graphs, reflecting the distribution of mannan on different yeast cells. Each adjacent pair represents two subsequent force maps for the same area with the same functionalized tip. The height and color codes at each pixel reflect the distribution of mannan. The blue color represents areas at which mannan polymers concentrated the most.

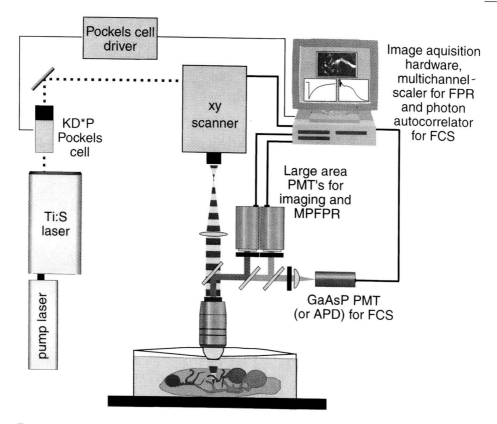

FIGURE 13–5 Integration of MPFPR and MPFCS into a multiphoton microscope instrument.

4. IN VIVO MPFPR AND MPFCS MEASUREMENTS

4.1 MPFPR

The most convenient instrument for in vivo diffusion measurements by MPFPR and MPFCS is a modified multiphoton microscope that includes the additional instrumentation outlined above. A schematic of a typical system is shown in Figure 13–5.

MPFPR is a simple technique to use in vitro and can yield accurate measurements of 3D diffusion coefficients. Good in vivo measurements, however, are often more difficult to obtain and interpret. Figure 13–6 shows the difference in the diffusion of a small dye (calcein) inside and outside of a cell.

In general, in vivo measurements require more complex diffusion models, and the data in Figure 13–6B are fit to an anomalous diffusion model. The same data could also be well-fit to a multiple diffusion component model; however, anomalous diffusion has the advantage of having fewer free parameters. The best-fit line for a single-component, normal-diffusion model is also shown in Figure 13–6B for comparison. The biological cause of anomalous diffusion is binding interactions and/or other forms of spatial hindrance within the cell, and an understanding of these interactions is often the primary goal of in vivo

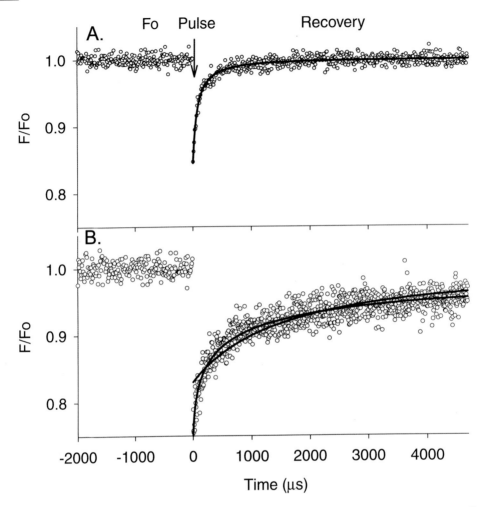

FIGURE 13–6 MPFPR measurements of calcein in solution and in AM-loaded RBL-2H3 cells. *A*, 10 μM calcein in water. Bleaching pulse length was 10 μs at 10 mW; monitoring power was 0.3 mW. Excitation was at 780 nm delivered through a ×40, 1.35 NA objective. Solid line indicates fit to a normal diffusion model yielding $D = 2.4 \times 10^{-6}$ cm^2 s^{-1}. Data are the average of 50 repeated pulses at 10 Hz. *B*, In vivo MPFPR curves measured in calcein–AM-loaded RBL cells. A condition similar to that in *A* exists except that the pulse length was 20 μs and the data are the average of 10 traces. Solid lines indicate the best-fit line results from the application of an anomalous diffusion model ($\alpha = 0.6$, $\Gamma = 5 \times 10^{-8}$ cm^2 s$^{-\alpha}$); poorer fit is from a single-component, normal-diffusion model.

diffusion measurements. Other causes of (apparent) anomalous diffusion can also arise, however, and should be avoided. Three common sources of artifactual complexity in vivo MPFPR measurements are as follows:

1. If the bleaching pulse is too deep or too long relative to the recovery time, the trace will not appear to be normal diffusion. This can occur in vitro as well as in vivo and is a common source of measurement error. Figure 13–7A shows the fractional error in the observed diffusion

coefficient as a function of the ratio of bleaching pulse length to the half time of the recovery. An in vitro (fluorescein in water) MPFPR curve taken using a bleaching pulse 10 times longer than the half time is better fit by a model assuming anomalous diffusion than normal diffusion (Fig. 13–7B).

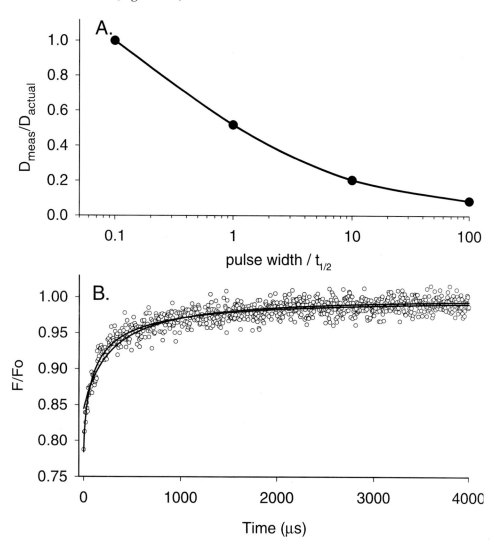

FIGURE 13–7 Error in diffusion (D) coefficient measurement due to overbleaching. *A*, Fractional error in the measured diffusion coefficient as bleaching pulse width is lengthened. Measurements were made on 2 μM fluorescein in water with 760 nm excitation through a ×40, 1.15 NA water immersion objective. The bleaching pulse was varied from 10 μs to 10 ms, and pulse intensity was set so that the same depth of bleach was reached at each pulse length. The value of D obtained using the 10 μs bleaching pulse was 2.6×10^{-6} cm^2 s^{-1}, and the recovery half time ($t_{1/2}$) from these data was 100 μs. *B*, Data taken at a bleaching pulse length of 1 ms (10 times longer then the actual half time). The single coefficient fit (normal diffusion) returns a diffusion constant of 4.2×10^{-7} cm^2 s^{-1}. A statistically better fit is obtained using an anomalous diffusion model with $\alpha = 0.73$ and $\Gamma = 9.4 \times 10^{-8}$ cm^2 s$^{-\alpha}$.

2. In MPFPR, photobleaching is usually assumed to be proportional to the intensity squared, although the equation given by Brown et al. (1999) is general. Equation 4 in this text assumes both two-photon excitation and photobleaching. Recent work has shown that multiphoton photobleaching can be higher order than 2 under certain conditions (Patterson and Piston, 2000). Higher order bleaching could be an additional source of measurement error in an MPFPR experiment if present and not specifically accounted for. In many cases, however, we find that bleaching does depend on the square of the intensity and appears to be a first-order kinetic process (over a reasonable intensity range). A check for the excitation order of photobleaching of a particular fluorophore is to measure the bleach depth (the first few points immediately after the bleaching pulse) as a function of pulse power. The length of the bleaching pulse must be much shorter than the diffusion time. The assumption that the photobleaching mechanism is both two-photon dependent and well approximated by first-order kinetics leads to the form $1 - \exp(cI^2)$ (Brown et al., 1999). Figure 13–8 shows the bleach depth of Alexa-488 bound to uridine triphosphate (UTP) and calmodulin as a function of the intensity of a 10 μs bleaching pulse fit to the assumed model: $A(1 - \exp[BI^2])$. Although the photobleaching efficiency varies between the Alexa-488 UTP and Alexa-488 calmodulin by a factor of two, both forms of conjugated Alexa-488 exhibit two-photon bleaching. Higher power for long dwell times and different chemical environments can produce more complex photobleaching mechanisms, as in Patterson and Piston (2000), and it is often important to check that bleaching is occurring with the assumed kinetics.

3. A third source of error is bleaching and recovery of intrinsic cellular fluorescence. Control (unloaded) cells should be imaged at the monitoring power being used for MPFPR measurements of the loaded cells to ensure that intrinsic fluorescence recovery is not being recorded along with the labeled probe's FPR signal.

Another variable with MPFPR measurements is the photobleaching rate of the fluorescent label being used. A probe should be selected that can be bleached with a pulse width of around 10% of the recovery half time at an average power that does not damage the cell. The bleaching efficiency of a given fluorophore depends on the wavelength and the environment. For example, Alexa-488, a probe usually considered to be difficult to photobleach, bleaches more significantly at short wavelengths (<880 nm). Furthermore, as shown in Figure 13–8, conjugated Alexa-488 bleaches differently depending on what it is bound to. For many probes the oxygen tension has a significant effect, and for intracellular measurements [O_2] can be readily modulated by the glucose concentration of the media, making a given fluorophore more or less "bleachable."

Cell damage due to the parked beam in MPFPR is a concern; however, with MPFPR the high-intensity pulse duration is on the order of fractions of a millisecond (although repeated several times). Typical in vivo measurement might

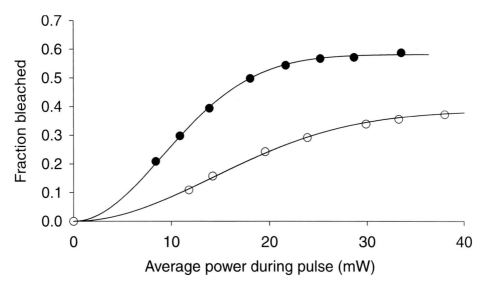

FIGURE 13–8 Photobleaching depth as a function of pulse intensity for Alexa-488 uridine triphosphate (open circles) and Alexa-488 calmodulin (closed circles). Bleach depth was determined from the average value of F/F_o during the first 20 μs after the bleaching pulse. Data were fit to a model that assumes two-photon bleaching and first-order bleaching kinetics.

use a bleaching power of 10–15 mW for 30 pulses of 100 μs duration, with a monitoring power of 1 mW and a total monitoring duration of a few seconds.

4.2 MPFCS

For concentrations of labeled molecules below about 1 μM, MPFCS can be used to measure diffusion constants in cells. As with in vivo MPFPR, there are several caveats:

1. Photon counts from light leakage, or those due to fluorescence from immobile or extremely slowly diffusion species, will not correlate, but introduce error in $G(0)$. If the number of molecules in the focal volume is of interest, for example, to calculate the concentration, $G(0)$ should be corrected if possible. Figure 13–9 illustrates the effect of noncorrelating background on the MPFCS curves of eGFP in solution. The inset in Figure 13–9 shows the relative change in $G(0)$ as a function of the fluorescence (F) to background (B) ratio. These data were obtained by placing an LED near the objective to provide controllable constant background counts. The solid line is the function $1/(1 + 2B/F + [B/F]^2)$, which can be used as a correction factor (Thompson, 1991). Noncorrelating background counts can be determined by measuring the counts from an unlabeled specimen at the same power used for the MPFCS measurement. The primary cause of noncorrelating background is room light or excitation light leakage onto the detector. Remarkably, diffusion coefficients obtained

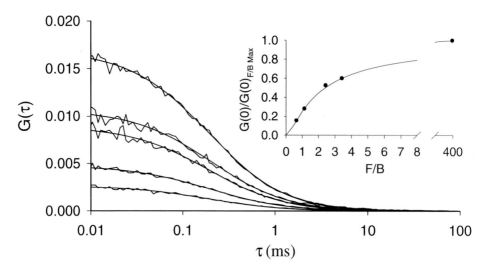

FIGURE 13–9 Effect of noncorrelating background (shot noise) counts on eGFP MPFCS curves. Background counts were added by the use of a green LED placed various distances from the objective lens during autocorrelation measurements. The fluorescence (F) to background (B) ratio was varied from 390 (no added counts, dark noise = 100 cps) to 0.6. $G(0)$ decreases with increasing noncorrelating counts and follows the predicted model shown in the inserted plot (see text for details); however, diffusion time (τ_D) remains constant (calculated $D = 6 \times 10^{-7}$ cm^2 s^{-1}).

from FCS are unaffected by high noncorrelating background counts. The data in Figure 13–9 yield the same diffusion coefficient at all levels of F/B.

2. Correlatable cellular autofluorescence is the major obstacle to measuring in vivo diffusion with FCS. Because low concentrations of probe are being used, the autofluorescence contribution to the total signal can be large. One of the major advantages of MPE for in vivo FCS is the fact that there is no out-of-focus autofluorescence generated (Schwille et al., 1999) that can be scattered into the detector. With MPFCS there are often less total counts due to autofluorescence, but still enough to potentially cause artifacts. Once the illumination intensity is determined based on the concentration of probe in the loaded cells, autocorrelation curves of unloaded cells should be measured. The amount of autofluorescence varies with cell type and culture conditions, but ideally it should be impossible to measure an autocorrelation curve under the illumination conditions used for the experiment. In practice, autofluorescence can often be transient, caused by slow-moving autofluorescent species. Figure 13–10 shows an MPFCS curve of eGFP in RBL cells taken at 920 nm with 9 mW at the sample. Autocorrelation traces (30 s per trace) made in unloaded cells revealed occasional bursts of correlatable signals. Even at the same position within a cell, a portion of the traces would show an autocorrelation indicative of sparse, slowly diffusing species (see Fig. 13–10B, C). For this reason, the data in Figure 13–10A were best fit to two diffusion components: a fast component due to eGFP

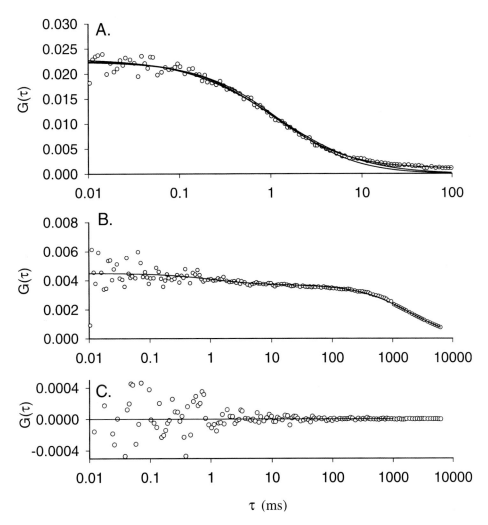

FIGURE 13–10 In vivo MPFCS of cytosolic, soluble eGFP in RBL-2H3 cells. RBL cells were loaded with recombinant eGFP (27 kDa) by electroporation, and 9 mW of excitation at 920 nm was delivered through a ×40, 1.15 NA Olympus objective. The detector was a Hamamatsu H7421-40 GaAsP photomultiplier tube; the dichroic/emission filter was 670DCLP/550DF150 (Chroma Technology Corp., Brattleboro, VT). *A*, In vivo MPFCS curve fit to three different models. a two-component, normal-diffusion model (best fit line), an anomalous diffusion model (next best fit), and a single-component, normal-diffusion model. Trace is the average of five 30 s autocorrelations. The calculated diffusion coefficient for eGFP in RBL cells from the data in *A* is $8 \times 10^{-8} \, cm^2 \, s^{-1}$. Representative MPFCS traces of autofluorescence from the same position in an unlabeled RBL cell are shown in *B* and *C*. Correlatable signal as shown in *B* appeared approximately 30% of the time; the remaining traces showed no correlation out to 10 s (as shown in *C*), even though count rates were more than 50 kHz.

diffusion and a small-amplitude component presumably due to auto-fluorescence. In this particular case (eGFP), a two-component normal diffusion model yielded a better fit than an anomalous diffusion model, presumably because there is minimal intracellular binding of the eGFP.

3. Photobleaching of the probe during the MPFCS measurement can cause correlation artifacts. The photobleaching behavior of a fluorophore in solution may be very different from its in vivo behavior, and probe selection for in vivo measurements are always ultimately based on testing in the cells under study.

A final, but critical concern when making in vivo MPFCS measurements is cell damage from the parked laser. With a Ti:Sapphire laser, enough power is available to cause obvious damage to a living specimen while a measurement is being made; however, less obvious levels of damage can be more difficult to detect. Shorter wavelengths (<850 nm) are especially lethal because they are equivalent to ultraviolet excitation. Because FCS requires many seconds of constant illumination at a relatively high laser power, it is best carried out at wavelengths greater than 880 nm.

5. CONCLUSION

The use of MPE extends the techniques of FPR and FCS. It moves FPR onto three dimensions and aids FCS by producing a well-defined focal volume with no out-of-focus excitation. If applied with care, MPFPR and MPFCS will greatly enhance our ability to measure and understand intracellular diffusion.

The authors thank Rebecca Williams, Jonas Korlach, Dan Larson, and Petra Schwille for their helpful discussions and Sally Kim and Neal Waxham for the Alexa-488–labeled calmodulin; and they acknowledge the financial support of the National Institutes of Health/National Centre for Research Resources (P41-RR0422) and the National Science Foundation (DIR 88002787).

REFERENCES

Axelrod, D., D. Koppel, J. Schlessinger, E. Elson, and W. Webb. Mobility measurement by analysis of fluorescence photobleaching recovery kinetics. *Biophys. J.* 16:1055–1069, 1976.

Berland, K. M., P. T. So, and E. Gratton. Two-photon fluorescence correlation spectroscopy method and application to the cellular environment. *Biophys. J.* 68:694–701, 1995.

Bouchard, J., and A. Georges. Anomalous diffusion in disordered media; statistical mechanisms, models and physical applications. *Phys. Rep.* 195(4&5):127–293, 1990.

Brown, E., E. Wu, W. R. Zipfel, and W. W Webb. Measurement of molecular diffusion in solution by multiphoton fluorescence photobleaching recovery. *Biophys. J.* 77:2837–2849, 1999.

Eigen, M., and R. Riger. Sorting single molecules: Applications to diagnostics and evolutionary biotechnology. *Proc. Natl. Acad. Sci. U.S.A.* 91:5740–5747, 1994.

Feder, T., I. Brust-Mascher, J. Slattery, B. Baird, and W. Webb. Constrained diffusion or immobile fraction on cell surfaces: A new interpretation. *Biophys. J.* 70(6):2767–2773, 1996.

Haupts, U., S. Maiti, P. Schwille, and W. Webb. Dynamics of fluorescence fluctuations in green fluorescent protein observed by fluorescence correlation spectroscopy. *Proc. Natl. Acad. Sci. U.S.A.* 95:13573–13578, 1998.

Koppel, D., D. Axelrod, J. Schlessinger, E. Elson, and W. Webb. Dynamics of fluorescence marker concentration as a probe of mobilty. *Biophys. J.* 16:1315–1369, 1976.

Madge, D., E. Elson, and W. Webb. Thermodynamic fluctuations in a reacting system—Measurement by fluorescence correlation spectroscopy. *Phys. Rev. Lett.* 29:705–708, 1972.

Madge, D., E. Elson, and W. Webb. Fluorescence correlation spectroscopy. II. An experimental realization *Biopolymers* 13:29–61, 1974.

Maiti, S., U. Haupts, and W. Webb. Fluorescence correlation spectroscopy: Diagnostics for sparse molecules. *Proc. Natl. Acad. Sci. U.S.A.* 94:11753–11757, 1997a.

Mertz, J., C. Xu, and W. Webb. Single-molecule detection by two-photon excited fluorescence. *Opt. Lett.* 20(24):2532–2534, 1995.

Patterson, G. H., and D. Piston. Photobleaching in two photon excitation microscopy. *Biophys. J.* 78:2159–2162, 2000.

Periasamy, N., and A. S. Verkman. Analysis of fluorophore diffusion by continuous distributions of diffusion coefficients. Applications of photobleaching measurements of multicomponent and anomalous diffusion. *Biophys. J.* 75(1):557–567, 1998.

Richards, B., and E. Wolf. Electromagnetic diffraction in the optical systems II. Structure of the image field in a planatic system. *Proc. R. Soc. Lond. Ser. A* 253:358–379, 1959.

Schätzel, K., M. Drewel, and S. Stimac. Photon correlation measurements at large lag times. *J. Mod. Opt.* 35(4):711–718, 1988.

Schwille, P., U. Haupts, S. Maiti, and W. Webb. Molecular dynamics in living cells observed by fluorescence correlation spectroscopy with one- and two-photon excitation. *Biophys. J.* 77:2251–2265, 1999.

Seksek, O., J. Biwersi, and A. S. Verkman. Translational diffusion of macromolecule-size solutes in cytoplasm and nucleus. *J. Cell Biol.* 138:131–142, 1997.

Thompson, N. L. Fluorescence correlation spectroscopy. In: *Topics in Fluorescence Spectroscopy*, Vol. 1, edited by J. R. Lakowicz. New York: Plenum Press, 1991, pp. 337–378.

Widengren, J., U. Mets, and R. Rigler. Fluorescence correlation spectroscopy of triplet states in solution: A theoretical and experimental study. *J. Chem. Phys.* 99:13368–13379, 1995.

Xu, C., and W. W. Webb. Multiphoton excitation of molecular fluorophores and nonlinear laser microscopy. *Topics in Fluorescence Spectroscopy, Volume 5: Nonlinear and Two-Photon Fluorescence,* edited by J. Lakowicz. New York: Plenum Press, 1997.

14

Cellular Response to Laser Radiation in Fluorescence Microscopes

Karsten König

1. Introduction

The integration of ultraviolet (UV) and visible (VIS) laser radiation in optical microscopes has led to significant advances in diagnostic methods in life sciences, including three-dimensional (3D) fluorescence imaging by confocal laser scanning microscopy (CLSM) (see Chapter 5), high-resolution fluorescence imaging by scanning near-field optical microscopy (SNOM), and fluorescence lifetime imaging (FLIM) (see Chapters 15–19) with subnanosecond temporal resolution. Further applications of these laser microscopes include intracellular noncontact microsurgery by powerful laser pulses and microphotochemistry, such as UV-induced release of caged compounds.

A novel approach in modern live cell microscopy is near-infrared (NIR) laser microscopy, employing 700–1200 nm wavelengths. The two major applications are (1) optical trapping for cellular force measurements and cellular/molecular micromanipulations using continuous wave (CW) laser tweezers (Ashkin et al., 1986; see also Chapter 22) and (2) multiphoton microscopy for fluorescence imaging based on nonresonant, nonlinear excitation of cellular fluorophores (see Chapters 9–12). In contrast to conventional one-photon microscopy, endogenous and exogenous fluorophores with UV and VIS electronic transitions are excited in multiphoton microscopes by the simultaneous absorption of two or more NIR photons of intense laser beams (Denk et al., 1990). With high numerical aperture (NA) objectives, the required high NIR intensities in the range of MW/cm^2 and GW/cm^2 are attained in the minute focal volume. Cellular components in out-of-focus regions do not experience the intense NIR radiation and hence remain unaffected. When NIR radiation is used in the range of 700–1200 nm, cells and tissues possess only low absorption and scattering coefficients. In most cell types, water with a molar absorption coefficient of less than 0.1 cm^{-1} is considered to be the major absorber. Exceptions are cells containing melanin, hemoglobin, and chlorophyll. Therefore, nonpigmented animal cells appear as almost transparent to low-intensity NIR radiation. Because there is hardly any efficient absorption in the out-of-focus regions, there will also be no out-of-focus bleaching and cell damage. Photodestruction can, however, potentially occur in the focal plane.

This chapter focuses primarily on the interaction between fluorescence excitation radiation and single living cells in laser microscopes. In particular, the influence of continuous wave UVA and NIR microbeams, as well as of femtosecond and picosecond NIR laser pulses, on cellular metabolism, ultrastructure, and viability is described.

2. Potential Causes of Photodamage

There are a number of potential photo-induced effects that may cause cellular damage. These include photothermal, photochemical, and photomechanical effects. Except for the possible damage resulting from light pressure and certain optical breakdown phenomena, photo-induced damage is virtually based on light absorption. There are two types of absorbers, the endogenous absorbers and the exogenous absorbers. Major endogenous absorbers with their one-photon absorption maxima are listed in Table 14–1. In most instances, the two-photon and three-photon absorption maxima and absorption coefficients of these molecules have not be determined. Interestingly, a variety of the endogenous absorbers emit fluorescence (Table 14–1). Particularly, the emission of endogenous absorbers can be used to obtain information on the metabolic state of the vital cells and organelles (König and Schneckenburger, 1994). For instance, the reduced coenzymes NADH and NADPH emit, in contrast to their oxidized forms NAD and NADP.

Otherwise, exogenous absorbers, such as certain fluorophores, applied to living cells to realize morphological or functional imaging may induce phototoxic reactions following linear or nonlinear light excitation. Exogenous absorbers may include fluorophore precursors (e.g., aminolevulinic acid for porphyrin synthesis), as well as caged compounds, stains, and ingredients in the medium. As discussed throughout this chapter, photochemical-induced cytotoxic reactions after absorption by endogenous absorbers play a major role among the different photodamaging processes.

TABLE 14–1 Absorption and fluorescence maxima of important endogenous absorbers

Absorber	Absorption Maxima (nm)	Fluorescence Maximum (nm)
Water	1500, 1900, 2900	
Melanin	<300	
Hemoglobin	410, 540, 580, 760	
Tyrosin	220, 275	305
Tryptophan	220, 280–290	320–350
NADH/NADPH	340	450–470
Flavins	370, 450	530
Zn-protoporphyrin	420, 550, 585	590
Coproporphyrin	400, 500, 530, 565, 620	625
Protoporphyrin	405, 505, 540, 575, 630	635

3. LIGHT EXCITATION PARAMETERS AND LASER MICROSCOPES

Most laser microscopes for fluorescence detection are based on the application of highly focused laser beams, typically referred as *microbeams,* in combination with scanning units (Pawley, 1995). Among the two modes of scanning, beam scanning and stage scanning (motorized stage movement), beam scanning is routinely used. It is based on the movement of a diffraction-limited microbeam across the sample with a typical beam dwell time per pixel of microseconds.

In multiphoton laser scanning fluorescence microscopy with high NA objectives, fluorescence photons are released from a minute subfemtoliter focal volume. In contrast, the excitation of molecules in one-photon fluorescence microscopes might occur in the entire volume of light exposure and therefore also in the out-of-focus regions. The employment of pinholes in CLSMs enables detection of fluorescence from the focal volume, but cannot exclude the out-of-focus photo-induced cytotoxic reactions.

Most laser scanning microscopes are based on the use of VIS fluorescence excitation radiation, such as 488 and 514 nm from an argon ion laser or 543 and 633 nm from a helium neon laser. A wide range of important fluorophores used in the life sciences, however, such as NADH, NADPH, DAPI, Hoechst 33342, and Fura-2 have only absorption bands in the UV region. The most common excitation source is a standard 50 W high-pressure mercury lamp of fluorescence microscopes in combination with bandpass filters to use the 365 nm line. For instance, UV laser microscopy is possible at 363.8 nm when an Ar^+ laser is used.

Multiphoton fluorescence excitation with red/NIR microbeams is based either on CW laser radiation at 100–300 mW power (Hänninen et al., 1994; König et al., 1995) or on pulsed radiation with Watt and kiloWatt powers (Denk et al., 1990). Appropriate CW laser sources are CW Ti:Sapphire ring lasers, Nd:YAG lasers, krypton ion lasers (647 nm), and laser diodes. For fast fluorescence imaging, however, high-repetition ultrashort laser pulses in the picosecond and femtosecond ranges are used with mean powers in the microWatt and milliWatt range. Whereas the pioneers of two-photon microscopy used a colliding-pulse mode-locked (CPM) dye laser, laser sources in multiphoton microscopes include mode-locked Ti:Sapphire lasers, Cr:LiSaF lasers, and frequency-doubled fiber lasers. Compact solid-state turnkey femtosecond laser systems are now commercially available.

Typically, the parallel laser beam is expanded to fill the back aperture of a high NA objective (NA > 1). The laser beam can be launched into either slightly modified conventional laser scanning microscopes or into a fluorescence microscope with an external x,y-scanner. Fluorescence imaging is performed by the internal photomultiplier tubes (de-scanned mode) of the standard microscopes, by a base-port photomultiplier tube, or with camera systems like slow-scan cooled charge-coupled device cameras.

Transmittance (T) and pulse broadening (M) due to group velocity dispersion of the optical systems without objective can be determined easily by power and pulse width measurements of the transmitted beam using conventional power meters and autocorrelators. If the transmittance and the dispersion values of the microscope objective in connection with the cell chamber are not known, the measurement of a parallel beam is recommended, which leaves a sandwich-setup consisting of a thin cell chamber between two identical objectives with nearly the same focal point. This process will circumvent the problem encountered with divergent beams (König et al.,1996c). For direct measurement of the pulse width at the fluorescent sample (*in situ* pulse width), the laser radiation should be split into two beams with a tunable time delay between and should be focused in the same focal volume of the fluorescent probe. The multiphoton excited target fluorescence signal versus time delay represents the desired autocorrelation function. Typical NIR values of a whole optical illumination system for multiphoton microscopy, including a high NA objective and cell chamber, are T = 30% (with photon loss more than 70% in the case of overillumination of the back aperture of the objective) and a 5000 fs^2 optical dispersion.

4. CELLS AND CELL HANDLING

Typically, in high-resolution fluorescence microscopy of living cells, a few single cells on the coverslip or within a special cell chamber are exposed to fluorescence excitation light to evoke emission of intracellular endogenous or exogenous fluorophores. This chapter describes the impact of this excitation light on the metabolism and vitality of single exposed cells compared with their nonilluminated neighboring cells.

To study long-term effects, light-exposed cells had to be traced for more than 1 day after laser exposure. For that purpose, cells were grown and laser-exposed in sterile round miniaturized cell chambers (MiniCem, JenLab GmbH, Jena, Germany). The plastic chamber consists of two round 0.17 μm thick windows for microscopy and a silicon ring. Injection needles were introduced to realize exchange of fluorophores, cells, and medium (Fig. 14–1). Single adherent cells grown on one of these windows can be differently marked by etching the glass with a Zeiss diamond-marking objective. The etched ring can be easily traced with a low-magnification objective, preferable a phase contrast objective. To visualize the general cell region containing the marked cell, another larger circle (about 3 mm in diameter) can be drawn with a waterproof pencil on the outside surface of the chamber window. After laser exposure, the cell chamber should be kept in the incubator.

Photodamage studies have been performed on a variety of cells, including human red blood cells, spermatozoa, Chinese hamster ovary cells (CHO, ATCC), and rat kangaroo kidney epithelium cells (*Potorous tridactylis*, PTK2, ATCC). CHO and PTK2 cells were maintained in Gibco-BRC minimum essential medium (MEM, 10% fetal bovine serum). Semen specimens were obtained from donors with normal semen parameters according to the World Health Organization's guidelines. Semen was diluted in HEPES-buffered isotonic saline solution containing 1% human serum albumin and was injected into microchambers. Peripheral human blood was freshly drawn from a finger of a healthy donor and diluted in a PBS–glucose–heparin solution.

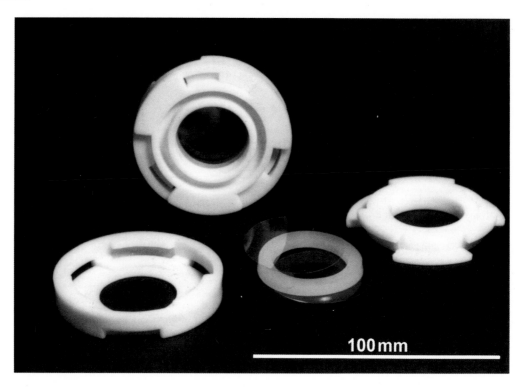

Figure 14–1 Photograph of the cell chamber for microscopy.

5. Single-Cell Studies of Photodamage

This section describes techniques used to study photodamage on single light-exposed cells in comparison with neighboring cells. The major prerequisite in such studies is to be able to trace the light-exposed cell among the surrounding non-exposed cells.

5.1 Single-Cell Studies of Morphology and Ultrastructure

Transmission light microscopy, electron microscopy, and force microscopes provide the opportunity to study photo-induced modifications on morphology and ultrastructure. For simultaneous monitoring of cellular response, the laser radiation can be used for cell exposure (fluorescence excitation) and for brightfield, phase contrast, or differential interference contrast microscopy to image photo-induced changes in morphology.

When NIR laser radiation is used at milliWatt powers, trapping effects may occur (Ashkin et al., 1986). When additional low-power white light sources are used, the laser-induced effects on the trapped cell can be imaged. For example, a NIR microbeam at a power of more than 5 mW and a halogen lamp can be used to confine a single erythrocyte and to monitor simultaneously the laser-induced effects on the cytoskeleton and the possible onset of hemolysis.

When the NIR microbeam powers are enhanced up to 100 mW, trapping forces in the range of 10–50 pN are induced, and even highly motile sperm cells can be confined. As a typical photodamage parameter, the exposure time when cells become immotile can be determined. In the case of long-term studies, adherent single cells, cell clusters, or cell monolayers should be used. It is highly recommended to involve the daughter cells of a laser-exposed cell and at least one further cell generation in the damage study. Typical long-term photodamage effects on the laser-exposed cell include membrane blebbing, cell shrinkage, loss of nucleus contours, impaired cell division, and uncontrolled cell growth. The photo-induced uncontrolled cell growth leads to the formation of giant cells with up to tenfold dimensions larger than the unexposed cell. Figure 14–2 shows the ultrastructure of a single giant CHO cell 48 hours after laser exposure with 150 fs laser pulses at 780 nm and 4 mW mean power compared with that of a normal nonexposed neighbor cell. The mitochondria-rich giant cell possesses a structured large nucleus and was not able to divide.

Scanning near-field microscopy (SNOM), force microscopes, and transmission electron microscopy allow studies of the influence of laser radiation on the cell membrane. To gain information on modification of intracellular ultrastructure at nanometer scale, however, transmission electron microscopy is the method of choice. A regular protocol is given by Liaw and Berns (1981), who describe a

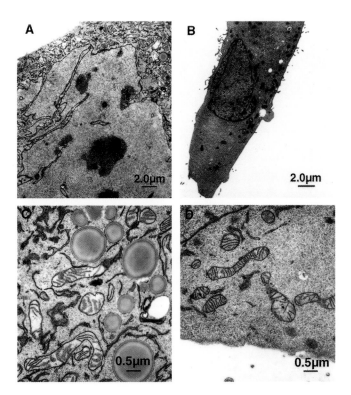

Figure 14–2 Ultrastructure of a giant CHO cell 48 hours after exposure to 150 fs laser pulses at 780 nm and 4 mW mean power (left) compared with a nonexposed control cell (right). Image: Oehring, Riemann, Halbhuber, König.

method to perform ultrastructure analysis of a particular laser-exposed single cell amidst the cell monolayer.

5.2 Application of Fluorescent Probes

There are a variety of fluorescent probes for studying photo-induced effects on cellular metabolism. For example, viability kits are available, which consist of a cell-permeable green-emitting "live cell" fluorophore, such as SYBR Green or ethidiumhomodimer (Molecular Probes), and a red fluorescent "dead cell" indicator, such as propidium iodide. Typically, these fluorophores can be excited by UVA radiation, blue-green excitation light, or NIR laser microbeams via a two-photon excitation process. Other indicators include DNA live cell stains (e.g., Hoechst 33342), calcium indicators (Fura, Indo), the pH indicator Snarf, and cytoskeleton probes. The membrane-specific fluorescent probe Laurdan (Molecular Probes, 365 nm absorption maximum) can be used to gain information on intracellular temperature changes due to its temperature-dependent fluorescence emission. Temperature-induced changes of membrane fluidity result in a Stokes' shift up to 50 nm in the blue-green spectral range (Liu et al., 1995).

Interestingly, endogenous fluorophores may also act as bioindicators of metabolic function (König and Schneckenburger, 1994). For example, the reduced pyridine coenzymes NADH and NADPH emit in the blue-green spectral range, whereas the oxidized forms do not fluoresce. Cellular damage, for example, by hypoxic conditions or light exposure, may have a significant impact on cellular NAD(P)H concentration and hence on autofluorescence. Monitoring autofluorescence modifications can therefore be a useful method to study photodamage.

5.3 Cytochemical Detection of Laser-Inflicted Damage

Yet another elegant means of understanding cellular damage induced by lasers is by performing in situ cytochemical localization studies, including immunochemistry. In contrast to routinely used conventional immunocytochemical procedures requiring the fixation of cells, cytochemical detection of reactive oxygen species (ROS) has been demonstrated in vivo. In nonlabeled living PtK2 cells exposed to 800 nm NIR laser irradiation, ROS have recently been localized using Ni-3,3'-diaminobenzidine (DAB) and Jenchrom *px blue* (Tirlapur et al., 2001). Similarly, Hockberger et al. (1999) have used DAB for cytochemical detection of H_2O_2 in cultured 3T3 cells exposed to blue light. Furthermore, the complimentary ultrastructural cytochemistry based on formation of electron-dense final products of DAB have revealed uniform staining of the peroxisomes in cells irradiated with blue light (450–490 nm).

5.4 Cloning Assay

A sensitive method to study the cellular response to laser light is to measure the reproduction behavior. It is recommended to also involve the cell division of the daughter cells and their further generations and to monitor cell clone formation. A well-known type of cells for cloning assays are the easy-to-handle CHO cells, with a normal cell division rate of about two per day. If a single cell remains unaffected by the laser exposure, the cell would divide within the next 12 h and would produce a clone of 4 cells after 24 h, of 16 cells after 2 days, and of 64 cells after 3 days. The cell growth (*CG*), which reflects the number of cell divisions, is therefore

$$CG(x) = \ln(n_x/n_0)/\ln 2 \tag{1}$$

where n_0 and n_x are the cell numbers at time of laser exposure and at a time period of x thereafter. Typical average values for nonexposed cells (control cells) are CG_0 (0–24 h) = 1.9 and CG_0 (0–48 h) = 3.8.

A relative cell growth value (*RCG*) can be defined as follows:

$$RCG = CG/CG_0 \tag{2}$$

where *CG* represents the laser-experienced cells and CG_0 the nonexposed control cells. As an example, Figure 14–3 demonstrates *RCG* values of CHO cells exposed to 250 fs pulses at 780 nm.

It is recommended to perform the exposure experiment after the first division to ensure the exposure of normal dividing cells. In our experience, most UVA- or NIR-exposed cells that are not significantly affected by the light may still have a delayed first division after exposure. Therefore, in our studies we consider clonal growth to be unaffected by photoradiation when the cell was able to form a clone

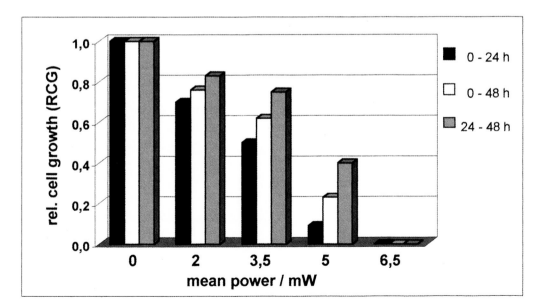

Figure 14–3 Relative cell growth values (RCG) of CHO cells exposed to 250 fs pulses at 780 nm versus mean laser power.

of at least eight cells 2 days after photoexposure ($CG[0–48$ h$] \geq 2$). A cloning efficiency of 100% means that all light-exposed cells demonstrated normal clonal growth (more than eight cells). In the case of 50% cloning efficiency (CE_{50}), half the light-exposed cells failed normal cell division. Zero percent cloning efficiency means that no cell revealed normal clonal growth. In this case, cells typically become large in size or die. For the cloning assay, we typically apply about 200 cells in the 2 ml cell chamber.

5.5 Comet Assay

Single-cell gel electrophoresis (comet assay) is a highly sensitive method to detect DNA strand breaks, to differentiate between single- and double-strand breaks, and to gain information on repair mechanisms. The assay is based on migration of the negatively charged DNA molecules in a weak electric field. When embedded in agarose gel, the migration distance is inversely proportional to the logarithm of the molecule length. Therefore, smaller DNA fragments migrate further toward the anode than larger fragments. Intact DNA remains in the nucleus. To visualize the migration of very few DNA fragments of a photodamaged cell, the DNA is typically labelled with the fluorophore propidium iodide and imaged with a sensitive camera. The DNA fluorescence pattern has the appearance of a comet with a head and a tail, indicating intact DNA and DNA fragments, respectively. Single-strand breaks can be visualized with the alkaline comet assay (alkaline conditions), whereas the neutral assay allows detection of double-strand breaks. The sensitivity of this method is about one single DNA break per 10^{10} dalton (Singh et al., 1988; Gedik et al., 1992).

To perform the single-cell alkaline comet assay, we add 1% low melting agarose at 45°C into the cell chamber, apply a lysis solution (2.5 M NaCl, 100 mM Na$_2$EDTA, 10 mM Tris, 1% sodium sarconisate, NaOH to adjust to pH = 10, 1% Triton X-100, 10% DMSO), and place the chamber in a horizontal electrophoresis tank (0.33 M NaOH, 1 mM Na$_2$EDTA). Electrophoresis is performed at 0.5 V/cm for 10–15 min. Afterwards, cells embedded in agarose are stained with propidium iodide and excited with 536 nm lamp light, 543 nm He-Ne laser, or 800 nm microbeams. Imaging is carried out with an intensified charge-coupled device camera or with the internal photomultiplier tube of a laser scanning microscope. Typically, the tail moment (tail length times tail fluorescence intensity) is used as a parameter to determine DNA damage.

6. CELL DAMAGE IN LIGHT MICROSCOPES

6.1 Cell Damage in UVA Microscopes

UVA radiation (315–400 nm) is not directly absorbed by the DNA. Severe photodamage, including DNA strand breaks may, however, occur by destructive photooxidation processes and oxidative stress, respectively (Tyrrell and Keyse, 1990). To study damage effects in CHO cells, highly focused CW UVA Ar$^+$ laser beams

($\times$63, NA = 1.25) at 364 nm, as well as the 365 nm radiation of a conventional mercury lamp of a UV Zeiss CLSM, have been employed.

Cells were scanned ten times with the laser microbeam in a 256 $\times$ 256 pixel field at 80 μs pixel dwell time. Starting with milliWatt powers, no UV-exposed cell survived. When exposed to laser powers above 20 μW, some cells remained alive but failed normal cell division (0% cloning efficiency). At a power of 6 μW, only 50% of the cells exhibited normal clonal growth (CE_{50}), and the other exposed cells died or produced giant cells. With a spot size of λ/NA $\approx$ 290 nm, the intensity and the fluence of the 6 μW microbeam was $\approx$9 kW/cm^2 and 7.2 J/cm^2 (ten scans), respectively.

For comparison, single CHO cells were exposed to 365 nm lamp light. A power of 1 mW ($\times$100, NA = 1.3) and a spot size of 190 μm resulted in 3.5 W/cm^2 intensity. A 10 s UVA exposure (35 J/cm^2) inhibited normal cell growth in all cells. A 50% cloning efficiency was obtained for about 1 s exposure time. Both the laser and lamp results confirm a severe damaging effect for 1–10 J/cm^2 fluences.

The UVA exposure induced a weak cellular autofluorescence in the blue-green spectral range due to NAD(P)H excitation. Fluorescence arose mainly in the mitochondria. The autofluorescence intensity decreased during exposure and reached a minimum after a fluence of 100 J/cm^2, likely due to auto-oxidation processes (König et al., 1996a). Further exposure, however, resulted in a strong increase in autofluorescence accompanied by an efflux of fluorophores into the cytoplasm, nucleus, and later into the extracellular medium.

The comet assay indicated photo-induced strand breaks. First breaks could be detected at about 10 J/cm^2 fluence. Severe damage was found in 95% of cells after 30 J/cm^2 UVA exposure (König et al., 1997). Spermatozoa exposed to 365 nm lamp radiation became immobile and revealed the onset of NADH-attributed autofluorescence in the sperm head, followed by cell death within 5 min of exposure.

6.2 Continuous Wave Near-Infrared Microbeam-Induced Cell Damage

It is obvious that high-intense NIR microbeams at a wavelength less than 800 nm may induce "UVA-like" destructive effects in the focal plane via a nonresonant two-photon excitation process. As indicated in Figure 14–4, two-photon excitation may result in fluorescence, but also in undesired intersystem crossing to a long-lived triplet state of endogenous or exogenous absorbers followed by photochemical reactions. Such reactions may include type I and type II photo-oxidation processes, leading to formation of toxic oxygen radicals and singlet oxygen (Hockberger et al., 1999).

Photodamage to living cells by CW NIR beams at 700–800 nm has been reported (König et al., 1995). For example, optical trapping of human spermatozoa with 0.1 W laser microbeams ($\approx$40 MW/cm^2) of the multimode CW Ti:Sapphire laser at wavelengths less than 800 nm resulted in trap-induced immobility within 1 min of exposure, disturbances in the intracellular redox state (autofluorescence modifications), and cell death. Longer wavelength CW microbeam exposure (800 nm, 1064 nm) showed no damaging effect during a 10 min trapping period and more than 20 GJ/cm^2 fluence, respectively.

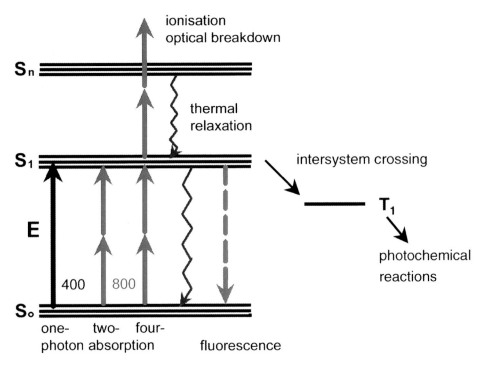

FIGURE 14–4 Principle of two-photon–induced fluorescence and photochemical damage and of photodamage by optical breakdown phenomena (see the text for details).

Using a live/dead viability kit, the NIR microbeam induced two-photon excited fluorescence of the dyes which enabled simultaneous monitoring of lethal effects during optical trapping (Fig. 14–5). Exposure of adherent CHO cells to 90 mW microbeams caused reduced cloning efficiency, giant cell formation, and cell death. When a CW Ti:Sapphire ring laser was used, a pronounced damaging effect was found at 760 nm. A 50% cloning efficiency was determined after an approximately 5 s exposure ($\approx$0.16 GJ/cm^2) at this wavelength. No cells were able to divide after 1 min exposure.

Interestingly, these damaging effects depend in part on laser output effects. In fact, the Ti:Sapphire ring laser used was not a single-frequency CW laser. Mode-beating effects resulted in the formation of unstable picosecond pulses with high transient peak powers and subsequent enhanced two-photon excitation. Transformation of the multimode CW laser into a single-frequency laser by introduction of an intracavity 20 MHz etalon reduced the damaging effect. For example, human spermatozoa could now be confined for as long as 400 ± 100 s in single-frequency traps (760 nm) compared with 60 ± 10 s in multimode traps before the onset of lethal effects. However, two-photon–induced photodamage could not be avoided (König et al., 1996b).

The action spectra of photodamage of CHO cells by CW NIR Ti:Sapphire ring laser radiation (no etalon) has been reported (Liang et al., 1996). It should be noted that the mentioned dramatic cellular response at certain wavelengths reflects not only the influence of endogenous absorbers (biological origin) but also the impact

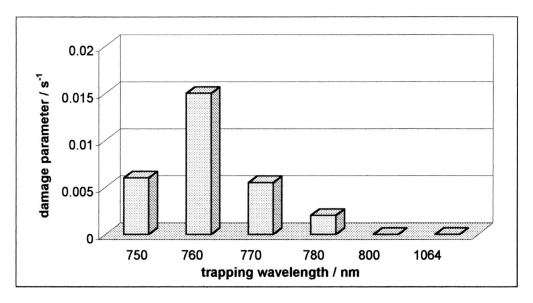

FIGURE 14–5 Loss of viability of optically confined human fluorescent spermatozoa by continuous wave near-infrared microbeams.

from the specific wavelength-dependent laser output (unstable picosecond laser pulses). The destructive effect of single-frequency CW lasers, as well as of multimode CW lasers, followed a power squared relation. There was no evidence of photodamage induced by significant cellular heating. Intracellular temperature measurements in 100 mW microbeam–exposed CHO cells and temperature calculations demonstrated only a weak temperature increase of less than 2 K (Liu et al., 1995; Schönle and Hell, 1998).

In contrast to the "transparent" spermatozoa and CHO cells, however, nontransparent erythrocytes may be damaged also by one-photon excitation of hemoglobin ($\alpha \approx 30$ cm^{-1} at 730 nm) and subsequent significant temperature increase. In particular, human erythrocytes confined in NIR traps may suffer from "heat-induced" membrane damage and hemolysis. The photodamage depends on wavelength, power, and trapping time. For example, fluences of approximately 1 GJ/cm^2 were required to induce damage at 800 nm and 50 mW power. Because low powers of less than 10 mW are sufficient to confine erythrocytes in the trap, nondestructive optical erythrocyte micromanipulation is nevertheless feasible within a wide power/energy range.

In conclusion, CW NIR laser microbeams, including laser tweezers, are sources of two-photon excitation and may act as a fluorescence excitation source and a source of "UV-like" photodamage. They can potentially induce destructive photochemical reactions, in particular when wavelengths in the 700–800 nm spectral range are used.

6.3 Cell Damage in Near-Infrared Microscopes by Ultrashort Laser Pulses

Compared with CW microbeam damage, possible multiphoton–induced destructive effects are enhanced in the case of ultrashort laser pulses due to the use of high peak powers. There is an optical window of safe multiphoton fluorescence imaging. Within this window, fluorescence microscopy can be performed on cells and embryos for hours without impact on cellular reproduction and vitality. For example, Chinese hamster embryos have been imaged with a two-photon scanning microscope for 24 hours without impact on *birth*; conventional one-photon laser scanning microscopy does not have this capability (Squirrel et al., 1999).

Photodamage does occur, however, above certain laser power (intensity) thresholds. For example, scanning of CHO cells more than 100 times (256×256 pixel field at 80 μs pixel dwell time) with an 80 MHz mode-locked Ti:Sapphire laser at 780 nm and 170 fs *in situ* pulse width reveals no changes in morphology or in cloning efficiency when the average power remained below 2 mW. In contrast, only 10 scans at higher powers of up to 8 mW lead to giant cell formation, reduced cloning efficiency, and cell death. The earliest discernible ultrastructural changes after NIR exposure were associated with mitochondria. Laser exposure results in swollen mitochondria in conjunction with loss of christae combined with the formation of electron-dense bodies in the mitochondrial matrix (Oehring et al., 2000).

The photodamage induced by NIR femtosecond laser pulses may lead to apoptosis-like cell death. In particular, changes in cytoplasmatic and nuclear calcium levels, fragmentation of F-actin cytoskeleton, disruption of plasma membrane integrity, and morphological deformations of the nuclei and nuclear envelope following exposure with 170 fs laser pulses with powers lower than 10 mW have been detected (Tirlapur et al., 2001).

To prove that water heating is not responsible for that damage, we monitored the effect of wavelength on cellular response. As expected, the damage potential was lower at longer wavelengths despite higher absorption coefficients of water.

At 780 nm, direct DNA damage could occur via a three-photon effect whereby amino acids would be excited. Microbeam exposure of out-of-nucleus regions only, however, also resulted in formation of giant CHO cells and impaired cellular reproduction.

To prove the influence of *in situ* pulse width, we varied the pulse width in the range of 80–1000 fs by a tunable pulse stretching unit. The unit consisted of either a folded SF14 prism pair at variable distance (fs range) or a pair of blazed gold-coated gratings (750 nm, 651 mm^{-1}). The gratings were used to obtain negatively chirped picosecond pulses in the range of 1–5 ps (König et al., 1999b).

For the same pulse energy, photodamage was found to be more pronounced for shorter pulses. When CHO cells were scanned ten times, 2 ps induced loss of viability at 22 mW mean power and impaired cell division at 11 mW (P_{50}), whereas 240 fs pulses induced the destructive effects at 7 mW and 3 mW, respectively. The photodamage process is likely based on a two-photon excitation process (Fig. 14–6). Because there is the same P^2/τ dependence of damage and fluorescence, the ratio of minimum laser power to induce cell damage over power to excite two-photon fluorescence is the same (typically 1–30) for picosecond as for femtosecond pulses. Both pulse width regions are appropriate in two-photon microscopy, so it makes no sense to use high-cost pulse compression units.

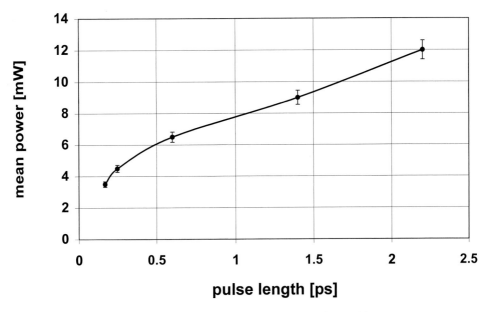

Figure 14–6 Plot showing that cell damage depends on pulse width.

7. Plasma-Induced Ablation

When power levels are higher than 10 mW, intensities rise to the TW/cm^2 region and a further destructive effect may occur. The extremely high light intensities induce intracellular optical breakdown and plasma formation accompanied by intense white luminescence and complex subnanosecond emission decay kinetics (König, 2000). Exposure at these light intensities results in severe shockwave-induced morphological damage, such as cell fragmentation. Typically, cells die immediately. During scanning, plasma formation occurs preferentially in the mitochondrial region. Onset of plasma does not typically start in the extracellular medium or in the nucleus.

Nevertheless, the immediate destructive plasma-mediated effect can be employed for highly precise processing of intracellular structures (König et al., 1999b). With 800 nm pulses at TW/cm^2 intensities in the line scan mode or for single point illumination, cutting and "drilling" can be performed with a cut size as low as 110 nm. The cut size is below the diffraction-limited spot size and reflects the effect of the highly intense core of the illumination spot. Femtosecond laser-based nanoprocessing has been used for highly localized intracellular chromosome dissection in living cells (König, 2000).

8. Conclusions and Recommendations

The UVA and NIR laser microbeams used in fluorescence microscopy can potentially harm vital cells. Ultraviolet A microbeams at 364/365 nm, used as excitation radiation in fluorescence microscopes, were found to damage cells at fluences of less than 10 J/cm². Such fluences are obtained during exposure times of seconds when high NA objectives are used. Ultraviolet A (320–400 nm) is well known to induce genetic, as well as nongenetic, damage via oxidative stress due to the formation of reactive oxygen species (i.e., singlet oxygen and oxygen radicals; Tyrrell and Keyse, 1990).

Multiphoton fluorescence imaging based on CW and ultrashort NIR microbeams appears to be a safe, novel optical section technique within a certain intensity window. The upper limit is determined by the onset of destructive effects, the lower one by the molecular multiphoton absorption cross sections of the chromophores and the detector sensitivity. Above the threshold, failed reproduction and apoptosis-like lethal effects are likely based on (1) two-photon photochemical photodamage and (2) immediate optomechanical, plasma-mediated cell damage.

The observed cell damage in the case of short-wavelength CW NIR microbeams, such as optical traps, is based on nonresonant two-photon excitation of endogenous chromophores. Due to the squared dependence of two-photon excitation on laser power, two-photon effects are amplified in multimode CW lasers where transient power fluctuations occur as a result of longitudinal mode beating. Laser tweezers at wavelengths longer than 800 nm are safer than traps in the 700–800 nm spectral range.

The transient intensity and photon flux densities of femtosecond microbeams in two-photon microscopes are on the order of 100 GW/cm^2 to TW/cm^2 and approximately 10^{31}–10^{32} photons cm^{-2} s^{-1}, respectively, for typical laser parameters (150 fs, 100 MHz, 1–10 mW average power). If possible, two-photon fluorescence microscopy in the femtosecond range should be performed with mean powers below 5–10 mW. More emphasis on sensitive fluorescence detection and the design of novel highly efficient two-photon fluorophores should be given in the future. On the other hand, highly localized destructive effects can be used to perform intracellular photochemistry and nanosurgery.

REFERENCES

Ashkin, A., J. M. Dziedzic, J. E. Bjorkholm, and S. Chu. Observation of a single-beam gradient force optical trap for dielectric particles. *Opt. Lett.* 11:288–290, 1986.

Denk, W., J. H. Strickler, and W. W. Webb. Two-photon laser scanning fluorescence microscopy. *Science* 248:73–76, 1990.

Gedik, C. M., S. W. B. Ewen, and A. R. Collins. Single-cell gel electrophoresis applied to the analysis of UV-C damage and its repair in human cells. *Int. J. Radiat. Biol.* 62:313–320, 1992.

Hänninen, P. E., E. Soini, and S. W. Hell. Continuous wave excitation two-photon fluorescence microscopy. *J. Microsc.* 176:222–225, 1994.

Hockberger, P. E., T. A. Skimina, V. E. Centonze, C. Lavin, S. Chu, S. Dadras, J. K. Reddy, and J. G. White. Activation of falvin-containing oxidases underlies light-induced production of H$_2$O$_2$ in mammalian cells. *Proc. Natl. Acad. Sci. U.S.A.* 96:6255–6260, 1999.

König, K. Invited review: Multiphoton microscopy in life sciences. *J. Microsc.* 200:83–104, 2000.

König, K., T. W. Becker, P. Fischer, I. Riemann, and K. J. Halbhuber. Pulse-length dependence of cellular response to intense near-infrared laser pulses in multiphoton microscopes. *Opt. Lett.* 24:113–115, 1999a.

König, K., M. W. Berns, and B. J. Tromberg. Time-resolved and steady-state fluorescence measurements of β-nicotinamide adenine dinucleotide–alcohol dehydrogenase complex during UVA exposure. *Photochem. Photobiol.* 37:91–95, 1997.

König, K., T. Krasieva, E. Bauer, U. Fiedler, M. W. Berns, B. J. Tromberg, and K. O. Greulich. Cell damage by UVA radiation of a mercury microscopy lamp probed by autofluorescence modifications, cloning assay, and comet assay. *J. Biomed. Opt.* 1:217–222, 1996a.

König, K., H. Liang, M. W. Berns, and B. J. Tromberg. Cell damage by near-IR beams. *Nature* 377:20–21, 1995.

König, K., H. Liang, M. W. Berns, and B. J. Tromberg. Cell damage in near infrared multimode optical traps as a result of multiphoton absorption. *Opt. Lett.* 21:1090–1092, 1996b.

König, K., I. Riemann, P. Fischer, and K. J. Halbhuber. Intracellular nanosurgery with near infrared femtosecond laser pulses. *Cell. Mol. Biol.* 45:195–201, 1999b.

König, K., and H. Schneckenburger. Laser-induced autofluorescence for medical diagnosis. *J. Fluorescence* 1:17–40, 1994.

König, K., L. Svaasand, Y. Liu, G. Sonek, P. Patrizio, Y. Tadir, M. W. Berns, and B. J. Tromberg. Determination of motility forces of human spermatozoa using an 800 nm optical trap. *Cell. Mol. Biol.* 42:501–509, 1996c.

Liang, H., K. T. Vu, P. Krishnan, T. C. Trang, D. Shin, S. Kimel, and M. W. Berns. Wavelength dependence of cell cloning efficiency after optical trapping. *Biophys. J.* 70:1529–1533, 1996.

Liaw, L. H. L., and M. W. Berns. Electron microscope autoradiography on serial sections of preselected living cells. *J. Ultrastruct. Res.* 75:187–194, 1981.

Liu, Y., D. Cheng, G. J. Sonek, M. W. Berns, C. F. Chapman, and B. J. Tromberg. Evidence for localized cell heating induced by infrared optical tweezers. *Biophys. J.* 68:2137–2144, 1995.

Oehring, H., I. Riemann, P. Fischer, K. J. Halbhuber, and K. König. Ultrastructure and reproduction behavior of single CHO-K1 cells exposed to near infrared femtosecond laser pulses. *Scanning.* 22:263–270, 2000.

Pawley, J. B. *Handbook of Biological Confocal Microscopy.* New York. Plenum Press, 1995.

Singh, N. P., M. T. McCoy, E. E. Tice, and E. L. Schneider. A simple technique for quantification of low levels of DNA damage in individual cells. *Exp. Cell. Res.* 175:184–191, 1988.

Squirrel, J. M., D. L. Wokosin, J. G. White, and B. D. Bavister. Long-term two-photon fluorescence imaging of mammalian embryos without compromising viability. *Nat. Biotechnol.* 17:763–762, 1999.

Tirlapur, U. K., K. König, C. Peuckert, R. Krieg, and K. J. Halbhuber,. Near infrared femtosecond laser pulses elicit generation of reactive oxygen species in mammalian cells leading to apoptosis-like cell death. *Exp. Cell Res.* 263:88–97, 2001.

Tyrrell, R. M., and S. M. Keyse. The interaction of UVA radiation with cultured cells. *J. Photochem. Photobiol. B* 4:349–361, 1990.

III

FLUORESCENCE RESONANCE ENERGY TRANSFER AND LIFETIME IMAGING MICROSCOPY

INTRODUCTION

The preceding chapters described microscopy methodologies to reveal the distribution of fluorescent stain in the cell and production of two- or three-dimensional images of intracellular structure based on that information. From advances in molecular biology during the past decade, we know that protein–protein association underlies the specificity of signal transduction. X-ray diffraction, nuclear magnetic resonance imaging, and electron microscopy methods have been used to study the structure and localization of proteins under nonphysiological conditions. We now need methods to study protein associations in living cells in three dimensions and in real time. Fluorescence resonance energy transfer (FRET) provides such a method.

The visualization and quantitation of protein associations under physiological conditions using steady-state fluorescence, two-photon excitation, and time-resolved, or lifetime imaging microscopy, methods are explained in the following four chapters. The fluorescence lifetime is defined as the average time that a molecule remains in an excited state before returning to the ground state. Two lifetime methods, time domain (Chapters 15, 17, and 18) and frequency domain (Chapters 16 and 19), are described in this section.

Chapter 15 outlines the basics of FRET, lifetime imaging microscopy, for various biological applications of FRET (monitoring of (1) protease activity of cellular proteins during apoptosis; (2) calcium, using calcium-sensitive cameleons; and (3) plasma membrane potential), and a FRET screening assay for quantification of gene expression using β-lactamase. The author also describes bioluminescence resonance energy transfer (BRET), which uses luciferase fusion protein as a donor and green fluorescent protein as an acceptor. This chapter introduces in the literature the correction for bleedthrough (or cross-talk), which is an inherent problem in steady-state (or digitized video) FRET microscopy. The author also explains the importance of lifetime imaging for FRET imaging, which provides dynamic events of the proteins (spatial and temporal) since the lifetime method does not depend on excitation intensity or fluorophore concentration targeted to the proteins. Theoretical and experimental techniques are described in Chapter 16 for frequency domain lifetime FRET imaging. Double-exponential decays are resolved using a single-frequency fluorescence lifetime imaging instrument for protein phosphorylation between a fused green fluorescent protein and an inodcyanine dye (Cy3) bound to an antibody against a phosphoamino acid.

Chapter 17 compares the various FRET microscopic techniques, such as wide-field, confocal, two-photon excitation, and double-exponential lifetime decay for FRET imaging. Moreover, this chapter describes the methodology of the correction for bleedthrough in intensity-based FRET imaging techniques.

The author also points out that it is possible to quantitate one or more protein associations with the lifetime method compared with the other three techniques. Lifetime FRET images are provided for the donor (CFP) in the presence and the absence of the acceptor (YFP) for enhancer binding protein-α (C/EBPα) dimerization in pituitary GHFT1-5 living cells.

Chapter 18 describes the time domain method of one- and two-photon excitation fluorescence lifetime imaging (FLIM) for calcium, pH, and oxygen concentration in biofilm. In the time-resolved, pump-probe microscopy (Chpater 19), two laser beams at different wavelengths and modulation frequencies are focused onto a diffraction-limited spot, and this allows recording of time-resolved images and lifetime measurements for cytoplasm and nuclear areas of the fibroblast cells.

15

Measurement of Fluorescence Resonance Energy Transfer in the Optical Microscope

Brian Herman, Gerald Gordon, Nupam Mahajan, and Victoria Centonze

1. Introduction

Fluorescence resonance energy transfer (FRET) can be used as a spectroscopic ruler to study and quantify the interactions of cellular components with each other, as well as the conformational changes within individual molecules at the molecular level (Herman, 1998). FRET is a process by which a fluorophore (donor) in an excited state may transfer its excitation energy to a neighboring chromophore (acceptor) nonradiatively through dipole–dipole interactions. This energy transfer manifests itself as both quenching of donor fluorescence intensity and lifetime (in the presence of acceptor) as well as an increase in the emission of acceptor fluorescence (sensitized emission). Because FRET decreases in proportion to the inverse sixth power of the distance between the donor and acceptor, this phenomenon is effective at measuring separation of the donor- and acceptor-labeled molecules when they are within 10–100 Å of each other.

FRET has long been used to examine inter- and intramolecular alterations in protein and lipid structure and function. More recently, the (re)discovery of green fluorescent protein (GFP), mutagenesis of GFP cDNA to yield numerous spectrally distinct mutant versions (mtGFPs) of this protein in conjunction with advances in optical imaging technologies, has enabled the measurement of interactions of intracellular molecular species in intact living cells, where the donor and acceptor fluorophores are actually part of the molecules themselves (Chalfie, 1995). In this chapter, we review the principles of FRET, the techniques used to measure FRET through an optical microscope, and recent applications of FRET measurements in the study of protein–protein interactions, calcium metabolism, alterations in membrane voltage potentials, protease activity, and high throughput screening assays.

2. PRINCIPLES OF FRET

In principle, if a donor molecule's fluorescence emission spectrum overlaps the absorbance spectrum of a fluorescent acceptor molecule and is spatially close enough to the acceptor molecule, the donor can transfer its excitation energy to the acceptor via dipole–dipole interactions (Fig. 15–1). Energy transfer efficiency varies most importantly as the inverse of the sixth power of the distance separating the donor and acceptor fluorophores. The energy transfer to the acceptor is nonradiative (is not mediated by a photon) and requires the distance between the chromophores to be relatively close (usually not exceeding 100 Å). The phenomenon can be detected by exciting the labeled specimen with light of wavelengths corresponding to the absorption maximum of the donor and detecting light emitted at the wavelengths corresponding to the emission maximum of the acceptor or by measuring the fluorescent lifetime of the donor in the presence and absence of the acceptor. The dependence of the energy transfer efficiency on the donor–acceptor separation provides the basis for the utility of this phenomenon as a molecular spectroscopic ruler.

Successful undertaking of FRET requires that several points be considered. (*1*) The concentrations of the donor and acceptor fluorophores need to be tightly controlled. Statistically, the highest probability of FRET occurs when multiple acceptor molecules surround a single donor. (*2*) Photobleaching needs to be prevented. Photobleaching can alter the donor–acceptor ratio and therefore the measured value of FRET. (*3*) The donor emission spectrum should substantially overlap the absorption spectrum of the acceptor. (*4*) There should be relatively little direct excitation of

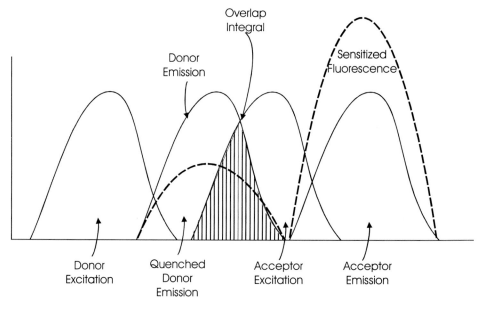

FIGURE 15–1 Principle of FRET. When donor and acceptor molecules are within 10–100 Å of one another, an excited donor can transfer its excitation energy to the acceptor in a nonradiative fashion. This leads to a decrease (quenching) in the emitted intensity of the donor and an increase in the emitted intensity of the acceptor (sensitized emission).

the acceptor at the excitation maximum of the donor. (5) The emission of both the donor and the acceptor must occur in a wavelength range where the detector has maximum sensitivity. (6) There should be little, if any, overlap of the donor absorption and emission spectrum, thus minimizing donor–donor self-transfer. (7) The emission of the donor should ideally result from several overlapping transitions and thus exhibit low polarization. (8) For FRET to occur, the donor must be fluorescent and have a sufficiently long lifetime. (9) It is necessary to demonstrate that the donor- and acceptor-conjugated reagents retain appropriate biological activity. (10) When undertaking FRET measurements, the molecular location of the donor and acceptor with respect to the tertiary structure of the reagent(s) that it is attached to needs to be considered. For example, donor–acceptor GFPs can be attached to either the carboxy or amino terminus of a protein. Because FRET requires the donor and acceptor fluorophores to be within 10–100 Å of each other (see later), and their dipoles are appropriately aligned with respect to one another, the potential exists that FRET might not be observed even though the proteins to which they are attached do interact if the donor and acceptor are on opposite ends (carboxy vs. amino terminus) of the interacting molecules.

The rate of energy transfer (K_T) and the energy transfer efficiency (E_T) are both related to the lifetime of the donor in the presence or absence of the acceptor:

$$K_T = \left(\frac{1}{\tau_D}\right)\left[\frac{R_0}{R}\right]^6 \tag{1}$$

where R_0 is the Förster critical distance (the donor–acceptor separation [in Angstroms] for which the transfer rate equals the donor de-excitation rate in the absence of acceptor), τ_D is the lifetime of the donor in the absence of the acceptor, and R is the distance separating the donor and acceptor molecules. R_0 is a measure of the maximal distance that can separate the donor and acceptor under which FRET will still occur. R_0 can be calculated from

$$R_0 = 0.211[K^2 \eta^{-4} Q_D J(\lambda)]^{1/6} \tag{2}$$

where K^2 is the orientation factor that describes the spatial relationship between the donor and acceptor absorption and emission dipoles in space ($K^2 = 2/3$ if the orientation of the donor and acceptor are random), η is the refractive index of the medium, Q_D is the quantum yield of the donor in the absence of acceptor, and $J(\lambda)$ is the overlap integral.

E_T is related to R by

$$R = R_0[(1/E_T) - 1]^{1/6} \tag{3}$$

where

$$E_T = 1 - (\tau_{DA}/\tau_D) \tag{4}$$

where τ_{DA} is the lifetime of the donor in the presence of the acceptor. Thus, by measuring the lifetime of the donor in the presence and absence of the acceptor, one can calculate the distance between the donor and acceptor.

3. TECHNIQUES TO MEASURE FRET IN THE OPTICAL MICROSCOPE

By combining optical microscopy with FRET, it is possible to obtain quantitative temporal and spatial information about the binding and interaction of proteins, lipids, enzymes, DNA, and RNA in vivo. Because energy transfer occurs over distances of 10–100 Å, a FRET signal corresponding to a particular field (or pixels) within a microscope image provides additional information beyond the optical microscopic limit of resolution. FRET microscopy is particularly useful when examining *temporal* and *spatial* changes in the distribution, interaction, and change in conformation of fluorescently conjugated biological molecules in living cells.

3.1 Intensity

Currently, most fluorescence microscopic imaging is performed as measurements of fluorescence emission. In FRET imaging, one typically measures changes in the relative amounts of emission intensity at two wavelengths, that of the donor and that of the acceptor. When FRET occurs, there is an increase in acceptor emission (I_A) and a concomitant decrease in donor emission (I_D). In a microscope, FRET can be detected by exciting the donor- and acceptor-labeled specimens with light of wavelengths corresponding to the absorption maximum of the donor and detecting light emitted at the wavelength corresponding to the maximum of the emission spectrum of the acceptor.

Although a change in the relative emission of either the donor or acceptor can be taken as indicative of FRET, it is customary to use the ratio of these two values (I_A/I_D) as a measure of FRET. The value of this ratio depends on the average distance between donor–acceptor pairs and allows corrections for differences in path length and accessible volume of the specimen, similar to the principle underlying ratio ion imaging (Diliberto et al., 1994). Because this method is simple and easy, it has been used in a variety of different FRET imaging applications (Adams et al., 1991; Periasamy et al., 1995).

Measurement of FRET using this intensity approach is also known as *steady-state FRET imaging* (or digitized video FRET imaging, see Chapter 17). There are a number of factors, however, that can affect the accuracy of this type of measurement that must be considered. We have discussed many of these sources of error (see earlier). Those specific to the optical microscope include the desire to obtain images with high signal-to-noise (S/N) ratio and the use of optical filters to spectrally discriminate donor and acceptor absorption and emission spectra. High S/N ratio images require a balance between time and intensity of excitation light exposure, optical efficiency of the microscope/detector system, and concentration of donor and acceptor fluorophores. Too high of a concentration of donor–acceptor fluorophores can cause self-quenching, which affects the accuracy of the FRET measurement. Photobleaching (which all fluorescent molecules are sensitive to) can alter the donor–acceptor ratio and therefore the value of FRET, as well as damage the specimen, especially if living cells or tissues are being observed.

In fixed specimens, however, it is possible to take advantage of photobleaching to measure FRET. This technique, known as *donor photobleaching fluorescence resonance energy transfer* (pbFRET) microscopy, has been used to measure proximity relationships between molecules, especially cell surface proteins, labeled with fluorophore-conjugated monoclonal antibodies, on a pixel-by-pixel base using digital imaging microscopy (Jovin et al., 1990). Photobleaching FRET is based on the theory that the fluorophore is sensitive to photodamage only when it is in its excited state. Because only a small percentage of fluorophores are in the excited state at any one time, fluorophores with longer lifetimes will have a higher probability of suffering photodamage and thus a faster rate of photobleaching. FRET, which shortens the fluorescent lifetime of the donor (see later) will protect against photobleaching. In fact, it has recently been shown that the photobleaching time of a fluorophore is inversely related to its excited state lifetime. This technique works well in fixed cells where photobleaching will not affect cell function and temporal resolution is not required (i.e., minutes are required for data acquisition using this approach).

For maximal FRET with any donor–acceptor pair, the donor emission spectrum should substantially overlap the absorption spectra of the acceptor, there should be relatively little direct excitation of the acceptor at the excitation maxima of the donor, and there should be little emission of the donor where the acceptor emission occurs. Unfortunately, in practice, it is difficult to identify donor–acceptor fluorophore pairs that meet these criteria. In most cases, overlap of the donor emission and acceptor absorbance is less than ideal, some direct excitation of the acceptor at the donor absorbance maximum occurs, and some emission of the donor occurs where the acceptor emission is maximal. This situation is further confounded in FRET microscopy because most of the commercially available optical filters used in fluorescent microscopes are not 100% efficient. Thus, these filters allow leakage of a small percentage of light with wavelengths outside of those that the filters are designed to pass. Unless one uses tightly regulated expression systems, it is often difficult to quantitatively titrate the concentration of the donor–acceptor fluorophores. Finally, corrections for background fluorescence, autofluorescence, and photobleaching may also be required.

Accurate FRET measurements in the microscope require that these potential sources of error be corrected. A simple method to correct for cross-talk (any detection of donor fluorescence with the acceptor emission filter and any detection of acceptor fluorescence with the donor emission filter) and for the dependence of FRET on the concentrations of the donor and acceptor has recently been developed (Gordon et al., 1998). The method requires a minimum of spectral information and can be readily implemented on a microscope or in a fluorometer.

The method uses three-filter sets: the donor, FRET, and acceptor filter sets. These filter sets are designed to isolate and maximize three signals: the donor fluorescence, the acceptor fluorescence due to FRET, and the directly excited acceptor fluorescence, respectively. The excitation filters for the donor filter set and the FRET filter set are either the same filter or two matched filters. The emission filters for the FRET and acceptor filter sets are either the same filter or matched filters. Microscopic samples containing just donor, just acceptor, or both donor and acceptor are examined with each of the filter sets, and the data are corrected for cross-talk and donor–acceptor concentrations. Neutral density filters may be used to match the intensity information using the different filter sets.

In practice, the fluorescence of the three different specimens is measured with the three separate filter sets: the donor filter set (*D*) excites the donor and detects primarily the donor emission, the acceptor filter set (*A*) excites the acceptor and detects primarily the acceptor emission, and the FRET filter set (*F*) excites the donor and detects primarily the acceptor emission. In the two-letter and three-letter symbols below, the first letter indicates the filter set applied (*D, A,* or *F*). The second letter indicates the fluorophores present (*a* for acceptor only, *d* for donor only, and *f* for both donor and acceptor). The third letter, if present, occurs only if the second letter is *f* and indicates that the symbol refers to the signal from only one of the fluorophores present (*a* for the signal from the acceptor or *d* for the signal from the donor). A bar over a symbol modifies the meaning of the symbol and indicates that the fluorescence would have occurred if there was no FRET.

The first step in the method is to express each of the three measured quantities as the sum of its contributions from the donor and acceptor fluorescence.

$$Df = Dfd + Dfa \tag{5a}$$

$$Ff = Ffd + Ffa \tag{5b}$$

$$Af = Afd + Afa \tag{5c}$$

Each of the donor contributions on the right side of Equations 5a, 5b, and 5c (*Dfd, Ffd,* and *Afd*) is then expressed as the difference between the fluorescence that would have occurred if there were no FRET and the fluorescence loss due to FRET. $\overline{Dfd}$ is the donor signal that would have occurred if there was no FRET using the donor filter set. $\overline{Dfd}$ is proportional to the total concentration of the donor labeled species. *FRET1* is the loss of donor fluorescence due to FRET using the donor filter set. *FRET1* is proportional to the concentration of interacting (bound) donor-labeled and acceptor-labeled species.

Each of the acceptor contributions on the right sides of Equations 5a, 5b, and 5c (*Dfa, Ffa,* and *Afa*) is then expressed as the sum of the fluorescence that would have occurred if there were no FRET and the fluorescence increase due to FRET. $\overline{Afa}$ is the acceptor signal that would have occurred if there was no FRET using the acceptor filter set. $\overline{Afa}$ is proportional to the total concentration of the acceptor-labeled species. The next step is to express all three equations in terms of $\overline{Dfd}$, $\overline{Afa}$, and *FRET1* with the result shown in Equations 6a, 6b, and 6c.

$$Df = \overline{Dfd} - FRET1 + \overline{Afa}\frac{Da}{Aa} + G \cdot FRET1\frac{Da}{Fa} \tag{6a}$$

$$Ff = (\overline{Dfd} - FRET1)\frac{Fd}{Dd} + \overline{Afa}\frac{Fa}{Aa} + G \cdot FRET1 \tag{6b}$$

$$Af = (\overline{Dfd} - FRET1)\frac{Ad}{Dd} + \overline{Afa} + G \cdot FRET1\frac{Ad}{Fd} \tag{6c}$$

There are seven new quantities introduced in equations 6a, 6b, and 6c. There are six two-letter symbols representing the fluorescence of specimens containing either only donor or only acceptor measured with each of the three-filter sets. The seventh is *G*, which is the factor relating the increase in acceptor signal due to FRET using the FRET filter set to the loss of donor signal due to FRET using the donor filter set. All of the terms with ratios of the two-letter symbols are due to cross-talk. Equations 6a, 6b, and 6c are three equations with three unknowns, $\overline{Dfd}$, $\overline{Afa}$ and *FRET1*. Solving for the unknowns yields the values needed to calculate *FRETN* defined as

$$FRETN = \frac{FRET1}{\overline{Dfd} \cdot \overline{Afa}} \propto \frac{[bound]}{[total\ d] \cdot [total\ a]} \qquad (7)$$

where [*total d*] is the total (free and bound) concentration of donor-labeled molecules and [*total a*] is the total concentration of acceptor-labeled molecules. We have incorporated these calculations into a simple EXCEL spreadsheet format that is available to interested parties.

3.2 Lifetime

As we can see, the use of steady-state FRET imaging requires the correction of many potential sources of errors. To address this situation, time-resolved FRET microscopy, which allows the quantitative measurement of donor–acceptor separation distances, has been developed. Time-resolved FRET is based on measuring the lifetime of the donor in the presence and absence of the acceptor (Herman et al., 1997; see also Chapter 17).

The fluorescence lifetime (τ) is the characteristic time that a molecule remains in an excited state before returning to the ground state. For single exponential decay of fluorescence, after a brief pulse of excitation light, the fluorescence intensity as a function of time is described as

$$I(t) = I_0 \exp(-t/\tau) \qquad (8)$$

where I_0 is the initial intensity immediately after the excitation pulse. The fluorescence lifetime (τ) is defined as the time it takes for the fluorescence intensity to decay to $1/e$ (37%) of the initial intensity (I_0).

Time-resolved FRET measurements have many advantages. Fluorescent lifetimes are independent of local intensity, concentration, and to a large extent photobleaching of the fluorophore. Fluorophores with similar spectra may have significant differences in their lifetimes, and the same fluorophore may display distinct lifetimes in different environments. Because fluorescent lifetimes are not affected by scattering, measurements of fluorescent lifetimes provide more sensitive and quantitative information from complex structures. The advantage of time-resolved FRET imaging is that the donor–acceptor distance can be mapped in a more accurate and quantitative manner.

There are two methods commonly employed to measure fluorescent lifetimes: time-domain pulsed methods (Periasamy et al., 1996) and frequency-domain or phase-resolved methods (Gadella et al., 1993; Lakowicz et al., 1991). Time-domain lifetime measurements employ pulsed excitation sources, and the fluorescent lifetime is determined directly from the fluorescence signal or by photon counting detection. Frequency-domain (phase-resolved) lifetime measurements utilize sinusoidally modulated light as an excitation source, and lifetimes are calculated from the phase shift or (de)modulation depth of the fluorescence emission signal. The relative merits of each of these approaches are currently a topic of debate.

Time-dependent (pulsed) lifetime measurements require detection systems with sufficient temporal resolution to capture most of the emitted photons from each excitation pulse. The extraction of complicated (multicomponent) lifetimes using phase modulation can require prolonged (i.e., excessive) exposure of the sample to damaging excitation energies and inadequate temporal resolution for biological processes. Both of these approaches can be used with wide-field, confocal, or multiphoton microscopy. Other chapters in this book deal with the use of time-domain methods to measure fluorescent lifetimes using multiphoton microscopy (Chapter 18) and FRET using phase-modulation lifetime imaging (Chapter 16).

We have developed instrumentation to image fluorescent lifetimes in single living cells with high sensitivity and temporal resolution (Wang et al., 1992). In wide-field fluorescent lifetime imaging microscopy (FLIM), parallel image detection using a high-speed two-dimensional gated image intensifier (with gating times from a subnanosecond to a few nanoseconds) is employed to acquire a series of images during the decay of the fluorescent lifetime, following pulsed excitation (Fig. 15–2).

A pulsed light source (a picosecond or femtosecond laser) is delivered to a fluorescence microscope through a multimode optical fiber system, which is focused to illuminate the entire field of view through a high numerical aperture objective lens. Time-resolved fluorescence microscopic images are obtained using a high-speed gated multichannel plate (MCP) image intensifier. The gated MCP is turned on for a very brief interval at some time interval (t_1) after the exciting pulse. The image acquired on the gated MCP image intensifier is focused at unity magnification onto a slow-scan cooled charge-coupled device (CCD) camera (which is continually left on).

This process is repeated many times at time t_1 with the emitted intensity from the gated MCP continually being accumulated on the CCD. The CCD is then read out, the timing of the gate window with respect to the excitation pulse is temporally shifted (t_2), and the whole process is repeated. Thus, time-resolved fluorescence images can be directly detected in short time and stored in a computer. In a given situation, the available fluorescence light level determines the integration time required to obtain acceptable S/N ratio images. Cooling the CCD to reduce dark current to negligible levels allows very long integration times and improves S/N.

Following the acquisition of a series of images during the decay of the fluorescent lifetime, a rapid calculation of fluorescent lifetimes is performed using a simple, compact, easy to implement algorithm. A single exponential decay of the fluorescence signal excited by a short-duration pulsed light source can be written as

$$F(t) = A \exp\left(-t / \langle \tau \rangle\right) \tag{9}$$

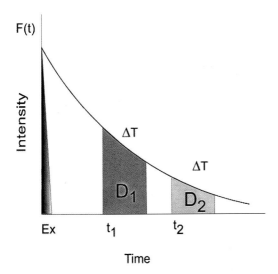

FIGURE 15–2 The principle of operation of the gated image intensifier used for time-resolved FLIM. At some time after the excitation pulse (t_1), the high voltage to the gated image intensifier is turned on for duration ΔT. This allows the intensifier to accumulate photons from the sample. The image accumulated on the intensifier at time t_1 (for the duration ΔT) is then read onto an integrating charge-coupled device (CCD) camera, and this process is repeated (using the same t_1 and ΔT) until the image integrated on the CCD is of a sufficient signal-to-noise ratio for accurate lifetime determination (~200 photons/pixel for 10% error in lifetime value). A second lifetime image (required for ratiometric lifetime determination) is then acquired in an identical fashion with the exception that the high voltage to the gated image intensifier is turned on at a time t_2 for duration ΔT. The lifetime ($\langle\langle \tau \rangle\rangle$) of the fluorophore can then be calculated as $\langle \tau \rangle = (t_2 - t_1)/$ natural log(D_1/D_2).

Because the decay of fluorescence is detected at two different delay times (t_1 and t_2) with the gate width ΔT, the fluorescence lifetime ($\langle\langle \tau \rangle\rangle$) can be calculated directly using only four parameter: the gated fluorescence signals, D_1 and D_2, and t_1 and t_2,

$$\langle \tau \rangle = (t_2 - t_1)/\text{natural log } (D_1/D_2) \tag{10}$$

4. BIOLOGICAL APPLICATIONS OF FRET

Within the past decade, more and more measurements of FRET in living cells or tissues have been made through the optical microscope. These include studies of growth factor and other cell surface receptor dimerization, the interaction of signal transduction components, endosomal fusion, the interaction of both endogenous and exogenous cellular proteins, protease activity, monitoring calcium and plasma membrane potential, and, most recently, high throughput screening assays.

4.1 Growth Factor and Other Cell Surface Receptor Dimerization

Studies of epidermal growth factor (EGF) receptor interaction have been carried out using frequency-domain lifetime imaging microscopy and FRET (Gadella and Jovin,

1995). EGF receptor clustering during signal transduction was monitored. This study demonstrated that the binding of EGF led to a fast and temperature-dependent microclustering of the EGF receptor. Also, the study showed that a concerted rotational rearrangement of the monomeric units comprising the dimeric receptor led to a potentiated mutual activation of the receptor tyrosine kinase domains.

Studies examining the effect of plasma membrane modification (by exogenous cholesterol and phosphatidylcholine) on the expression of free HLA heavy chains and β_2m-bound HLA-I molecules on JY human B lymphoblasts have also been performed using FRET microscopy (Bodnar et al., 1996). Both the cholesterol level and the lipid structure/fluidity of the plasma membrane in lymphoblastoid cells were found to regulate lateral organization and consequently the presentation efficiency of HLA-I molecules. CD8/MHC-1 and LFA-1/ICAM interactions in cytotoxic T lymphocytes were studied using FRET microscopy and suggested a critical role for CD8β-subunit in the signal transmission of peripheral T lymphocytes (Bacso et al., 1996). Finally, studies using pbFRET demonstrate that clustering of the mast cell function-associated antigen (MAFA) on the surface of rat mucosal-type mast cell line 2H3 (RBL-2H3) leads to suppression of the secretory response induced by the type I Fc epsilon receptor (Jurgens et al., 1996).

4.2 Endosomal Fusion

Time-resolved FRET imaging has been used to study the extent of membrane fusion of individual endosomes in single cells (Oida et al., 1993). With time-domain FLIM and FRET, the extent of fusion and the number of fused and unfused endosomes were clearly visualized and quantitated.

4.3 Interaction Signal Transduction Components

Recently, FRET and FLIM have been used to image protein phosphorylation in living cells following receptor activation (Wouters and Bastiaens, 1999). Ligand-induced activation of the EGF receptor via phosphorylation was visualized with microinjected Cy3 phosphorylation epitope-specific antibodies (as the acceptor) into cells expressing either a carboxy-terminal GFP ErbB1 receptor fusion construct or a GFP–protein kinase Cα (PKCα) fusion construct (as the donor). EGF binding led to ErbB1 activation, causing tyrosine phosphorylation of the ErbB1 receptor and recruitment of the Cy3-phosphotyrosine specific antibody to the activated ErbB1 receptor. Similarly, activation of PKCα led to autophosphorylation and recruitment of Cy3-labeled antibodies to PKCα, as the antibody in this case only recognizes the phosphorylated epitope on PKCα. This approach is useful for measuring the dynamics of receptor activation and phosphorylation-dependent activity in cells.

4.4 Interaction of Endogenous and Exogenous Cellular Proteins

One of the most useful applications of FRET microscopy is in the examination of dynamic interactions between cellular components with each other and/or with exogenous proteins. The literature is replete with examples of these types of studies, and we discuss only a few representative studies here. In these studies, donor and acceptor fluorophores may be bound to interacting molecules indirectly via antibody labeling or directly via chemical derivatization or as recombinant fluorescent fusion proteins (i.e., GFP). Regardless of the method used for labeling the molecules of interest, it is essential to determine whether binding of the antibody or derivatization of the cellular constituent with the fluorophore alters its biological activity.

FRET imaging has been used to determine whether human papillomavirus (HPV) 16 E6 protein and the tumor suppressor p53 protein are in close ($\leq$50 Å) physical proximity to each other in HPV-infected cells (Liang et al., 1993). In this study, FRET was quantitated as the ratio of acceptor emission to donor emission when the samples were excited at the maximal absorbance of the donor using p53–FITC (donor) and E6–rhodamine (acceptor) conjugated antibodies. The results of these studies documented, for the first time, that HPV 16 E6 is physically associated with p53 in HPV-expressing cells.

Apoptosis is a physiological process of cell death resulting from an intricate cascade of sequential protein–protein interactions. Specific proteins from the Bcl-2 multigene family are known to either positively or negatively influence apoptotic progression. One hypothesis that has been advanced to explain how this family of proteins exerts their pro- versus antiapoptotic activity is that the ratio of the expression of pro- versus antiapoptotic Bcl-2 family members serves as a rheostat to determine the cellular sensitivity to apoptosis. In particular, pro- and antiapoptotic family members can undergo homotypic, as well as heterotypic interactions, and through these interactions can modulate overall biological pro- and antiapoptotic activity.

To examine homotypic versus heterotypic interactions between Bcl-2 family members during apoptosis, EGFP–Bax and BFP–Bcl-2 or CFP–Bax and YFP–Bcl-2 fusion proteins were co-expressed in the same cell (Mahajan et al., 1998). FRET microscopy, performed with an Olympus IX70 and a Hamamatsu Orca-II camera interfaced to a Metamorph (Universal Imaging Corp.) image processing system, was used to obtain measurements of fluorescent intensities from regions of interest in cells using the three-filter set/three specimen FRET approach described earlier (Color Fig. 15–3). The data indicate that Bcl-2 and Bax interact in mitochondria of living cells and that apoptosis is associated with a decrease in the interaction of these two proteins (Fig. 15–4). These data are consistent with the hypothesis that apoptosis decreases the interaction of Bcl-2 with Bax, preventing Bcl-2 from inhibiting the pro-apoptotic activity of Bax, leading to programmed cell death.

4.5 Protease Activity

Caspase-mediated proteolysis of many cellular proteins is a critical event during apoptosis. To determine which, where, and when specific caspases are activated in mammalian cells upon receipt of specific death stimuli, various caspase-specific FRET fluorescent substrates, consisting of four amino acid long caspase-specific recognition sequences (e.g., YVAD for caspase-1 and DEVD for caspase-3), introduced between either BFP or CFP (donors) and GFP or YFP (acceptors) (Mahajan et al., 1999) have been developed. Following transfection and

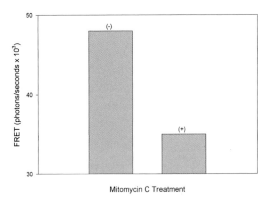

FIGURE 15–4 Alteration in Bcl-2/Bax interaction following stimulation of apoptosis by mitomycin C (MMC- 5 μg/ml for 12 hr). MMC treatment resulted in a decrease in the interaction of Bcl-2 with Bax as assessed by FRET microscopy.

expression in cells (in the absence of endogenous caspase activity), the proximity between the fluorophores, BFP and GFP or CFP and YFP, results in FRET. After induction of apoptosis, caspase-specific fluorescent substrates were cleaved, resulting in a loss of FRET. These targetable caspase-specific substrates are likely to prove highly valuable in elucidating the pathways regulating apoptosis.

4.6 Calcium

Cameleons are chimeric proteins consisting of BFP or CFP (donor), calmodulin (CaM), a glycylglycine linker, the CaM-binding domain of myosin light chain kinase (M13), and EGFP or YFP (acceptor) (Miyawaki et al., 1998). Ca^{2+} binding to CaM causes a change in the interaction of CaM with M13 and an increase in the efficiency of FRET. As with the caspase-specific FRET substrates just discussed, because cameleons can be targeted genetically they offer great promise for monitoring Ca^{2+} in whole organisms, tissues, organelles, and submicroscopic environments where such measurements were previously impossible.

The original cameleons were not particularly sensitive to changes in Ca^{2+} at resting cytoplasmic Ca^{2+} levels and were highly sensitive to quenching at physiological pH. Improved cameleons that have been recently developed are more resistant to quenching at physiological pH and display a higher sensitivity to Ca^{2+} at resting mammalian cytoplasmic Ca^{2+} levels (Miyawaki et al., 1999). Additional modifications to the GFP molecules themselves (i.e., circular permutation) may prove useful to enhancing the efficiency of the FRET process between donor and acceptor GFP fusion proteins (Baird et al., 1999).

4.7 Plasma Membrane Potential

Improved membrane potential sensors, based on FRET between voltage-sensing oxinol dyes and voltage-insensitive donor fluorophores associated with cell membranes, have been developed (Cacciatore et al., 1999). This assay retains the voltage sensitivity of oxonol probes while reporting real-time changes in membrane potential. In this approach, two dye molecules, a coumarin-linked phospholipid (CC2-DMPE) and an oxonol dye (DiSBAC$_2$[3]), are loaded into the plasma membrane of cells. CC2-DMPE partitions into the outer leaflet of the plasma membrane, where it acts as a fixed FRET donor to the mobile voltage-sensitive oxonol acceptor. A negative potential inside will push the negatively charged oxonol to the outer leaflet of the plasma membrane, resulting in FRET. Conversely, depolarization of the plasma membrane results in rapid translocation of the oxonol to the inner surface of the plasma membrane, decreasing FRET.

FRET is measured as a change in the ratio of the donor to acceptor emission when excited at the donor absorption maximum and has a reported sensitivity of 1% change in the FRET ratio/1 mV change in plasma membrane potential (for DiSBAC$_2$[3]). This FRET assay (using DiSBAC$_2$[3]) is 100 times faster than standard oxonol redistribution assays. Increasing the hydrophobicity of the oxonol can further increase the temporal resolution of this assay, allowing real-time measurement of action potentials.

4.8 High Throughput Screening

A decrease in reagent costs and an increase in assay numbers/unit time are two highly attractive features of high throughput screening. The recent developments of highly sensitive detection systems, miniaturization, and microfluidics have allowed the reduction of assay volumes from 100–200 μl to 1–2 μl and the ability to screen over 100,000 compounds a day. FRET is particularly well suited for high throughput screening because the results produced are in the format of a ratio between the two wavelengths. This greatly reduces the many potential artifacts and signal variability, such as variability in cell number, probe concentration, optical paths, or light fluctuations. For FRET high throughput screening applications, a multiwell plate format is preferred. These plates are typically constructed of a black polymer lining the wells and a low fluorescence clear bottom with good light transmission in the region of the ultraviolet to red wavelengths.

A number of examples of high throughput FRET assays have been recently described (Gonzalez et al., 1999; Mere et al., 1999; Zlokarnik et al., 1998). These include adaptation of the protease and plasma membrane voltage FRET assays previously described. One representative high throughput FRET screening assay is used for quantification of gene expression in single living cells using β-lactamase as a reporter of gene expression (Mere et al., 1999; Zlokarnik et al., 1998). The basic concept of this assay is as follows: The enzyme β-lactamase hydrolyzes the β-lactam ring of cephalosporin, creating a free amino group that triggers spontaneous elimination of any leaving group previously attached to the 3' position of cephalosporin. This leads to disruption of the normally close apposition of the 3' and 7' positions of the cephalosporin ring structure. 7-Hydroxycoumarin (the donor) is attached to the 7' position, while fluorescein (the acceptor) is attached to the 3' position. This donor–acceptor cephalosporin has been named CCF2, and a membrane-permeate ester derivative of this cephalosporin (CCF2/AM) can be added to cell cultures containing transfected gene constructs containing β-lactamase. In the intact cephalosporin the 3′ and 7′ positions are in close proximity and FRET occurs. Following the expression of the gene containing the β-lactamase, CCF2 is cleaved at the 3′ position, leading to a decrease in FRET. The extent by which FRET decreases is directly proportional to the level of expression of the specific gene reporter construct.

4.9 Bioluminescence Resonance Energy Transfer

Bioluminescence resonance energy transfer (BRET) is similar to FRET except that it uses a bioluminescent luciferase fusion protein as the donor and a GFP fusion protein as acceptor (Xu et al., 1999). The main advantage of BRET over FRET is that it avoids the consequences of fluorescence excitation. In BRET, the donor fluorophore of the FRET technique is replaced by a luciferase. In the presence of a substrate, bioluminescence from the luciferase excites the acceptor fluorophore through the same Förster resonance energy transfer mechanisms used by FRET. Because BRET does not involve optical excitation, all the light emitted by the fluorophore must result from resonance transfer. In this respect, BRET may be theoretically superior to FRET for quantifying resonance transfer. In fact, early development of high throughput screening assays using BRET is underway.

In principle, there may exist certain situations in which BRET may be more advantageous than FRET for measuring interacting partners. For example, BRET may be superior for cells that are either photoresponsive or damaged by the wavelength of light used to excite FRET. Cells or tissues that have substantial autofluorescence may be better assayed with BRET. Other advantages of BRET include its insensitivity to photobleaching and the facts that it does not suffer from complications due to simultaneous excitation of both donor and acceptor fluorophores as FRET does and that it does not require specific acceptor-only control experiments to be performed (as one must for FRET).

BRET is not without its own limitations, however. BRET requires the use of the substrate coelenterazine. Coelenterazine is hydrophobic, and while it permeates many cell types, there may be some cell types that will not. Coelenterazine can also exhibit autoluminescence in certain media. Lack of sensitivity may also hinder the use of BRET in some cases. Depending on the expression level of the luciferase, the observed luminescence can be dim, and, for ratio imaging, only a small proportion of the total light is collected. Therefore, a sensitive light-measuring device is required.

REFERENCES

Adams, S. R., A. T. Harootunian, Y. J. Buechler, S. S. Taylor, and R. Y. Tsien. Fluorescence ratio imaging of cyclic AMP in single cells. *Nature* 349:694–697, 1991.

Bacso, Z., L. Bene, A. Bodnar, J. Matko, and S. A. Damjanovich. Photobleaching energy transfer analysis of CD8/MHC-I and LFA 1/ICAM-1 interactions in CTL-target cell conjugates. *Immunol. Lett.* 54:151–156, 1996.

Baird, G. S., D. A. Zacharias, and R. Y. Tsien. Circular permutation and receptor insertion within green fluorescent proteins. *Proc. Natl. Acad. Sci. U.S.A.* 96:11241–11246, 1999.

Bodnar, A., A. Jenei, L. Bene, S. Damjanovich, and J. Matko. Modification of membrane cholesterol level affects expression and clustering of class I HLA molecules at the surface of JY human lymphoblasts. *Immunol. Lett.* 54:221–226, 1996.

Cacciatore, T. W., P. D. Brodfuehrer, J. E. Gonzalez, T. Jiang, S. R. Adams, R. Y. Tsien, W. B. Kristan, Jr., and D. Kleinfeld. Identification of neural circuits by imaging coherent electrical activity with FRET-based dyes. *Neuron* 123:449–459, 1999.

Chalfie, M. Green fluorescent protein. *Photochem. Photobiol.* 62:651–656, 1995.

Diliberto, P. A., X. F. Wang, and B. Herman. Confocal imaging of Ca^{2+} in cells. *Methods Cell Biol.* 40:244–262, 1994.

Gadella, T. W., Jr., and T. M. Jovin. Oligomerization of epidermal growth factor receptors on A431 cells studied by time-resolved fluorescence imaging microscopy. A stereochemical model for tyrosine kinase receptor activation. *J. Cell. Biol.* 129:1543–1558, 1995.

Gadella, T. W. Jr., T. M. Jovin, and R. M. Clegg. Fluorescence lifetime imaging microscopy (FLIM)—Spatial resolution of microstructures on the nanosecond time scale. *Biophys. Chem.* 48: 221–239, 1993.

Gonzalez, J. E., K. Oades, Y. Leychkis, A. Harootunian, and P. A. Negulescu. Cell-based assays and instrumentation for screening ion-channel targets. *Drug Discovery Today* 4:431–439, 1999.

Gordon, G. W., G. Berry, N. P. Mahajan, X. H. Liang, B. Levine, and B. Herman. Quantitative fluorescence resonance energy transfer measurements using fluorescence microscopy. *Biophys. J.* 74:2702–2713, 1998.

Herman, B. *Fluorescence Microscopy.* Oxford, England: Bios Scientific Publishers, 1998, 170 pp.

Herman, B., P. Wodnicki, S. Kwon, A. Periasamy, G. W. Gordon, N. P. Mahajan, and X. F. Wang. Recent developments in monitoring calcium and protein interactions in cells using florescence lifetime microscopy. *J. Fluorescence* 7:85–91, 1997.

Jovin, T. M., D. J. Arndt-Jovin, G. Marriott, R. M. Clegg, M. Robert Nicoud, and T. Schormann. Distance, wavelength and time: The versatile 3rd dimensions in light emission microscopy. In: *Optical Microscopy for Biology,* edited by B. Herman and K. Jacobson. New York: Wiley, 1990, pp. 575–602.

Jurgens, L., D. J. Arndt-Jovin, I. Pecht, and T. M. Jovin. Proximity relationships between the type I receptor for Fc epsilon (Fc epsilon RI) and the mast cell function–associated antigen (MAFA) studied by donor photobleaching fluorescence resonance energy transfer microscopy. *Eur. J. Immunol.* 26:84–91, 1996.

Lakowicz, J. R., and K. Berndt. Lifetime-selective fluorescence imaging using an rf phase-sensitive camera. *Rev. Sci. Instrum.* 62:1727–1734, 1991.

Liang, X. H., M. Volkmann, R. Klein, B. Herman, and S. J. Lockett. Co-localization of the tumor suppressor protein p53 and human papillomavirus E6 protein in human cervical carcinoma cell lines. *Oncogene* 8:2645–2652, 1993.

Mahajan, N. P., K. Linder, G. Berry, G. W. Gordon, R. Tsien, R. Heim, and B. Herman. Bcl-2 and Bax interactions in mitochondria probed with fluorescent protein and fluorescence resonance energy transfer. *Nat. Biotechnol.* 16:547–552, 1998.

Mahajan, N. P., D. C. Harrison-Shostak, J. Michaux, and B. Herman. Novel mutant green fluorescent protein protease substrates reveal the activation of specific caspases during apoptosis. *Chem. Biol.* 6:401–409, 1999.

Mere, L., T. Bennet, P. Cassin, P. England, B. Hamman, T. Rink, S. Zimermean, and P. Negulescu. Miniaturized FRET assays and microfluidics: Key components for ultra-high-throughput screening. *Drug Discovery Today* 4:363–369, 1999.

Miyawaki, A., O. Griesbeck, R. Heim, and R. Y. Tsien. Dynamic and quantitative Ca^{2+} measurements using improved cameleons. *Proc. Natl. Acad. Sci. U.S.A.* 96:2135–2140, 1999.

Miyawaki, A., J. Llopis, R. Heim, J. M. McCaffery, J. A. Adams, M. Ikura, and R. Y. Tsien. Fluorescent indicators for Ca^{2+} based on green fluorescent proteins and calmodulins. *Nature* 388:882–887, 1998.

Oida, T., Y. Sako, and A. Kusumi. Fluorescence lifetime imaging microscopy (flimscopy). Methodology development and application to studies of endosome fusion in single cells. *Biophys. J.* 64:676–685, 1993.

Periasamy, A., X. F. Wang, P. Wodnicki, G. W. Gordon, S. Kwon, P. A. Diliberto, and B. Herman. High speed fluorescence microscopy: Lifetime imaging in the biomedical sciences. *J. Microsc. Soc. Am.* 1:13–23, 1995.

Periasamy, A., P. Wodnicki, X. F. Wang, S. Kwon, G. W. Gordon, and B. Herman. Time resolved fluorescence lifetime imaging microscopy (TRFLIM) using a picosecond pulsed tunable dye laser system. *Rev. Sci. Instrum.* 67:3722–3731, 1996.

Wang, X. F., A. Periasamy, D. M. Coleman, and B. Herman. Fluorescence lifetime imaging microscopy: Instrumentation and application. *CRC Crit. Rev. Anal. Chem.* 23(5):1–26, 1992.

Wouters, F. S., and P. I. H. Bastiaens. Fluorescence lifetime imaging of receptor tyrosine kinase activity in cells. *Curr. Biol.* 9:1127–1130, 1999.

Xu, Y., D. W. Piston, and C. H. Johnson. A bioluminescence resonance energy transfer (BRET) system: Application to interacting circadian clock proteins. *Proc. Natl. Acad. Sci. U.S.A.* 96:151–156, 1999.

Zlokarnik, G., P. A. Negulescu, T. E. Knapp, L. Mere, N. Burres, L. Feng, M. Whitney, K. Roemer, and R. Y. Tsien. Quantification of transcription and clonal selection of single living cells with beta-lactamase as reporter. *Science* 279:84–88, 1998.

16

Frequency-Domain Fluorescence Lifetime Imaging Microscopy: A Window on the Biochemical Landscape of the Cell

PETER J. VERVEER, ANTHONY SQUIRE, AND PHILIPPE I. H. BASTIAENS

1. INTRODUCTION

Fluorescence microscopy is an established technique for determining the localization and properties of molecules in biological specimens. Obvious advantages of fluorescence are sensitivity, specificity, and spectral characteristics that depend on the environment of the probe. In addition, the low energy content of fluorescence photons in the visible part of the spectrum permits nondestructive measurements in living cells. Imaging the spatial distribution of a molecule using its fluorescence intensity has been complemented with (micro) spectroscopic techniques for studying the physical and chemical properties of the molecular environment of the fluorophore, which allow the observation of biochemical activity in cells. This has typically been achieved by exploiting the steady-state spectral characteristics of fluorescent probes that change their emission energy upon reaction with the environment. With such techniques, an image that is related to the physiological parameter of interest can be calculated from the ratio of intensities obtained at two excitation or emission wavelengths, eliminating the concentration and light path dependence of the fluorescence intensity. To quantify these images, the ratio as a function of the physiological parameter of interest has to be calibrated separately.

Most physical and chemical changes in the molecular environment of a dye, however, tend to affect its quantum yield, but not its emission–excitation spectrum. Therefore, quantitative imaging of processes related to changes in quantum yield is necessary, which is possible by measuring the excited state lifetime of fluorophores. Fluorescence lifetime imaging microscopy (FLIM) enables the spatial mapping of these nanosecond fluorescence kinetics in cells. The advantage of FLIM is that fluorescence lifetimes are independent of probe concentration or light path length, which are difficult parameters to control in a biological specimen such as a cell. Examples of physiological parameters that have been imaged quantitatively by FLIM are molecular associations (Gadella and Jovin, 1995), proteolytic processing (Bastiaens and Jovin, 1996), phosphorylation (Ng et al., 1999), intracellular calcium (Lakowicz et al., 1994; Szmancinski and Lakowicz, 1995), and pH (Sanders et al., 1995).

Photophysical processes that depopulate the excited state, and thereby alter the quantum yield, affect the fluorescence lifetimes. A typical example of such a process is fluorescence resonance energy transfer (FRET) (Clegg, 1996), which has been put to good use for measuring protein interactions in cells (Tsien et al., 1993; Tsien, 1998; see Chapters 15, 17). Proximity of fluorescent dyes on labeled biomolecules is manifested by quenching of the donor fluorescence (by nonradiatively transferring energy to the acceptor) and sensitized emission of the acceptor. In the microscope FRET can be detected by taking the ratio of the donor and acceptor fluorescence emissions. This causes problems, however, when the donor and acceptor are on separate molecules because the intensities are proportional to fluorophore concentrations, which cannot be easily controlled in a cell. Direct excitation of the acceptor and donor fluorescence bleedthrough in the acceptor channel are difficult to avoid. Therefore, the ratio depends on the local concentration of donor- and acceptor-tagged proteins in addition to the FRET efficiency. Measurement of the donor quenching with FLIM solves this problem because the lifetimes are independent of probe concentration and the acceptor fluorescence is easily filtered from that of the donor (Bastiaens and Squire, 1999). In addition, quantitative images of the population of protein states, such as bound and unbound, can be extracted from FLIM data without the need for external calibration.

In this chapter we describe the experimental setup and numerical analysis methods for frequency-domain FLIM. The experimental and computational procedures for basic single-frequency FLIM and multiple harmonic frequency detection (mfFLIM) are included. We end this chapter with an example an application of FLIM where protein phosphorylation is quantitatively imaged inside cells.

2. PRINCIPLES OF FLIM

2.1 General Approaches

Generally, the fluorescence response of a sample to a pulse excitation can be described by a sum of exponential decays. For a homogenous sample consisting of a single fluorophore, the lifetime is equal to the time required for the fluorescent intensity to fall to $1/e$ of its initial value after pulse excitation (assuming that the decay kinetics of the fluorophore are well described by a single exponential). In general, we are interested in heterogeneous samples containing different species of molecules, whose response is a sum of exponential decays and whose composition differs with spatial position in the sample.

Two general techniques exist for determining the decay kinetics of a sample. The first and most intuitive method involves a direct time-resolved measurement of the fluorescence intensity following excitation with a short pulse of light. Typically, delayed gating techniques are employed to record the fluorescence intensity at several time points, repeated over many excitation pulses in order to achieve a sufficient signal-to-noise ratio. For the complex decay kinetics of a heterogeneous sample, the measured fluorescence response must be fitted to a multiple exponential model. This method is generally denoted by *time-domain* FLIM (Draaijer et al., 1995; Periasamy et al., 1996; Scully et al., 1996; French et al., 1997; Sytsma et al., 1998).

The second method employs sinusoidally modulated light to continuously excite the sample. The fluorescence is also sinusoidally modulated with the same frequency, but is shifted in phase and has a reduced modulation depth relative to the average intensity. From this change in phase and modulation, the lifetime of a homogenous sample can be directly calculated. For measuring multiple lifetimes simultaneously, this method can be extended by exciting with a sum of sinusoidal modulated signals with differing frequencies. This approach is denoted by *frequency-domain* FLIM (Lakowicz and Berndt, 1991; Gadella et al., 1993; Clegg and Schneider, 1996; Carlsson and Liljeborg, 1997; Schneider and Clegg, 1997; Squire et al., 2000) and is described in more detail in the next sections.

2.2 Heterodyne and Homodyne Detection Techniques

The basic principle of phase and modulation fluorimetry with sinusoidal excitation light is illustrated in Figure 16–1a. Excitation with sinusoidal modulated light results in sinusoidal modulated fluorescence emission at the same frequency as the excitation but is shifted in phase and has a reduced relative modulation depth. In general, the reciprocal of the frequencies is chosen so as to span the full lifetime range in the sample (typically tens to hundreds of MHz for nanosecond fluorescence lifetimes). Accurate determination of the fluorescence waveform at these high frequencies can be achieved using heterodyne or homodyne detection methods, where a high-frequency input signal (tens to hundreds MHz) is transformed into a low-frequency signal (Hz to tens of Hz) or a static phase-dependent signal, respectively. The basic principle of heterodyne/homodyne detection is shown in Figure 16–1b.

The technique involves multiplying the signal of interest with a reference waveform. It is essential that the reference waveform maintains phase coherence with the input signal, and consequently they are derived from a common modulation source. In heterodyne detection the reference waveform is chosen with a frequency close to that of the signal. This results in additional sum and difference frequency components at the output of the multiplier (Piston et al., 1989). This output is fed into a lowpass filter with a cut-off frequency well below that of either input, giving a signal oscillating at the difference frequency only. This slowly oscillating signal contains information on the modulation and phase of the input signal and can be time sampled with a high degree of accuracy. In the homodyne detection mode, the reference waveform has a frequency equal to the input signal and gives a filtered signal dependent only on the phase difference of the input signal and the reference waveform. This may be phase sampled with high accuracy by stepping the phase of the reference waveform and recording the filtered signal amplitude. The resulting phase-dependent signal is also periodic and contains information on the phase and modulation of the input signal.

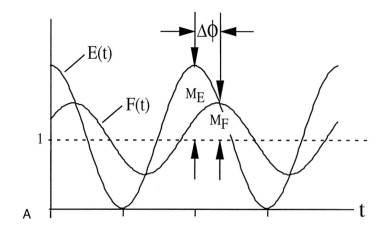

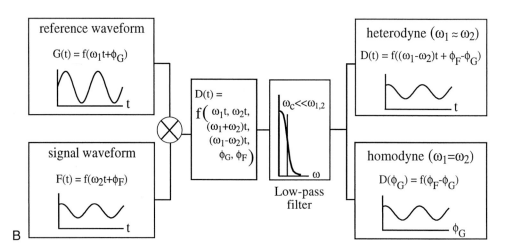

FIGURE 16–1 *a,* Basic principle of FLIM in the frequency domain, showing the normalized excitation and fluorescence signals. The excitation light $E(t)$ is sinusoidally modulated at a high frequency. The fluorescence light $F(t)$ is modulated at the same frequency, but it is phase shifted by $\Delta\phi$, and its modulation depth M_F is decreased, from which decay parameters can be derived. *b,* Principle of homodyne and heterodyne detection. The fluorescence signal is multiplied with a reference waveform that has a frequency equal (homodyne) or close (heterodyne) to that of the signal waveform. The result is lowpass filtered to obtain a signal that oscillates slowly with time (heterodyne) or that is constant in time (homodyne). The former can be accurately sampled in time, and the latter can be sampled by varying the phase ϕ_G of the reference waveform.

2.3 Fluorescence Lifetime Imaging in a Light Microscope

Fluorescence lifetime imaging in a light microscope can be achieved using image intensifiers (Kume et al., 1988) as mixing devices for heterodyne/homodyne detection. Such a device converts a light image incident upon its photocathode surface into photoelectrons that are then amplified within the capillaries of a microchannel plate (MCP). These are then converted to an amplified light image by excitation of

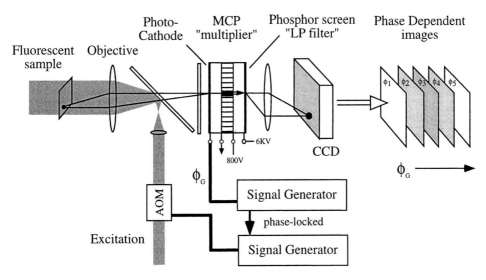

Figure 16–2 Homodyne detection using an image intensifier in a light microscope. Excitation light is sinusoidally modulated using a standing wave acousto-optic modulator (AOM). An image of the sample fluorescence is formed on the photocathode of the intensifier. Homodyne detection is achieved by modulating the gain of the image intensifier using a signal generator phase-locked to the signal generator driving the AOM. The amplified image is integrated on the phosphor screen, which effectively acts as a lowpass (LP) filter. The resulting phase-dependent image is recorded with a charge-coupled device (CCD) camera. MCP, microchannel plate.

a photoluminescence surface. A high-frequency voltage modulation across either the MCP or photocathode acts as the reference waveform. This results in a repetitive modulation of the gain characteristics at every pixel of the detector. Thus, the MCP behaves as a multiplier because the response is proportional to the incident image intensity multiplied by the gain. The slow response time of the phosphor screen at the output of the imaging device acts as a lowpass filter.

Figure 16–2 specifically illustrates homodyne detection using an image intensifier in a light microscope, for the simplest case of single-frequency sinusoidal excitation. Here the excitation light is passed through a standing wave acousto-optic modulator (SW-AOM) in order to obtain sinusoidal modulation, which is driven using a high-frequency signal generator. The excitation light is focused at the back focal plane of a microscope objective in order to provide uniform (Koehler) illumination of the fluorescence sample. The same objective is used to form an image of the sample fluorescence on the photocathode of the image intensifier.

Homodyne detection is performed by modulating the gain of the image intensifier at the same frequency as the excitation using a second signal generator phase-locked to the first. The resulting phase-dependent image at the phosphor screen output of the image intensifier is recorded with a charge-coupled device (CCD) camera. The signal at each pixel of the image is a sinusoidal function of the phase of the intensifier modulation and can therefore be sampled by recording a sequence of images at sequential phase settings of the gain modulation.

3. THEORY OF FREQUENCY DOMAIN LIFETIME MEASUREMENTS

The theory of fluorescence lifetime imaging in the frequency domain has been described extensively in the literature (Gratton and Limkeman, 1983; Gratton et al., 1994; Gadella et al., 1994; Clegg and Schneider, 1996; Squire and Bastiaens, 1999; Squire et al., 2000). For a general description of frequency-domain FLIM, we consider excitation with any repetitively modulated excitation source and a sample composed of Q fluorescent species. In general, the response of a fluorescent system to any form of excitation (repetitive or not) is given by the convolution of the excitation waveform $E(t)$ with the response of the system to impulse excitation $F_\delta(t)$:

$$F(t) = QE \int_{-\infty}^{t} E(u) F_\delta(t - u) \, du \tag{1}$$

where the term QE accounts for the quantum efficiency of the fluorophore and any instrumentation factors in the excitation and detection. The excitation light can be any repetitively modulated light source, which can be represented as a Fourier series:

$$E(t) = E_0 + \sum_{n=1}^{\infty} E_n \cos(n\omega t + \phi_{E,n}) \tag{2}$$

where ω is the fundamental circular frequency of the modulation, E_0 is the time-independent average intensity, and E_n is the intensity amplitude of the nth harmonic frequency component with phase $\phi_{E,n}$.

The fluorescent response $F_\delta(t)$ of the sample to impulse excitation is given by an exponential decay summed over all species:

$$F_\delta(t) = \sum_{q=1}^{Q} a_q \exp(-t/\tau_q) \tag{3}$$

where a_q is the amplitude of the qth fluorescent species and τ_q is the corresponding lifetime. Thus, substituting Equations 2 and 3 into Equation 1, we obtain the following fluorescence response:

$$F(t) = QE\left(F_0 + \sum_{n=1}^{\infty} E_n M_n \cos(n\omega t + \phi_{E,n} - \Delta\phi_n) \right) \tag{4}$$

where $F_0 = E_0 \sum_{q=1}^{Q} a_q \tau_q$ is the average fluorescence. The relationships for the phase shift $(\Delta\phi_n)$ and modulation (M_n) terms are

$$\Delta\phi_n = \arctan(A_n/B_n) \tag{5a}$$

$$M_n = \sqrt{A_n^2 + B_n^2} \tag{5b}$$

where

$$A_n = \sum_{q=1}^{Q} \frac{\alpha_q n \omega \tau_q}{1 + (n \omega \tau_q)^2} \tag{6a}$$

$$B_n = \sum_{q=1}^{Q} \frac{\alpha_q}{1 + (n \omega \tau_q)^2} \tag{6b}$$

In the above equations we have made the substitution

$$\sum_{q=1}^{Q} \alpha_q = \sum_{q=1}^{Q} a_q \tau_q = 1$$

Thus α_q gives the fractional contribution of the qth species to the average fluorescence.

By experimentally determining the frequency dependence of the phase shift ($\Delta\phi_n$) and modulation (M_n) of a fluorescent sample and fitting to the dispersion relationships given by Equations 5 and 6, the lifetime parameters can be estimated (see Section 5, later). To experimentally determine the modulation (M_n) and phase shift ($\Delta\phi_n$) of the fluorescence signal with a high degree of accuracy, the time-dependent fluorescence response of the sample is mapped using the heterodyne or homodyne detection method. In FLIM this is achieved by modulating the gain of an image intensifier. Here again we consider any repetitively modulated source for the gain characteristics of the image intensifier $G(t)$, which can thus be expanded as a Fourier series:

$$G(t) = G_0 + \sum_{m=1}^{M} G_m \cos(m\omega't + \phi_{G,m} + mk\Delta\psi) \tag{7}$$

where ω' is the fundamental frequency of the gain modulation, G_0 is the average gain, G_m is the amplitude of the mth frequency component with phase $\phi_{G,m}$, and $k\Delta\psi$ is an additional adjustable phase of the source of modulation in the gain.

The response of the MCP is proportional to the gain multiplied by the fluorescent signal. When these are given by general Equations 4 and 7, the product results in a signal composed of (1) a time invariant component; (2) a component oscillating at the harmonic frequencies in the gain; (3) a component oscillating at the harmonic frequencies in the fluorescence; and (4) components oscillating at the sums and differences of the harmonic frequencies in the gain and fluorescence. The slow response time of the phosphor screen at the output of the imaging device gives an integrated signal $D(t,k)$ consisting only of the slowly varying signals corresponding to the closely matched difference frequency components:

$$D(t,k) = D_0\left(1 + \sum_{n=1}^{N} M_{R,n} M_n \cos(n\Delta\omega t + nk\Delta\psi - \phi_{R,n} - \Delta\phi_n)\right) \tag{8}$$

where $\Delta\omega = |\omega - \omega'|$ is the difference in the fundamental frequencies of the gain and fluorescence signal, $M_{R,n} = E_n G_n / 2E_0 G_0$ is the modulation measured for a reference sample with no intrinsic lifetime, $\phi_{R,n} = \phi_{G,n} - \phi_{E,n}$ is the phase measured from a reference sample, and $D_0 = QE\,G_0E_0$ is the average intensity, which also accounts for the quantum efficiency term described above. The fundamental frequency difference is chosen to be small so that the heterodyne signal can be accurately sampled in time (typically the heterodyne signal will oscillate with a frequency from Hz to tens of Hz).

For the homodyne detection mode, the source of the gain modulation is chosen with a frequency equal to the fundamental of the fluorescent signal (i.e., $\Delta\omega = 0$), which results in a time invariant signal dependent only on the phase difference between the gain modulation source and the fluorescence:

$$D(k) = D_0\left(1 + \sum_{n=1}^{N} M_{R,n}M_n\cos(nk\Delta\psi - \phi_{R,n} - \Delta\phi_n)\right) \qquad (9)$$

This signal can be sampled by acquiring an image at each sequential step ($k = 0, 1, 2, 3 \ldots K - 1$) of $\Delta\psi$ in the phase of the gain modulation. Once sampled, both the amplitudes and phases of each harmonic component in the signal can be determined from a Fourier analysis. If a reference signal is similarly sampled, then both experimental observables M_n and $\Delta\phi_n$ can be calculated (see Section 5).

The number of sampled time or phase points that need to be measured is constrained by the Nyquist theorem and should be more than twice the frequency content of the detected signal ($K > 2N$) to avoid aliasing effects. In biological applications, the microscopic samples are sensitive to photobleaching effects, and with live cells we are often required to observe rapid dynamics. This places constraints on the number of images that can be acquired and therefore on the number of harmonics in the measured signal. This complicates the numerical analysis and requires advanced methods for analyzing complex decay kinetics (see Section 5).

4. Experimental Setup for Frequency-Domain FLIM

4.1 Single-Frequency FLIM

Figure 16–3 shows an overall schematic for the FLIM configuration in our laboratory (Squire and Bastiaens, 1999; Squire et al., 2000). The setup for obtaining FLIM data at multiple frequencies requires some additional components, which are indicated in Figure 16–3 by dashed outlines and are discussed later.

The FLIM instrument is based around an inverted microscope (Zeiss Axiovert 135 TV). For a light source we employ an argon/krypton continuous wave (cw) mixed gas laser (Coherent Innova 70C Spectrum), which provides selection of discrete laser lines over the visible region from 457.9 to 647.1 nm. Sinusoidal modulation of the output of the argon/krypton laser is achieved by the use of SW-AOMs (Intra-Action Corp. Belwood). In our setup, any one of three SW-AOMs are

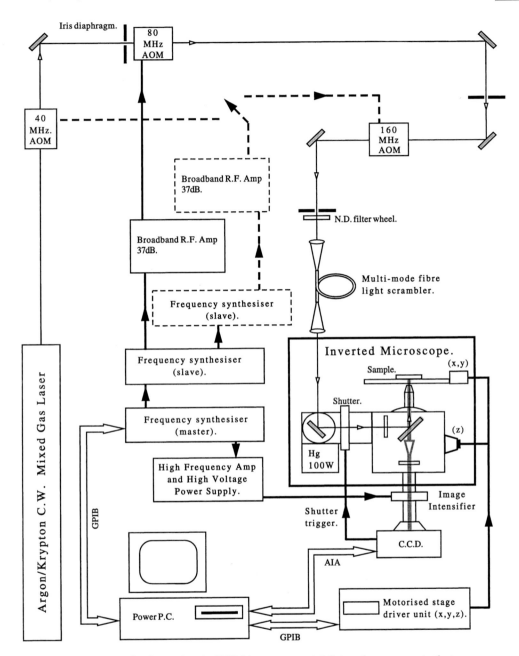

FIGURE 16–3 Overall schematic of a FLIM instrument. Additional components that are necessary for multiple-frequency FLIM are indicated by dashed outlines. See text for a detailed description of the different components. AOM, acousto-optic modulator; C.C.D., charge-coupled device camera.

used, which have about ±25% tunability with central resonant frequencies of 40, 80, and 160 MHz.

To remove the effects of laser speckle at the sample plane, the temporal and spatial coherence properties of the laser are disrupted by launching the main beam into a 1.5 m step index silica fiber having a 1 mm core and numerical aperture of 0.37 (Technical Video Ltd) using a 12 cm focal length lens. The fiber converts the Gaussian intensity profile of the main laser beam into a uniform cone of light, and by vibrating it at frequencies of 100 Hz a moving random speckle pattern is generated in the illumination, which is integrated during detection. An 8 cm focal length achromatic lens is used to collect and collimate the laser light just before the epi-illumination port of the microscope in order to obtain Koehler illumination at the sample plane. Attached to the second epi-illumination port is a 100 W mercury arc lamp (Zeiss HBO 100 W/2) with a controllable intensity output (Zeiss AttoArc). The mode of illumination can be quickly switched between the laser and lamp via a rotating mirror. Light from the sample is imaged directly onto the photocathode of a high-frequency-modulated image intensifier head (C5825, Hamamatsu) via the TV port situated directly below the sample.

For heterodyne/homodyne detection the effective gain of the image intensifier can be modulated at a frequency between 300 kHz and 300 MHz by the application of a biased sinusoidal voltage to the photocathode. The amplified phase-locked outputs from high-frequency synthesizers (2023 Marconi) provide a highly stable sinusoidal voltage source for modulating both the excitation field via the SW-AOMs and the gain characteristics of the image intensifier unit.

The amplified image at the phosphor screen is projected onto the chip of a scientific-grade CCD camera (Quantix, Photometrics) using a telescopic lens with a magnification of 0.5. The 12 bit camera houses a Kodak KAF1400 chip with a 1317×1035 array of 6.8 μm square pixels. With a maximal readout rate of 5 Mpixels s^{-1} a full frame can be read in 0.5 s. At modulation frequencies of 100 MHz the image intensifier has a modulation transfer function (MTF) of about 12 lp/mm, which corresponds to a line spacing of 41.7 μm on the CCD. A sampling criterion of 2.5 per resolve corresponds to a pixel dimension of 16.7 μm; thus 2×2 binning of the CCD chip is employed for the collection of FLIM image sequences.

A BNC output on the CCD indicating shutter status provides synchronous triggering of an external high-speed shutter (Uniblitz VS25 and D122 shutter and driver, Vincent Associates), mounted in the optical path between the epi-illumination port of the microscope and the filter block. The phase setting of the frequency synthesizer modulating the image intensifier gain is controlled via commands sent over a GPIB interface using a high-performance PCI-GPIB card housed in a Macintosh 8600 PowerPC. Software extensions were written in C for the image analysis program IPLab Spectrum (Signal Analytics Corp.), which enabled control of the phase of the frequency synthesizer output to the image intensifier. Additional extensions were written for processing and analyzing FLIM sequences within IPLab.

4.2 Multiple-Frequency FLIM

In multiple-frequency FLIM, heterodyne/homodyne detection at a harmonic set of frequencies is performed simultaneously as theoretically described above. The basis of the theory implies the presence of higher harmonic content in the excitation (Eq. 2) and the gain characteristics of the detector (Eq. 7). Such harmonic content must be experimentally introduced. A repetitively pulsed light source offers one means of introducing higher harmonics into an excitation source. The use of a mode-locked laser in combination with pulse pickers to obtain a sufficiently low fundamental frequency has previously been reported (Alcala et al., 1985; Watkins et al., 1998). Alternatively, a CW laser can be pulsed by passing the light through a Pockels cell pulsed by a high-voltage comb generator (Verkman et al., 1991). Both methods generate pulses whose widths are sufficiently short compared with their modulation period, giving a wide spectrum of harmonic frequencies.

We employ combinations of SW-AOMs (see Fig. 16–3) in order to introduce a limited set of frequencies into the excitation light (Piston et al., 1989). For example, two SW-AOMs in series will modulate the light at the frequencies of each of the SW-AOMs and the sum and difference frequencies. The frequency of each of the SW-AOMs has to be carefully chosen such that the sum and difference frequencies form part of a harmonic set. In practice this can be difficult because each of the SW-AOMs operates efficiently at well-defined discrete frequencies. For the 40 MHz SW-AOM, the resonances are about 200 kHz apart with a finesse of about 2.66. Furthermore, each SW-AOM introduces a loss of about 50% in the excitation light. For these reasons we used combinations of only two SW-AOMs, either the 40 or 160 MHz SW-AOMs in combination with the 80 MHz SW-AOM, offering a minimum of four harmonic frequencies from 20 up to 360 MHz.

Higher harmonic content in the gain characteristics of an image intensifier can be achieved by the application of a repetitive-voltage pulse either to the photocathode or the MCP. A square-pulse modulation of the gain characteristics of the image intensifier is achieved by the application of a biased sinusoidal voltage to the photocathode. This arises because the application of a positive voltage at the photocathode prevents the transfer of negatively charged photoelectrons to the MCP, and the device is effectively in an "off" state. This situation is rapidly reversed and the device is "on" for small negative voltages. Thus, the application of a zero mean sinusoidal voltage effectively leads to square-wave modulation as the voltage swings from negative (on) to positive (off) voltages every 180°.

By increasing the voltage offset of the sinusoidal voltage, the duration of the on state of the device relative to the off state can be reduced, leading to square-pulse modulation and an increased amplitude in the higher harmonic content. Although this provides a simple means of pulsing the image intensifier, the subsequent reduction in the amplitude of the on voltage with increasing offset at the photocathode leads to some reduction of image resolution. This occurs because photoelectrons are less strongly accelerated in the weaker electric field between the photocathode and MCP and are thus capable of larger deviations from a direct trajectory.

A frequency synthesizer (2023, Marconi) is used to provide a highly stable, high-frequency sinusoidal voltage source, which is phase-locked with the frequency synthesizers used to drive the SW-AOMs. Homodyne detection at each harmonic frequency in the fluorescence signal was performed by matching the frequency of the gain modulation source with the fundamental frequency component in the fluorescence excitation wave.

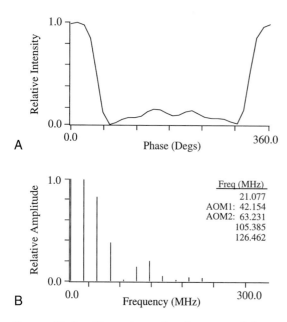

FIGURE 16–4 *a,* Phase-dependent intensity of the excitation light for a multiple-frequency FLIM instrument, using homodyne detection. Values were obtained by spatially averaging an image sequence of a reflective sample. *b,* Fourier analysis of Figure 16.4a, showing the amplitude of each harmonic. AOM, acousto-optic modulator.

For a typical homodyne single- or multiple-frequency FLIM experiment a reference signal must be obtained, using a piece of reflecting aluminum, in order to determine a reference phase and modulation at each frequency. The results for a typical multiple frequency measurement are shown in Figure 16–4. Here the excitation light was modulated using the 40 and 80 SW-AOMs set at 42.154 and 63.231 MHz, respectively, in order to form part of a harmonic set with the difference (21.077 MHz) and sum (105.385 MHz) frequencies. A phase-sampled reference signal is obtained by averaging each image in the sequence, as shown in Figure 16–4a. Figure 16–4b shows the spectral content in the homodyne-detected excitation light, calculated from a Fourier analysis of Figure 16–4a. Some additional harmonics are seen to arise, probably induced by anharmonic modulations in each of the SW-AOMs.

5. Numerical Analysis of FLIM Data

5.1 Introduction

This section describes the numerical analysis techniques that are used to obtain information about the fluorescence decay kinetics at each resolvable image element (pixel) from phase-dependent image data acquired by FLIM or multiple-frequency FLIM. Figure 16–5 shows a schematic representation of the steps involved, divided in two main parts: (*1*) harmonic analysis of the phase-dependent

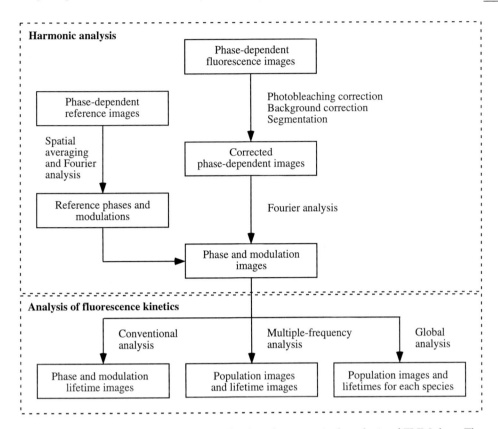

FIGURE 16–5 Schematic of the steps involved in the numerical analysis of FLIM data. The analysis is divided in two steps: harmonic analysis of the phase-dependent image sequence (top) and decay analysis of the resulting phase and modulation images (bottom). The latter can be done in several ways, whereas the first step is always the same.

image sequence to obtain phase and modulation images and (2) analysis of the phase and modulation images to obtain information about the fluorescence decay kinetics of the sample. The harmonic analysis is the same in all cases, whereas the decay analysis can be done in several ways.

5.2 Harmonic Analysis of the Phase-Dependent Images

A number of pre-processing steps are applied to the phase-dependent image sequence before the phase and modulation images are calculated. Optionally, a photobleaching correction is applied, in which case the phase-dependent series of images must be acquired twice where the second sequence is phase sampled in the reverse direction. A first-order correction for photobleaching is then applied by reversing the second sequence of images and adding it to the first-sequence (Gadella et al., 1994). This first order correction is adequate in most applications. The resulting sequence is then corrected for background fluorescence. A small region in a flat area outside the cell is interactively selected, and in each phase-dependent image the mean intensity value inside that region is calculated and subtracted. To select the objects of interest, a mask is then created from the average of all images by thresholding the intensities.

Each image in the phase-dependent sequence is described by Equation 9. At each pixel, the phase shifts ($\Delta\phi_n$) and modulations (M_n) are the parameters of interest because they depend on the decay kinetics of the sample. Equation 9 can be rewritten in a linear form:

$$D(k) = D_0 + \sum_{n=1}^{N} [v_n \cos(kn\Delta\psi) + w_n \sin(kn\Delta\psi)] \tag{10}$$

This equation has the form of a discrete Fourier transformation, and its parameters can be estimated using discrete sine and cosine transformations (Clegg and Schneider, 1996):

$$D_0 = \frac{1}{K} \sum_{k=0}^{K-1} D(k) \tag{11a}$$

$$v_n = \frac{2}{K} \sum_{k=0}^{K-1} D(k) \cos(kn\Delta\psi) \tag{11b}$$

$$w_n = \frac{2}{K} \sum_{k=0}^{K-1} D(k) \sin(kn\Delta\psi) \tag{11c}$$

where K is the number of phase steps at which measurements are made. Equation 11 is the simplest approach to estimate the parameters of Equation 10. However, any other method for estimating the parameters of a linear equation is applicable. In particular, we favor the use of a singular value decomposition (SVD) because it allows estimation of the variances of the parameters (Press et al., 1992).

Given the parameters of Equation 10, the phase shifts and modulations can be found from

$$\Delta\phi_n = \arctan(w_n/v_n) - \Delta\phi_{R,n} \tag{12a}$$

$$M_n = \sqrt{v_n^2 + w_n^2}/D_0 M_{R,n} \tag{12b}$$

Using error propagation, the variances of $\Delta\phi_n$ and M_n can be calculated:

$$\sigma_{\Delta\phi_n}^2 = \frac{w_n^2 \sigma_{v_n}^2 + v_n^2 \sigma_{w_n}^2}{(v_n^2 + w_n^2)^2} + \sigma_{\Delta\phi_{R,n}}^2 \tag{13a}$$

$$\sigma_{M_n}^2 = \frac{v_n^2 + w_n^2}{D_0^4 M_{R,n}^2} \sigma_{D_0}^2 + \frac{v_n^2 \sigma_{v_n}^2 + w_n^2 \sigma_{w_n}^2}{D_0^2 M_{R,n}^2 (v_n^2 + w_n^2)} + \frac{v_n^2 + w_n^2}{D_0^2 M_{R,n}^4} \sigma_{M_{R,n}}^2 \tag{13b}$$

To calculate the phases ($\Delta\phi_n$) and modulations (M_n), the reference phases ($\Delta\phi_{R,n}$) and reference modulations ($M_{R,n}$) must be known. They can be obtained by imaging a reflective sample, which has a lifetime equal to zero. Each image in the phase-dependent image sequence is averaged to obtain a single sequence with an improved signal-to-noise ratio. The reference phases and modulations are then obtained from this averaged sequence as described before, with the substitution $\Delta\phi_{R,n} = 0$, $M_{R,n} = 1$, and $\sigma^2_{\Delta\phi_{R,n}} = \sigma^2_{M_{R,n}} = 0$ in the formulas above.

5.3 Analysis of the Fluorescence Kinetics Parameters

In each pixel, information about the fluorescence decay kinetics of the sample is extracted from the phase and modulation images using Equation 6. For multiple-component decay kinetics ($Q \geqslant 2$), the sum of the fractional contributions (α_q) is equal to one, and we can eliminate one fraction by substituting $\alpha_Q = 1 - \sum_{q=1}^{Q-1} \alpha_q$ into Equation 6:

$$A_n = \frac{n\omega\tau_Q}{1 + (n\omega\tau_Q)^2} + \sum_{q=1}^{Q-1}\left(\frac{n\omega\tau_q}{1 + (n\omega\tau_q)^2} - \frac{n\omega\tau_Q}{1 + (n\omega\tau_Q)^2}\right)\alpha_q \qquad (14a)$$

$$B_n = \frac{1}{1 + (n\omega\tau_Q)^2} + \sum_{q=1}^{Q-1}\left(\frac{1}{1 + (n\omega\tau_q)^2} - \frac{1}{1 + (n\omega\tau_Q)^2}\right)\alpha_q \qquad (14b)$$

As described above, estimated values for the phase shifts and modulations are computed from the data. Estimated values of A_n and B_n can be calculated from those by

$$\hat{A}_n = M_n \sin(\Delta\phi_n) \qquad (15a)$$
$$\hat{B}_n = M_n \cos(\Delta\phi_n) \qquad (15b)$$

with variances

$$\sigma^2_{A_n} = \sin^2(\Delta\phi_n)\sigma^2_{M_n} + M_n^2\cos^2(\Delta\phi_n)\sigma^2_{\Delta\phi_n} \qquad (16a)$$
$$\sigma^2_{B_n} = \cos^2(\Delta\phi_n)\sigma^2_{M_n} + M_n^2\sin^2(\Delta\phi_n)\sigma^2_{\Delta\phi_n} \qquad (16b)$$

Information about the decay kinetics can be retrieved in several ways. We describe three different approaches in the next sections.

5.3.1 Phase and modulation lifetimes

The most common approach to obtain decay information from the measured phases and modulations is to assume $Q = 1$ and calculate phase and modulation lifetime values from Equation 6 (Clegg and Schneider, 1996):

$$\tau_{\Delta\phi_n} = \tan(\Delta\phi_n)/n\omega \qquad (17a)$$
$$\tau_{M_n} = \sqrt{1/M_n^2 - 1}/n^2\omega^2 \qquad (17b)$$

with variances calculated by error propagation:

$$\sigma^2_{\tau_{\Delta\phi,n}} = (1 + \tan^2(\Delta\phi_n))^2 \sigma^2_{\Delta\phi_n}/n\omega \qquad (18a)$$
$$\sigma^2_{\tau_{M,n}} = \sigma^2_{M_n}/n^2\omega^2 M_n^4(1 - M_n^2) \qquad (18b)$$

If the sample is homogeneous, both calculations have the same value, which is independent of the frequency $n\omega$. In most samples, however, each resolvable image element contains a mixture of fluorophores with different lifetimes, and the decay kinetics are not mono-exponential. In such a case, the phase and modulation lifetimes will differ and change with frequency.

5.3.2 Multiple frequency analysis

If the decay kinetics of the sample at each pixel are heterogeneous, the decay parameters can be retrieved either by direct fitting to the measured phases and modulations (Eq. 9) (Gratton et al., 1994; Lakowicz et al., 1984; Squire et al., 2000) or by fitting the images of $\hat{A}_n$ and $\hat{B}_n$, calculated using Equation 15, to Equation 6 or Equation 14. The last two equations are linear in the fractions α_q, which makes them easier to fit. We will therefore assume a fit either to Equation 6 or Equation 14. The fractions α_q and lifetimes τ_q are found by minimizing a χ^2 error measure at each pixel (Gratton et al., 1994):

$$\chi^2(\alpha_q, \tau_q) = \sum_{n=1}^{N} \left[\left(\frac{\hat{A}_n - A_n(\alpha_q, \tau_q)}{\sigma_{A_n}} \right)^2 + \left(\frac{\hat{B}_n - B_n(\alpha_q, \tau_q)}{\sigma_{B_n}} \right)^2 \right] \quad (19)$$

where $A(\alpha_q, \tau_q)$ and $B(\alpha_q, \tau_q)$ are the functions given by Equation 6 or Equation 14. Minimization of Equation 19 can be achieved by a Levenberg-Marquardt nonlinear fitting routine (Press et al., 1992). A Levenberg-Marquardt type of routine may return a covariance matrix as a result. The values on the diagonal of the covariance matrix can be used as estimations for the variances of the lifetime and fraction parameters. It is important to note that, in order to successfully fit the data the number of lifetimes in the model must be equal to or less than the number of harmonic frequencies ($Q \leq N$). This follows from the fact that, for a successful fit, the number of parameters (Q lifetimes + Q or $Q - 1$ fractions) must be smaller than or equal to the number of independent measurements (N phase shifts + N modulations).

5.3.3 Global analysis

Although the multiple-frequency analysis enables the estimation of lifetimes and fractional contributions to the steady-state fluorescence on a pixel-by-pixel basis, such a fit may be difficult for typical FLIM data. To minimize sample exposure and acquisition time, the number of harmonics (N) in the data must be minimized (typically $N = 4$). Consequently, the signal-to-noise ratio of the data may be too low for a reliable fit. A solution for this problem is to exploit prior knowledge about the system under study by using global analysis techniques (Beechem, 1992; Verveer et al., 2000). In particular, one can assume that there is a limited number of molecule species that each have a discrete fluorescence lifetime τ_q that does not vary spatially. The imaging of FRET in a protein complex is an example of a case where this is a reasonable assumption, as is depicted graphically in Figure 16–6.

By fitting all pixels simultaneously, using the assumption of lifetime invariance, the number of parameters that must be estimated can be greatly reduced. This can be achieved by minimizing the χ^2 measure:

$$\chi^2(\alpha_{q,m}, \tau_q) = \sum_{m=1}^{M} \sum_{n=1}^{N} \left[\left(\frac{\hat{A}_{n,m} - A_{n,m}(\alpha_{q,m}, \tau_q)}{\sigma_{A_{n,m}}} \right)^2 + \left(\frac{\hat{B}_{n,m} - B_{n,m}(\alpha_{q,m}, \tau_q)}{\sigma_{B_{n,m}}} \right)^2 \right] \quad (20)$$

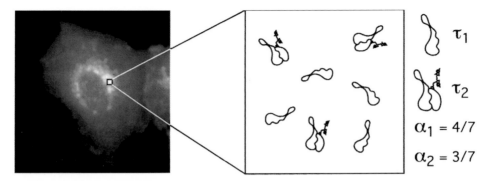

Figure 16–6 Principle of global analysis, illustrated for the imaging of fluorescence resonance energy transfer (FRET) in a protein complex. In an unbound state, the donor molecule has a lifetime value τ_1, which does not vary spatially. On the formation of a protein complex the lifetime of the donor decreases to a value $\tau_2 < \tau_1$, due to the transfer of energy to an acceptor. Because the spatial configuration of the complex is fixed, τ_2 is also spatially invariant. Thus the decay parameters of interest are the lifetimes τ_1 and τ_2 of the two species of molecule (global parameters), and the fractional amount of molecules α of each species in each pixel (local parameters).

This equation is similar to Equation 19, but instead of a repeated pixel-by-pixel minimization, an additional summation is done over the M pixels in the image, and a single minimization is performed. A subscript m is added to indicate a dependence on spatial position. The parameters that are simultaneously varied to minimize this χ^2 measure are the Q spatially invariant lifetime values (τ_q) and the M sets of fractions ($\alpha_{q,m}$) at each pixel. We use a truncated Newton algorithm (Schlick and Fogelson, 1992) to minimize Equation 20.

The error analysis for global analysis is much more complicated than that for a multiple-frequency analysis. Because the number of parameters that must be estimated can be very large, the memory requirements are too high for an algorithm that returns a covariance matrix. Several other, more rigorous techniques are available to obtain errors on some or all of the parameters. These are computationally demanding, however, because they require a sequence of minimizations to be performed (Straume et al., 1991).

The global analysis approach can lead to dramatic improvements in the precision and accuracy of the estimated lifetime and fraction parameters. It can also be shown that the number of lifetimes that can be fit with the global approach is about twice the number of harmonics in the data if Equation 14 is used (Verveer et al., 2000). This follows from the requirement that the number of estimated parameters is smaller than or equal to the number of measurements:

$$M(Q - 1) + Q \leq 2MN \Rightarrow Q \leq 2N, \quad \text{if } M \geq Q \tag{21}$$

Thus, for a given number of harmonics, twice the number of lifetimes can be fit by a global analysis compared with a conventional analysis. An important consequence is the possibility of fitting a two-component model to single-frequency data, which is impossible with conventional analysis approaches.

5.4 Calculation of Species Populations from Fluorescence Fractions

Images of the fluorescence of each species can be obtained by multiplying the fractions (α_q) with the average fluorescence (D_0). Each of these is proportional to the population (number of excited molecules) of the corresponding species. They can only be compared with each other, however, if they are corrected for differences in the quantum yield and measured spectral properties of the chromophore. The quantum efficiencies are proportional to the lifetimes of the chromophore, and the spectral properties depend on the detector and filter set used. Absolute values for these correction factors are difficult to obtain, but it is sufficient to obtain relative values, for example, by measuring the intensities of an equimolar sample of the molecular species involved with the same detector and filter set. In the particular case of imaging FRET, the spectral properties of all species are the same. Thus, in that case the populations can be corrected using the lifetime values (τ_q), and no separate measurements of the spectral properties are necessary. From the corrected populations, fractional populations of all species can be obtained by normalizing their sum to one at each pixel.

6. Applications of FLIM to Protein Phosphorylation Imaging

In this section, we describe a FLIM assay for protein phosphorylation based on FRET (Bastiaens and Squire, 1999) between a fused green fluorescent protein (GFP) and an indocyanine dye (Cy3) bound to an antibody against a phospho-amino acid. Efficient FRET will only occur when the Cy3 chromophore is within nanometer range of the GFP by binding of the antibody to the phosphorylated polypeptide or protein. This imaging assay is generic for protein covalent modification when an antibody is available against the modified epitope. In this system, it is essential to determine FRET exclusively via changes in the donor quantum yield because the acceptor-carrying antibody is usually present in excess. Also, FRET detection by changes in donor quantum yield removes the necessity for absolute specificity of the antibody for its epitope, because the proximity of the donor and acceptor dyes is the only criterion for reaction of the protein. Detection of antibody binding to the phosphorylated epitope exclusively through donor lifetime changes requires stringent optical filtering so that only the donor fluorescence is detected. Here we look at the phosphorylation of the receptor tyrosine kinase ErbB-1 in cells after stimulation with epidermal growth factor (EGF) (Wouters and Bastiaens, 1999).

A cytosolic C–terminal fusion of EGFP with ErbB1 has been constructed for this purpose. Binding of EGF to the receptor induces dimerization and trans-autophosphorylation of tyrosine residue catalyzed by its intrinsic tyrosine kinase activity. This phosphorylation on tyrosine residues is detected by a Cy3-conjugated antiphosphotyrosine antibody (PY72-Cy3). Activation of the receptor is measured in MCF7 breast carcinoma cells after 5 minutes incubation with EGF.

In Color Figure 16–7a,b the GFP fluorescence lifetime maps calculated from the average of phase and modulation lifetimes at a single frequency (80 MHz) are shown before (Color Fig. 16–7a) and after (Color Fig. 16–7b) photobleaching of the acceptor Cy3. In Color Figure 16–7c histograms are shown for the lifetime values in Color Figure 16–7a, b. The large difference in lifetimes before and after photobleaching at the membrane show the phosphorylation of the receptor. In this conventional analysis, a fluorescence lifetime map is obtained where the lifetime in each pixel depends on the relative population of molecules exhibiting FRET. These lifetime values are a frequency-dependent function of the true lifetimes of each species, and therefore a broad distribution of apparent values is found (Color Figure 16–7c). From these data apparent FRET efficiencies E_i at each pixel i can be calculated by

$$E_i = 1 - \tau_i^{da}/\tau_i^d \tag{22}$$

where τ_i^{da} is the apparent lifetime at each pixel i in the sample where the donor and acceptor are present, and τ_i^d is the corresponding fluorescence lifetime of the donor in the absence of the acceptor obtained from the photobleached specimen (Color Fig. 16–7b). From this apparent efficiency map, the population of phosphorylated molecules could be approximated by assuming a linear relationship with E_i. This approximation would require the FRET efficiency for the fully phosphorylated receptor to be known.

A more quantitative image of phosphorylation populations emerges when a global analysis is applied to the same data set as shown in Color Figure 16–7d–f. Here, the prior knowledge about the system was encoded so that only two species were present: phosphorylated and unphosphorylated receptors. The two lifetimes obtained reflect the quantum yields of GFP in the absence ($\tau = 2.26$ ns) and presence ($\tau = 0.76$ ns) of FRET, and one minus their division gives the true FRET efficiency in the ErbB1–GFP/PY72–Cy3 complex ($E = 66\%$). The fractional population image in Color Figure 16–7d shows high levels of phosphorylated receptor in membranes after stimulation with EGF (up to 100% phosphorylation). The global analysis of the FLIM data set where the acceptor was photobleached gave a low fluorescence fractional contribution of a short lifetime component, demonstrating essentially single-component fluorescence decay kinetics of all GFPs (Color Fig. 16–7e).

Comparison of the fluorescence lifetime histograms of the conventional (Color Fig. 16–7c) and global (Color Fig. 16–7f) analysis further clarifies the difference between the two data analysis approaches. In the global analysis the histogram consists of two peaks at the discrete values of the lifetimes that were obtained, and thus the spatial information is contained in the *population* of phosphorylated states (Color Fig. 16–7d). In contrast to the conventional approach described above, true populations with their associated discrete lifetimes are obtained. These lifetimes fully define the FRET efficiency in the protein complex so that no prior knowledge about the FRET efficiency in the fully phosphorylated state is required, nor is it necessary to measure a sample in the absence of acceptor (e.g., after photobleaching). The multiplication of the fractional populations (Color Fig. 16–7d) with the steady-state fluorescence (Color Fig. 16–7g) gives images that are proportional to the concentrations of phosphorylated and unphosphorylated receptors (Color Fig. 16–7h, i). Separation of populations in this way shows that the receptor in the secretory pathway is not phosphorylated in contrast to the receptor at the plasma membrane.

7. CONCLUSION

We have described the theoretical background and experimental setups for both single-frequency and multiple-frequency FLIM. The development of multiple-frequency FLIM allows us to spatially map the complex decay kinetics that arise in biological samples by simultaneous measurements at several frequencies. A global analysis approach was described that allows precise quantitative analysis of the data using a small number of frequencies. A remarkable and useful result is that double-exponential decays can be resolved using a standard single-frequency FLIM instrument. These developments enabled a quantitative FLIM assay for protein phosphorylation in cells based on FRET. The combination of the several technical and theoretical advancements that are described in this chapter enables quantitative measurements and thus provides a powerful tool for investigating biochemical reactions, such as protein interactions in cells.

REFERENCES

Alcala, J. R., E. Gratton, and D. M. Jameson. A multifrequency phase fluorometer using the harmonic content of a mode-locked laser. *Anal. Instrum.* 14:225–250, 1985.

Bastiaens, P. I. H., and T. M. Jovin. Microspectroscopic imaging tracks the intracellular processing of a signal transduction protein: Fluorescent labeled protein kinase C bI. *Proc. Natl. Acad. Sci. U.S.A.* 93:8407–8412, 1996.

Bastiaens, P. I. H. and A. Squire. Fluorescence lifetime imaging microscopy: Spatial resolution of biochemical processes in the cell. *Trends Cell Biol.* 9:48–52, 1999.

Beechem, J. M. Global analysis of biochemical and biophysical data. *Methods Enzymol.* 210:37–54, 1992.

Carlsson, K., and A. Liljeborg. Confocal fluorescence microscopy using spectral and lifetime information to simultaneously record four fluorophores with high channel separation. *J. Microsc.* 185:37–46, 1997.

Clegg, R. M. Fluorescence resonance energy transfer spectroscopy and microscopy. In: *Fluorescence Imaging Spectroscopy and Microscopy,* edited by X. F. Wang and B. Herman. New York: Wiley, 1996, pp. 179–251.

Clegg, R. M., and P. C. Schneider. Fluorescence lifetime resolved imaging microscopy: A general description of lifetime-resolved imaging measurements. In: *Fluorescence Microscopy and Fluorescence Probes,* edited by J. Slavik. New York: Plenum Press, 1996, pp. 15–33.

Draaijer, A., R. Sanders, and H. C. Gerritsen. Fluorescence lifetime imaging, a new tool in confocal microscopy. In: *Handbook of Biological Confocal Microscopy,* edited by J. B. Pawley. New York: Plenum Press, 1995, pp. 491–505.

French, T., P. T. C. So, D. J. Weaver, T. Coelho-Sampaio, E. Gratton, E. W. Voss, and J. Carrero. Two-photon fluorescence lifetime imaging microscopy of macrophage-mediated antigen processing. *J. Microsc.* 185:339–353, 1997.

Gadella, T. W. J., Jr., R. M. Clegg, and T. M. Jovin. Fluorescence lifetime imaging microscopy: Pixel-by-pixel analysis of phase modulation data. *Bioimaging* 2:139–159, 1994.

Gadella, T. W. J., Jr., and T. M. Jovin. Oligomerization of epidermal growth-factor receptors on A431 cells studied by time resolved fluorescence imaging microscopy—A stereochemical model for tyrosine kinase receptor activation. *J. Cell Biol.* 129:1543–1558, 1995.

Gadella, T. W. J., Jr., T. M. Jovin, and R. M. Clegg. Fluorescence lifetime imaging microscopy (FLIM)—Spatial-resolution of microstructures on the nanosecond time-scale. *Biophys. Chem.* 48:221–239, 1993.

Gratton, E., and M. Limkeman. A continuously variable frequency cross-correlation phase fluorometer with picosecond resolution. *Biophys. J.* 44:315–324, 1983.

Gratton, E., M. Limkeman, J. R. Lakowicz, B. P. Maliwa, H. Cherek, and G. Laczko. Resolution of mixtures of fluorophores using variable-frequency phase and modulation data. *Biophys. J.* 4:479–486, 1984.

Kume, H., K. Koyama, K. Nakatsugawa, S. Suzuki, and D. Fatlowitz. Ultrafast microchannel plate photomultipliers. *Appl. Opt.* 27:1170–1178, 1988.

Lakowicz, J. R., and K. Berndt. Lifetime-selective fluorescence imaging using an rf phase sensitive camera. *Rev. Sci. Instrum.* 62:1727–1734, 1991.

Lakowicz, J. R., G. Laczko, H. Cherec, E. Gratton, and M. Limkeman. Analysis of fluorescence decay kinetics from variable-frequency phase shift and modulation data. *Biophys. J.* 46:463–477, 1984.

Lakowicz, J. R., H. Szmacinski, W. J. Lederer, M. S. Kirby, M. L. Johnson, and K. Nowaczyk. Fluorescence lifetime imaging of intracellular calcium in COS cells using Quin-2. *Cell Calcium* 15:7–27, 1994.

Ng, T., A. Squire, G. Hansra, F. Bornancin, C. Prevostel, A. Hanby, W. Harris, D. Barnes, S. Schmidt, H. Mellor, P. I. H. Bastiaens, and P. J. Parker. Imaging protein kinase Ca activation in cells. *Science* 283:2085–2089, 1999.

Periasamy, A., P. Wodnicki, X. F. Wang, S. Kwon, G. W. Gordon, and B. Herman. Time resolved fluorescence lifetime imaging microscopy using a picosecond pulsed tunable dye-laser system. *Rev. Sci. Instrum.* 67:3722–3731, 1996.

Piston, D. W., G. Marriott, T. Radivoyevich, R. M. Clegg, T. M. Jovin, and E. Gratton. Wideband acoustooptic light-modulator for frequency-domain fluorometry and phosphorimetry. *Rev. Sci. Instrum.* 60:2596–2600, 1989.

Press, W. H., S. A. Teukolky, and W. T. Vetterling. *Numerical Recipes in C—The Art of Scientific Computing*, 2nd ed. Cambridge: Cambridge University Press, 1992.

Sanders, R., A. Draaijer, H. C. Gerritsen, P. M. Houpt, and Y. K. Levine. Quantitative Ph imaging in cells using confocal fluorescence lifetime imaging microscopy. *Anal. Biochem.* 227:302–308, 1995.

Schlick, T., and A. Fogelson. TNPACK—A truncated Newton minimization package for large scale problems: I. Algorithm and usage. *ACM Trans. Math. Soft.* 18:46–70, 1992.

Schneider, P. C., and R. M. Clegg. Rapid acquisition, analysis, and display of fluorescence lifetime—resolved images for real-time applications. *Rev. Sci. Instrum.* 68:4107–4119, 1997.

Scully, A. D., A. J. MacRobert, S. Botchway, P. O'Neill, A. W. Parker, R. B. Ostler, and D. Phillips Development of a laser-based fluorescence microscope with subnanosecond time resolution. *J. Fluoresc.* 6:119–125, 1996.

Squire, A., and P. I. H. Bastiaens. Three dimensional image restoration in fluorescence lifetime imaging microscopy. *J. Microsc.* 193:36–49, 1999.

Squire, A., P. J. Verveer, and P. I. H. Bastiaens. Multiple frequency fluorescence lifetime imaging microscopy. *J. Microsc.* 197:136–149, 2000.

Straume, M., S. G. Frasier-Cadore, and M. L. Johnson. Least-squares analysis of fluorescence data. In: *Topics in Fluorescence Spectroscopy*, edited by J. R. Lakowicz. New York: Plenum Press, 1991.

Sytsma, J., J. M. Vroom, C. J. Degrauw, and H. C. Gerritsen. Time gated fluorescence life-time imaging and microvolume spectroscopy using two-photon excitation. *J. Microsc.* 191:39–51, 1998.

Szmancinski, H., and J. R. Lakowicz. Possibility of simultaneously measuring low and high calcium concentrations using Fura-2 and lifetime-based sensing. *Cell Calcium* 18:64–75, 1995.

Tsien, R. Y. The green fluorescent protein. *Annu. Rev. Biochem.* 76:509–538, 1998.

Tsien, R. Y., B. J. Bacskai, and S. R. Adams. FRET for studying intracellular signaling. *Trends Cell Biol.* 3:242–245, 1993.

Verkman, A. S., M. Armijo, and K. Fushimi. Construction and evaluation of a frequency domain epifluorescence microscope for lifetime and anisotropy decay measurements in subcellular domains. *Biophys. Chem.* 40:117–125, 1991.

Verveer, P. J., A. Squire, and P. I. H. Bastiaens. Global analysis of fluorescence lifetime imaging microscopy data. *Biophys. J.* 78:2127–2137, 2000.

Watkins, A. N., C. M. Ingersoll, G. A. Baker, and F. V. Bright. A parallel multiharmonic frequency-domain fluorometer for measuring excited-state decay kinetics following one-, two-, or three-photon excitation. *Anal. Chem.* 70:3384–3396, 1998.

Wouters, F. S., P. I. H. Bastiaens. Fluorescence lifetime imaging of receptor tyrosine kinase activity in cells. *Curr. Biol.* 9:1127–1130, 1999.

17

Wide-Field, Confocal, Two-Photon, and Lifetime Resonance Energy Transfer Imaging Microscopy

Ammasi Periasamy, Masilamani Elangovan, Horst Wallrabe, Magarida Barroso, James N. Demas, David L. Brautigan, and Richard N. Day

1. Introduction

The light microscope has been used for almost a century to produce images of cells, and this approach has contributed enormously to our understanding of cellular structure and function (Bright and Taylor, 1986; Herman, 1998; Inoué and Spring, 1997; Pawley, 1995; Periasamy and Herman, 1994). In turn, molecular biological studies over the past few decades have shown that cellular events, such as signal transduction and gene transcription, require the assembly of proteins into specific macromolecular complexes. What we require now are methods to visualize these protein–protein associations as they occur in the living cell. Recent advances in digital imaging coupled with the development of new fluorescent fluorophores now provide the tools to begin the study of protein-protein interactions in the intact cell. In this chapter we describe four different imaging techniques that apply the method of fluorescence resonance energy transfer (FRET) to surpass the optical limitations of the light microscope, allowing detection of the physical interactions of proteins in the living cell.

The physical basis for FRET has been understood both in theory and in experimental applications for many years (Förster, 1965; Lakowicz, 1999; Stryer, 1978), but it is only with recent technical advances that FRET microscopy has become generally applicable to biomedical research (Clegg, 1996; Day, 1998; Gordon et al., 1998; Heim and Tsien, 1996; Jurgens et al., 1996; Periasamy and Day, 1998, 1999). The FRET imaging approach has a significant advantage over methods that have the same spatial resolution such as X-ray diffraction, nuclear magnetic resonance, and electron microscopy.

FRET detects localized molecular interactions in their natural environment within the living cell (Guo et al., 1995). For example, the application of digitized video (DV)-FRET imaging can reveal the two-dimensional spatial distribution of steady-state interactions between two protein partners in intact cells. Laser scanning confocal FRET microscopy (C-FRET), when used in combination with optical sectioning, allows the addition of the third dimension of detection of where steady-state protein associations occur inside the cell. The use of fluorophores that are excited by near-ultraviolet (UV) light in these live cell-imaging studies is limited, however, by problems associated with photobleaching and photodamage. This limitation can be overcome through application of the technique of two-photon FRET (2P-FRET), which uses red light to excite these near-UV fluorophores (Periasamy, 2000).

The information obtained using these methods regarding steady-state protein interactions in intact cells, although valuable, falls short of revealing a key feature of the living system—its dynamic organization. The temporal resolution of the imaging modalities described above is not sufficiently high to reveal protein interactions in real time. This degree of temporal resolution can be achieved using the technique of fluorescence lifetime imaging (FLIM) (see Section 3.4; Periasamy et al., 1996, 1999a,b; Wang et al., 1991). The FLIM method detects the nanosecond decay kinetics of a fluorophore and provides a spatial lifetime map for the fluorophore within the cell.

The local environment that surrounds a fluorophore influences its fluorescence lifetime. Thus, an event like the transfer of excitation energy from a donor fluorophore to acceptor molecules in its local environment results in measurable changes in the donor fluorescence lifetime. An important advantage to measuring energy transfer by donor FLIM is that only the donor fluorophore is relevant; the acceptor fluorescence need not be measured. This reduces background interference and enhances the precision of the measurement of protein interactions to a temporal resolution of subnanoseconds. Use of this technology will significantly improve and expand existing capabilities for studying dynamic events inside the living cell. In this chapter we discuss the methodology for monitoring protein associations in living cells using each of the techniques described above with both conventional (fluorescein isothiocyanate [FITC] and Cy3 pair) fluorophores and combinations of color variants of green fluorescent proteins (GFPs).

2. SPECTRAL ILLUSTRATIONS FOR DIFFERENT FLUOROPHORE PAIRS AND THE FILTER CONFIGURATIONS

FRET is the quantum mechanical process for the direct radiationless transfer of energy from a donor (D) to an acceptor (A) molecule. For FRET to occur between a donor and acceptor, the following four conditions must be fulfilled (for more detailed discussion of FRET theory, see Chapter 15 and Lakowicz, 1999). First, the donor emission spectrum must significantly overlap the absorption spectrum of the acceptor. We illustrate in Color Figure 17–1 many different combinations of fluorophores with spectral characteristics that could be suitable as donor–acceptor combinations for energy transfer imaging. Second, the distance between the donor and acceptor fluorophores must fall within the range of approximately 1–10 nm. Third, the donor emission dipole moment, the acceptor absorption dipole moment, and their separation vectors must be in favorable mutual orientation. Finally, the emission of the donor should have a high quantum yield.

TABLE 17–1 Filter configuration for FRET imaging for the fluorophore pair shown in Color Figure 17–1

Pair		Excitation (nm)	Emission (nm)		Dichroic (nm)
Donor	Acceptor	(Donor)	Donor	Acceptor	
BFP	GFP	365/15	460/50	535/50	390
BFP	YFP	365/15	460/50	535/26	390
BFP	RFP	365/15	460/50	610/60	390
CFP	RFP	440/21	480/30	610/60	455
CFP	YFP	440/21	480/30	535/26	455
GFP	Rhod-2	488/20	535/50	595/60	505
FITC	Rhod-2	488/20	535/45	595/60	505
FITC	CY3	488/20	535/45	595/60	505
Cy3	CY5	525/45	595/60	695/55	560
Alexa-488	Rhod-2	488/20	535/45	595/60	505

The first condition for energy transfer dictates that the efficiency of FRET will improve with increased overlap of the donor fluorophore emission spectra with the absorption spectra for the acceptor fluorophore. The problem we encounter, however, is that if the spectral overlap of donor and acceptor fluorophores is increased, there is an increased spectral bleedthrough signal detected in the acceptor channel. The spectral bleedthrough background results from a combination of both acceptor fluorophore excited by the donor excitation light plus donor emission detected in the acceptor channel. The result is a high and variable background signal from which a weak FRET-based acceptor emission must be extracted. Fortunately, there are methods for correcting for this spectral bleedthrough, and these are discussed below.

It is important to carefully select filter combinations that reduce the spectral bleedthrough background to improve the signal-to-noise (S/N) ratio for the FRET signals. In Table 17–1, we suggest some possible filter combinations for FRET imaging using several of the different fluorophore combinations shown in Color Figure 17–1. These filters are commercially available from either Chroma Technology Corporation (www.chroma.com) or Omega Optical, Inc. (www. omegafilters.com). Some of these filter combinations were used for the FRET imaging experiments described in this chapter. For the studies described here, the same dichroic mirror was used to acquire both the donor and acceptor images. This is important because there can be small differences in the mechanical position of the dichroic mirror from one filter cube to another, and the use of different cubes can introduce artifacts in the processed FRET images.

3. THE ACQUISITION AND PROCESSING OF FRET IMAGES

3.1 Digitized Video FRET

Any conventional microscope can be used for DV-FRET (for more detail, see Chapter 4 and Periasamy and Herman, 1994). The DV-FRET imaging system used in our studies consists of an IX-70 Olympus (Melville, NY) inverted microscope equipped with epifluorescence and transmitted illumination optics (Periasamy and Day, 1999). The epifluorescent light source was a 100 W Hg-Xe combination arc lamp (Hamamatsu Corp., Middlesex, NY) and a halogen lamp for transillumination with a long working distance condenser (numerical aperture [NA] = 0.55).

A Plan Apo 60× (NA = 1.2) water-immersion objective lens was used for these studies. A Ludl (Ludl Electronic products Ltd., Hawthrone, NY) motorized microscope stage and focus system, along with the excitation, emission, and neutral density (ND) filter wheels, was interfaced to the SGI INDY (Silicon Graphics Inc.) computer through a controller driven by Inovision Corporation ISEE software (RTP, Raleigh, NC). The excitation and emission filters were installed in their respective filter wheels and were paired with a dichroic mirror mounted in a cube.

The donor and acceptor images were collected sequentially using the same excitation filter and dichroic mirror, but different emission filters. The charge-coupled device (CCD) camera used for DV-FRET image acquisition was a Hamamatsu Orca-200 CCD camera system. This camera has high quantum efficiency in the visible spectrum. For further information regarding the sensitivity and other parameters of the detectors for fluorescence imaging, see Chapter 3.

Cell preparation for DV-FRET is as follows. The HeLa cells shown in Color Figure 17–2 were transfected with expression plasmids encoding GFP– and BFP–Pit-1 as described previously (Day, 1998; Periasamy and Day, 1999). The transiently transfected cells were diluted in phenol red-free culture medium containing serum and used to inoculate culture dishes containing 25 mm round glass coverslips. The cells were kept at 33°C in a humidified 5% CO_2 incubator for 24–48 hours before FRET imaging. A coverslip with a monolayer of transfected cells was placed in a specifically designed chamber for the microscope stage (Periasamy and Herman, 1994).

The cells expressing the GFP–fusion protein were initially identified using the GFP filter set (see Table 17–1), and a reference image was obtained. A second image was then acquired for donor (D) fluorescence (BFP) from the same field of cells using the donor filter set, followed by acquisition of a third image of the same field using the acceptor (A) filter set. The camera gain, ND filter, and image acquisition time were kept constant for the donor and acceptor images. Camera dark-current noise (see Chapter 3) and background contributions due to medium, cells, and optics (Aubin, 1979) were digitally subtracted from donor (I_D) and acceptor (I_A) images. With the ISEE software, a pixel-by-pixel ratio of the background-subtracted acceptor to donor images yielded the FRET image (I_A/I_D) (Color Fig. 17–2a).

Some of the signal in Color Figure 17–2a results from the spectral bleedthrough background for the combination of the BFP and GFP fluorophores. To obtain the optimal FRET signal, it is important to correct for the bleedthrough background (or cross-talk) signal in the acceptor channel. Recently, Gordon and colleagues (1998) reported a method for cross-talk correction (see also Chapter 15). Their approach required the collection of a series of donor and acceptor images to correct for the cross-talk signal. This approach can be problematic for fluorophores that are sensitive to photobleaching, such as BFP. As an alternative method, we have used a very narrow-band filter to reduce the cross-talk and then increased the exposure time to improve the S/N ratio in the acquired images.

Computer software (Inovision Corporation, RTP, NC) designed to correct for the emission bleedthrough signal was then used to process the FRET image (Color Fig. 17–2b). This correction requires that a series of images from cells labeled with either the donor or acceptor proteins alone be obtained using the same filter configuration that is to be used for FRET data acquisition. The computer algorithm then determines the percentage of cross-talk in each channel from the donor and acceptor control images, and this data set is then used to correct the bleedthrough in the FRET images. As shown in Color Figure 17–2, there is a significant improvement in the localized FRET signal after the bleedthrough correction (red color). This correction technique, however, is not capable of completely removing the bleedthrough signal. Moreover, because the bleedthrough background varies depending on the excitation intensity and fluorophore concentration, there can be error in the percentage of bleedthrough estimation for a given fluorophore.

3.2. Laser Scanning Confocal FRET Microscopy

A major limitation to the DV-FRET microscopy technique is that the emission signals originating from above and below the focal plane contribute to out-of-focus signal that reduces the contrast and seriously degrades the image. Laser scanning confocal FRET (C-FRET) microscopy (see Chapter 5) can overcome this limitation owing to its capability of rejecting signals from outside the focal plane. This capability provides a significant improvement in lateral resolution and allows the use of serial optical sectioning of the living specimen.

A disadvantage of this technique is that the wavelengths available for excitation of different fluorophore pairs is limited to standard laser lines. If a confocal microscope is also equipped with UV laser and UV optics, it is possible to use the BFP fluorophore combinations, shown in Color Figure 17–1. As mentioned above, however, this fluorophore is sensitive to photobleaching, and this is exacerbated by the intense laser source. Standard laser lines do allow C-FRET to be used for many fluorophore combinations, including CFP–RFP, GFP–rhodamine or –Cy3, FITC– or Alexa 488–Cy3, or rhodamine and Cy3–Cy5.

3.2.1 Cell preparation

We developed C-FRET microscopy techniques to quantify the level of co-localization (1–10 nm range) between apically internalized Alexa 488–labeled pIgA-R ligand with basolaterally internalized Cy3-labeled ligand using the internalization protocol described below. Co-localization of specific receptor–ligand complexes is used to define endosomal compartments. In particular, we are interested in studying transcytosis, which is a multistep vesicular pathway that connects the apical and the basolateral plasma membrane domains of polarized epithelial cells through an apical endosome compartment (Barroso and Sztul, 1994). Our work uses polarized MDCK cells that are stably transfected with polymeric IgA receptor (pIgA-R) and an anti-pIgA-R–IgG ligand, labeled with Alexa- 488 or Cy3.

MDCK cells stably transfected with pIgA-R were grown to confluence on Transwell Clear (Corning-Costar) inserts for 3 days to achieve full polarization. pIgA-R ligand–Alexa 488 and –Cy3 were co-internalized apically and basolaterally, respectively, for 4 hours at 17°C, washed with PBS to remove unbound ligand, and immediately fixed with 4% paraformaldehyde/PBS. After 4 hours internalization, co-localization is observed predominantly in the subapical region (Barroso and Sztul, 1994). We used Alexa 488 as the donor fluorochrome paired with Cy3 as the acceptor fluorochrome. Negative controls comprise cells internalized from the apical membrane with Alexa 488–labeled pIgA-R ligand (donor only) or internalized from the basolateral membrane with Cy3-labeled pIgA-R ligand (acceptor only).

3.2.2 Data collection

A Nikon PCM 2000 laser scanning confocal microscope, a 60× water immersion lens, and Argon and green He-Ne lasers were used. Images under FRET conditions were taken with an Argon laser (488 nm), a 10% ND filter, and a 590 nm long-pass filter. We acquired and analyzed images for different labeling conditions: a single-labeled Cy3 (acceptor) reference sample, a single-labeled Alexa 488 (donor) reference, and a double-labeled Alexa 488–Cy3 (donor–acceptor) sample. It is important to note that we also scanned the double-labeled and the single-labeled specimens with the green He-Ne laser alone (data not shown). This establishes the Cy3 (acceptor) base levels without background fluorescence caused by the donor excitation wavelength. These images allowed us to select cells with similar acceptor intensities for our FRET analysis (see below).

3.2.3 FRET analysis

The spectral bleedthrough is the major problem in obtaining a true FRET signal in C-FRET technology (Color Fig. 17–3F). This is the same issue encountered with DV-FRET as discussed in Section 3.1. We have dissected this signal into three components: first, excitation of the acceptor molecule by the incident light intended for the donor (Color Fig. 17–3D); second, overlap between donor and acceptor emission in the red channel (Color Fig. 17–3B); and, third, actual FRET. All of these elements were incorporated into a specially developed algorithm (see below) applied to the contaminated FRET image (Color Fig. 17–3F); the result is a corrected FRET image as shown in Color Figure 17–3G. The algorithm considers the fluorescence intensity values of seven different images: the two single-labeled reference specimens (Color Fig. 17–3A–D) and the one double-labeled specimen (Color Fig. 17–3E,F), excited separately by the donor (Color Fig. 17–3) and acceptor (data not shown) wavelengths, with the images collected in the donor channel (Color Fig. 17–3A,C,E) and the acceptor channel (Color Fig. 17–3B,D,F).

The algorithm works on the assumption that there is a linear relationship between the fluorescence intensity levels of donor and acceptor and their bleedthrough. By establishing these levels in the two-reference specimens pixel and finding matching pixel ranges in the double-labeled specimen, a correction factor is obtained. Then, this correction factor is subtracted from the uncorrected FRET signal, resulting in the corrected FRET image (Color Fig. 17–3G).

We believe that this approach has significant advantages over the standard method of completely bleaching the acceptor and then comparing the prebleach and postbleach images (Kenworthy et al., 2000). Our program will allow a quicker method to analyze FRET in live as well as fixed cells without the use of photobleaching.

3.2.4 Data analysis

As shown in Color Figure 17–3A, B, we show an image collected from the donor-only specimen (Alexa 488, single labeled, excitation 488 nm) in both the green (Color Fig. 17–3A) and red (Color Fig. 17–3B) channels. In Color Figure 17–3C,D, we show the image collected from the acceptor-only specimen (Cy3, single labeled, excitation 488 nm), again in the green (Color Fig. 17–3C) and the red (Color Fig. 17–3D) channels. As expected, we did not see any fluorescence in the green channel. The noticeable background fluorescence in the red channel (Color Fig. 17–3B, D) is caused by the Donor and acceptor emission bleed-through into the red channel. This is a major issue in interpreting the FRET signal in a double-labeled specimen.

In Color Figure 17–3E, F, we show the image collected from a donor- and acceptor-containing specimen (Alexa 488–Cy3, double labeled, excitation 488 nm). The green channel (Color Fig. 17–3E) shows the donor labeling (Color Fig. 17–3A). Fluorescence in the red channel (Color Fig. 17–3F) is assumed to be the accumulation of FRET plus background.

As mentioned earlier, the algorithm modifies (Elangovan and Periasamy, 2000) the image in Color Figure 17–3F and produces the corrected FRET image in Color Figure 17–3G. A significant reduction in the fluorescence intensity level is observed between Color Figure 17–3F and 17–3G which is due to the removal of bleedthrough between Alexa 488 and Cy3 fluorochromes. The method is particularly suitable for the type of cellular analysis described here, where adjacent colocalization between basolaterally and apically internalized pIgA-R molecules within the membranes of the apical endocytic compartment can be demonstrated through C-FRET. This method is also ideal for biological analysis that is sensitive to photobleaching. We will use this method to investigate the nature of submicron domains within the endosomal membranes by monitoring whether an endosomal transmembrane protein, the polymeric IgA receptor (pIgA-R), is sorted into domains (<10 nm) from which other proteins are excluded.

3.3 Two-Photon Excitation FRET Microscopy

As mentioned earlier, the advantage of C-FRET over DV-FRET lies in the ability to reject the out-of-focus signal that originates from outside the focal plane. A significant improvement over C-FRET can be achieved by eliminating the out-of-focus signal altogether by limiting excitation to only the fluorophore at the focal plane. This is precisely what two-photon excitation microscopy does.

Two-photon absorption was theoretically predicted by Göppert-Mayer in 1931, and it was experimentally observed for the first time in 1961 by using a ruby laser as the light source (Kaiser and Garrett, 1961). The original idea of two-photon fluorescence scanning microscopy was experimentally demonstrated in biological samples by Denk and others (1990; see also Chapters 9–13). It is explained in Chapters 9–13 that the rate of excitation is proportional to the square of the instantaneous intensity.

With extremely high instantaneous intensity, two photons of light (at approximately twice the wavelength normally required to excite a fluorophore) can simultaneously occupy the fluorophore absorption cross section. The energies of these photons are combined to excite the fluorophore. The extremely high instantaneous intensity required for two-photon excitation is made possible by focusing a femtosecond (fs) pulsed, mode-locked laser on the specimen to yield on the order of 10^{-50} to 10^{-49} cm^4 s/photon/molecule. Because two-photon excitation occurs only in the focal volume, the detected emission signal is exclusively in-focus light. Furthermore, because two-photon excitation uses longer wavelength light, it is less damaging to living cells, thus limiting problems associated with fluorophore photobleaching and photodamage as well as intrinsic fluorescence of cellular components.

3.3.1 Cell preparation

The cDNA sequence for the transcription factor C/EBPα was fused in frame with the sequence encoding either the BFP or ds-RFP color variants. For transfection, pituitary GHFT1-5 cells were harvested and transfected with 3–10 μg of purified expression plasmid encoding the BFP or RFP fusion proteins by electroporation, as described previously Day, 1998. For imaging, a coverslip with a monolayer of transfected cells was placed in a specifically designed chamber for the microscope stage (Periasamy and Herman, 1994).

3.3.2 2P-FRET imaging system

A Verdi pumped tunable ti:sapphire laser (Coherent, Mira 900, Tunable 700–1000, 76 MHz) was coupled to the laser port of a Biorad MRC600 (Biorad, Hercules, CA) laser scanning confocal scan head through beam steering optics. The MRC600 scan head was coupled to a Nikon TE300 (Eclipse) epifluorescent microscope (Nikon, Melville, NY). A 60× Plan Apo, 1.2 NA, water immersion objective lens was used for these experiments. It is important to use a non-phase objective lens to minimize light scattering in laser coupling to the microscope (Periasamy, 2000).

3.3.3 2P-FRET imaging

Cells expressing the RFP tagged C/EBPα protein were identified by 535 nm excitation using an arc lamp light source coupled to the microscope. The Biorad scan head was coupled at the side port of the microscope. For 2P excitation, the Ti:Sapphire laser wavelength was tuned from 870 nm down to 700 nm, in order to see minimum or no signal from the RFP–C/EBPα-only labeled proteins. We found that no RFP signal was seen at 740 nm. The BFP–C/EBPα protein molecule was excited with an infrared wavelength from the Ti:Sapphire laser, which caused a considerable amount of signals at 740 nm. The 740 nm was used as an excitation wavelength to acquire the donor and acceptor images. Before using the BFP– and

RFP–C/EBPα combination, we used the BFP–C/EBPα alone to monitor the bleedthrough signal in the RFP channel. We acquired the images with BFP alone on both the channels, and the BFP bleedthrough was corrected, as shown in Color Figure 17–4. The acquired image was background subtracted and bleedthrough corrected to obtain a 2P-FRET image (Elangovan and Periasamy, 2000).

3.4. Fluorescence Lifetime Imaging FRET Microscopy

Each of the fluorescence microscopy techniques described uses intensity measurements to reveal fluorophore concentration and distribution in the cell. Recent advances in camera sensitivities and resolutions have improved the capability of these techniques to detect dynamic cellular events. Unfortunately, even with the improvements in technology, these fluorescence microscopic techniques do not have high-speed (<s) time resolution to fully characterize the organization and dynamics of complex cellular structures.

In contrast, the time-resolved fluorescence microscopic technique (see Chapter 15) allows the measurement of dynamic events at very high temporal resolution. Combined with FRET, this approach can monitor interactions between cellular components with very high spatial resolution. Moreover, one need not implement spectral bleedthrough correction in FLIM images. To date, most measurements of fluorescence lifetimes have been performed in solution or in cells in suspension (Lakowicz, 1999). Fluorescence lifetime imaging microscopy (FLIM) was developed to overcome this limitation to allow lifetime measurements to be acquired from living cells in culture (Periasamy et al., 1996; see also Chapters 15–16 and 18). Moreover, the combination of FLIM and FRET provides a means for analyzing dynamic protein interactions in four dimensions (three-dimensional space and time).

3.4.1 Theoretical simulation and experimental verification of double-exponential decay

In quantitative analysis, it is important to estimate measurement precision in the presence of noise. We evaluated the performance over a wide range of experimental conditions in order to assess the optimum conditions and the theoretical limitations for contiguous and overlap gating procedures for single- and double-exponential decay (Periasamy et al., 1999; Sharman et al., 1999). In particular, we examined the effect of shot noise associated with photon counting.

The rapid lifetime determination (RLD) method is a family of data analysis techniques for fitting experimental data that conform to single- and multiple-exponential decays with or without baseline contribution (Sharman et al., 1999). We have already devised theoretical simulations of the RLD method for single-exponential decay using Monte Carlo simulations (Ballew and Demas, 1989), which were experimentally demonstrated using a gated image intensifier charge-coupled device (CCD) camera for live cellular imaging (Periasamy et al., 1996, 1999a,b). Because our main interest is to explore the multicomponent analysis of the fluorescence decay in FLIM–FRET data in order to determine the protein associations in a complex biological environment, we extended our theoretical simulations from single-exponential to double-exponential decay.

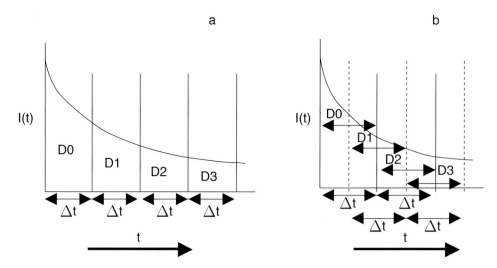

FIGURE 17–5 Illustration of contiguous (*a*) and overlap (*b*) gating. Four images were acquired during the decay after laser pulse excitation.

A range of values for each variable was considered in order to test the optimal conditions for each form of RLD. Theoretical simulations are primarily used to determine pre-exponential factors and lifetime, and Monte Carlo theoretical simulations were used to estimate the precision and accuracy of the data reduction method. The contiguous-gate method shows an increase in percent relative standard deviation at lower lifetime ratios compared with the overlap-gated case. For example (Sharman et al., 1999), when the ratio of the lifetimes equals 1.4 (meaning that the lifetimes are very similar), the optimal relative standard deviation is 10% for the contiguous-gated method and 5% for the overlap-gated method (the difference is a longer time interval/lifetime ratio). Clearly, the overlap-gated method is an improvement over contiguous gating.

To demonstrate this methodology we used BHK21 cells labeled with GFP (not targeted to any proteins) and acquired and then processed the images using contiguous and overlap gating. The data acquisition of four gates during the decay after laser pulse excitation is shown in Figure 17–5 (Sharman et al., 1999). The overlap gating FLIM image (OG1) provides a better S/N ratio than the contiguous gating (CG1). We also demonstrated that in the overlap-gated mode, there were no signals in the second component lifetime image (OG2) (see Color Fig. 17–6) because the GFP was not targeted to any proteins of the living cell (excitation, 460 nm; emission, 535/50 nm).

3.4.2. FLIM–FRET imaging of protein associations

Figure 17–7 shows the FLIM–FRET imaging system. The Coherent Verdi (532 nm; 5 W) pumped Ti:Sapphire pulsed infrared laser (76 MHz; 150 fs; tunable wavelength, 700–1100 nm) is tuned using x-wave optics for the 880 nm and then doubled to obtain a 440 nm pulsed laser line to excite the donor CFP–C/EBPα. The 440 nm pulsed laser beam was coupled to the microscope epifluorescence port through the motorized beam expander to illuminate the live specimen.

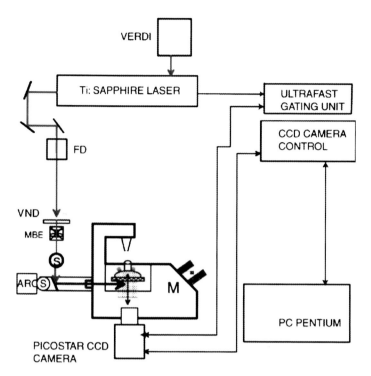

FIGURE 17–7 Schematic diagram of FLIM–FRET imaging system (see text for an explanation of the system). FD, Frequency doubler; VND, variable neutral density filter; M, microscope; ARC, arc lamp; S, shutter.

To acquire the time-resolved images of the donor in the presence and absence of the acceptor, we used dichroic filter (455 nm) and an emission filter (480/30 nm) provided by Chroma in the Nikon filter cube. We implemented an overlap-gated scheme (see Fig. 17–5) to acquire and process the images for single- and double-exponential decays. First, we used the CFP–C/EBPα living cells to acquire the donor-only time-resolved images (DD_0 and DD_1), and then we replaced the chamber with CFP– and YFP–C/EBPα cells and repeated the same procedure to collect overlap-gated time-resolved images (AD_0, AD_1, AD_2, AD_3). These donor images in the absence (τ_D) and in the presence of the acceptor (τ_{DA}) were processed for lifetime images. As shown in Color Figure 17–8, we selected the same protein complex area in τ_D and τ_{DA1} to compare its lifetime distribution. If the energy was transferred from one protein to another, then $\tau_{DA1} < \tau_D$ ($\tau_D = 2.18$ ns and $\tau_{DA1} = 1.6$ ns (Table 17–2). The energy transfer efficiency ($E = 1 - [\tau_{DA1}/\tau_D]$) of this event is about 22%.

There is no second component lifetime donor image in the absence of the acceptor, but the lifetime image of the second component in the presence of the acceptor indicates (see Color Fig. 17–8) that there is energy transfer (or distance distribution) distribution. Also, the lifetime is zero away from the area of the protein complex. There are a number of proteins that did not significantly take part in the energy transfer process when compared with τ_D.

TABLE 17–2 Lifetime distribution of the donor in the absence (τ_D) and presence (τ_{DA}) of the acceptor in a single living cell

P*	τ_D (ns)	τ_{DA1} (ns)	τ_{DA2} (ns)	E (%)
a	2.18	1.8	0.57	17
b	2.18	1.92	0.31	12
c	2.18	1.69	0.47	22
d	2.18	1.30	0.36	40

*P indicates position marked on the image in Color Figure 17.8.

4. CONCLUSIONS

We have described in this chapter the application of FRET using the microscopic techniques of DV-FRET, C-FRET, 2P-FRET, and FLIM–FRET to study protein–protein interactions in living cells. For each technique we outlined some important factors one should consider for imaging. The spectral overlap is the major problem with DV-FRET, so we addressed how to correct the bleedthrough signal from the donor channel to the acceptor channel. Moreover, C-FRET provides better discrimination of out-of-focus information at different sections of the cell, which allowed us to acquire the FRET signal in order to quantify the organization and dynamics of endosomes in epithelial cells using Alexa 488 and Cy3 as a fluorophore pair. We also demonstrated that one could gain a tremendous advantage from the 2P-FRET technique by reducing the photobleaching of BFP molecules by using infrared wavelength (740 nm) excitation.

The error involved in quantitating the FRET signal in intensity FRET imaging (DV-FRET, C-FRET, and 2P-FRET) is a major problem. The FLIM–FRET images are independent of fluorophore concentration and excitation intensity and provided two- and three-dimensional lifetime distributions of FLIM–FRET images of CFP– and YFP-C/EBPα protein dimerization. The decrease in the donor lifetime in the presence of the acceptor indicates the occurrence of energy transfer. Moreover, the efficiency and the distance between donor and acceptor of energy transfer can be easily calculated using the lifetime. This novel FLIM–FRET imaging technology with very high temporal and spatial resolution could revolutionize the monitoring of protein associations in living cells.

This work was supported by grants from the W. M. Keck Foundation and the Academic Enhancement Program of the University of Virginia.

REFERENCES

Aubin, J. E. Autofluorescence of viable cultured mammalian cells. *J. Histochem. Cytochem.* 27:36–43, 1979.

Ballew, R. M., and J. N. Demas. An error analysis of the rapid lifetime determination method for the evaluation of single exponential decays. *Anal. Chem.* 61:30–33, 1989.

Barroso, M., and E. S. Sztul. Basolateral to apical transcytosis is indirect and involves BFA and trimeric G protein sensitive passage through the apical endosome. *J. Cell Biol.* 124:83–100, 1994.

Bright, G. R., and D. L. Taylor. Imaging at low light level in fluorescence microscopy. In: *Applications of Fluorescence in the Biomedical Sciences,* edited by D. L. Taylor, F. Lanni, A. S. Waggoner, R. F. Murphy, and R. R. Birge. New York: Alan R. Liss, 1986, pp. 257–288.

Clegg, R. M. Fluorescence resonance energy transfer. In: *Fluorescence Imaging Spectroscopy and Microscopy, Chemical Analysis Series,* Vol. 137, edited by X. F. Wang and B. Herman. New York: John Wiley & Sons, 1996, pp. 179–251.

Day, R. N. Visualization of Pit-1 transcription factor interactions in the living cell nucleus by fluorescence resonance energy transfer microscopy. *Mol. Endocrinol.* 12:1410–1419, 1998.

Denk, W., J. H. Strickler, and W. W. Webb. Two-photon laser scanning fluorescence microscopy. *Science* 248:73–76, 1990.

Elangovan, M., and A. Periasamy. Bleed-through and photobleaching corection in multiphoton FRET microscopy. *SPIE Proc.* 4262: In Press, 2001.

Förster, T. Delocalized excitation and excitation transfer. In: *Modern Quantum Chemistry,* Vol. 3, edited by O. Sinanoglu. New York: Academic Press, 1965, pp. 93–137.

Goppert-Mayer, M. Ueber Elementarakte mit zwei Quantenspruengen. *Ann. Phys.* 9:273–295, 1931.

Gordon, G. W., G. Berry, X. H. Liang, B. Levine, and B. Herman. Quantitative fluorescence resonance energy transfer measurements using fluorescence microscopy. *Biophys. J.* 74:2702–2713, 1998.

Guo, C., S. K. Dower, D. Holowka, and B. Baird. Fluorescence resonance energy transfer reveals interleukin (IL)-1–dependent aggregation of IL-1 type I receptors that correlates with receptor activation. *J. Biol. Chem.* 270:27562–27568, 1995.

Heim, R., and R. Y. Tsien. Engineering green fluorescent protein for improved brightness, longer wavelengths and fluorescence resonance energy transfer. *Curr. Biol.* 6:178–182, 1996.

Herman, B. *Fluorescence Microscopy,* 2nd ed. New York: Springer-Verlag, 1998.

Inoué, S., and K. Spring. *Video Microscopy,* 2nd ed. New York: Plenum Press, 1997.

Jurgens, L., D. Arndt-Jovin, I. Pecht, and T. M. Jovin. Proximity relationships between the type I receptor for Fc epsilon (Fc epsilon RI) and the mast cell function-associated antigen (MAFA) studied by donor photobleaching fluorescence resonance energy transfer microscopy. *Eur. J. Immunol.* 26:84–91, 1996.

Kaiser, W., and C. G. B. Garrett. Two-photon excitation in CaF2:Eu2+. *Phys. Rev. Lett.* 7:229–231, 1961.

Kenworthy, K. A., N. Petranova, and M. Edidin. High-resolution FRET microscopy of cholera toxin B-subunit and GPI-anchored proteins in cell plasma membranes. *Mol. Biol. Cell* 11:1645–1655, 2000.

Lakowicz, J. R. *Principles of Fluorescence Spectroscopy,* 2nd ed. New York: Plenum Press, 1999.

Pawley, J. *Handbook of Biological Confocal Microscopy,* 2nd ed. New York: Plenum Press, 1995.

Periasamy, A. Two-photon excitation energy transfer microscopy. *SPIE Proc.* 3921:299–304, 2000.

Periasamy, A., and R. N. Day. FRET imaging of Pit-1 protein interactions in living cells. *J. Biomed. Opt.* 3:154–160, 1998.

Periasamy, A., and R. N. Day. Visualizing protein interactions in living cells using digitized GFP imaging and FRET microscopy. *Methods Cell Biol.* 58:293–314, 1999.

Periasamy, A., and B. Herman. Computerized fluorescence microscopic vision in the biomedical sciences. *J. Comput. Assist. Microsc.* 6:1–26, 1994.

Periasamy, A., K. K. Sharman, R. Ahuja, I. Eto, and D. L. Brautigan. Fluorescence lifetime imaging of green fluorescent protein of living cells. *SPIE Proc.* 3604: 6–12, 1999a.

Periasamy, A., K. K. Sharman, and J. N. Demas. Fluorescence lifetime imaging microscopy using rapid lifetime determination method: Theory and applications. *Biophys. J.* 76:A10, 1999b.

Periasamy, A., P. Wodnicki, X. F. Wang, S. Kwon, G. W. Gordon, and B. Herman. Time resolved fluorescence lifetime imaging microscopy using picosecond pulsed tunable dye laser system. *Rev. Sci. Instrum.* 67:3722–3731, 1996.

Sharman, K. K., J. N. Demas, H. Asworth, and A. Periasamy. Error analysis of the rapid lifetime determination (RLD) method for double exponential decays: Evaluating different window systems. *Anal. Chem.* 71:947–952, 1999.

Stryer, L. Fluorescence energy transfer as a spectroscopic ruler. *Annu. Rev. Biochem.* 47:819–846, 1978.

Wang, X. F., T. Uchida, D. M. Coleman, and S. Minami. A two-dimensional fluorescence lifetime imaging system using a gated image intensifier. *Appl. Spectrosc.* 45:360–366, 1991.

18

One- and Two-Photon Confocal Fluorescence Lifetime Imaging and Its Applications

Hans C. Gerritsen and Kees de Grauw

1. Introduction

Fluorescence microscopy is a commonly used tool in biological and biophysical research. The great power of fluorescence microscopy lies in the excellent contrast, selectivity, and sensitivity that can be achieved with this technique. The high contrast stems from the shift between the wavelength at which the specimen is excited and the wavelength of the emitted fluorescence light. This shift, the Stokes' shift, enables the efficient suppression of scattered excitation light by means of simple optical cut-off or bandpass filters. The contrast and sensitivity that can be realized are so high that even fluorescence imaging of single molecules can be accomplished.

The fluorescent molecules can be easily conjugated to chemically selective groups or antibodies such that specific sites in a specimen can be selectively labeled. As a result, it is possible to selectively stain, for instance, specific parts of cells. This makes fluorescence microscopy a popular tool in morphological studies. The number of probes that can be distinguished in fluorescence microscopy is limited by overlap of the comparatively broad emission spectra of the fluorescent labels. In general, the maximum number of probes that can be imaged in a single wavelength excitation experiment is limited to two or three.

Fluorescence microscopy can also be employed for the imaging (quantification) of concentrations of molecules. A good example of this is the imaging of ion concentrations using fluorescent probes that alter their properties in the presence of specific ions. These probes have a high selectivity for specific ions and exhibit distinct changes in their photophysical properties in the presence of ions. Fluorescent ion indicators can be classified in two groups: intensity probes and ratio probes. Intensity probes exhibit a marked increase in the fluorescence intensity with an increasing ion concentration. They are not, however, suitable for quantitative imaging of ion concentrations because the observed fluorescence intensity depends on both the ion concentration and the probe concentration. The main problems with the use of intensity probes arise from inhomogeneous partitioning of the probe within or between cells, probe leakage, and photobleaching. All these factors affect the fluorescence intensity, which complicates the interpretation of the fluorescence images.

Using ratio probes can circumvent these drawbacks (Tsien and Poenie, 1986; Grynkiewicz et al., 1985). Here, the probes exhibit a characteristic shift in the excitation or emission spectrum in the presence of the ions. In this case, the dependence on the probe concentration can be eliminated by taking the ratio of the fluorescence intensities recorded at two different emission wavelengths for excitation at a given wavelength (emission ratio). Alternatively, the detection takes place at one emission wavelength while exciting the probe at two different wavelengths (excitation ratio). Although the ratio method cancels out probe concentration and other intensity effects, the local chemical environment of the probe molecules can compromise the spectral properties of the ratio probes. Therefore, this method requires an in situ calibration procedure, which is particularly tedious when working with cells. An additional complication of this method is that many of the ratio probes, calcium ratio probes in particular, have absorption bands in the ultraviolet (UV) spectrum. Ultraviolet radiation is not only potentially harmful to biological cells, but also gives rise to an autofluorescence background in the images.

Many of the limitations of conventional fluorescence microscopy can be overcome by using the fluorescence lifetime of a molecule as a contrast mechanism. The fluorescence lifetime of a molecule is independent of the fluorescence intensity and differs per type of fluorescent molecule. In addition, the fluorescence lifetime can be sensitive for the chemical environment of the fluorescent molecule. Fluorescence lifetime imaging (FLIM) (Bastiaens and Squire, 1999; Buurman et al., 1992; French et al., 1997; Gadella et al., 1993; Lakowicz et al., 1992; Ni et al., 1991; Periasamy et al., 1996; Wang et al., 1991; see also Chapters 15–17) can be used in many different imaging experiments, including multilabeling experiments, the quantitative imaging of ion concentrations, the quantitative imaging of other molecules (such as oxygen), and the quantification of the energy transfer efficiency in fluorescence resonance energy transfer (FRET) experiments.

In this chapter we give a brief theoretical description of the fluorescence lifetime, treat several practical aspects related to the implementation of FLIM, and illustrate the potential of FLIM with several examples. The experiments described here were carried out using a confocal and a multiphoton excitation microscope.

2. BASICS OF FLUORESCENCE

For an extensive description of the fundamentals of fluorescence, we refer the reader to Chapter 1. Here, only a brief summary will be given. The fundamental difference between phosphorescence and fluorescence is the excited state from which emission takes place. In practice, the difference shows up as a difference in the time it takes for emission to occur after the absorption of light. Fluorescence typically transpires on a time scale of a few nanoseconds, while phosphorescence takes place on a much longer time scale, typically microseconds to seconds.

The fluorescence lifetime, τ, of a molecule is one of the most important parameters of a fluorescent molecule, and it is particularly useful when we are interested in acquiring quantitative information. From Equations 1 and 2 in Chapter 1 and the definition of the natural lifetime ($\tau_n = 1/\Gamma$), we find that the quantum efficiency of a fluorescent molecule can be expressed as $Q = \tau/\tau_n$. Due to interactions between the fluorescent molecule and its environment, additional nonradiative decay channels that quench the fluorescence may be introduced. Consequently, the nonradiative rate (k_{nr}) increases and the fluorescence lifetime reduces to τ'. Most importantly, the quantum efficiency reduces to $Q' = \tau'/\tau_n$, and the observed fluorescence intensity goes down. For this reason intensity images are not reliable for the imaging of concentrations.

By employing fluorescence lifetime imaging, however, the change in quantum efficiency can, in principle, be monitored and corrected for. The correction is accomplished simply by dividing the fluorescence intensities by the measured fluorescence lifetimes. Alternatively, the fluorescence lifetime yields information about the interactions between the fluorescent probe and its environment. For instance, the polarity of the solvent has a strong influence on the fluorescent properties of a fluorescent probe, including the fluorescence lifetime. Another well-known effect is the reduction of the fluorescence lifetime due to collisional quenching. Selective quenching by collisions with molecular oxygen may be used to quantify free oxygen concentrations.

3. THREE-DIMENSIONAL QUANTITATIVE IMAGING

Confocal imaging and multiphoton excitation imaging (Denk et al., 1990) dramatically enhance the suitability of fluorescence microscopy for investigations in three dimensions. This rests in the fact that these arrangements avoid the detection of a general background signal due to fluorescence from out-of-focus planes. Thus, it makes possible the noninvasive imaging of three-dimensional structures with exceptional contrast and sensitivity. The combination of three-dimensional imaging with quantitative imaging is particularly challenging. An important consequence of high-resolution three-dimensional imaging is that only a small volume is imaged, and, in general, a small number of fluorescent molecules are present in this volume element. On assuming a typical dye concentration of tens of micrometers, we find that about 100 dye molecules are present in the focus of a high numerical aperture (NA) microscope objective.

For conventional confocal and multiphoton fluorescence intensity imaging, images of reasonable quality can be acquired with as few as 20 detected photons per pixel. For quantitative images, however, this is not sufficient. Even a quantitative comparison based on the ratio of two intensity images would require a much higher signal level. Based on photon statistics alone, about 200 photons per pixel would be required to achieve an accuracy of 10% in this ratio. To realize this level of accuracy throughout the whole image, including the dim regions, the average intensity should be much higher. From the above line of reasoning we see that quantitative imaging requires roughly two orders of magnitude more signal than simple intensity imaging experiments.

Photobleaching constrains the maximum number of photons that can be emitted per fluorescent molecule. Good fluorophores may emit 10^4–10^5 photons before photobleaching takes place. Based on the above numbers and assuming an optimistic overall detection efficiency of 1% for the microscope, we find that a maximum of 10^4–10^5 detected photons can be expected from this volume element. Consequently, photobleaching effects may show up after only a few frames have been recorded. Moreover, the limited number of available molecules and the high signal requirement also put a limit on the maximum acquisition (pixel) rate. This example clearly stresses the importance of the overall efficiency of the microscope and of the method employed for quantitative imaging.

At this point it is important to emphasize a key difference between confocal and multiphoton excitation imaging. In confocal microscopy, axial resolution is obtained by spatial filtering. A detection pinhole is used in combination with point illumination. In multiphoton excitation microscopy, spatial selection of a volume element is not achieved by spatial filtering but by the nonlinear dependence of the multiphoton absorption process itself (see Chapters 9 and 10). The two imaging methods differ significantly in the spatial extent of photobleaching. In confocal microscopy, fluorescent molecules are excited throughout the whole illumination cone that hits the sample. Importantly, the bleach rate of planes above and below the focal plane is equal to the bleach rate in the focal plane. Therefore, the total amount of photodamage to the specimen due to photobleaching is proportional to the number of planes recorded. This effect becomes particularly important when large three-dimensional data sets are recorded and ultimately limits the amount of information that can be obtained from thick specimens.

In multiphoton microscopy the excitation, and thus the photobleaching, is restricted to the focal volume of the objective. Consequently, the amount of photobleaching of one plane in a three-dimensional stack is independent of the total number of planes in the stack. Instead it depends on the total number of three-dimensional stacks that is recorded.

4. FLUORESCENCE LIFETIME IMAGING

The fluorescence decay time of a molecule can be employed for fluorescence imaging. One of the experimental complications in FLIM is the fact that fluorescence lifetimes are generally on the order of a few nanoseconds. Therefore, fast detection schemes are required for FLIM. At present two methods are in use for fluorescence lifetime imaging. The first, based on phase fluorometry, is a frequency domain method and relies on the measurement of the phase shift between the pulsed or modulated excitation light and the fluorescence. The phase shift between excitation and emission is due to the delay (fluorescence lifetime) of the emission with respect to the excitation light. This method is extensively described in Chapter 16 and will not be further discussed here. The second method relies on straightforward time-gated detection of fluorescence after pulsed excitation of the specimen (Buurman et al., 1992; Ni et al., 1991).

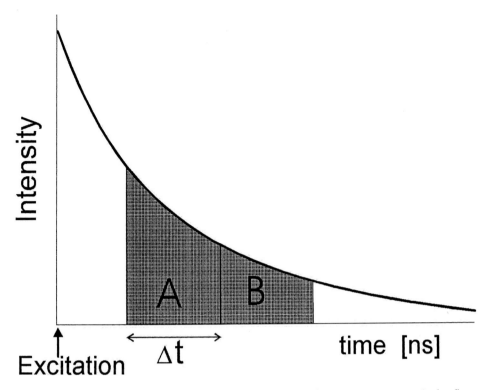

FIGURE 18–1 The principle of time gating. After exciting the specimen at $t = 0$, the fluorescence is collected in two (or more) time windows.

The fluorescence emission is detected sequentially in two or more time gates, each delayed by a different time relative to the excitation pulse (see Fig. 18–1). In the case of a two-gate detection scheme, the ratio of the signal acquired in the two gates, here referred to as the *window ratio*, is a measure of the fluorescence lifetime. For a monoexponential fluorescence intensity decay, the fluorescence lifetime is given by

$$\tau = \Delta t / \ln (I_A / I_B) \qquad (1)$$

where Δt is the time-offset between the two windows and I_A and I_B are the corresponding integrated fluorescence intensities. In deriving Equation 1 it is assumed that the two gates are of the same width. In the case of a multiexponential fluorescence decay the window ratio yields only an effective fluorescence lifetime. The number of time gates can, however, be easily extended to enable the recording of multiexponential decays (Scully et al., 1997). The multiexponential analysis requires a comparatively slow fitting procedure and is, at present, not suitable for visualizing lifetime information during the image acquisition. In contrast, the window ratio can be displayed in real time, even at high frame rates.

We have implemented time-gated detection schemes in both a confocal microscope (Buurman et al., 1992) and in a multiphoton excitation microscope (Sytsma et al., 1998). The implementation of time-gated detection in scanning microscopes is straightforward because only one pixel is acquired at a time. In both setups we incorporated a very efficient detection scheme that relies on the use of time-gated single-photon counting.

Briefly, the pulses that are coming out of the photomultiplier are first amplified and then fed into a discriminator. The output of the discriminator is fed into gated fast counters that are enabled for a specific time after a preset delay with respect to the excitation pulse. A computer sets the delay with respect to the excitation pulses and the open time of each gate. All the gates open sequentially after each and every excitation pulse. Therefore, the window ratio is generally not affected by fading of the fluorescence due to photobleaching.

For the two-gate system, Ballew and Demas (1989) demonstrated that the optimum gate width in a two-gate system with equal gate widths amounts to 2.5 times the fluorescence lifetime. Under these conditions, the two-gate system that we employ collects the fluorescence for a period $5 \cdot \tau$. On assuming that the first gate opens at time zero, this corresponds to a total of 99.3% of the total fluorescence emission. This makes the method very efficient. Moreover, theoretical calculations (Köllner and Wolfrum, 1992), as well as computer simulations (Gerritsen et al., 1996), demonstrated that at the optimum gate width a two-gate system requires only 225 detected photons for an accuracy of 10% in the fluorescence lifetime. Recently, we demonstrated that this accuracy could indeed be realized in practice, thus proving that the accuracy is determined by photon statistics alone (Sytsma et al., 1998).

Time-gated detection is also an effective method for suppressing background signals correlated with the excitation pulse. Scattered light, including direct and multiple scattered excitation light and Raman scattering, reaches the detector at $t \approx 0$ and can be very effectively suppressed by opening the first gate a few hundred picoseconds after $t = 0$. This can improve the signal-to-background ratio in the images without a significant loss of signal. Furthermore, time gating enables the suppression of autofluorescence in biological specimen. In general, autofluorescence has a comparatively short fluorescence lifetime, and a significant improvement in signal-to-background ratio can again be realized by offsetting the first gate with respect to the excitation pulse.

5. Examples

5.1 Quantitative Ion-Concentration Imaging

A common feature of fluorescent ion indicators is that they show changes not only in intensity or spectral properties in the presence of specific ions but also in their fluorescence decay behavior. Therefore, many of the present ion indicators can be used for FLIM of ion concentrations (Lakowicz et al., 1992; Sanders et al., 1994, 1995). Often the ion indicator exists in two forms: one bound to the ion and one free, with each form exhibiting a distinct fluorescence lifetime. A change in the ion concentration shifts the equilibrium between the bound and free probe concentrations. The shift between the two states is reflected in the amplitudes of the fluorescence intensity decay components. This behavior can be used as an alternative method for the quantitative imaging of ion concentrations. This is particularly interesting because the decay behavior is generally independent of variations in the excitation intensity, probe concentration, photobleaching, and absorption effects.

5.2 Calcium Imaging

CalciumGreen-1 (Molecular Probes) is a popular calcium indicator that exhibits a strong increase of its fluorescence intensity upon binding to calcium. This property and the fact that it can be excited with visible light make it attractive for the imaging of free calcium. The use of this and other single wavelength probes is hampered by inhomogeneous probe partitioning within the cells, leakage of the probe, and fading of the fluorescence due to photobleaching. All these factors affect the fluorescence intensity and complicate the quantification of the ion concentrations. Therefore, quantitative calcium imaging is usually carried out using dual wavelength probes like Indo and Fura. These probes, however, require excitation in the ultraviolet, and in addition a cumbersome calibration procedure is needed in order to convert the images into quantitative calcium images.

CalciumGreen-1 exhibits strong differences between the fluorescence lifetimes of the free probe and the probe–calcium complex. This opens up the possibility to use this probe as a fluorescence lifetime probe. Here, we employed CalciumGreen-1 for confocal fluorescence lifetime imaging of calcium concentrations in rat myocytes. This example is described in detail by Sanders et al. (1994). The microscope response was calibrated using probes dissolved in buffers containing known ion concentrations. The calibration curve, the window ratio as a function of the calcium concentration in the calibration buffers, is shown in Figure 18–2. The sigmoidally shaped calibration curve shows a useable Ca^{2+} range from about pCa 6.5 (300 nM) to 8.5 (3 nM). This makes the probe particularly suitable for lifetime imaging at the low end of the physiological free Ca^{2+} concentration range.

The shape of the calibration curve is consistent with the existence of two distinct states of the probe, one bound to Ca^{2+} and one free, each state with its

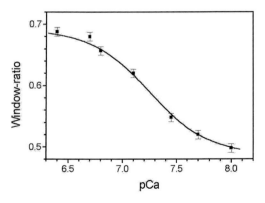

FIGURE 18–2 The window ratio as a function of pCa for CalciumGreen (in buffers).

own fluorescence lifetime. Moreover, the validity of this description has been confirmed by standard time-correlated single-photon counting measurements (Sanders et al., 1994). These indeed yielded monoexponential decays for calcium-free and calcium-saturated buffers with decay times of 0.46 and 3.53 ns, respectively. Intermediate calcium concentrations, on the other hand, resulted in a biexponential decay with components corresponding to the bound and free states, but with amplitudes depending on the calcium concentration.

The fluorescence intensity image of a CalciumGreen-stained cardiac rat myocyte is shown in Figure 18–3A. The fluorescence intensity in the nucleus is clearly much higher than that in the cytoplasm. This suggests a higher Ca^{2+} concentration in the nucleus of the myocyte. The fluorescence lifetime image (window-ratio image) of the same cell (Fig. 18–3B), however, shows a uniform gray value. Here, the gray value is a direct measure of the fluorescence lifetime. Therefore, we may conclude that the calcium concentration is constant. From the calibration curve and the gray values of the image a concentration of 20 nM $\pm$ 8% in the nucleus and 18 nM $\pm$ 18% in the cytoplasm can be found. Ratio imaging experiments using Fura-2 indeed confirm the presence of equal Ca^{2+} concentrations of this order in the nucleus and cytoplasm of this type of myocyte. The difference between the intensity image and the window-ratio image can be explained by the preferential partitioning of CalciumGreen into the nucleus.

It should be noted that the calibration in this experiment is based on probes dissolved in calcium buffers. The calibration may not be transferable to cells because the chemical environment inside the cells is different from that in the buffers. Therefore, the absolute value of the calcium concentrations found here might not be very accurate. The observation that the concentrations are constant is, however, independent of the absolute value of the concentration.

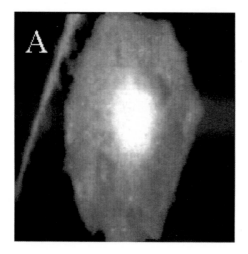 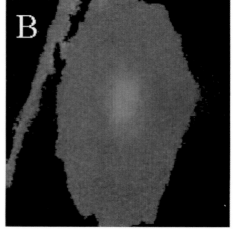

FIGURE 18–3 The confocal fluorescence intensity image (*a*) and lifetime image (*b*) of a CalciumGreen-stained cardiac rat myocyte. In *b*, the gray value is a measure of the Ca^{2+} concentration.

5.3 pH Imaging

Reliably verifying the absolute value of the calcium concentration with other fluorescence imaging methods is not straightforward. Such a comparison can, however, be easily carried out in the case of pH measurements. Therefore, we compared the results of FLIM with those of conventional emission-ratio imaging using the pH probe carboxy-SNAFL-1. A detailed description of this example is given by Sanders et al. (1995). This probe shows a 60 nm shift in the position of the emission band in the pH range 7–9 (Rink et al., 1982), and, in addition, it shows significant changes in the fluorescence decay behavior in this pH range. Moreover, the calibration of the pH response inside cells can be easily accomplished by means of the nigericin high potassium method (Rink et al., 1982).

The calibration curve of the window ratio versus the pH recorded in buffer and in cells overlap almost perfectly (Fig. 18–4a), indicating that the calibration of the window ratio versus pH is hardly sensitive to the presence of cellular components. Moreover, calibration curves recorded on buffers containing additives that mimic the cellular environment showed little or no influence of the additives. In all the experiments, the maximum deviation between the calibration curves amounted to about 0.2 pH units at high pH. The calibration curves of the emission ratio versus pH on the same buffers and cells are shown in Figure 18–4b. Significant deviations are found between the two calibration methods. The emission ratios recorded in cells are systematically lower than those recorded in buffer. A one pH unit lower plateau is observed at the low end of the pH range. These results clearly demonstrate that the fluorescence emission ratio of carboxy-SNAFL-1 is sensitive to the details in the chemical environment and therefore requires a calibration procedure on cells. On the other hand, the probe can be used as a fluorescence lifetime probe with only a simple calibration on buffers.

Herman et al. (1997) carried out experiments showing the effect of similar additives as in the above experiments, but they used the Ca probe Calcium Crimson. These experiments revealed a comparatively small influence of the additives on the fluorescence lifetime behavior of this probe. These findings suggest that fluorescence lifetime-based imaging is less sensitive than conventional methods for the

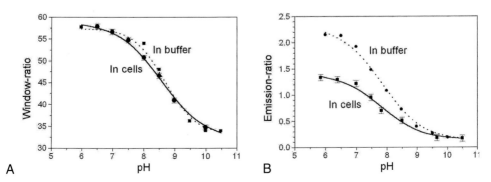

FIGURE 18–4 The C-SNAFL1 window ratio calibration curves (*a*) and the emission ratio calibration curves (*b*) recorded in buffer and in cells.

details in the chemical environment of the probe. This makes FLIM an attractive tool for quantitative ion concentration imaging.

5.4 pH Imaging in Biofilm

Recently, we employed FLIM in combination with multiphoton excitation microscopy for the quantitative imaging of pH in dental biofilm (Vroom et al., 1999). The biofilm consists of different types of bacteria with a typical size of 1 μm and can be up to several hundred micrometers thick. The strong scattering properties of this specimen make a high penetration with confocal microscopy difficult. A high penetration into this sample can, however, be easily achieved by using multiphoton excitation imaging.

Two parameters of great interest in biofilm are the oxygen and pH distributions. The fermentation of sugars by the bacteria leads to a pH drop in the biofilm that can result in very low pH values. The conventional method for the quantification of pH in the acidic range is excitation ratio imaging. The implementation of excitation ratio imaging in multiphoton excitation microscopy is not straightforward and requires the use of two pulsed (femtosecond) lasers. FLIM is an interesting alternative that requires only single wavelength excitation. Here, we employed the pH sensitivity of carboxy-fluorescein for the quantitative imaging of pH inside biofilm. This probe can be employed for (multiphoton) FLIM-based pH imaging in the acidic range from pH 3.5 to 7.

Two-photon excitation FLIM experiments were carried out on biofilm at an excitation wavelength of 800 nm using a time-gated detection system with four time gates. The gates had widths of 1, 1, 3, and 5 ns, respectively, and the opening of the first gate was delayed by 1 ns with respect to the excitation pulse. The latter delay was introduced in order to suppress the autofluorescence from the biofilm. Typically, the decay times of the autofluorescence are less than 1 ns, while the (average) decay of the carboxy-fluorescein is in the 3–4 ns range. In Figure 18–5 a typical pH calibration curve (window ratio 1/4 as a function of pH) is shown. The curve was recorded on "inactive" biofilm immersed in pH buffer without any

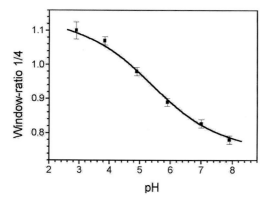

Figure 18–5 The pH calibration curve for carboxy-fluorescein. The window ratios were recorded in biofilm immersed in pH buffer.

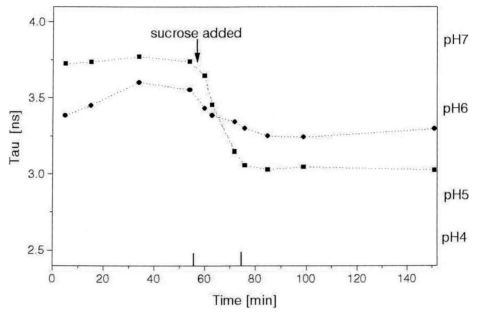

Figure 18–7 The average pH as a function of time in two different regions in the biofilm (the boxed areas in Color Fig. 18–6B).

nutrients. The pH values used in the calibration are the average pH value over one calibration image.

Color Figure 18–6 shows some typical intensity (Color Fig. 18–6A) and lifetime (Color 18–6B, C) images of the biofilm. The $100 \times 100 \ \mu m^2$ images were recorded 5 μm below the biofilm surface. In the intensity image individual bacteria are discernible, indicating that some of the carboxy-fluorescein has accumulated on the surface of the bacteria. However, the lifetime images (Color Fig. 18–6B, C) do not show the individual bacteria any longer. Color Figure 18–6B was recorded before the addition of 14 mM sucrose and shows regions of different pH in the pH range 7–5.8. Color Figure 18–6C shows the same field, but was recorded 20 minutes after the addition of the sucrose. The addition of the sucrose clearly lowered the pH throughout the whole image. Now the pH is in the pH 6–4 range.

In Figure 18–7 the average pH as a function of time in two different regions in the biofilm is shown. The two regions are indicated in Color Figure 18–6B by the black boxes. At a time of 55 min, the sucrose was added to the specimen. This results in a rapid lowering of the pH in both the selected regions; however, the magnitude of the pH drop differs for the two regions. This difference is likely to be due to differences in the types of bacteria present in the two regions.

5.5 Oxygen Imaging

Fluorescent probes can be used for the quantitative sensing and imaging of oxygen (Bambot et al., 1994; Gerritsen et al., 1997). Here, the quenching of the excited state of the probe by collisions with O_2 molecules is utilized. The quenching decreases both the fluorescence lifetime and the fluorescence intensity. Dynamic quenching of fluorescence can be conveniently described by the Stern-Volmer equation:

$$\frac{I_0}{I} = \frac{\tau_0}{\tau} = 1 + K_D[O_2] \qquad (2)$$

where (I_0, τ_0) and (I, τ) are the fluorescence intensities and lifetimes in the absence and presence of oxygen, respectively. $K_D = k_q\tau_0$ is the Stern-Volmer quenching coefficient. Equation 2 shows that the ratio τ_0/τ linearly depends on the oxygen concentration. The value of K_D is determined by the temperature, collision radius, and diffusion coefficients of the probe and O_2 molecules (Lakowicz, 1986). K_D can be determined from a calibration procedure in which the ratio τ_0/τ is measured as a function of dissolved oxygen concentration.

The experiments described here were carried out with the oxygen-sensitive probe TRIS (1, 10-phenanthroline) ruthenium (II) chloride hydrate (TPR, Aldrich Chemicals). This probe has a long fluorescence lifetime of about 1 μs in the absence of oxygen. Therefore, a pulse picker was used to reduce the 82 MHz repetition rate of the laser output to 160 kHz. Two time gates of 1200 ns each were used, of which the first was delayed by 30 ns with respect to the laser pulse to suppress autofluorescence. These settings result in a comparatively low duty cycle of the experiment and pixel dwell times of 8 ms. In Figure 18–8a the calibration curve of TPR is shown. The window ratio shows a nonlinear dependence on the oxygen concentration and varies from 3 at zero oxygen to approximately 11 at 500 μM dissolved oxygen. The window ratios can be easily converted to lifetimes by employing Equation 1. At zero oxygen an unquenched lifetime, τ_0, of 1060 ns is found, and at 500 μM dissolved oxygen it has decreased to 500 ns.

In Figure 18–8b the ratio of the unquenched and the quenched lifetimes is plotted as a function of oxygen concentration. In agreement with Equation 2, a linear relation is found, proving that dynamic quenching occurs. Linear regression of the data yields a value of $K_D = 2.35 \pm 0.05$ (mM)$^{-1}$ for the Stern-Volmer coefficient ($R = 0.997$).

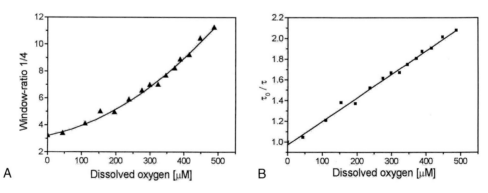

A

B

FIGURE 18–8 The window ratio (a) and the ratio of the unquenched and the quenched lifetimes (b) as a function of the oxygen concentration for a TPR solution.

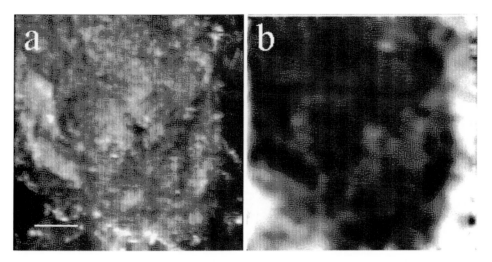

FIGURE 18–9 Intensity (*a*) and window-ratio image (*b*) of TPR-stained biofilm. The gray value in *b* is a measure of the oxygen concentration, and the scale bar denotes 5 μm.

Quantitative oxygen images were recorded in biofilm material that was deposited on a cover glass and submersed in a buffer solution containing 100 μM TPR. In Figure 18–9, 30 $\times$ 30 μm^2 intensity and window-ratio images are shown. In the intensity image (Fig. 18–9a) a nonuniform fluorescence intensity is observed. The outline of a compact microcolony of approximately 25 $\times$ 25 μm^2 of bead-like bacteria is observed, with borders that appear dark. The window-ratio image (Fig. 18–9b) shows a homogeneous gray value over the microcolony. This indicates that the oxygen concentration within the microcolony is more or less constant. The borders appear bright, corresponding with a higher oxygen concentration. When converting the window ratios to dissolved oxygen concentrations, an oxygen concentration of approximately 300 μM was found at the borders. This corresponds with the value expected for a solution that is in contact with air. Therefore, this signal is ascribed to the dye solution in which the biofilm material is submersed. In the microcolony, an average value of 50 ($\pm$20) μM is found.

Dissolved oxygen concentrations in biofilm have also been determined using microelectrodes (Costerton et al., 1994). These measurements showed that, despite the permeable water channels in the biofilm, highly anaerobic conditions exist within the bacterial microcolonies, which corresponds with our observation.

6. CONCLUSION

Fluorescence lifetime imaging is a powerful imaging method that is very well suited for quantitative imaging. It can be used for quantitative imaging of ion concentrations without using UV probes and dual excitation methods. There are clear indications that fluorescence lifetime–based ion concentration imaging is less sensitive to the precise details in the chemical environment of the fluorescent probe. In addition, FLIM allows the quantification of small molecules like oxygen.

In FLIM, background signals arising from autofluorescence that also scattered light can be easily reduced or suppressed. Fluorescence lifetime imaging can be implemented in a straightforward manner in scanning microscopes like multiphoton and confocal microscopes.

REFERENCES

Ballew, R. M., and J. N. Demas. An error analysis of the rapid lifetime determination method for the evaluation of single exponential decays. *Anal. Chem.* 61:30–33, 1989.

Bambot, S. B., R. Holavanahali, J. R. Lakowicz, G. M. Carter, and G. Rao. Phase fluorometric sterilizable optical oxygen sensor. *Biotech. Bioeng.* 43:1139–1145, 1994.

Bastiaens, P. I. H., and A. Squire. Fluorescence lifetime imaging microscopy: Spatial resolution of biochemical processes in the cell. *Trends Cell Biol.* 9:48–52, 1999.

Berlin, J. R., and M. Konishi. Ca^{2+} transients in cardiac myocytes measured with high and low affinity Ca^{2+} indicators. *Biophys. J.* 65:1632–1647, 1993.

Buurman, E. P., R. Sanders, A. Draaijer, H. C. Gerritsen, J. J. F. van Veen, P. M. Houpt, and Y. K. Levine. Fluorescence lifetime imaging using a confocal laser scanning microscope. *Scanning* 14:155, 1992.

Costerton, J. W., Z. Lewandowski, D. DeBeer, D. E. Caldwell, D. R. Korber, and G. James. Biofilms, the customized microniche. *J. Bacteriol.* 176:2137–2142, 1994.

Denk, W., J. H. Strickler, and W. W. Webb. Two-photon laser scanning fluorescence microscopy. *Sci. Rep.* 248:73, 1990.

French, T., P. T. C. So, D. J. Weaver, T. Coehlo-Sampaio, E. Gratton, E. W. Voss, and J. Carrero. Two-photon fluorescence lifetime imaging microscopy of macrophage-mediated antigen processing. *J. Microsc.* 185:339–353, 1997.

Gadella, T. W. J., T. M. Jovin, and R. M. Clegg. Fluorescence lifetime imaging microscopy (FLIM)—Spatial resolutions of microstructures on the nanosecond time scale. *Biophys. Chem.* 48:221–239, 1993.

Gerritsen, H. C., R. Sanders, A. Draaijer, and Y. K. Levine. The photon economy of fluorescence lifetime imaging. *Scanning* 18:55–56, 1996.

Gerritsen, H. C., R. Sanders, A. Draaijer and Y. K. Levine. Fluorescence lifetime imaging of oxygen in living cells. *J. Fluorescence* 7(1):11–16, 1997.

Grynkiewicz, G., M. Poenie, and R. Y. Tsien. A new generation of Ca^{2+} indicators with greatly improved fluorescence properties. *J. Biol. Chem.* 260:3340–3450, 1985.

Herman, B., P. Wodnicki, S. Kwon, A. Periasamy, G. W. Gordon, N. Mahajan, and X. F. Wang. Recent developments in monitoring calcium and protein interactions in cells using fluorescence lifetime microscopy. *J. Fluorescence* 7(1):85–92, 1997.

Köllner, M., and J. Wolfrum. How many photons are necessary for fluorescence-lifetime measurements? *Chem Phys. Lett.* 200(1,2):199–204, 1992.

Lakowicz, J. R. *Principles of Fluorescence Spectroscopy.* New York: Plenum Press, 1986.

Lakowicz, J. R., and K. Berndt. Lifetime-selective fluorescence imaging using an rf phase-sensitive camera. *Rev. Sci. Instrum.* 62:1727–1734, 1991.

Lakowicz, J. R., H. Szmacinski, and K. Nowaczyk. Fluorescence lifetime imaging of calcium using Quin-2. *Cell Calcium* 13:131–147, 1992.

Ni, T., and L. A. Melton. Fluorescence lifetime imaging: An approach for fuel equivalence ratio imaging. *Appl. Spectrosc.* 45(6):938, 1991.

Periasamy, A., P. Wodnicki, X. F. Wang, S. Kwon, G. W. Gordon, and B. Herman. Time-resolved fluorescence lifetime imaging microscopy using a picosecond pulsed tunable dye-laser system. *Rev. Sci. Instrum.* 67:3722–3731, 1996.

Rink, T. J., R. Y. Tsien, and T. Pozzan. Cytoplasmic pH and free Mg^{2+} in lymphocytes. *J. Cell Biol.* 95:189, 1982.

Sanders, R., A. Draaijer, H. C. Gerritsen, P. Houpt, and Y. K. Levine. Quantitative pH imaging in cells using confocal fluorescence lifetime imaging microscopy. *Anal. Biochem.* 227:302–308, 1995.

Sanders, R., H. C. Gerritsen, A. Draaijer, P. Houpt, and Y. K. Levine. Fluorescence lifetime imaging of free calcium in single cells. *Bioimaging* 2(3):131–138, 1994.

Scully, A. D., R. B. Ostler, D. Phillips, P. O. O'Neill, K. M. S. Townsend, A. W. Parker, and A. J. MacRobert. Application of fluorescence lifetime imaging microscopy to the investigation of intracellular PDT mechanisms. *Bioimaging* 5(1):9–18, 1997.

Sytsma, J., J. M. Vroom, C. J. de Grauw, and H. C. Gerritsen. Time gated fluorescence lifetime imaging and micro-volume spectroscopy using two-photon excitation. *J. Microsc.* 191(1):39–51, 1998.

Tsien, R. Y., and M. Poenie. Fluorescence ratio imaging: A new window into intracellular ionic signaling. *Trends Biochem. Sci.* 11:450–455, 1986.

Vroom, J. M., K. J. de Grauw, H. C. Gerritsen, D. J. Bradshaw, P. D. Marsh, G. K. Watson, J. J. Birmingham, and C. Allison. Depth penetration and detection of pH gradients in biofilms by two-photon excitation microscopy. *Appl. Environ. Microbiol.* 65(8):3502–3511, 1999.

Wang, X. F., T. Uchida, D. M. Coleman, and S. Minami. A two-dimensional fluorescence lifetime imaging system using a gated image intensifier. *Appl. Spectrosc.* 45(3):360, 1991.

19

Biological Applications of Time-Resolved, Pump-Probe Fluorescence Microscopy and Spectroscopy in the Frequency Domain

CHEN-YUAN DONG, CHRISTOF BUEHLER, PETER T. C. SO, TODD FRENCH, AND ENRICO GRATTON

1. INTRODUCTION

In biological applications of optical microscopy, technical developments often lead to novel imaging modalities with significant applications. For example, the development of confocal microscopy led to microscopic imaging with enhanced contrast (see Chapter 5), and the more recent development of two-photon fluorescence microscopy revolutionized fluorescence microscopy by providing an imaging modality capable of high image contrast, reduced photodamage, and exciting possibilities in controlling localized photochemical reactions in three dimensions (see Chapters 9–13).

In fluorescence microscopy, an exciting development has led to a new imaging methodology based on the frequency-domain, pump-probe technique. By focusing two laser sources at different wavelengths and modulation frequencies to a diffraction-limited spot, we have demonstrated the unique capabilities of this technology in biological applications. Localization of the pump-probe volume results in time-resolved images at high spatial contrast. Furthermore, unlike other intensity and time-resolved techniques, the pump-probe technology can provide time-resolved, high-frequency information without a fast photodetector. In this chapter, we address both the technical background and the biological applications of this technique. We discuss the theoretical background and experimental details and show examples of biological imaging and spectroscopic applications of this novel modality in microscopy.

2. Principles of the Pump-Probe Technique

2.1 One-Photon Absorption, Spontaneous Emission, and Stimulated Emission

The interaction of radiation with matter has long been a subject of intense study. It is well known that individual photons of the right energy can shift molecules from the ground state to the excited state. In contrast, a molecule in the excited state can return to the ground state by emitting a photon spontaneously, in a process called *spontaneous emission*. Both one-photon absorption and spontaneous emission transitions involve the interaction of one photon with one molecule. The transition probabilities between ground state $|0\rangle$ and excited state $|n\rangle$ can be represented mathematically as

$$\Gamma^{abs} \sim |\langle n|\vec{R} \cdot \vec{\gamma}|0\rangle|^2 \quad \text{and} \quad \Gamma^{em} \sim |\langle 0|\vec{R} \cdot \vec{\gamma}|n\rangle|^2 \tag{1}$$

where Γ^{abs} and Γ^{em} represent the absorption and spontaneous emission transition rate, respectively. The term $\vec{R} \cdot \vec{\gamma}$ represents the interaction between the electric field represented by the polarization vector $\vec{\gamma}$ and the position operator $\vec{R}$. In addition to spontaneous fluorescence emission, excited state molecules can also be induced to emit photons and in the process, return to the ground state. The photons in this process of stimulated emission have some unique properties. In particular, the stimulated emission photons have the same optical properties as the photons inducing such transitions. Mathematically, the stimulated emission transition rate has the same form as given for the spontaneous emission process $\Gamma' \sim |\langle 0|\vec{R} \cdot \vec{\gamma}|n\rangle|^2$ (Baym, 1973).

The principle of stimulated emission has been used to generate coherent radiation and is the key working concept behind laser actions (Siegman, 1986). As we will see later, the combination of fluorescence excitation and stimulated emission can be applied in frequency-domain, pump-probe techniques for interesting applications in fluorescence microscopy and spectroscopy.

2.2 Application of Time-Domain, Pump-Probe Technique in Ultrafast Spectroscopy

Spectroscopic techniques based on the pump-probe methodology have been useful in understanding many fundamental processes. The inherent ultrafast feature of the technique makes it the ideal tool for studying picosecond or subpicosecond phenomena in biology, chemistry, and condensed matter. In a common implementation, a pulsed laser is split by a beam and recombined at the sample. By introducing an optical path length difference between the split pulses, a temporal delay between the pump and probe beams can be derived to probe ultrafast molecular dynamics. The strong pump beam excites the ground state molecules, and the weaker probe source monitors the relaxation of excited state molecules. To obtain a signal of high signal-to-background ratio, the pump beam must be sufficiently strong to deplete a substantial portion of the ground state molecules. As a result, photodamage to the specimen can be significant. Different implementations of the pump-probe methodology exist, but the key concept of using an optical delay line for probing ultrafast molecular dynamics remains invariant (Evans, 1989; Fleming, 1986; Lytle et al., 1985). Ultrafast phenomena in biological systems, such as rhodopsin, heme proteins, and photosynthetic reaction centers, have been studied with pump-probe techniques (Hochstrasser and Johnson, 1988).

Ultimately, the temporal resolution of time-domain pump-probe spectroscopy is determined by the pulse width of the laser source. The narrowest pulse width achievable is less than 10 fs (Zhou et al., 1994). Recently, there have been reports of the generation of attosecond laser pulses, opening more exciting possibilities in probing ultrafast phenomena (Papadogiannis et al., 1999).

2.3 Pump-Probe Technique in the Frequency Domain

In addition to time-domain implementation, pump-probe techniques can also be implemented in the frequency domain. The Lytle group first proposed the frequency-domain technique in the asynchronous sampling approach (Elzinga et al., 1987a,b). In this approach, a pulsed pump laser operated at a high repetition rate is focused for molecular excitation. The probe laser is overlapped at the sample to monitor ground state molecules or to induce stimulated emission from excited state molecules. The key difference in the frequency-domain approach is that, instead of mechanically varying the optical path length difference, temporal evolution of molecular dynamics is studied by offsetting the pulse repetition frequencies of the two lasers. The difference in pulse repetition frequencies is equivalent to introducing a variable delay between the pump and probe sources. As a result, molecular excited states are repeatedly sampled after excitation by the pump beam.

Effectively, the modulation frequencies of the pump and probe lasers are mixed at the sample, generating cross-correlation signals at both the sum and difference frequencies. In the case of pulsed laser systems, cross-correlations between higher frequency pump and probe harmonics also exist. These cross-correlation signals can be analyzed for time-resolved information in the system of interest. Naturally, it is the low-frequency, cross-correlation signals that are analyzed for time-resolved information (Dong et al., 1995; Buehler et al., 1999, 2000).

A variation of pump-probe spectroscopy is the stimulated emission approach. In this case, the wavelength of the probe laser is chosen to induce stimulated emission from the excited state molecules. Instead of monitoring the probe beam, this approach obtains time-resolved information by measuring the fluorescence directly. Because the ground state population need not be saturated, as is done in the transient absorption approach, photobleaching in the stimulated emission approach is reduced. In spectroscopic studies, the stimulated emission approach has been successfully demonstrated (Kusba et al., 1994; Lakowicz et al., 1994). In microscopy, the concept of using simulated emission to result in superior resolution for microscopic imaging has been proposed and realized (Hell and

Wichmann, 1994; Klar and Hell, 1999). In the pump-probe case, both time-resolved microscopic imaging and spectroscopic applications have been demonstrated in a fluorescence microscope (Dong et al., 1995; Buehler et al., 1999, 2000).

2.4 Generation of the Frequency-Domain, Pump-Probe, Fluorescence Signal

In conventional frequency-domain spectroscopy, the fluorescence decay $F(t) = F_0 e^{-t/\tau}$ characterized by a lifetime of τ and amptitude F_0 is studied by measuring the frequency response of the fluorescent sample to intensity-modulated excitation light sources. Due to the sample's lifetime, the fluorescence responds with a phase-shifted and demodulated signal relative to that of the excitation source. Time-resolved information can be obtained from the fluorescent sample by measuring the phase shift and demodulation factor. In the case of a single-exponential species, the phase shift ϕ and modulation m responding to the modulation frequency of ω, are given by

$$\tan(\phi) = \omega\tau_\phi \qquad (2a)$$

$$m = \frac{1}{\sqrt{1 + \omega^2\tau_m^2}} \qquad (2b)$$

where τ_ϕ and τ_m represent the respective lifetimes determined from ϕ and m (Lakowicz, 1999). Time-resolved polarization information of a rotational species can also be obtained by measuring the phase shift $\Delta\phi = \phi_\cdot - \phi_\parallel$ and modulation ratio Y between the parallel and perpendicular components of the fluorescence decay (Gratton and Limkeman, 1983; Gratton et al., 1984; Alcala and Gratton, 1985; Lakowicz, 1999).

In the case of pump-probe methodology, the equation of motion governing the excited state molecular population $N(\vec{r},t)$ is given by

$$\frac{dN(\vec{r},t)}{dt} = -\frac{1}{\tau}N(\vec{r},t) + \sigma I(\vec{r},t)[c - N(\vec{r},t)] - \sigma'I'(\vec{r},t)N(\vec{r},t) \qquad (3)$$

where σ and σ' are the absorption and stimulated emission cross sections, respectively; $I(\vec{r},t)$ and $I'(\vec{r},t)$ are the respective pump and probe fluxes, and c is the total molecular concentration. In Equation 3, the excited state population is determined by three processes. First, there is the natural decay of the excited state population represented by $(-1/\tau)N(\vec{r},t)$. There is also the excitation process given by $\sigma I(\vec{r},t)[c - N(\vec{r},t)]$. Finally, the probe beams de-excite the excited state population in a term given by $-\sigma'I'(\vec{r},t)N(\vec{r},t)$ (Buehler et al., 2000).

In the case of sinusoidal pump and probe profiles (including laser pulse train), an approximate solution for Equation 3 can be obtained. In this case, the pump and probe fluxes can be represented as

$$I(\vec{r},t) = I(\vec{r}) \sum_{n=0}^{\infty} L(n\omega)\cos(n\omega t) \quad \text{and}$$

$$I'(\vec{r},t) = I'(\vec{r}) \sum_{n=0}^{\infty} L'(n\omega')\cos(n\omega' t) \qquad (4)$$

where $L'(n\omega')$ and $L(n\omega)$ are the line shape functions of the pump and probe sources at angular modulation frequencies ω and ω', respectively. Because fluorescence is related to the decay of the excited state population and the fluorescence quantum yield (q) via the relation $F(\vec{r},t) = -q\, dN(\vec{r},t)/dt$, differential equation theory can be used along with Equations 3 and 4 to obtain a solution for fluorescence density $F(\vec{r},t)$ (Boyce and DiPrima, 1977; Buehler et al., 2000). After simplification, the first-order solution for spatially integrated fluorescence is given by

$$F(t) = \frac{qc\tau\sigma\sigma'}{2} \sum_{n=0}^{\infty} \frac{J(n\omega,n\omega')L(n\omega)L'(n\omega')}{\sqrt{1 + (n\omega\tau)^2}} \cos\left[n(\omega' - \omega)t - \phi_n\right]\int I(\vec{r})I'(\vec{r})d^3\vec{r} \quad (5)$$

Note that in Equation 5 an arbitrary frequency-dependent factor, $J(n\omega,n\omega')$, is introduced. This J factor is intended to account for the additional decay in the frequency spectrum due to relative jitter between the pump and probe laser sources (Buehler et al., 2000).

3. ADVANTAGES OF FREQUENCY-DOMAIN METHODOLOGY IN PUMP-PROBE FLUORESCENCE MICROSCOPY AND SPECTROSCOPY

There are two key advantages in applying frequency-domain, pump-probe methodology to fluorescence microscopy and spectroscopy. First, because both the pump and probe lasers are needed in generating the cross-correlation signal, the signal is generated most strongly at the focal volume where both laser sources are high in intensity. The result is the localization of observation volume, thus resulting in confocal-like imaging quality for microscopy. The second significant consequence of the frequency-domain technique is the availability of ultrafast temporal information from the low-frequency, cross-correlation signal without using high-speed photodetectors. Both of these advantages are inherent in Equation 5 and are discussed later (Dong et al., 1995, 1997; Buehler et al., 1999, 2000).

3.1 Confocal-Like Imaging Quality

The confocal-like feature of pump-probe microscope is illustrated in the spatial dependence of the pump-probe signal. As Equation 5 shows, the magnitude of the pump-probe signal depends on the spatial integral of the product of pump and probe fluxes $\int I(\vec{r})I'(\vec{r})d^3\vec{r}$. The integrand represents the point spread function (PSF) of a pump-probe imaging system based on a one-photon excitation and a one-photon stimulated emission process:

$$PSF_{pp} = I(\vec{r})I'(\vec{r}) \quad (6)$$

For focusing light with a wavelength of λ using a circular objective with a numerical aperture (NA) of $\sin(\alpha)$, the intensity distribution is well characterized from diffraction theory and is given by

$$I(u,v) = \left| 2 \int_0^1 J_0(v\rho)e^{-\frac{1}{2}iu\rho^2}\rho d\rho \right|^2 \tag{7}$$

where $u = 4k(\sin[\alpha/2])^2 z$ and $v = k(\sin[\alpha])r$ are the dimensionless axial and radial coordinates, respectively, and $k = 2\pi/\lambda$ is the wave number (Born and Wolf, 1985; Sheppard and Gu, 1990; Dong et al., 1997).

In comparison, the pump-probe PSF is remarkably similar to those of other depth discrimination microscopic techniques, such as confocal and two-photon microscopy. The confocal and two-photon PSFs are given, respectively, as

$$PSF_c = I(u,v)I(u'',v'') \tag{8}$$

$$PSF_{2P} = I^2(u/2,v/2) \tag{9}$$

where the double prime in the confocal case represents the detection wavelength. For the two-photon case, the excitation wavelength is assumed to be double that for one-photon excitation (Sheppard and Gu, 1990; Dong et al., 1997).

As is well established, the physical origins in image contrast enhancement are very different for confocal and two-photon microscopy. For confocal microscopy, a detection pinhole physically blocks out-of-focus photons from reaching the optical detector to result in point-like detection. In two-photon microscopy, requirements of high excitation intensity in nonlinear excitation limits molecular absorption to the focal region. The point-like excitation is the origin of depth discrimination for two-photon microscopic imaging. The pump-probe method offers the same type of depth discrimination. The physical origin leading to image contrast enhancement in the pump-probe technique is, however, very different from its confocal and two-photon cousins. The generation of the pump-probe signal depends on the excitation and stimulated emission processes acting on a common set of fluorescent molecules. This interaction dictates that the cross-correlation signal comes mainly from the focal volume where both laser sources have high intensities. In fact, examination of Equations 6 and 8 shows that the pump-probe and confocal PSFs are the same if probe and confocal detection wavelengths are identical (Dong et al., 1997).

Although the shape of the PSF gives an intuitive characterization of the imaging quality, an alternative formulation is to examine the Fourier transforms of the PSFs. The spatial frequency content derived from the resulting optical transfer function (OTF) provides further quantitative information on a technique's image resolution. In general, the higher the frequency content of a technique, the higher resolution it possesses (Dong et al., 1997).

In addition to PSFs and OTFs, an important parameter to quantify is the amount of fluorescence a given technique observes away from the focal regions. A suitable parameter for that purpose is the normalized, integrated fluorescence intensity at different axial planes, $F_{int}(u)$, defined as (Sheppard and Gu, 1990; Dong et al., 1997)

$$F_{int}(u) = \frac{\int_0^\infty I(u,v)v\,dv}{\int_0^\infty I(0,v)v\,dv} \qquad (10)$$

To summarize the similarities and differences between pump-probe and other techniques, the optical parameters discussed are listed in Table 19–1.

3.2 Ultrafast Spectroscopy from the Low-Frequency, Cross-Correlation

The second important feature of the pump-probe technique is its inherent capability for ultrafast spectroscopic studies without using high speed optical detectors. Recall from Equation 5 that, under generalized periodic excitation, the temporal dependence of the low-frequency, cross-correlation signal is of the form

$$F(t) \sim \sum_{n=0}^\infty \frac{J(n\omega,n\omega')L(n\omega)L'(n\omega')}{\sqrt{1+(n\omega\tau)^2}} \cos[n(\omega'-\omega)t - \phi_n] \qquad (11)$$

Because the frequency difference $\omega' - \omega$ can be chosen to be small, high-frequency temporal information given by $1/\sqrt{1+(n\omega\tau)^2}$ and ϕ_n may be obtained from a much lower frequency, cross-correlation signal at frequency $n(\omega' - \omega)$ (assuming $\omega' > \omega$). Therefore, high-speed detectors are not necessary to extract high-frequency information (Buehler et al., 2000).

There are three factors that affect the temporal resolution of pump-probe techniques in the frequency domain. First, the temporal resolution can be adversely affected by the line shape function of the light sources given by $L(n\omega)$ and $L'(n\omega')$. For commonly used pulsed systems, the pulse train can be modeled as Lorentzian in shape with FWHM $\Delta t = 2b$ and period T (Alcala and Gratton, 1985):

$$L(t) = \sum_{n=-\infty}^\infty 1/[b^2 + (t - nT)^2] \qquad (12)$$

TABLE 19–1 Comparison of optical parameters between one-photon, two-photon, one-photon pump-probe, and one-photon confocal microscopy

	Radial FWHM at $u = 0$ (v)	Axial FWHM at $v = 0$ (u)	Radial Cut-off Frequency (Normalized)	Axial Cut-off Frequency (Normalized)	Axial Depth Discrimination (FWHM, u)
One-photon	1.617	11.140	2	0	None
Two-photon	2.320	16.040	2	0.5	17.020
One-photon pump/probe	1.160	8.020	4	1	8.510
One-photon confocal	1.160	8.020	4	1	8.510

Data are from Dong et al. (1997). FWHM, full width half maximum.

and the corresponding frequency-domain representation is given by

$$L(\omega) = \frac{\pi e^{-b\omega}}{b} \sum_{n=0}^{\infty} \delta(\omega - n\omega_0) \tag{13}$$

where $\omega_0 = 2\pi/T$. Therefore, the nth frequency component of the frequency spectrum has amplitude $\pi e^{-b\omega}/b$ and the -3 dB bandwidth is given by $f_{3dB} = \ln2/(\pi\Delta t)$ (Alcala and Gratton, 1985). For a pulsed laser system with a 100 ps pulse width, the -3 dB point is about 2.2 GHz, which can distort the pump-probe fluorescence spectrum. The second demodulating factor affecting the system frequency response is the natural decay term $1/\sqrt{1 + (n\omega\tau)^2}$. The -3 dB point of the decay is $\sqrt{3}/\tau$. Assuming a typical fluorescence lifetime in the range of 1–10 ns, the -3 dB point of this process can vary from 1.7 to about 0.17 GHz. For a shorter lifetime down to 100 ps, the -3 dB bandwidth is at 17 GHz. Therefore, a substantial demodulation in fluorescence can be due to the fluorophores' natural lifetime.

Finally, there can be demodulation due to relative jitter between the two lasers. The factor $J(n\omega, n\omega')$ can depend on the type of laser systems and may not be so straightforward to quantify. Nonetheless, it is important to keep in mind that any relative drift in the timing of laser pulses between the pump and probe sources may be regarded as additional broadening of the laser pulse width and can result in further demodulation of the system frequency response. Therefore, to optimize the performance of a pump and probe system, lasers with narrow pulses and a mutually stable phase relationship should be used.

4. Methods of Implementation

Different implementations of pump-probe techniques in the frequency-domain are possible. A generalized instrument diagram is shown in Figure 19–1. There are three key parts in a pump-probe instrument: the frequency-modulated light sources, microscope optics setup, and synchronization of laser sources with detection electronics and signal processing apparatus.

4.1 Frequency-Modulated Light Sources

4.1.1 Pulsed laser systems
Because the effectiveness of pump-probe techniques depends critically on how well the two light sources are overlapped, well-collimated light sources are the natural choice for these studies. Therefore, pulsed laser systems represent a natural light source for frequency-domain, pump-probe applications. Furthermore, pulsed laser systems have inherent frequency harmonics ideal for pump-probe applications. Furthermore, many commercial mode-locked laser systems accept external trigger signals, and two such systems can be easily phase locked for frequency-domain experiments.

In the work by the Gratton group, a 10 MHz clock is used to phase lock two synthesizers (PTS500, Programmed Test Sources, Inc., Littleton, MA). Each of these

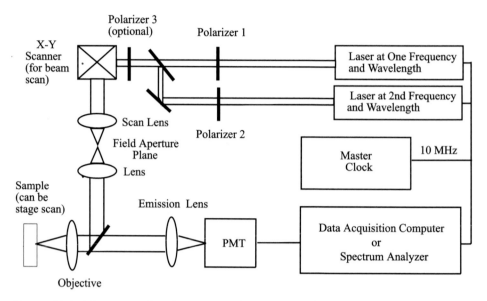

FIGURE 19–1 A frequency-domain, pump-probe instrument for microscopy and spectroscopy. PMT, photomultiplier tube.

synthesizers controls the base repetition frequency of a mode-locked, neodymium (Nd) laser system (Nd-YAG and Nd-YLF, Antares, Coherent Inc., Santa Clara, CA). The frequency-doubled 532 nm output of one of the Nd lasers can be used to excite the fluorescent molecules (such as rhodamine B). The frequency-doubled output of the other Nd laser can be used to pump a dye laser (such as DCM dye laser, Model 700, Coherent Inc.). The longer lasing wavelength of the dye laser can be used to induce stimulated emission from the rhodamine B molecules already in the excited state.

The typical laser pulse width of the Nd system is about 100 ps and that of the dye laser is around 10 ps. Power and polarization of the laser sources can be controlled by polarizers. Typical laser powers at the sample are in the 100 μW range for the pump source and in a few mW range for the probe laser. For fast data acquisition in imaging applications, the two pulsed laser systems are offset in frequency by 5 KHz. The pump laser is at 76.2 MHz–2.5 KHz, and the probe source has a base repetition frequency of 76.2 MHz + 2.5 KHz. For high-frequency spectroscopy where a smaller frequency difference is desired, the difference in base frequencies can be reduced to 210 Hz (pump at 76.2 MHz and the probe at 76.2 MHz + 210 Hz). The choice of 210 Hz moves the cross-correlation frequency away from both the ubiquitous 60 Hz harmonic and $1/f$ noises. This arrangement of laser source choice has been used successfully for both microscopic imaging and spectroscopic studies (Dong et al., 1995; Buehler et al., 2000).

4.1.2 Sinusoidally modulated laser diodes
In addition to pulsed laser systems, sinusoidally modulated lasers can also be used for frequency-domain, pump-probe applications. Compared with other laser

sources, such as the Nd-based systems previously described, laser diodes are advantagous in that they are economical, compact, and stable. The drawback is that laser diodes do not have much flexibility in tunability.

In an implementation previously described (Buehler et al., 1999; Dong et al., 2001), a 10 MHz line is used to synchronize two synthesizers independently modulated to drive two laser diodes. A laser diode operating at 635 nm (APM08 [635-05], Power Technology Inc., Mabelvale, AR) is used for excitation, and another laser diode emitting at 680 nm (APM08 [690-40], Power Technology Inc.) is used for stimulating emission. Two home-built current sources provide the DC biasing of the laser diodes. The modulation signal from the synthesizer is combined with the DC bias using a bias-T (PBTC-1G, Mini-Circuits, Brooklyn, NY). Typical modulation frequencies of the pump and probe sources are 80 MHz and 80 MHz + 5 KHz, respectively.

4.2 Microscope Optics Setup

One key factor for successful pump-probe experiments is to deliver and overlap frequency-modulated laser sources; the two laser sources need to be recombined and focused at the sample. A commercial research microscope or a home-built apparatus can be used. For microscopic imaging, the key is to use a high-quality, high NA objective (such as a Zeiss Plan-Neofluar 63×, NA 1.25, Thornwood, NY). As Color Figure 19–2 shows, the two lasers are typically combined outside of the microscope system. If needed, the power and polarization of each laser source can be independently controlled with polarizers. The combined lasers are then guided into the microscope. A beam expander combination ensures filling of the objective's back aperture for tight focusing. In a beam-scanning arrangement, the scan lens used ensures proper translation of angular deviation from the x–y scanner (Model 6350, Cambridge Technology, Watertown, MA) into linear translation at the sample. If the stage-scanning is utilized, piezoelectric transducers are then used for physical movement of the sample stage.

4.3 Synchronization of Laser Sources with Detection Electronics

Because the pump-probe methodology is essentially a lock-in technique, sample scanning and A/D signal processing need to be synchronized to the laser's cross-correlation frequency. A signal clock from the same 10 MHz synchronization clock for the lasers is used to synchronize both the scanning and digitization cycle. Pixel residence time is determined by the cross-correlation frequency. At the cross-correlation frequency of 5 KHz, the shortest pixel residence time (corresponding to one A/D waveform per pixel) is 0.2 ms. The fluorescence generated at the sample is collected in an epi-illuminated fashion and optically filtered before reaching the photomultiplier tubes (PMTs).

Both the DC and cross-correlation signals are present at the PMT, and additional electronic filtering is needed to isolate the cross-correlation signal. This can be accomplished by a series of current amplifier (SR570, Stanford Research,

Sunnyvale, CA), bandpass filters at 5 KHz and another stage of amplification (SR560, Stanford Research). After these signal processing steps, the signal is then fed into a 12 bit digitizer (A2D-160, DRA Laboratories, Sterling, VA).

To reduce harmonic noise, a four-point-per-waveform digitization scheme is used. The scanner mode typically used raster scans in a 256 $\times$ 256 pixel area for imaging, and the digitized images are recorded and stored by custom software. For spectroscopic studies, a spectrum analyzer (Hewlett Packard 35665A, Rolling Meadows, IL) is used instead to display, acquire, and store the frequency spectrum for analysis (Dong et al., 1995; Buehler et al., 1999, 2000).

5. Applications for Microscopic Imaging and Spectroscopic Measurements

5.1 High-Resolution Microscopic Imaging of Biological Specimens

5.1.1 Lifetime-resolved imaging

In this section, we discuss examples of pump-probe microscopic imaging along with its unique features. First, enhanced image contrast resulting from superior axial depth discrimination is demonstrated, and a comparison between pump-probe and one-photon fluorescence imaging is made. Furthermore, because pump-probe microscopy is inherently a time-resolved technique, we show that different fluorescent species with similar emission spectra but different lifetimes can be easily identified.

To show the superior imaging characteristics of pump-probe microscopy, two systems were labeled and imaged under pump-probe and conventional one-photon microscopy: erythrocytes and mouse fibroblasts. In both samples, a membrane probe rhodamine-DHPE (Molecular Probes, Eugene, OR) was used for labeling. In the case of the erythrocytes, the dye was directly applied to the sample for about 30 min. The cells were spun down and washed before imaging procedures commenced. In the case of mouse fibroblasts, the cells, grown on coverslips, were first fixed in acetone. Again, the fixed cells were labeled with rhodamine-DHPE for about 30 min before viewing. The laser sources used were the pulsed Nd dye laser systems described in Section 4.1.1.

Both the pump-probe and one-photon images of erythrocytes and mouse fibroblasts are shown in Color Figure 19–2. One-photon figures represent images obtained by scanning only the pump beam and recording of the intensity. A comparison between the pump-probe and one-photon images clearly shows the superior image quality of pump-probe microscopy. In the case of erythrocytes, the pump-probe image shows the ring structure in the periphery of the cells much better than that from one-photon imaging. One-photon imaging clearly registers more of the fluorescence from off-focal regions. A similar observation is made for the mouse fibroblast sample. In this example, the enhanced depth discrimination shows superior rejection of off-focal fluorescence and results in a sharper image quality of the cellular structures.

Another unique feature of the pump-probe technique stems from its time-resolved nature. It is a method capable of offering lifetime imaging as additional contrast in distinguishing fluorescent species. Fluorescent molecules similar in emission properties, but with different lifetimes, may be resolved. To demonstrate this characteristic, mouse fibroblasts were fixed and doubly labeled with the nucleic acid stain ethidium bromide and the membrane probe rhodamine-DHPE.

The first harmonic amplitude and phase images are shown in Color Figure 19–3. While the harmonic amplitude at one excitation frequency (76.2 MHz) offers no time-resolved information, the phase image clearly shows that the nuclear and cytoplasmic regions have very different lifetimes. To measure the actual lifetime, the phase data needed to be determined relative to a reference compound, aqueous rhodamine B. The average lifetime obtained from the lifetime histogram reveals that the membrane and cytoplasmic region has an average lifetime of 2.0 ± 0.5 ns, and the nuclear region has a characteristic lifetime of 6.6 ± 4.8 ns. In comparison, aqueous rhodamine B has a lifetime of 1.5 ns, as determined from frequency-domain phase fluorometry. Also, the lifetimes of free ethidium bromide and that bound to nucleic acids are 1.7 and 24 ns, respectively. The phase image shows that the cytoplasmic region is substantially labeled by rhodamine-DHPE. Furthermore, the fact that the average lifetime of the nuclear region lies in between those of bound and unbound nucleic acid shows that, most likely, both fluorescent species are present in the nuclear region (Dong et al., 1995).

In addition to pulsed laser systems, frequency-modulated laser diodes can also be used as laser sources for pump-probe imaging. Color Figure 19–4 shows time-resolved, pump-probe images of nuclear labeled mouse STO cells. Due to wavelengths of the laser diodes chosen (in the 600–700 nm range), the fluorescent dyes used had to be excitable and be able to emit in the red spectral range. TOTO-3 (Molecular Probes), a nucleic acid stain, satisfies the wavelength requirement. Nile Blue was used as a lifetime reference. As a cellular example, laser diode systems were applied to TOTO-3–labeled mouse STO cells. The phase image was analyzed, and a corresponding lifetime of 2.82 ns was obtained.

5.1.2 Time-resolved polarization imaging

While the magic angle condition for lifetime imaging can be satisfied by setting the pump and probe beam polarization 54.7° apart, time-resolved polarization can also be satisfied by setting the probe polarization parallel and perpendicular to that of the pump beam. In this fashion, the parallel and perpendicular populations of the excited states can be sampled. Two examples of time-resolved imaging are shown in Color Figures 19–5 and 19–6. Color Figure 19–5 shows an example of depolarization inside a fluorescent latex sphere, while Color Figure 19–6 shows depolarization inside a mouse fibroblast labeled with CellTracker™ CMTMR (Molecular Probes). To demonstrate depolarization effects of the dye inside the fluorescent latex sphere, both the parallel (pump∥probe) and the perpendicular (pump ⊥ probe) orientation images were measured. For the parallel orientation, the average phase inside a square region was determined to be 68.3°. In the perpendicular orientation, a larger phase of 74.1° was measured. The phase difference of 5.8° between the two orientations illustrates the depolarization effect of the dye inside the sphere. The depolarization phenomenon observed can be due to either molecular rotation or energy transfer. To distinguish the two effects, polarization measurements at −10°C were conducted using a phase fluorometer (K2, ISS, Champaign, IL). Here, the measured polarization value of 0.025 shows that depolarization may be due predominantly to energy transfer. In fact, it has been shown that fluorescent spheres from the same manufacturer are 95% efficient in excitation energy transfer (Roberts et al., 1998).

Time-resolved polarization studies can also be conducted in cellular systems. Mouse fibroblasts labeled with a rhodamine-based stain, CellTracker™ CMTMR (Molecular Probes), were imaged in the fashion described above. The chloromethyl group from the dye reacts with thiols inside cells. The dye binds to glutathione and results in a complex that is cell impermeant and concentrated in the cytoplasm (Molecular Probes product information sheet). Time-resolved polarization images of the CellTracker™ CMTMR–glutathione complex are shown in Color Figure 19–6. Lifetime images (at 76.2 MHz) are shown in Color Figure 19–6A, and polarization images at the same frequency are shown in Color Figure 19–6B. Aqueous rhodamine B was used as the phase reference. The lifetime in most of the cell is uniform. For regions I and III, however, the respective lifetimes of 1.93 and 1.75 ns are quite distinct from the values in region II (2.75 ns) and region IV (2.85 ns).

Rotation correlation times can also be determined. Assuming initial anisotropy of 0.39, 27 ps (I), 104 ps (II), 107 ps (III), and 151 ps (IV) are the determined rotation anisotropies from the phase data. These fast rotation correlation times reveal interesting properties of the cytoplasm. Because a CMTMR molecule is about twice the size of a glutathione molecule, the rotational effect is primarily determined by the CMTMR molecules. The fast rotational correlation times indicate that the molecules see an aqueous environment inside the cytoplasm (Buehler et al., 2000). This result is consistent with previous studies that examined cytoplasmic microviscosity using small probes (Periasamy et al., 1991; Swaminathan et al., 1997).

The significance of these images is that time-resolved polarization information can be resolved across the entire cell. Unlike single-point measurements, local variations can be revealed and time-resolved global information can be obtained in cellular systems.

5.2 High-Frequency Fluorescence Spectroscopy: Lifetime and Rotation

In addition to time-resolved imaging, frequency-domain, pump-probe techniques can also be used for high-frequency spectroscopic measurements. Both ultrafast lifetime and rotation correlation time measurements can be performed.

Figure 19–7 is a plot of the power spectrum of 280 μM rhodamine B in TRIS showing that the frequency response of fluorescence decay can be easily expanded into the GHz range. Harmonics can be measured up to 4.572 GHz. The frequency harmonics (peaks) can be fitted to obtain the lifetime of the rhodamine sample. To properly fit for the lifetime, additional decay due to the lasers' frequency responses and jitters must be taken into account. A model presented in Equations 5, 11, and 13 was used for fitting, and the resulting single exponential fit yields

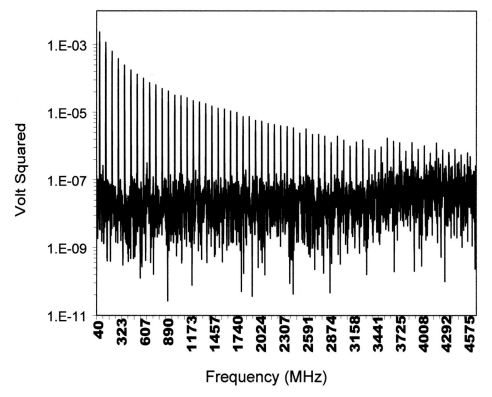

FIGURE 19–7 Pump-probe frequency response of 280 μM rhodamine B in TRIS. A single-exponential fit yields a lifetime of 1.54 ns for the sample. [Adapted from *Biophys. J.* 79: 536–549, 2000.]

a lifetime of 1.54 ns for the rhodamine B sample. The fit also yields a *b* factor of 61.8 ps, which corresponds to a FWHM of the laser pulse at 123.6 ps. Note that this is a simplified fit because the effects of the relative jitter between the two lasers were not considered. Nonetheless, the fitted pulse width of the laser is a reasonable value, considering that the broader pulse width of the Nd-based and dye lasers described previously is around 100 ps.

The time-resolved polarization spectrum of rhodamine B in TRIS can also be determined. Both the modulation ratio and the differential phase are plotted in Figure 19–8. Similar to the imaging example, parallel and perpendicular polarization data are obtained by changing the relative polarization between the pump and probe beams into the parallel and perpendicular configurations, respectively. The data were fit by using an isotropic rotator model with the fitting software Globals Unlimited (LFD, Urbana, IL). With a single-exponential lifetime of 1.54 ns and an initial anisotropy of 0.39, the rotational correlation time was determined to be 88 ps.

The key point demonstrated here is that ultrafast fluorescence harmonics can be studied without fast detectors. Compared with conventional phase fluorometers (LFD, Physics Department, UIUC), the frequency response of frequency-domain, pump-probe spectroscopy can be easily extended into the GHz range.

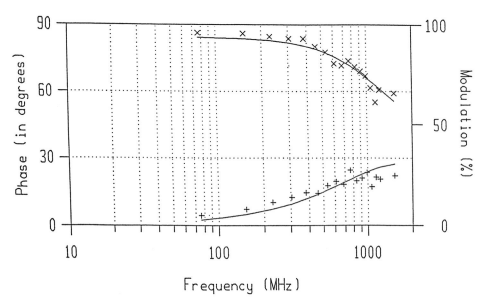

FIGURE 19–8 Time-resolved polarization spectrum of rhodamine B in TRIS. The modulation ratio (×) and differential phase (+) are plotted along with the data fit. The rotational correlation time was determined to be 88 ps. [Adapted from *Biophys. J.* 79: 536–549, 2000.]

6. CONCLUSION

We have presented the development of frequency-domain techniques for applications in fluorescence microscopy and spectroscopy. This technology is unique in that it is capable of achieving both confocal-like image quality for microscopic imaging and ultrafast temporal resolution characteristics of the pump-probe techniques. Operated in the stimulated-emission mode, ground state depletion is not necessary for pump-probe studies, and photobleaching is greatly reduced.

Three areas hold great promise for the future of this novel technology. First, with the expanding future of blue-green laser diode technology, there is a significant potential for broadening this technology with existing microscope systems into an ultrafast, time-resolved imaging technique with excellent axial depth discrimination capability. Furthermore, pump-probe technology has the potential to revolutionize time-resolved fluorescence spectroscopy. Not only can simultaneous harmonics be acquired, thus greatly reducing data acquisition time, but it also allows ultrafast, time-resolved spectroscopic studies to be performed in point-like domains inside cellular systems. Finally, the principles of pump-probe methodology can be implemented with other novel forms of microscopy, such as total internal reflection microscopy (see Chapter 21), to be developed into new tools for microscopy and spectroscopy.

REFERENCES

Alcala, J. R., and E. Gratton. A multi-frequency phase fluorometer using the harmonic content of a mode-locked laser. *Anal. Instrum.* 14:225–250, 1985.

Baym, G. *Lectures on Quantum Mechanics.* Menlo Park: The Benjamin/Cummings Publishing Company, 1973.

Born, M., and E. Wolf. *Principles of Optics*, 5th ed. Oxford: Pergrmom Press, 1985.

Boyce, W. E., and R. C. DiPrima. *Elementary Differential Equations and Boundary Value Problems*, 3rd. New York: John Wiley & Sons, 1977.

Buehler, C., C. Y. Dong, P. T. C. So, and E. Gratton. Frequency-domain, pump-probe microscopic imaging using intensity modulated laser diodes. *Proc. SPIE* 3603:262–217, 1999.

Buehler, C. H., C. Y. Dong, P. T. C. So, and E. Gratton. Time-resolved polarization imaging by pump-probe (stimulated emission) fluorescence microscopy. *Biophys. J.* 79:536–549, 2000.

Dong, C. Y., C. Buehler, P. T. C. So, T. French, and E. Gratton. Implementation of intensity-modulated laser diodes in time-resolved, pump-probe fluorescence microscopy. *Appl. Opt.* 40(7):1109–1115, 2001.

Dong, C. Y., P. T. C. So, C. Buehler, and E. Gratton. Spatial resolution in scanning pump-probe fluorescence microscopy. *Optik* 106(1):7–14, 1997.

Dong, C. Y., P. T. C. So, T. French, and E. Gratton. Fluorescence lifetime imaging by asynchronous pump-probe microscopy: Application to cellular systems. *Biophys. J.* 69:2234–2242, 1995.

Elzinga, P. A., R. J. Kneisler, F. E. Lytle, G. B. King, and N. M. Laurendeau. Pump/probe method for fast analysis of visible spectral signatures utilizing asynchronous optical sampling. *Appl. Opt.* 26(19):4303–4309, 1987a.

Elzinga, P. A., F. E. Lytle, Y. Jian, G. B. King, and N. M. Laurendeau. Pump/probe spectroscopy by asynchronous optical sampling. *Appl. Spectrosc.* 41(1):2–4, 1987b.

Evans, D. K., editor. *Laser Applications in Physical Chemistry.* New York: Marcel Dekker, 1989.

Fleming, G. R. *Chemical Applications of Ultrafast Spectroscopy.* New York: Oxford University Press, 1986.

Gratton, E., D. M. Jameson, and R. D. Hall. Multifrequency phase and modulation fluorometry. *Annu. Rev. Biophys. Bioeng.* 13:105–124, 1984.

Gratton, E., and M. Limkeman. A continuously variable frequency cross-correlation phase fluorometer with picosecond resolution. *Biophys. J.* 44:315–324, 1983.

Hell, S. W., and J. Wichmann. Breaking the diffraction resolution limit by stimulated-emission-depletion fluorescence microscopy. *Opt. Lett.* 19(11):780–782, 1994.

Hochstrasser, R. M., and C. K. Johnson. Biological processes studied by ultrafast laser techniques. In: *Ultrashort Laser Pulses,* edited by W. Kaiser. New York: Springer-Verlag, 1988, pp. 357–417.

Klar, T., and S. W. Hell. Subdiffraction resolution in far-field fluorescence microscopy. *Opt. Lett.* 24(14):954–956, 1999.

Kusba, J., V. Bogdanov, I. Gryczynski, and J. R. Lakowicz. Theory of light quenching: Effects on fluorescence polarization, intensity, and anisotropy decays. *Biophys. J.* 67(5):2024–2040, 1994.

Lakowicz, J. R. *Principles of Fluorescence Spectroscopy,* 2nd ed. New York: Kluwer Academic/Plenum Publishers, 1999.

Lakowicz, J. R., I. Gryczynski, V. Bogdanov, and J. Kusba. Light quenching and fluorescence depolarization of rhodamine B. *J. Phys. Chem.* 98:334–342, 1994.

Lytle, F. E., R. M. Parrish, and W. T. Barnes. An introduction to time-resolved pump/probe spectroscopy. *Appl. Spectrose.* 39(3):444–451, 1985.

Papadogiannis, N. A., B. Witzel, C. Kalpouzos, and D. Charalambidis. Observation of attosecond light localization in higher order harmonic generation. *Phys. Rev. Lett.* 83(21):4289–4292, 1999.

Periasamy, N., M. Armijo, and S. A. Verkman. Picosecond rotation of small polar fluorophores in the cytosol of sea urchin eggs. *Biochemistry* 30(51):11836–11841, 1991.

Roberts, D. V., B. P. Wittmerhaus, Y. Z. Zhang, S. Swan, and M. P. Klinkosky. Efficient excitation energy transfer among multiple dyes in polystyrene microspheres. *J. Luminescence.* 79:225–231, 1998.

Sheppard, C. J. R. and M. Gu. Image formation in two-photon fluorescence microscopy. *Optik* 86(3):104–106, 1990.

Siegman, A. E. *Lasers*. Mill Valley: University Science Books, 1986.

Swaminathan, R., C. P. Hoang, and A. S. Verkman. Photobleaching recovery and anisotropy decay of green fluorescent protein GFP-S65T in solution and cells: cytoplasmic viscosity probed by green fluorescent protein translational and rotational diffusion. *Biophys. J.* 72:1900–1907, 1997.

Zhou, J., G. Taft, C. P. Huang, I. P. Christov, H. C. Kapteyn, and M. M. Murnane. Sub-10 fs pulse generation in Ti:Sapphire: Capabilities and ultimate limits. In: *Ultrafast Phenomena IX*, edited by P. F. Barbara, W. H. Knox, G. A. Mourou, and A. H. Zewail. Berlin: Springer-Verlag, 1994, pp. 39–40.

IV

OTHER ADVANCED METHODS IN CELLULAR IMAGING

INTRODUCTION

We have seen in the previous chapters that the fluorescence microscopy methods become an esssential tool for cellular imaging and for studying their functional activities with high temporal and spatial resolution. The microscopic techniques discussed in this section are used for different biological investigation other than what has been described in Part IV. Multicolor fluorescence should become routine, especially with the development of novel fluorescent reagents with reduced photobleaching and large and variable stoke shifts. Chapter 20 describes recent technological improvements to image biological specimens by spectral methods with microscopic spatial resolution. Spectral imaging could improve on molecular–biological procedures by automatic detection such as mapping brain oxygenation and targeting cellular components such as actin, endosomes, mitochondria in mouse 3T3 fibroblast cells.

Total internal reflection fluorescence (TIRF) microscopy provides a means to selectively excite in an aqueous or cellular environment without exciting fluorescence from regions further from the surface . Theory and optical configurations of TIRF are discussed in Chapter 21. TIRF can be used for many biological applications such as observing the position, extent, composition, and motion of cell substrate contact regions; single molecule studies; and measurements of the kinetic rates of binding of extra-cellular and intracellular proteins to cell surface receptors and artificial membranes.

Laser trap is a beam of laser light brought to a diffraction-limited focus through a microscope objective. At the focal point, a translucent partical or a living cell may be captured, held and manipulated. Chapter 22 clearly describes about the mechanics of laser traps to show when the technique is appropriate for certain applications, and to point out common problems and their solutions. The physics and fundamental limitations of laser traps in cell biology are also discussed.

Bioluminescence method is widely used for monitoring gene expression in living cells and tissues. Bioluminescence provides additional advantages in the ability to couple enzyme reactions to luciferase. Chapter 23 outlines several bioluminescent reporter genes currently used in live cells, and mammalian nervous system. Also, the author explains the instrumentation involved for bioluminescence imaging.

Chapter 24 describes the instrumentation involved in atomic force microscopy (AFM) imaging and the details of cellular and molecular level applications. This AFM provides higher resolution in the atomic scale than the other microscopic techniques described in this book for cellular imaging.

20

Spectral Microscopy for Quantitative Cell and Tissue Imaging

Daniel L. Farkas

1. Introduction

Light is a most versatile tool for the study of biological systems and phenomena; the variety, range, nondestructiveness, and spatial resolution of optical imaging are all important for investigating *structure* in biological systems, all the way down to the subcellular level. In imaging of live specimens, biological *function* can also be explored (at the cell, tissue, organ, or even whole-organism level). To achieve this more challenging task, technological requirements are heightened and additional features of light such as speed, coherence, polarization and wavelength can be used to generate the contrast and discrimination needed (Farkas et al., 1995, 1997). This chapter concentrates on recent improvements in our ability to image biological specimens by spectral methods, with microscopic spatial resolution and without sacrificing temporal resolution, where relevant.

Imaging in biology, as in any field, involves detection, recognition, quantitation, and reconstruction. There are, however, many features specific only to biological specimens, and the resulting difficulties may not be fully appreciated by the engineers and physicists who, as developers of new methods of imaging, often expect these to behave in the biological world with the precision they show in the laboratory. This hope is often confounded by the inherent variability of living organisms; nevertheless, when combined with necessary insights into the physical complexity of biological samples, imaging methods like the ones described here will become increasingly valuable in biology and, in time, in other domains such as clinical practice.

2. Light Microscopy as a Research Tool

Light microscopy is one of the most firmly established research methods, both historically and in breadth of use, to the extent that the microscope has become an icon of the sciences. Modern, digital light microscopy workstations allow detailed, quantitative imaging of structures within cells (Arndt-Jovin et al., 1985; Giuliano et al., 1995), while automation and optimization of these instruments make live monitoring of cellular dynamics possible. The information content derived is further enhanced if several modes of microscopy are used simultaneously (Farkas et al., 1993; Taylor et al., 1997), and these methodologies have yielded significant advances in the understanding of cellular functions and dynamics. Additionally, this basic biological information can be useful in more applied fields, such as genetics (Eastmond et al., 1995), toxicology (Taylor et al., 1994), pathology (Taylor, 1994) and cancer research (Folli et al., 1994; Fox et al., 1995; Replogle-Schwab et al., 1996).

The vast majority of current microscopic bioimaging is intensity-based (whether transmitted/reflected light, fluorescence, or luminescence is used). If experimental conditions are carefully controlled, this makes quantitation not only possible but also relatively easy, provided that a limited number of the system's components are monitored at one time.

More recently, lifetime-based imaging has been introduced (Lakowicz et al., 1992) and applied to interesting cases. There are, however, many other optical parameters and contrast mechanisms to consider; spectral imaging and coherence (Fujimoto et al., 1995) are such cases, and we address spectral imaging here.

3. OPTICAL IMAGING IN THE SPECTRAL DOMAIN

When light impinges upon an object, it can be re-emitted, reflected/scattered, transmitted, or absorbed. Spectroscopy measures these phenomena in the temporal or frequency domain in order to determine important physical properties of the object being probed. Depending on the energy of the light and the nature of the object, some of these interactions, such as absorption by particular chemical bonds, can reflect the presence and quantity of certain well-understood constituents. The addition of imaging to spectroscopy can trace its origins to pioneering efforts in airborne satellite-based remote sensing, which was developed on the premise that information present in the optical properties of natural and manmade objects could be used to detect and monitor these entities from a distance. Similar premises are at the basis of noncontact studies by biological spectral imaging, particularly with "optical biopsy" (Levenson and Farkas, 1997; Flotte, 1998).

Optical imaging uses light to probe a scene in order to determine structure and organization. Unlike spectroscopy, in its simplest form it does not probe fundamental chemical or physical properties; rather, it presents data to our human visual system for perception and understanding. Thus, while there is a connection between image and content, it is not easy to combine the two rigorously, and often analysis may address spectral and spatial content sequentially rather than simultaneously. Although spectroscopy and imaging have been coupled in the past, usually this has involved obtaining a point or a line of spectroscopic information out of an entire two-dimensional image. Spectral imaging has been brought to biology in an effort to determine, with high spatial resolution, not only what a scene "looks like" but also what it contains and where the various classes of objects are located.

Human color vision is, of course, a form of imaging spectroscopy by which we determine the intensity and proportion of wavelengths present in our environment. Spectral imaging improves on the eye in that it can break up the light content of an image not just into red, green, and blue but into an arbitrarily large number of wavelength classes (see Color Fig. 20–1). Furthermore, it can extend the range to include the invisible ultraviolet and infrared regions of the spectrum denied to the unaided eye; this type of imaging is usually known as *hyperspectral.* The result of (hyper)spectral imaging is a data set, known as a *data cube,* in which spectral information is present at every picture element (pixel) of a digitally acquired image. Integration of spectral and spatial data in scene analysis remains a challenge.

The information conveyed by color (or indeed spectral features in general) varies. In some instances, color reflects the presence of a single, important chromophore that can be detected and quantified by virtue of its absorbing certain wavelengths. Hemoglobin, chlorophyll, bilirubin, and so forth, are such molecules. Furthermore, some substances will reflect changes in their functional status with absorbance shifts, the obvious example being the color shift involved in going from oxy- to deoxyhemoglobin (the basis for pulse oximetry used clinically to monitor oxygenation). In other, more complex cases, color can function as a signature, signifying an identity, even when we do not understand all the physical bases for the perceived color composition. In contrast to intrinsically colored objects, when histology or pathology samples are examined in transmission under a microscope, they generally appear almost clear and, especially when single cells are examined, will be virtually invisible unless optical techniques such as phase or differential interference contrast microscopy are used. In other words, in these cases, the specimens are optically weak, there is almost no useable spectral information in the visible range, and staining is required before viewing. The most commonly used stains provide intense color signals, but the relationship between the observed colors and the cellular constituents is obscure at best and variable at worst.

Finally, exogenous, spectrally defined chromophores or fluorophores can be coupled to agents that bind to cellular components with great specificity in such techniques as ion-sensitive signaling (Moreton, 1994), immunohistochemistry (Taylor and Cote, 1997), immunofluorescence (Brelje et al., 1993), and in situ hybridization (Speicher et al., 1996; Herrington, 1998). In these cases, the colors usually bear only an arbitrary connection to the molecular structures they are coupled to, although in some cases spectral shifts and/or intensity changes can be used to detect intra- and intermolecular interactions.

4. Spectral Imaging Technologies in Microscopy

4.1 Filter Changers

While not providing spectral imaging in the strict sense of more or less continuous, high-resolution spectral content, filter changers represent the main competition to all more advanced spectral imaging devices. The term filter changers refers to various means of interposing a number of (usually) bandpass interference filters into the imaging light path and, in the case of fluorescence, of having a set of filter cubes each comprising an excitation filter, a dichroic mirror, and an emission filter. With suitable selection of dyes and filter combinations, as many as five to eight different fluorescent dyes (all emitting in the visible region) can be imaged simultaneously (DeBiasio et al., 1987; Reid et al., 1992). Cross-talk correction is required to compensate for emission from one dye entering into another dye's main collection channel, but such algorithms have been well worked out (Galbraith et al., 1989, 1991; Morrison, 1998). An advantage of this approach is its basic simplicity and relatively low cost. Of all the technologies discussed here, fixed-wavelength filter-based systems are the most likely to find their way into turnkey systems (including clinical).

Potential disadvantages include their inflexibility: New dyes may be difficult to add to the repertoire, and there remains the possibility of image shift associated with each change of filter, although with careful engineering this can be minimized or, if necessary, corrected for ex post facto. Such systems are not well suited for spectral analysis of complex samples whose signals are not made up of combinations of well-characterized fluorophores or chromophores. If sufficient wavelengths are sampled, however, and if the spectral behavior of the specimen has few high-frequency features, multiple filters can extract useful information (see Color Figure 20–2). For example, for reflectance/absorbance imaging of skin, a filter wheel containing as many as 17 different filters (each with a bandpass of 40 nm) spanning the region from 420 to 1040 nm has been used to obtain broad-range, discontinuous spectral data (Tomatis et al., 1998).

4.2 Excitation Wavelength Selection

There are a number of different techniques that permit the acquisition of spectrally resolved images. In remote sensing, the illumination source is the sun, whose spectral qualities are constant; consequently, all spectral discrimination is effected on the returned light. In contrast, in "nonremote" sensing, the illumination arm can be varied spectrally, providing information, in reflection or absorbance, very similar to that which can be obtained by spectrally filtering the remitted light. Thus, with a tunable light source, the illuminating light is scanned continuously or discontinuously through a number of wavelengths. A gray-scale image is taken at each desired wavelength; the resulting image stack constitutes a spectral cube.

This method benefits from simplicity, (relatively) low cost, and the fact that no additional optical or mechanical elements are interposed in the imaging light path, resulting in minimal image degradation. It does require rigorously achromatic optics in the objectives and transfer optics to prevent image blurring at the spectral extremes. It may be nearly impossible to design transmission optics that will behave well in the extended range of 350 to 1100 nm (which can be captured by some charge-coupled device [CCD] cameras), and, for some applications, resort to reflective (Cassegrainian) optics (which are intrinsically achromatic) may be necessary. The light sources can be tuned using either diffraction gratings, as in most monochromators, or tunable filters, such as acousto-optic tunable filters (AOTFs) or liquid crystal tunable filters (LCTFs), in case there is need for greater speed in switching wavelengths than can be achieved with a monochromator, which must physically shift the grating position.

4.3 Tunable Filters in the Imaging Path

Tunable filters, as their name implies, can be tuned to permit the transmission of narrow, preselected bands of light, and these can be rapidly and randomly switched. A "continuous" spectrum can be obtained by collecting a gray-scale image at each contiguous spectral band. Thus, these devices are "band sequential" in operation. It should be noted that only a portion of the photons emerging from the imaged specimen will be passed to the collector because the filter is only transparent to a narrow wavelength region at any one time.

4.3.1 Liquid crystal tunable filters

LCTFs consist of a number of liquid crystal layers each of which passes a number of different frequencies; stacking them results in a single dominant transmission band, along with much smaller side bands (and in some cases, unfortunately, additional major transmission bands far from the desired spectral region). They can be switched from wavelength to wavelength in about 50 ms and are optically well behaved in that they do not seem to induce image distortion or shift. Each filter assembly can span approximately 1 octave of wavelength (e.g., 400–800 nm), and their useful range can extend into the near infrared. As noted above, they can be introduced into either the illumination (excitation) or emission pathways or both.

Throughput is a problem in that half the light corresponding to one polarization state is lost automatically, and peak transmission of the other half probably does not exceed 40% at best. Out-of-band rejection is not sufficient to prevent excitation light from leaking into the emission channel without the use of a dichroic mirror or cross-polarization (Hoyt, 1996). A small controller box in addition to a PC is needed to drive an LCTF assembly.

4.3.2 Acousto-optic tunable filters

In an AOTF, filtered light of narrow spectral bandwidth is angularly deflected away from the incident beam at the output of the crystal. The central wavelength of this filtered beam is determined by the acoustic frequency of the AOTF; this wavelength can be changed within approximately 25 μs to any other wavelength. High-end AOTFs involve additional optics compared with LCTFs and require higher power and more involved electronics. The payoff is much faster switching and the ability to vary not only the wavelength but also the bandwidth and the intensity of the transmitted light. Thus, experiments involving luminescence lifetimes or very rapid acquisition of multiple wavelengths are possible using this technology.

Over the past few years, we overcame some of the shortcomings of AOTFs that have impeded their use in high-resolution imaging: (*1*) As with LCTFs, usually only one polarization state is available; (2) there is image shift when wavelengths are changed; (3) intrinsic image blur is present; and (4) out-of-band rejection is no greater than 10^{-2}–10^{-3}. Approaches and solutions to some of these problems have been described previously (Wachman et al., 1997, 1998a,b). Recently, improvements in transducer apodization (in the emission path) and the use of two AOTFs in tandem (in the excitation path) have been adopted to further improve out-of-band rejection (Farkas et al., 1998).

Apodization refers to a technique in which the transducer is sectioned into a number of discrete slices with the relative amplitudes of the electronic signal used to drive each arranged to generate an acoustic spatial profile within the crystal that optimizes the shape of the filter spectral bandpass. According to our calculations, a critical parameter for fabricating a crystal for apodization is the distance separating individual transducer slices. In particular, minimizing this distance can provide the greatest reduction in the out-of-band light passing through the filter—a point that has not been generally appreciated. Based on this we (Wachman et al., unpublished data) fabricated an imaging AOTF crystal for the current study with a transducer consisting of 11 electronically isolated, individually driven slices with a slice-to-slice separation of approximately 50 μm. This design results in a greater than 10 dB decrease in the out-of-band light between the primary side lobes with even greater decreases obtainable in specific regions.

In addition, we have developed an AOTF illumination system that can be fiber-coupled to any microscope (or other optical system) to provide speed and spectral versatility for excitation as well as detection. This source consists of a 500 W short-arc Xenon lamp followed by a pair of identical AOTFs placed in series, with the output of the second AOTF coupled into a fiber. This crystal configuration enables both polarizations to be filtered twice and recombined with a minimum of extra optics. The double filtering greatly reduces both the out-of-band light levels and the diffuse broadband scattered light from the output face of the first crystal. In a single-AOTF system with a white-light source, these can degrade the output tremendously, making it unusable for most applications. In contrast, our double-AOTF system has out-of-band levels lower than 10^{-4} relative to the peak, with low levels of diffuse scatter. This system has been tested for its suitability for both bright-field and fluorescence applications.

4.4 Fourier Transform Imaging Spectroscopy

Fourier transform interference spectroscopy (FTIS) is the technique most widely used for generating spectral information in the infrared region (Buican, 1990; Lewis et al., 1995; Jackson et al., 1997). The method takes advantage of the principle that when light is allowed to interfere with itself at a number of optical path lengths, the resulting interferogram reflects its spectral constitution. Monochromatic light generates a pure cosine wave, and multispectral light, which consists of a mixture of wavelengths, will generate interferograms that can be modeled as a sum of their respective cosine waves. An inverse Fourier transform of such an interferogram will regenerate the presence and intensities of all the contributing wavelengths. The classic Michelson interferometer, which used two separate light paths to generate an interference pattern, is usable in the infrared region of the optical spectrum because mechanical tolerances at those wavelengths are not unreasonable. With the much shorter wavelengths in the visible region, however, the required optical path length differences are much smaller, and vibration and mechanical imprecision become limiting. The Sagnac interferometer design, which directs the interfering light beams in opposite directions around a common path, overcomes many of these problems and has been successfully adopted for use for visible light imaging spectroscopy by Applied Spectral Imaging, Inc. (Cabib et al., 1996; Garini et al., 1996).

There are advantages and drawbacks to Fourier transform techniques. As with tunable filters, and in contrast to fixed interference filters, wavelength ranges can be flexibly tailored to the exact spectral properties of the image. Compared with monochromator- or prism-based devices, FTIS enjoys a throughput advantage, as it does not require a narrow-slit aperture that would reduce the signal-to-noise ratio. The so-called *Fellgett advantage* refers to the fact that the interferometer can collect all the emitted or transmitted photons simultaneously (subject, of course, to normal losses) rather than collecting photons in a single narrow wavelength band at any one time, as with the band sequential devices. If photobleaching of labile fluorescent labels occurs during imaging, the Fellgett advantage could be significant.

On the other hand, unlike a band-sequential device, FTIS is unable to variably alter its sensitivity (or imaging time) in different spectral regions to compensate for changes in detector sensitivity or light flux (Farkas et al., 1998; Levenson et al., 1999). A major disadvantage of FTIS is that an interferogram must be converted by mathematical operations into a spectral data cube. This can require large working memory and considerable processing time (although of course both of these considerations become less important almost daily). Maximum image size is also currently limited by these considerations to about 500 × 500 pixels in the instruments currently mounted onto light microscopes.

4.5 Spectrotomography

Spectral imaging can also be accomplished using a number of techniques that disperse the spectral information onto detectors either in sequential mode or, more recently, simultaneously. For example, a scene can be imaged repeatedly while a direct vision prism is rotated, multiplexing the spatial and spectral information in the focal plane. A computer demultiplexes the result to obtain a spectral image (Descour et al., 1997). Alternatively, a computer-generated hologram disperser can be used to distribute various diffraction orders of the primary scene over a (much larger) two-dimensional CCD array. The position and intensity of the resulting multiple image mosaic reflect the spectral content of the original scene; the spectral content of the image is reconstructed using an iterative multiplicative algebraic reconstruction algorithm (De la Iglesia et al., 2000).

The advantage of this approach is that spatial and spectral information can be acquired with a single image exposure without any necessity for scanning in either the spatial or wavelength domain. Thus, spectral images can be acquired in a single exposure, and, depending on signal intensity, the sampling can be accomplished at video rate or even faster. This advantage is "paid for" with much higher CCD array size requirements because the CCD must capture not only the primary image but all of its higher order diffraction images. With current CCDs, a 128 × 128 pixel primary image is a realistic goal. In addition, the reconstruction of the spectral data cube from the dispersed images is computationally intense. Thus, while images may be acquired rapidly, the spectral information will have to be calculated off-line for the foreseeable future.

4.6 Prisms

A form of microscope-based spectral imaging that resembles the techniques used in many so-called push-broom remote sensing devices is represented by the PARISS system (Warren and Hackwell, 1989). A prism diffracts light collected through a narrow slit, dispersing its spectral components onto a CCD array. In the current system, the imaged strip is stationary, and the sample is moved beneath it. The resulting spectral information from each pixel along the strip being analyzed can be classified rapidly using neural-net software that accompanies the commercial instrument (Lightforms, Inc.). The vertical and horizontal resolution of the unit is not as high as that of true imaging spectrometers, and a motorized stage is necessary to build up a complete scan. The unit is compact and relatively inexpensive, however, and has already been used for a number of interesting applications (Jeremy Lerner, personal communication).

5. ANALYSIS OF SPECTRAL DATA

As can be seen, there are a large and growing number of spectral imaging techniques; spectral discrimination technology is no longer really the challenge. While a large body of theoretical and practical work on multispectral analysis already exists, it is naturally mainly directed toward remote sensing applications (Descour et al., 1995; Landgrebe, 1998). Unfortunately, many of the highly tuned algorithms contain constraints specific to ground reflectance imaging and cannot be directly applied to biological imaging data sets (Boardman, 1989). The next step, then, for biological spectral imaging will be to develop robust software that will permit quantitative, intuitive analysis. Of course, the tools will vary with the sample and the question being asked. Generally, the tasks involve visualization, classification, and quantitation.

At the simplest level, visualization merely involves the representation of the spectral data as a color image. When imaging is confined to the visible wavelengths, it is straightforward to generate an RGB version of the data set. If nonvisible wavelength bands are acquired, then some form of mapping the data into the visible range is needed. Classification is equivalent to spectral segmentation and is an exclusive operation in which a pixel or object is ultimately assigned to a single class. Midway between visualization and classification is spectral similarity mapping, which assigns an intensity to each pixel reflecting how closely its spectrum matches that of a reference spectrum. This operation can reveal subtle morphological detail (Rothman et al., 1997).

Finally, quantitation or determination of the concentration of each class at every pixel is either a stepping stone to classification (e.g., when combinatorial dye mixtures are used) or an end in itself. Overall, the investigative steps consist of (1) detection and/or selection of appropriate spectra for subsequent analysis and (2) the analysis of the image itself. More sophisticated considerations include determination of the optimal dimensionality of the data set, the number of training sets available, the appropriateness of parametric versus nonparametric classifiers, and so on. These go beyond our scope here and the reader is directed to a useful overview (Landgrebe, 1998).

5.1 Selection of Reference Spectra

Fluorescence and immunohistochemistry-based applications involve the detection of (presumably) spectrally distinct signals. In the analysis, either stored spectra from a spectral library can be used or classification spectra can be directly derived from the sample itself, assuming that pure or unmixed spectral data for each color are available. When complex samples, stained with histological dye combinations, are to be analyzed, the concept of "pure" spectra does not really apply, and the use of empirically derived spectral signatures from histological or cytological features may be necessary. Again, these spectra can be derived from a previously accumulated spectral library or from the sample itself.

We have had the most experience with totally manual spectral selection. This straightforward but subjective maneuver involves examining an RGB image of a spectral data cube and, using more or less expert histological insight, choosing a few pixels or areas to provide reference spectra for subsequent analysis. While this is a useful exploratory procedure, and may be necessary for some applications, it would be desirable to have the instrument automatically detect and apply the most appropriate reference spectra. In this way, variations in technique, illumination, chemistry, filter integrity, and so forth can be dynamically adjusted for. We have made a beginning along these lines, attempting to automatically select appropriate spectra for the classification and pixel-unmixing of immunohistochemically triple-stained samples (Levenson et al., 1999).

5.2 Spectral Classification

A large number of classification algorithms have been devised for analysis of hyperspectral sets, but it is not yet clear which strategies are most suitable for which biological samples. A great deal depends on whether the subject is spectrally complex, as in absorbance/scattering images of live tissues or transmission images of histology, cytology, or pathology specimens, or well-defined, containing only known chromophores or fluorophores. Classification techniques include deterministic approaches, fuzzy set theory, fractals, stochastic models, neural nets, and others.

The most intuitively obvious approach of evaluating each pixel in terms of its similarity (in Euclidean minimum distance) with all the reference spectra and then selecting the "nearest" one proves to be unfortunately sensitive to noise; approaches using a stochastic process and second-order statistics appear to be more robust (Swain and Davis, 1978). Such statistical methods underscore the importance of having sufficient "training sets," or examples of classifying spectra, and point to a counterintuitive finding. It appears that, contrary to what would be expected, increasing the spectral resolution (dimensionality) of the data set does not indefinitely improve classification accuracy. Instead, after a point, determined by the richness of the training set, increasing dimensionality actually decreases class separability (Landgrebe, 1998).

5.3 Pixel-Unmixing

It is often the case that areas in the sample corresponding to individual detector pixels are not spectrally homogeneous. This may be deliberately encountered, as when a cell nucleus is probed for the simultaneous expression of a number of proteins using multiple fluorescent antibodies. The task would be the detection and possibly the quantitation of each species, even in the presence of a high degree of spatial overlap. A chemometric model can be used to resolve the resulting spectral mixture, using a constrained linear unmixing approach (Boardman, 1989; Farkas et al., 1998) or more sophisticated techniques. We have found that the choice of endmember spectra for the unmixing, not unexpectedly, significantly affects the output, adding to the impetus to develop semiautomated spectral selection techniques as a preamble to unmixing or classification operations.

6. SELECTED EXAMPLES OF MICROSCOPIC SPECTRAL IMAGING IN BIOLOGY AND MEDICINE

6.1 Cellular Imaging in Automated Microscopy

The spectral range available for light microscopy was traditionally limited to the visible region, but solid-state imaging devices have extended it into the near infrared. The FTIS instrument described above has been used successfully for quantitative cytology (Malik et al., 1996a,b; Levenson and Farkas, 1997). The ability to unambiguously differentiate five to six fluorophores by filter-based multiwavelength imaging has also made it possible to label monitor *independently* a corresponding number of components in the same sample, even within living, moving cells (DeBiasio et al., 1987). This constitutes a unique way of mapping not only the location and interrelationships of cellular components but also their dynamic reorganization and interactions upon performing a certain physiological function. In the automated microscopy setting (Farkas et al., 1993) this is easier to achieve than fully spectral imaging, but it is unlikely that more than six cellular components (Color Fig. 20–2) can be simultaneously quantitated and tracked. De la Iglesia and coworkers have applied this approach to more clinically relevant cells (Plymale et al., 1999) and are currently extending the AOTF and spectrotomography technologies discussed above to the same applications (de la Iglesia et al., 2000).

6.2 Imaging of Conventionally Stained Pathology Samples

More than a century ago, advances in pathology, along with microbiology, established the central underpinnings for our modern understanding of tissues, cells, and disease. The recognition that the cell formed the basic building block of organisms and that *cellular* disorders underlay such important processes as cancer, inflammation, and so forth, came some 200 years after the discovery of the microscope. Some of this remarkable delay is due to the fact that unstained tissue sections are extremely difficult to visualize with conventional microscopic optics.

To a great extent, subsequent progress in pathology relied on developments in the dye industry, which created a torrent of coal-based dyes intended for fabrics. A large number of these dyes were tested on biological samples, and ultimately detailed techniques were elaborated for specific and nonspecific staining of different cellular and extracellular constituents (DeMay, 1996). For reasons of simplicity, cost, and, of course, utility, a pair of dyes, hematoxylin and eosin, became by far the most widely used set of stains for standard pathology specimens, hematoxylin staining largely nuclear and eosin extranuclear structures. Other stains used for cytology specimens include Giemsa and the Papanicolaou stain (used, of course, for the Pap smear; Papanicolau, 1942). Of relevance here, all of these general stains appear to exhibit spectral complexity when bound to tissue that can be of use in the analysis of clinical material (Levenson and Farkas, 1997).

Stained tissue sections are typically evaluated by pathologists with the naked eye, using a process of pattern recognition, for disorders of cellular size, shape, and organization, in combination with structural evaluation of overall tissue architecture. Architectural clues usually are paramount, and specific color information is rarely necessary. In fact, until recently, most pathology texts were illustrated almost exclusively with gray-scale photomicrographs; however, no pathologist would be satisfied with viewing specimens in black and white, so color definitely contributes to the overall assessment. It should be stressed that spectral analysis depends on the choice of stains themselves. While the use of hematoxylin and eosin is hallowed by tradition, it may not provide the most useful spectral information. Spectra from cells stained with the Papanicolaou stain, which in addition to hematoxylin and eosin also contains an orange and a green stain, exhibit considerably more spectral complexity than cells stained only with hematoxylin and eosin (Levenson, Wachman, and Farkas, unpublished data). No systematic re-examination of the suitability of various stain combinations for spectral imaging has been done. Thus, spectral imaging may require quality assurance steps above those normally associated with standard histology; conversely, it is hoped that it will lead to better standardization efforts.

The application of spectral imaging in the analysis of pathology and cytology specimens was recently reviewed by our group (Levenson et al., 1999).

6.3 Immunohistochemistry

Immunohistochemistry is widely used clinically for the detection of diagnostically or prognostically significant molecules in or on cells (Fritz et al., 1992).

Typically, one color is used at a time, and if more than one antigen is to be analyzed, serial sections are made and a different antibody is applied to each one. This procedure results in more slides and more preparation steps than if the reagents could be multiplexed on a single slide. If co-expression of antigens in a single cell is a question, as it might be in some lymphomas, or in breast cancer prognosis determinations (Shackney et al., 1996), then multiply staining single slides with different chromophore-coupled antibodies may be helpful.

Triple-staining procedures are not often performed because of technical difficulties and because it can be difficult to determine where and to what extent the different stains may physically overlap due to co-expression of two or more of the antigens in the same cellular compartment. Spectral imaging can be used to examine such double- and triple-stained specimens and to tease out (potentially quantitatively) patterns of expression.

6.4 Immunofluorescence and In Situ Hybridization

Immunofluorescence involves the detection of antigens (usually proteins) by means of antibodies directly or indirectly coupled to fluorophores, whereas in situ hybridization uses nucleic acids or nucleic acid analogues (PNAs), similarly coupled to fluorescent labels, to detect DNA or RNA species. It is often desirable to image a number of markers simultaneously; spectral techniques are well suited to accomplishing such multiplexing, although, as noted above, fixed filter combinations may prove to be the most useful in a clinical setting. Multicolor immunofluorescence still awaits the appropriate combination of robust reagents and versatile instruments.

As many as 10 or more signals can be expected to be separable when these conditions are fully met. If the signals are guaranteed not to overlap, as in the case of "painted" chromosomes in a metaphase spread, then combinatorial approaches to sample classification can be used (Speicher et al., 1996; Schröck et al., 1996, 1997), permitting detection of 24 or more spectrally distinct classes of signal (as in automated spectral karyotyping, for example).

6.5 In Vivo Fluorescence and Elastic-Scattering Spectroscopy

We will touch lightly on the topic of in vivo imaging (the so-called optical biopsy) because a recent symposium presented a number of approaches in this area (*Ann. N.Y. Acad. Sci.*, Vol. 838, 1998). The essential advantage of spectral imaging over conventional biopsy is that the instrument (e.g., endoscope) can be brought to the patient. The technique relies on the assumption that tissue abnormalities, even if invisible to the eye, can be made evident due to distinct autofluorescence or scattering behavior. The promise of autofluorescence in the detection of colonic mucosal premalignant lesions was elaborated on by Zonios et al. (1998). While their report dealt only with single-point spectroscopy, their results provide empirical and theoretical support for the clinical use of imaging fluorescence spectroscopy.

Reflectance/absorbance spectroscopy complements fluorescence spectroscopy for in vivo diagnostic applications. Also known as elastic-scattering spectroscopy, the technique relies on the wavelength dependence of both scattering and absorption (spectral reflections are typically suppressed) to detect composite spectral signatures of normal and diseased tissue (Bigio and Mourant, 1997). The optical behavior of tissue and particularly skin has been studied in detail (Anderson and Parrish, 1981), and recently spectral information has been used in such advanced methods as two-photon microscopy (Masters et al., 1997). Efforts to develop clinically acceptable diagnostic imaging systems for distinguishing benign nevi, atypical nevi, and malignant melanomas in skin have previously combined morphometric with three-color (RGB)–based analysis, without achieving sufficiently high performance. More recently, systems that add imaging in 7 (Gutkowicz-Krusin et al., 1997) or 17 (Tomatis et al., 1998) wavelength bands, extending into the near infrared, are showing considerable promise. We have recently applied spectral microscopy to both the assessment of clinical treatments in melanoma patients (Kirkwood et al., 1999) and the mesoscopic imaging of in vivo lesions, with the goal of improved diagnosis and correlation with gene expression (Yang et al., 1999).

6.6 Functional Imaging of Brain Oxygenation by Intravital Spectral Microscopy

We recently reported using a microscope equipped with AOTFs (Wachman et al., 1997) to take images of physiologically relevant parameters in a manner that would not be possible with conventional instruments. Two-dimensional maps of both oxygen saturation and oxygen tension were obtained through a cranial window in a living mouse. Oxygen saturation (SO_2) data were obtained by imaging reflectance spectroscopy; oxygen tension (PO_2) data were determined by frequency-domain lifetime measurements of palladium-porphyrin, an oxygen-sensitive phosphorescent probe. These images could be superimposed on a white-light image of the brain to show functional variations with brain activity. These studies were pursued further into the functional imaging realm by adding variables to effect changes in brain oxygenation (Shonat et al., 1998). An interesting example is shown in Color Figure 20–3, illustrating the effect of amphetamine administration on oxygen saturation of hemoglobin and oxygen tension in a living mouse.

7. Outlook

Spectral imaging is appealing for many reasons, not the least of which are its intuitive nature (because it is similar to human vision) and its ability to assist with one of the most important tasks in digital microscopy: image segmentation. It is our hope and prediction that wavelength discrimination-based methods of optical imaging will play an ever-increasing role in biological research and medical practice. Multicolor fluorescence should become routine, especially with the development of novel fluorescent reagents with reduced photobleaching and large and variable Stokes' shifts. Fully spectral imaging techniques will need to speed up considerably if they are to be used in any kind of automated screening. Video-rate

AOTF or computed spectral tomography systems may find their niche in such applications.

Robust, automated software tools will need to be developed. One can even speculate that some functions of a pathologist could eventually be automated or at least assisted. Finally, spectral imaging could improve on molecular–biological procedures by automatically detecting and targeting cells for analysis and manipulation.

I would like to thank my colleagues, and particularly my collaborators Drs. Felix De la Iglesia, Alan Koretsky, Richard Levenson, Ross Shonat, and Elliot Wachman, for permission to give examples of their work and for valuable discussions. Support from the National Science Foundation (through grant NSF-MCB 8920118 to our Science and Technology Center) and from corporate partners (Bio-Rad, Carl Zeiss, ChromoDynamics, NEOS, Nikon, Olympus, and Parke-Davis) is much appreciated.

REFERENCES

Anderson, R. R., and J. A. Parrish. The optics of human skin. *J. Invest. Dermatol.* 77:13–19, 1981.

Arndt-Jovin, D. J., M. Robert-Nicoud, S. J. Kaufman, and T. M. Jovin. Fluorescence digital microscopy in cell biology. *Science* 230:247–256, 1985.

Bigio, I. J., and J. R. Mourant. Ultraviolet and visible spectroscopies for tissue diagnostics: Fluorescence spectroscopy and elastic-scattering spectroscopy. *Phys. Med. Biol.* 42:803–814, 1997.

Boardman, J. W. Inversion of imaging spectrometry data using singular value decomposition. *Proc. IGARSS 89* 4:2069–2072, 1989.

Brelje, T. C., M. W. Wessendorf, and R. L. Sorenso. Multicolor laser scanning confocal immunofluorescence microscopy: Practical application and limitations. In: *Cell Biological Applications of Confocal Microscopy. Methods Cell Biol.* San Diego: Academic Press, 1993, pp. 97–181.

Buican, T. N. Real-time Fourier transform spectroscopy of fluorescence imaging and flow cytometry. *Proc. SPIE* 1250:126–133, 1990.

Cabib, D., R. A. Buchwald, Y. Garini, and D. G. Soenksen. Spatially resolved Fourier transform spectroscopy: A powerful tool for quantitative analytical microscopy. *Proc. SPIE* 2687:278–291, 1996.

DeBiasio, R., G. R. Bright, L. A. Ernst, A. S. Waggoner, and D. L. Taylor. Five parameter fluorescence imaging: Wound healing of living Swiss 3T3 cells. *J. Cell Biol.* 105:1613–1622, 1987.

De la Iglesia, F. A., J. R. Haskins, D. L. Farkas, and G. H. Bearman. Coherent multi-probes and quantitative spectroscopic multimode microscopy for the study of simultaneous intracellular events. *Proceedings of the International Society for Analytical Cytometry Conference,* Montpellier, France, 2000 (in press).

DeMay, R. M. *The Art and Science of Cytopathology,* Vol. 2. Chicago: ASCP Press, 1996.

Descour, M. R, J. M. Mooney, D. L. Perry, and L. R. Illing. Imaging spectrometry. *Proc. SPIE* 2480, 1995.

Descour, M. R., C. E. Volin, E. L. Dereniak, K. J. Thome, A. B. Schumacher, D. W. Wilson, and P. D. Maker. Demonstration of a high-speed non-scanning imaging spectrometer. *Opt. Lett.* 22:1271–1273, 1997.

Eastmond, D. A., M. Schuler, and D. S. Rupa. Advantages and limitations of using fluorescence in situ hybridization for the detection of aneuploidy in interphase human cells. *Mutat. Res.* 348:153–162, 1995.

Farkas, D. L., B. T. Ballou, G. W. Fisher, and D. L. Taylor. From in vitro to in vivo by dynamic multiwavelength imaging. *Proc. SPIE* 2386:138–149, 1995.

Farkas, D. L., B. Ballou, G. W. Fisher, C. Lau, W. Niu, E. S. Wachman, and R. M. Levenson. Optical image acquisition, analysis and processing for biomedical applications. *Springer Lecture Notes Comput. Sci.* 1311:663–671, 1997.

Farkas, D. L., G. Baxter, R. L. DeBiasio, A. Gough, M. A. Nederlof, D. Pane, J. Pane, D. R. Patek, K. W. Ryan, and D. L. Taylor. Multimode light microscopy and the dynamics of molecules. *Annu. Rev. Physiol.* 55:785–817, 1993.

Farkas, D. L., R. M. Levenson, B. C. Ballou, C. Du, C. Lau, E. S. Wachman, and G. W. Fisher. Non-invasive image acquisition and advanced processing in optical bioimaging. *Comput. Med. Imaging Graphics* 22:89–102, 1998.

Flotte, T. J. Pathology correlations with optical biopsy techniques, *Ann. N.Y. Acad. Sci.* 838:143–149, 1998.

Folli, S., P. Westerman, D. Braichotte, A. Pelegrin, G. Wagnieres, H. Van Der Bergh, and J. P. Mach. Antibody–indocyanin conjugates for immunophotodetection of human squamous cell carcinoma in nude mice. *Cancer Res.* 54:2643–2649, 1994.

Fox, J. L., P. H. Hsu, M. S. Legator, L. E. Morrison, and S. A. Seelig. Fluorescence in situ hybridization: Powerful molecular tool for cancer prognosis. *Clin. Chem.* 41:1554–1559, 1995.

Fritz, P., H. Multhaupt, J. Hoenes, D. Lutz, R. Doerrer, P. Schwarzmann, and H. V. Tuczek. Quantitative immunohistochemistry: Theoretical background and its application in biology and surgical pathology. *Prog. Histochem. Cytochem.* 24:1–53, 1992.

Fujimoto, J. G., M. E. Brezinski, G. J. Tearney, S. A. Boppart, B. Bouma, M. R. Hee, J. F. Southern, and E. A. Swanson. Optical biopsy and imaging using optical coherence tomography. *Nat. Med.* 1:970–972, 1995.

Galbraith, W., L. A. Ernst, D. L. Taylor, and A. S. Waggoner. Multiparameter fluorescence and selection of optimal filter sets: Mathematics and computer program. *Proc. SPIE* 1063:74–122, 1989.

Galbraith, W., M. C. Wagner, J. Chao, M. Abaza, L. A. Ernst, L. A., Nederlof, M. A., Hartsock, R. J., Taylor, and D. L. Waggoner. Imaging cytometry by multiparameter fluorescence. *Cytometry* 12:579–596, 1991.

Garini, Y., N. Katzir, D. Cabib, R. A. Buckwald, D. G. Soenksen, and Z. Malik. Spectral bio-imaging. In: *Fluorescence Imaging Spectroscopy and Microscopy, Chemical Analysis Series*, Vol. 137, edited by X. F. Wang and B. Herman. New York: John Wiley & Sons, Inc., 1996.

Giuliano, K. A., P. L. Post, K. M. Hahn, and D. L. Taylor. Fluorescent protein biosensors: Measurement of molecular dynamics in living cells. *Annu. Rev. Biophys. Biomol. Struct.* 24:405–434, 1995.

Gutkowicz-Krusin, D., M. Elbaum, P. Szwaykowski, and A. W. Kopf. Can early malignant melanoma be differentiated from atypical melanocytic nevus by in vivo techniques? Part II. Automatic machine vision classification. *Skin Res. Technol.* 3:15–22, 1997.

Herrington, C. S. Demystified . . . in situ hybridization. *Mol. Pathol.* 51:8–13, 1998.

Hoyt, C. Liquid crystal tunable filters clear the way for imaging multiprobe fluorescence. *Biophot. Int.* July/August:49–51, 1996.

Jackson, M., M. G. Sowa, and H. H. Mantsch. Infrared spectroscopy: A new frontier in medicine. *Biophys. Chem.* 68:109–125, 1997.

Kirkwood, J. M., D. L. Farkas, A. Chakraborty, K. F. Dyer, D. J. Tweardy, J. L. Abernethy, H. D. Edington, S. S. Donnelly, and D. Becker. Systemic interferon—Treatment leads to Stat3 inactivation in melanoma precursor lesions. *Mol. Med.* 5:11–20, 1999.

Lakowicz, J. R., H. Szmacinski, K. Nowaczyk, and M. L. Johnson. Fluorescence lifetime imaging of free and protein-bound NADH. *Proc. Natl. Acad. Sci. U.S.A.* 89:1271–1275, 1992.

Landgrebe, D. On information extraction principles for hyperspectral data: A white paper. 1998. Available at: http://dynamo.ecn.purdue.edu/~biehl/MultiSpec/documentation.html. Accessed 12/29/00.

Levenson, R. M., E. M. Balestreire, and D. L. Farkas. Spectral imaging: Prospects for pathology. In: *Applications of Optical Engineering to the Study of Cellular Pathology*, edited by E. Kohen, 1999, pp. 133–149.

Levenson, R. M., and D. L. Farkas. Digital spectral imaging for histopathology and cytopathology. *Proc. SPIE* 2983:123–135, 1997.

Lewis, E. N., P. J. Treado, R. C. Reeder, G. M. Story, A. E. Dowrey, C. Marcott, and I. W. Levin. Fourier transform spectroscopic imaging using an infrared focal-plane array detector. *Anal. Chem.* 67:3377–3381, 1995.

Malik, Z., D. Cabib, R. A. Buckwald, A. Talmi, Y. Garini, and S. G. Lipson. Fourier transform multipixel spectroscopy for quantitative cytology. *J. Microsc.* 182:133–140, 1996a.

Malik, Z., M. Dishi, and Y. Garini. Fourier transform multipixel spectroscopy and spectral imaging of protoporphyrin in single melanoma cells. *Photochem. Photobiol.* 63:608–614, 1996b.

Masters B. R., P. T. So, and E. Gratton. Multiphoton excitation fluorescence microscopy and spectroscopy of in vivo human skin. *Biophys. J.* 72:2405–2412, 1997.

Moreton, R. B. Optical methods for imaging ionic activities. *Scanning Microsc. Suppl.* 8:371–390, 1994.

Morrison, L. Spectral overlap corrections in fluorescence in situ hybridizations employing high fluorophore densities. *Cytometry Suppl.* 9:CT108, 1998.

Papanicolaou, G. N. A new procedure for staining vaginal smears. *Science* 95:438–439, 1942.

Plymale, D. R., J. R. Haskins, and F. A. de la Iglesia. Monitoring simultaneous subcellular events in vitro by means of coherent multiprobe fluorescence. *Nat. Med.* 5:351–355, 1999.

Reid, T., A. Baldini, T. C. Rand, and D. C. Ward. Simultaneous visualization of seven different DNA probes by in situ hybridization using combinatorial fluorescence and digital imaging microscopy. *Proc. Natl. Acad. Sci. U.S.A.* 89:1388–1392, 1992.

Replogle-Schwab, R., K. J. Pienta, and R. H. Getzenberg. The utilization of nuclear matrix proteins for cancer diagnosis. *Crit. Rev. Eukaryot. Gene Expr.* 6:103–113, 1996.

Rothmann, C., A. M. Cohen, and Z. Malik. Chromatin condensation in erythropoiesis resolved by multipixel spectral imaging: Differentiation versus apoptosis. *J. Histochem. Cytochem.* 45:1097–1108, 1997.

Schröck, E., S. du Manoir, T. Veldman, B. Schoell, J. Wienberg, M. A. Ferguson-Smith, Y. Ning, D. H. Ledbetter, I. Bar-Am, D. Soenksen, Y. Garini, and T. Ried. Multicolor spectral karyotyping of human chromosomes. *Science* 273:494–497, 1996.

Schröck, E., T. Veldman, H. Padilla-Nash, Y. Ning, J. Spurbeck, S. Jalal, L. G. Shaffer, P. Papenhausen, C. Kozma, M. C. Phelan, E. Kjeldsen, S. A. Schonberg, P. O'Brien, L. Biesecker, S. du Manoir, and T. Ried. Spectral karyotyping refines cytogenetic diagnostics of constitutional chromosomal abnormalities. *Hum. Genet.* 101:255–262, 1997.

Shackney, S. E., A. A. Pollice, C. A. Smith, L. Alston, S. G. Singh, L. E. Janocko, K. A. Brown, S. Petruolo, D. W. Groft, R. Yakulis, and R. J. Hartsock. The accumulation of multiple genetic abnormalities in individual tumor cells in human breast cancers: Clinical prognostic implications. *Cancer J. Sci. Am.* 2:106–113, 1996.

Shonat, R. D., E. S. Wachman, W. Niu, A. P. Koretsky, and D. L. Farkas. Near-simultaneous hemoglobin saturation and oxygen tension maps in mouse brain using an AOTF microscope. *Biophys. J.* 73:1223–1231, 1997.

Shonat, R. D., E. S. Wachman, W. Niu, A. P. Koretsky, and D. L. Farkas. Near-simultaneous hemoglobin saturation and oxygen tension maps in mouse cortex during amphetamine stimulation. *Adv. Exp. Biol. Med.* XX:149–158, 1998.

Speicher, M. R., S. Gwyn Ballard, and D. C. Ward. Karyotyping human chromosomes by combinatorial multi-fluor FISH. *Nat. Gen.* 12:368–375, 1996.

Swain, P. H. and S. M. Davis. *Remote Sensing: The Quantitative Approach.* New York: McGraw Hill, 1978.

Taylor, C. R. The current role of immunohistochemistry in diagnostic pathology. *Adv. Pathol. Lab. Med.* 7:59–105, 1994.

Taylor, D. L., K. Burton, R. L. DeBiasio, K. A. Giuliano, A. H. Gough, T. Leonardo, J. A. Pollock, and D. L. Farkas. Automated light microscopy for the study of the brain:

Cellular and molecular dynamics, development and tumorigenesis. *Ann. N.Y. Acad. Sci.* 820:208–228, 1997.

Taylor, C. R., and R. J. Cote. Immunohistochemical markers of prognostic value in surgical pathology. *Histol. Histopathol.* 12:1039–1055, 1997.

Taylor, D. L., R. DeBiasio, G. LaRocca, D. Pane, P. Post, J. Kolega, K. Giuliano, K. Burton, A. Gough, A. Dow, H. Yu, A. Waggoner, and D. L. Farkas. Potential of machine-vision light microscopy in toxicologic pathology. *Toxicol. Pathol.* 22:145–159, 1994.

Tomatis, S., C. Bartoli, A. Bono, N. Cascinelli, C. Clemente, and R. Marchesini. Spectrophotometric imaging of cutaneous pigmented lesions: Discriminant analysis, optical properties and histological characteristics. *J. Photochem. Photobiol. B.* 42:32–39, 1998.

Wachman E. S., D. L. Farkas, and W. Niu. Submicron Imaging System Having an Acousto-Optic Tunable Filter. U.S. Patent 5,796,512 (24 claims), 1998b.

Wachman E. S., D. L. Farkas, and W. Niu. Light Microscope Having Acousto-Optic Tunable Filters. U.S. Patent 5,841,577 (42 claims), 1998b.

Wachman, E. S., W. H. Niu, and D. L. Farkas. Imaging acousto-optic tunable filter with 0.35-micrometer spatial resolution. *Appl. Opt.* 35:5520–5526, 1996.

Wachman, E. S., W. Niu, and D. L. Farkas. AOTF microscope for imaging with increased speed and spectral versatility. *Biophys. J.* 73:1215–1222, 1997.

Warren, D. W., and J. Hackwell. A compact prism spectrograph for broadband infrared spectral surveys with array detectors. *Proc. SPIE* 1055:314–321, 1989.

Yang, P., D. L. Farkas, J. M. Kirkwood, J. L. Abernethy, H. D. Edington, and D. Becker. Macroscopic spectral imaging and gene expression analysis of the early stages of melanoma. *Mol. Med.* 5:785–794, 1999.

Zonios, G., R. Cothren, J. M. Crawford, M. Fitzmaurice, R. Manoharan, J. Van Dam, and M. S. Feld. Spectral pathology. *Ann. N.Y. Acad. Sci.* 838:108–115, 1998.

21

Total Internal Reflection Fluorescence Microscopy

Daniel Axelrod

1. Introduction

Total internal reflection fluorescence (TIRF) microscopy provides a means to selectively excite fluorophores in an aqueous or cellular environment that is very near a solid surface (within ≤100 nm: less than one-fifth the thickness of a confocal microscopy section) without exciting fluorescence from regions further from the surface. Fluorescence excitation by this thin zone of electromagnetic energy (called an *evanescent field*) results in images with very low background fluorescence, virtually no out-of-focus fluorescence, and minimal exposure of cells to light in any other planes in the sample. Figure 21–1 shows an example of TIRF on intact living cells in a culture compared with standard epifluorescence. The above features have led to numerous applications, including the following:

1. Selective visualization of cell–substrate contact regions: TIRF can be used qualitatively to observe the position, extent, composition, and motion of these contact regions, even in samples in which fluorescence elsewhere would normally obscure the fluorescent pattern in the contact regions. A variation of TIRF, used to identify cell–substrate contacts, involves doping the solution surrounding the cells with a nonadsorbing and nonpermeable fluorescent volume marker; focal contacts then appear relatively dark (Gingell et al., 1987; Todd et al., 1988).
2. Visualization and spectroscopy of single-molecule fluorescence near a surface (Vale et al., 1996; Ha et al., 1999; Dickson et al., 1996, 1998; Sako et al., 2000): The purpose here is to observe the properties of individual molecules without the ensemble averaging inherent in standard spectroscopies on bulk materials and thereby detect kinetic features and states that otherwise would be obscured. Related to single-molecule detection is the capability of seeing fluorescence fluctuations as fluorescent molecules enter and leave the thin evanescent field region in the bulk. These fluctuations (visually obvious in TIRF) can be quantitatively autocorrelated to obtain kinetic information about the molecular motion ("fluorescence correlation spectroscopy").

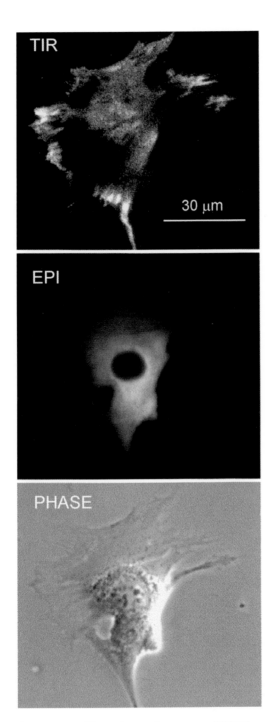

FIGURE 21–1 Two views of a mouse BC3H1 smooth muscle cell microinjected with rhodamine dextran: total internal reflection fluorescence (TIR; using configuration E shown in Fig. 21–6); and epi-illumination (EPI; with a much shorter exposure time than the TIR because of the much larger quantity of rhodamine dextran through the bulk of the cell). The objective was a Zeiss ×40 water immersion, NA = 0.75 achromat, used on a Leitz Diavert microscope. The images were recorded on a 576 × 384 pixel slow-scan cooled charge-coupled device camera (Photometrics Star-1).

3. Tracking of secretory granules in intact cells before and during the secretory process: The thin evanescent field allows small intensity changes to be interpreted as small motions of granules in the direction normal to the substrate with accuracy as small as 2 nm, much smaller than the light microscope resolution limit (Lang et al., 1997; Steyer and Almers, 1999; Oheim et al., 1998, 1999; Han et al., 1999; Schmoranzer et al., 2000).

4. Measurements of the kinetic rates of binding of extracellular and intracellular proteins to cell surface receptors and artificial membranes (Hellen and Axelrod, 1991; McKiernan et al., 1997; Lagerholm et al., 2000; Fulbright and Axelrod, 1993; Sund et al., 1999): Some of these studies combine TIRF with fluorescence recovery after photobleaching (FRAP). TIRF–FRAP can be used additionally to measure lateral surface diffusion coefficients along with the on/off kinetics of reversibly adsorbed fluorescent molecules.

5. Visualization of membrane micromorphological structures and dynamics of living cells: By utilizing the unique polarization properties of the evanescent field of TIRF, endocytotic or exocytotic sites, ruffles, and other submicroscopic irregularities can be highlighted (Sund et al., 1999).

6. High-contrast visualization of submembrane cytoskeletal structure on thick cells: Although TIRF cannot view deeply into thick cells, it can display the fluorescence-marked submembrane filament structure with high contrast in thick cells without distortion from out-of-focus planes.

7. Long-term fluorescence movies of cells during development in culture (Wang and Axelrod, 1994): Because the cells are exposed to excitation light only at their cell–substrate contact regions and not through their bulk, they tend to survive longer under observation, thereby enabling time-lapse recording of a week in duration. During this time, newly appearing cell surface receptors can be immediately marked by fluorescent ligands that are continually present in the full cell culture medium while maintaining a low fluorescence background.

8. Comparison of membrane-proximal ionic transients with simultaneous transients deeper in the cytoplasm (Omann and Axelrod, 1996): Because TIRF is completely compatible with standard epifluorescence, brightfield, dark-field, or phase contrast illumination, these methods of illumination can be switched back and forth rapidly by electro-optic devices.

2. THEORY OF TIRF

The thin layer of illumination is an "evanescent field" that exponentially decays in intensity with increasing distance normal to the surface. The evanescent field is produced by an excitation light beam traveling in a solid (e.g., a glass coverslip or tissue culture plastic) incident at a high angle θ upon the solid–liquid surface at which the sample (e.g., single molecules or cells) adhere. That angle θ, measured from the normal, must be large enough for the beam to totally internally reflect (TIR) rather than refract through the interface, a condition that occurs above some "critical angle." TIR generates a very thin electromagnetic field in the liquid with the same frequency as the incident light, exponentially decaying in intensity with distance from the surface. This field is capable of exciting fluorophores near the surface, while avoiding excitation of a much larger number of fluorophores further out in the liquid.

2.1 Infinite Plane Waves

The simplest case of TIR is that of an "infinitely" extended plane wave incident upon a single interface. When a light beam propagating through a transparent medium 3 of high index of refraction (e.g., glass) encounters a planar interface with medium 1 of lower index of refraction (e.g., water), it undergoes TIR for incidence angles (measured from the normal to the interface) greater than the "critical angle." The critical angle θ_c for TIR is given by

$$\theta_c = \sin^{-1}\left(\frac{n_1}{n_3}\right) = \sin^{-1} n \tag{1}$$

where n_1 and n_3 are the refractive indices of the liquid and the solid, respectively, and $n = n_1/n_3$. Ratio n must be less than unity for TIR to occur. (A refractive index n_2 will correspond to an optional intermediate layer, to be discussed below.)

For incidence angle $\theta < \theta_c$, most of the light propagates through the interface with a refraction angle (also measured from the normal) given by Snell's law. (Some of the incident light internally reflects back into the solid.) For $\theta > \theta_c$, all of the light reflects back into the solid. Even with TIR, however, some of the incident energy penetrates through the interface and propagates parallel to the surface in the plane of incidence. The field in the liquid, called the *evanescent field* (or *wave*), is capable of exciting fluorescent molecules that might be present near the surface.

For an infinitely wide beam (i.e., a beam width many times the wavelength of the light, which is a good approximation for unfocused or weakly focused light), the intensity of the evanescent wave (measured in units of energy/area/s) exponentially decays with perpendicular distance z from the interface:

$$I(z) = I(0)e^{-z/d} \tag{2}$$

where

$$d = \frac{\lambda_0}{4\pi}(n_3^2 \sin^2\theta - n_1^2)^{-1/2} \tag{3}$$

λ_0 is the wavelength of the incident light in vacuum. Depth d is independent of the polarization of the incident light and decreases with increasing θ. Except for $\theta \to \theta_c$ (where $d \to \infty$), d is in the order of λ_0 or smaller. A physical picture of refraction at an interface shows TIR to be part of a continuum rather than a sudden new phenomenon appearing at $\theta = \theta_c$. For small θ, the refracted light waves in the liquid are sinusoidal, with a certain characteristic period noted as one moves normally away from the surface. As θ approaches θ_c, that period becomes longer as the refracted rays propagate increasingly parallel to the surface. At exactly $\theta = \theta_c$, that period is infinite because the wavefronts of the refracted light are normal to the surface themselves. This situation corresponds to $d = \infty$. As θ increases beyond θ_c, the period becomes mathematically imaginary. Physically, this corresponds to the exponential decay of Equation 2.

The intensity of the evanescent field at any position is the squared amplitude of the complex electric field vector at that position:

$$I(z) = \mathbf{E}(z) \cdot \mathbf{E}^*(z) \tag{4}$$

The intensity as defined here is proportional to the probability rate of energy absorption by a fluorophore in the evanescent wave. The polarization (i.e., the vector direction) of the electric field of the evanescent wave depends on the incident light polarization, which can be either "s" (polarized normal to the plane of incidence formed by the incident and reflected rays) or "p" (polarized in the plane of incidence).

For s-polarized incident light, the evanescent electric field vector direction remains purely normal to the plane of incidence. For p-polarized incident light, the evanescent electric field vector direction remains in the plane of incidence, but it "cartwheels" along the surface with a nonzero longitudinal component (Fig. 21–2). This feature distinguishes evanescent light from freely propagating subcritical refracted light, which has no longitudinal component. The longitudinal component

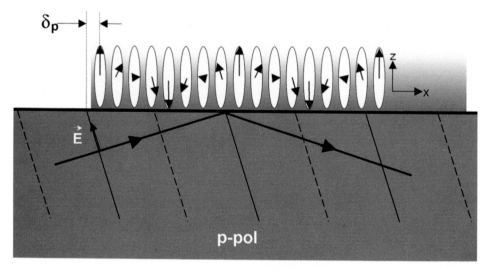

FIGURE 21–2 Schematic drawing of the evanescent polarization resulting from p-polarized (p-pol) (in the plane of incidence) incident light. The incident light wavefronts (with the intervals from solid to dashed wavefront lines representing one-half of a wavelength in the glass) define the spacing of the spatial period along the interface, and reflected wavefronts are not shown. The p-pol evanescent field is elliptically polarized in the x–z plane (primarily z-pol with a weaker x-component at a relative phase of $\pi/2$). For pictorial clarity, only two periods of evanescent electric field oscillation are shown; in reality, the evanescent region is much more extended and contains many more periods of oscillation in the x-direction. The exact phase relationship between the incident field and the evanescent field is a function of incidence angle and is represented here by δ_p (see Eq. 6).

approaches zero as the incidence angle is reduced from the supercritical range back toward the critical angle.

All these properties are implicit in the complete mathematical expressions for the amplitude and phase of the evanescent electric field:

$$\mathbf{E}_s(z) = 2 \cos\theta(1 - n^2)^{-1/2} e^{-i\delta_s}e^{-z/2d} \,\hat{\mathbf{y}}$$
$$\mathbf{E}_p(z) = 2 \cos\theta(n^4 \cos^2\theta + \sin^2\theta - n^2)^{-1/2}e^{-i\delta_p}e^{-z/2d}[-i(\sin^2\theta - n^2)^{1/2}\hat{\mathbf{x}} + \sin\theta\,\hat{\mathbf{z}}]$$
$$(5)$$

where the plane of incidence is the x–z plane, the incident electric field amplitude in the substrate is normalized to unity for each polarization, and the phase lags relative to the incident light are:

$$\delta_p = \tan^{-1}\left[\frac{(\sin^2\theta - n^2)^{1/2}}{n^2\cos\theta}\right]$$

$$\delta_s = \tan^{-1}\left[\frac{(\sin^2\theta - n^2)^{1/2}}{\cos\theta}\right] \tag{6}$$

A p-polarized evanescent field can be uniquely utilized to highlight submicroscopic irregularities in the plasma membrane of carbocyanine dye–labeled living cells (Sund et al., 1999), as shown schematically in Figure 21–3.

The corresponding evanescent intensities in the two polarizations (assuming incident intensities normalized to unity) are

$$I_p(z) = \frac{(4 \cos^2\theta)(2 \sin^2\theta - n^2)}{n^4 \cos^2\theta + \sin^2\theta - n^2}e^{-z/d}$$

$$I_s(z) = \frac{(4 \cos^2\theta)}{1 - n^2}e^{-z/d} \tag{7}$$

FIGURE 21–3 Schematic drawing of the excitation probability of oriented carbocyanine fluorophores embedded in a membrane in a z-polarized evanescent field (the dominant direction of a p-polarized evanescent field). The membrane is depicted in cross section with a curved region corresponding to a bleb or an invagination. The direction of the absorption dipole of the fluorophores is known to be parallel to the local plane of the membrane and free to rotate in it (Axelrod, 1979) and is shown with bidirectional arrows. Higher excitation probability is depicted by lighter shades. The z-component of the electric field selectively excites the regions of oblique membrane orientation.

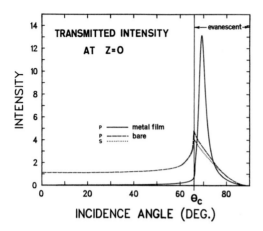

Figure 21–4 Evanescent intensities $I_{p,s}$ at $z = 0$ versus θ, assuming that the incident intensities in the glass are set equal to unity. At angles $\theta > \theta_c$, the transmitted light is evanescent; at angles $\theta < \theta_c$, it is propagating. Both s- and p-polarizations are shown. Refractive indices $n_3 = 1.46$ (fused silica) and $n_1 = 1.33$ are assumed here, corresponding to $\theta_c = 65.70$. Also shown is the evanescent intensity that would be obtained with a thin (20 nm) aluminum film coating.

Intensities $I_{p,s}(0)$ are plotted versus θ in Figure 21–4. The evanescent intensity approaches zero as $\theta \rightarrow 90°$. On the other hand, for supercritical angles within 10° of θ_c, the evanescent intensity is as great or greater than the incident light intensity. The plots can be extended without breaks to the subcritical angle range, where the intensity is that of the freely propagating refracted light in medium 1. One might at first expect the subcritical intensity to be slightly *less* than the incident intensity (accounting for some reflection at the interface), but certainly not *more* as shown. The discrepancy arises because the intensity in Figure 21–4 refers to EE^* alone rather than to the actual energy flux of the light, which involves a product of EE^* with the refractive index of the medium in which the light propagates.

Regardless of polarization, the spatial period of the evanescent electric field is $\lambda_0/(n_3 \sin \theta)$ as it propagates along the surface. Unlike the case of freely propagating light, the evanescent spatial period is not at all affected by the medium 1 in which it resides. It is determined only by the spacing of the incident light wavefronts in medium 3 as they intersect the interface. This spacing can be experimentally important because it determines the spacing of interference fringes produced when two coherent TIR beams are made to intersect at the surface (Fulbright and Axelrod, 1993).

2.2 Finite-Width Incident Beams

For a finite-width beam, the evanescent wave can be pictured as the beam's partial emergence from the solid into the liquid, travel for some finite distance along the surface, and then re-entrance into the solid. The distance of propagation along the surface is measurable for a finite-width beam and is called the *Goos-Hanchen shift*. The Goos-Hanchen shift ranges from a fraction of a wavelength at $\theta = 90°$ to infinite at $\theta = \theta_c$, which corresponds to the refracted beam skimming along the interface. A finite beam can be expressed as an integral of infinite plane waves approaching at a range of incidence angles. In general, the intensity profile of the finite beam evanescent field can be calculated from the mathematical form for the evanescent wave at each infinite plane wave incidence angle, integrated over all the constituent incident plane wave angles.

For a TIR Gaussian laser beam focused with a narrow angle of convergence, the experimentally observed evanescent illumination is approximately an elliptical Gaussian profile, and the polarization and penetration depth are approximately equal to those of a single infinite plane wave. If the angle of convergence is greater and the mean angle is within a few degrees of the critical angle, however, the evanescent field tends to become a long thin stripe.

2.3 Intermediate Layers

In actual experiments in biophysics, the interface may not be a simple interface between two media, but rather a stratified multilayer system. One example is the case of a biological membrane or lipid bilayer interposed between glass and an aqueous medium. Another example is a thin metal film coating, which can be used to quench fluorescence within the first approximately 10 nm of the surface. We discuss here the TIR evanescent wave in a three-layer system in which incident light travels from medium 3 (refractive index n_3) through the intermediate layer (n_2) toward medium 1 (n_1). Qualitatively, several features can be noted:

1. Insertion of an intermediate layer never thwarts TIR, regardless of the intermediate layer's refractive index n_2. The only question is whether TIR takes place at the n_3:n_2 interface or the n_2:n_1 interface. Because the intermediate layer is likely to be very thin (no deeper than several tens of nanometers) in many applications, precisely which interface supports TIR is not important for qualitative studies.
2. Regardless of n_2 and the thickness of the intermediate layer, the evanescent wave's profile in medium 1 will exponentially decay with a characteristic decay distance given by Equation 3. The overall distance of penetration of the field measured from the surface of medium 3 is, however, affected by the intermediate layer.
3. Irregularities in the intermediate layer can cause scattering of incident light, which then propagates in all directions in medium 1. Experimentally, scattering appears not be a problem on samples even as inhomogeneous as biological cells. Direct viewing of incident light scattered by a cell surface lying between the glass substrate and an aqueous medium confirms that scattering is many orders of magnitude dimmer than the incident or evanescent intensity and will thereby excite a correspondingly dim contribution to the fluorescence.

A particularly interesting kind of intermediate layer is a metal film. Classic electromagnetic theory (Hellen and Axelrod, 1987) shows that such a film will reduce the s-polarized evanescent intensity to nearly zero at all incidence angles.

The p-polarized behavior is, however, quite different. At a certain sharply defined angle of incidence θ_p ("the surface plasmon angle"), the p-polarized evanescent intensity becomes an order of magnitude brighter than the incident light at the peak (see Fig. 21–4). This strongly peaked effect is due to a resonant excitation of electron oscillations at the metal–water interface. For an aluminum film at a glass–water interface, θ_p is greater than the critical angle θ_c for TIR. The intensity enhancement is rather remarkable because a 20 nm thick metal film is almost opaque to the eye.

There are some potentially useful experimental consequences of TIR excitation through a thin metal film coated on glass:

1. The metal film will almost totally quench fluorescence within the first 10 nm of the surface, and the quenching effect is virtually gone at a distance of 100 nm. Therefore, TIR with a metal film–coated glass can be used to selectively excite fluorophores in the 10–200 nm distance range.
2. A light beam incident upon a 20 nm aluminum film from the glass side at a glass–aluminum and film–water interface does not have to be collimated to produce TIR. Those rays that are incident at the surface plasma angle will create a strong evanescent wave; those rays that are too low or high in incidence angle will create a negligible field in the water. This phenomenon may ease the practical requirement for a collimated incident beam in TIR and make it easier to set up TIR with a conventional arc light source.
3. The metal film leads to a highly polarized evanescent wave, regardless of the purity of the incident polarization.

3. OPTICAL CONFIGURATIONS

A wide variety of optical arrangements for TIRF have been employed. In general, an inverted microscope is most convenient because it provides more room to add TIR optics above rather than below the stage. However, some upright microscope configurations are still very workable. Most configurations use an added prism to direct the light toward the TIR interface, but it is also possible to use a high numerical aperture (NA > 1.4) microscope objective for this purpose. This section gives examples of these arrangements, partly as a guide to simple systems that work and partly as a basis for creative variations. For concreteness in the descriptions, we assume that the sample consists of fluorescence-labeled cells in culture adhered to a glass coverslip (Axelrod, 1979).

In all cases, the critical angle must be considered in the design. The index of refraction of the standard glass coverslip upon which cells are grown is about $n_1 = 1.52$. The index of refraction of the intact cell interior can be as high as $n_3 = 1.38$. Therefore, to obtain TIR at this interface, the angle of incidence must be larger than the critical angle of 65°. If the cells are not intact (e.g., permeabilized, hemolyzed, or fixed) so that the lower refractive index is that of aqueous buffer ($n_1 = 1.33$) instead of cytoplasm, then the critical incidence angle is 61°.

3.1 "Prismless" TIR Through a High Aperture Objective

By using an objective with a sufficiently high NA, supercritical angle incident light can be cast upon the sample by epi-illumination through the objective. The incident beam must be constrained to pass through the periphery of the objective's pupil and must emerge with only a narrow spread of angles. This can be accomplished by setting the incident beam to be focused off-axis at the objective's back focal plane. The further the beam focus is positioned radially off-axis, the larger the angle that the beam will emerge from the objective. It emerges into the immersion oil (refractive index n_3) at a maximum angle θ (measured from the optical axis) given by

$$\mathrm{NA} = n_3 \sin \theta \tag{6}$$

Because $n_i \sin \theta_i$ is conserved (by Snell's law) in each material i as the beam traverses into the solid substrate supporting the sample, the right side of Equation 6 can also refer to the refractive index and angle in the substrate. For total internal reflection to take place at the sample surface, θ must be greater than the critical angle θ_c given by

$$n_1 = n_3 \sin \theta_c \tag{7}$$

where n_1 is the refractive index of the medium in which the sample resides. From Equations 6 and 7, it is evident that the NA must be greater than n_1, preferably by a substantial margin. This is no problem for an interface with water with $n_1 = 1.33$ and an NA = 1.4 objective. For viewing the inside of a cell at $n_1 = 1.38$, however, an NA = 1.4 objective will produce TIR at barely above the critical angle. The evanescent field in this case will be quite deep, and dense nonhomogeneities in the sample (such as cellular organelles) will convert some of the evanescent field into scattered propagating light.

Fortunately, objectives are now available or planned that have an NA of more than 1.4. One such objective is an Olympus ×100 NA = 1.65; this works very well for through-the-lens TIR on living cells. The drawbacks are that the objective (1) requires expensive $n_i = 1.78$ coverslips made of either LAFN21 or SF11 glass (custom cut by VA Optical Co, San Anselmo, CA); (2) requires special $n_i = 1.78$ oil, which is volatile and leaves a crystalline residue; and (3) is expensive. A newer objective circumvents all these problems: a ×60 NA = 1.45 (available from Olympus and Zeiss). It uses standard glass ($n_i = 1.52$) coverslips and oil, and yet has an aperture adequate for TIR on cells (see the example in Fig. 21–5). The 1.45 objective is probably the method of choice for TIR (Axelrod, 2001).

The advantages of prismless inverted TIR include the possibility of viewing a sample that is completely accessible from the top and the compatibility (in fact necessity) for using the highest aperture, highest resolution, brightest objectives available. The arrangement is also easily compatible with intersecting beams to produce interference fringes: The high mechanical stability enables one to achieve interfringe spacings of about 0.3 μm without the blurring effects of small vibrations. Such fringes can be useful in studies of surface diffusion by combing TIR with fluorescence recovery after photobleaching (e.g., Fulbright and Axelrod, 1993).

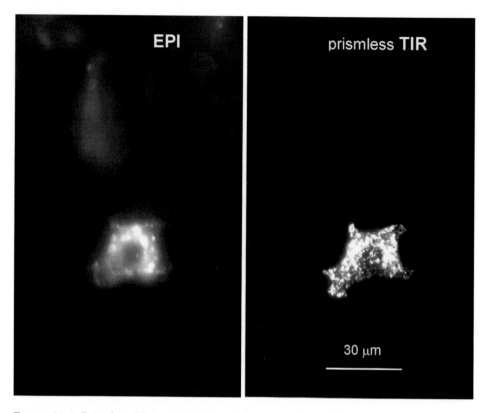

FIGURE 21–5 Prismless TIRF and EPI laser illumination (λ = 488 nm) through an Olympus ×60 NA = 1.45 objective of a bovine chromaffin cell marked with GFP coupled to the secretory granule–specific atrial natriuretic protein. The two views are obtained simply by adjusting the off-axis radial position of the laser beam focus at the objective's back focal plane. Note that TIRF produces a much higher contrast view of membrane-proximal secretory granules.

Two possible arrangements for prismless TIR are shown in Figure 21–6: one with a laser illumination and one that utilizes a conventional arc source rather than a laser beam. In the case of the laser beam, the illumination can be continuously switched from standard epi to TIR simply by increasing the off-axis position of the beam focus at the objective's back focal plane. Once TIR is achieved, further increases in the off-axis position serve to increase the incidence angle at the sample and thereby decrease the depth of the evanescent field. In the case of arc-lamp illumination, switching back and forth between epi and TIR is done by placing or removing the opaque disk, as shown. The size of the TIRF area on the sample is directly proportional to the angle of convergence of the rays at the back focal plane. In the case of TIR illumination, this is easily increased by expanding the beam width at the focusing lens just before the beam enters the microscope.

A disadvantage of prismless TIR is that the evanescent illumination is not pure: a small fraction of the illumination of the sample results from excitation light scattered within the objective, and a small fraction of the observed fluorescence arises from luminescence of the objective's internal elements.

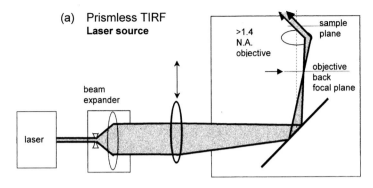

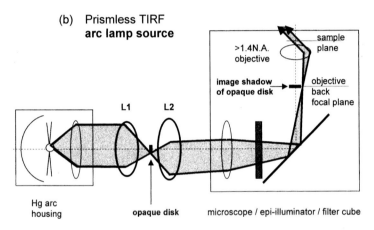

FIGURE 21–6 Two arrangements for prismless TIRF in an inverted microscope. *a*, Laser illumination. The beam is focused at the back focal plane at a radial position sufficient to lead to supercritical angle propagation into the coverslip. *b*, Arc lamp illumination. The goal is to produce a sharp-edged shadow image of an opaque circular disk at the objective back focal plane such that only supercritical light passes through the objective. The actual physical opaque disk must be positioned at an equivalent upbeam optical plane, which, in Köhler illumination, must also contain a real image of the arc. This goal is accomplished by creating such a plane in the arc light path with an extra lens L1, along with a following lens L2 to refocus that plane in the objective. The illumination at the back focal plane is a circular annulus; it is shown as a point on one side of the optical axis for pictorial clarity only.

3.2 TIRF with a Prism

Although a prism may restrict sample accessibility or choice of objectives in some cases, prism-based TIR is very inexpensive to set up and produces a "cleaner" evanescent-excited fluorescence than prismless TIR. Figure 21–7 shows several schematic drawings for setting up laser/prism-based TIR in an inverted microscope. In configurations A–D (Fig. 21–7A–D), the buffer-filled sample chamber consists of a lower bare glass coverslip, a spacer ring (often made of 60 μm thick Teflon), and an inverted coverslip so that the cells face down. The upper surface of the cell coverslip is put in optical contact with the prism lowered from above

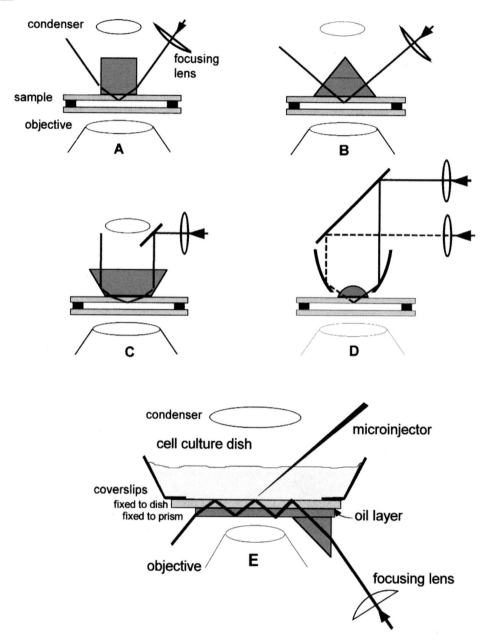

FIGURE 21–7 Schematic drawings for prism-based TIR in an inverted microscope, all using a laser as a light source. The vertical distances are exaggerated for clarity. The first four configurations (*A–D*) use a TIR prism above the sample; the fifth (*E*) places the prism below the sample. Details are provide in the text.

by a layer of immersion oil or glycerol. The lateral position of the prism is fixed, but the sample can be translated while still maintaining optical contact. The lower coverslip can be oversized, and the Teflon spacer can be cut with gaps so that solutions can be changed by capillary action with entrance and exit ports. Alternatively, commercially available solution-changing cell chambers (e.g., Sykes-Moore chamber from Bellco Glass Co. or rectangular cross-sectional microcapillary tubes from Wilmad Glass Co.) can be used. Configuration E (Fig. 21–7E) allows complete access to the sample from above for solution changing and/or electrophysiology studies.

In configuration A, the square prism allows simultaneous use of transmitted light illumination for phase contrast and dark field. In configuration B, custom truncation and polishing of the top of the triangular prism (shown as an option with dashed lines in Fig. 21–7B) also provides compatibility with phase contrast and dark field. In configuration C, with its 60° trapezoidal prism (inexpensively manufactured by truncating and polishing the apex of a commercially available triangular prism), the incoming beam is vertical so that the TIR spot shifts laterally very little when the prism is raised and re-lowered during changes of sample. To achieve TIR with an incidence angle of only 60°, the prism *must* be made of higher than standard refractive index material (e.g., flint glass at $n_i = 1.64$ available from Rolyn Optics). In configuration D, the parabolic mirror and hemispherical prism are positioned so that the beam traverses a radius of the prism toward a TIR spot at the focus of the parabola. In this manner, a lateral shift of the vertical incoming beam will always focus at the same spot. Substantial changes of incident angle can thereby be accessed quite conveniently. In addition, interference fringes in the TIR evanescent field can be created by splitting the incoming beam into two beams (as indicated by the additional dashed beam); each reflecting at different azimuthal positions in the parabola, but recombining at the same parabola focus (Fulbright and Axelrod, 1993). The spacing of the fringes can be adjusted by varying the relative azimuthal positions of the two beams.

Configuration E utilizes a prism below the microscope stage and multiple TIR reflections in the substrate coverslip (which acts as a waveguide). This configuration is completely open from above and is compatible with simultaneous microinjection or microelectrophysiology. In the simplest form, a small commercially available triangular prism is placed in optical contact (via oil) with the bottom of the cell-containing coverslip. The sample can be translated while the prism remains laterally fixed. Only water (or air) immersion objectives are compatible with the multiple internal reflections at the lower surface of the coverslip in this configuration.

Figure 21–8 shows an exceptionally convenient (and low-cost) prism-based TIRF setup for an upright microscope. The laser beam is introduced in the same port in the microscope base that is intended for the transmitted light illuminator (which should be removed), thereby utilizing the microscope's own in-base optics to direct the beam vertically upward. An extra lens just upbeam from the microscope base weakly focuses the TIR spot and permits adjustment of its position. This system gives particularly high-quality images if a water immersion objective is employed and submerged directly into the buffer solution in an uncovered cell chamber. This system is also easily used with cells adhering directly on tissue culture plastic dishes rather than on coverslips; the plastic–cell interface is then the site of TIR.

If the objective has a long enough working distance, reasonable accessibility to micropipettes is possible. In this configuration, flexibility in incidence angles (to obtain a range of evanescent field depths) is sacrificed in exchange for convenience. A set of various-angled trapezoids will, however, allow one to

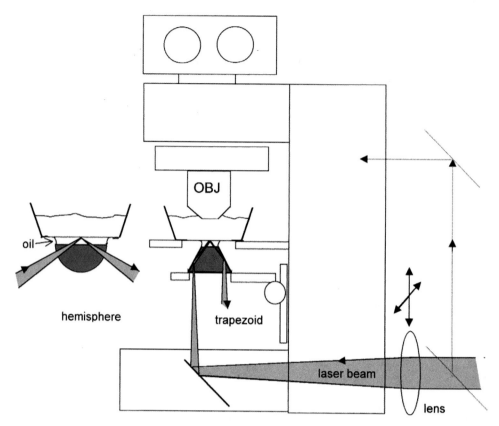

FIGURE 21–8 TIRF for an upright microscope utilizing the integral optics in the microscope base and a trapezoidal prism on the condenser mount and movable up and down. The position of the beam is adjustable by moving the external lens. An alternative hemispheric prism configuration for variable incidence angle is also indicated to the left. Vertical distances are exaggerated for clarity. An extra set of mirrors can be installed to deflect the beam into an epi-illumination light path (shown with dashed lines).

employ various discrete incidence angles. In an alternative approach for varying incidence angles over a continuous range, a hemispherical prism can be substituted for the trapezoidal prism (Loerke et al., 2000). The incident laser beam is directed along a radius line at an angle set by external optical elements.

In all these prism-based methods, choice of optical materials is somewhat flexible, as follows:

1. The prism used to couple the light into the system and the (usually disposable) slide or coverslip in which TIR takes place need not be matched exactly in refractive index.
2. The prism and slide may be optically coupled with glycerol, cyclohexanol, or microscope immersion oil, among other liquids. Immersion oil has a higher refractive index (thereby avoiding possible TIR at the prism–coupling liquid interface at low incidence angles), but it tends to be more autofluorescent (even the "extremely low" fluorescence types).

3. The prism and slide can both be made of ordinary optical glass for many applications, unless shorter penetration depths arising from higher refractive indices are desired.

Optical glass does not transmit light below about 310 nm and also has a dim autoluminescence with a long (several hundred microsecond) decay time, which can be a problem in some fluorescence recovery after photobleaching (FRAP) experiments. The autoluminescence of high-quality fused silica (often called "quartz") is much lower. Tissue culture dish plastic (particularly convenient as a substrate in the upright microscope setup) is also suitable, but tends to have a significant autofluorescence compared with ordinary glass. Different brands of tissue culture plastic have significantly different amounts of autofluorescence (Corning brand is one of the least fluorescent). More exotic high n_3 materials, such as sapphire, titanium dioxide, and strontium titanate, can yield exponential decay depths (d) as low as $\lambda_0/20$.

In all the prism-based methods, the TIRF spot should be focused to a width no larger than the field of view. The larger the spot, the more that spurious scattering and out-of-focus fluorescence from the immersion oil layer between the prism and coverslip will increase the (generally) low fluorescence background attainable by TIRF. Also, the incidence angle should exceed the critical angle by at least a couple of degrees. At incidence angles very near the critical angle, the cells cast a noticeable "shadow" along the surface.

4. General Experimental Considerations

4.1 Light Source

Virtually any laser with a total visible output in or above the 0.5 W range should be adequate. The most popular laser for cell biology work with a microscope appears to be a 3 W continuous wave argon laser. Air-cooled argon or diode lasers of less than 100 mW are probably marginal in power, especially for dim samples. Laser illumination produces interference fringes, which are manifested as intensity variations over the sample area. For critical applications, it may be advisable to rapidly jiggle the beam (e.g., in a commercially available optical fiber phase scrambler) or to compute a normalization of sample digital images against a control digital image of a uniform concentration of fluorophores. Image reconstruction by scanning a focused TIR spot in a raster pattern, an approach that might both permit a lower laser power and avoid interference fringes, is not straightforward. A small spot requires a large convergence angle of the focused beam directed toward the TIR interface. Part of this large angular range may overlap into subcritical angle incidence and thereby result in some propagating refracted light. Arc illumination is feasible with prismless TIR and has the advantages of easy selection of excitation colors with filters and freedom from coherent light interference fringes. It is somewhat dimmer, however, because much of the arc-lamp power directed toward the sample at subcritical angles is necessarily blocked.

4.2 Functionalized Substrates

TIRF experiments often involve specially coated substrates. A glass surface can be chemically derived to yield special physical or chemical absorptive properties. Covalent attachments of certain specific chemicals are particularly useful in cell biology and biophysics, including poly-L-lysine for enhanced adherence of cells, hydrocarbon chains for hydrophobicizing the surface in preparation for lipid monolayer adsorption, and antibodies, antigens, or lectins for producing specific reactivities. Derivation generally involves pretreatment of the glass by an organosilane.

A planar phospholipid coating (possibly with incorporated proteins) on glass can be used as a model of a biological membrane. Methods for preparing such model membranes on planar surfaces suitable for TIR are reviewed by Thompson and Palmer (1988).

Aluminum coating (for surface fluorescence quenching) can be accomplished in a standard vacuum evaporator; the amount of deposition can be made reproducible by completely evaporating a premeasured constant amount of aluminum. After deposition, the upper surface of the aluminum film spontaneously oxidizes in air. This aluminum oxide layer appears to have some similar chemical properties to the silicon dioxide of a glass surface; it can be derived by organosilanes in much the same manner.

4.3 Photochemistry at the Surface

Illumination of surface adsorbed proteins can lead to apparent photochemically induced cross-linking. This effect is observed as a slow, continual, illumination-dependent increase in the observed fluorescence. It can be inhibited by deoxygenation (aided by the use of an O_2-consuming enzyme/substrate system, such as protocatachuic deoxygenase/protocatachuic acid or a glucose/glucose oxidase system) or by 0.05 M of cysteamine.

5. TIRF VERSUS OTHER OPTICAL SECTIONING MICROSCOPY TECHNIQUES

Confocal microscopy (CM) is another technique for apparent optical sectioning, achieved by exclusion of out-of-focus emitted light with a set of image plane pinholes (see Chapter 5). Confocal microscopy has the clear advantage in versatility. That method of optical sectioning works at any plane of the sample, not just at an interface between dissimilar refractive indices. Other differences exist, however, that in some special applications are favorable for the use of TIRF:

1. The depth of the optical section in TIRF is approximately 0.1 μm, whereas in CM it is a relatively thick at approximately 0.6 μm.
2. In some applications (e.g., FRAP, FCS, or on cells whose viability is damaged by light), illumination and not just detected emission is best restricted to a thin section; this is possible only with TIRF.

3. Because TIRF can be adapted to and made interchangeable with existing standard microscope optics, even with "home-made" components, it is much less expensive than CM. TIRF microscopy kits are commercially available from Olympus and from TILL-Photonics companies; the mirrors and prisms can also be purchased separately from optical supply companies and configured by the end-user.

Two-photon microcopy (TPM) has many desirable features, including true optical sectioning, whereby the plane of interest is the only one that is actually excited (as in TIRF). Two-photon microscopy is not restricted to the proximity of an interface (see Chapters 9–13). The optical section of TPM is still, however, much thicker than that of TIRF.

Cell–substrate contacts can be located by a nonfluorescence technique completely distinct from TIRF, known as *internal reflection microscopy* (IRM). Using conventional illumination sources, IRM visualizes cell–substrate contacts as dark regions. Internal reflection microscopy has the advantage that it does not require the cells to be labeled. On the other hand, the disadvantages are that it contains no information of biochemical specificities in the contact regions, and that it is less sensitive to changes in contact distance (relative to TIRF) within the first critical 100 nm of the surface.

This work was supported by NIH grant 5 R01 NS38129. The author also wishes to thank all the past graduate students who have contributed to aspects of the TIRF work described here: Thomas Burghardt, Nancy Thompson, Edward Hellen, Andrea Stout, Ariane McKiernan, Michelle Dong Wang, Robert Fulbright, Laura Johns, and Susan Sund.

REFERENCES

Axelrod, D. Carbocyanine dye orientation in red cell membrane studied by microscopic fluorescence polarization. *Biophys. J.* 26:557–574, 1979.

Axelrod, D. Selective imaging of surface fluorescence with very high aperture microscope objectives. *J. Biomed. Opt.* 6:6–13, 2001.

Dickson, R. M., D. J. Norris, and W. E. Moerner. Simultaneous imaging of individual molecules aligned both parallel and perpendicular to the optic axis. *Phys. Rev. Lett.* 81:5322–5325, 1998.

Dickson, R. M., D. J. Norris, Y.-L. Tzeng, and W. E. Moerner. Three-dimensional imaging of single molecules solvated in pores of poly(acrylamide) gels. *Science* 274:966–969, 1996.

Fulbright, R. M., and D. Axelrod. Dynamics of nonspecific adsorption of insulin to erythrocyte membrane. *J. Fluorescence* 3:1–16, 1993.

Gingell, D., O. S. Heavens, and J. S. Mellor. General electromagnetic theory of internal reflection fluorescence: The quantitative basis for mapping cell-substratum topography. *J. Cell Sci.* 87:677–693, 1987.

Ha, T. J., A. Y. Ting, J. Liang, W. B. Caldwell, A. A. Deniz, D. S. Chemla, P. G. Schultz, S. Weiss. Single-molecule fluorescence spectroscopy of enzyme conformational dynamics and cleavage mechanism. *Proc. Natl. Acad. Sci. U.S.A.* 96:893–898, 1999.

Han, W., Y.-K. Ng, D. Axelrod, and E. S. Levitan. Neuronal peptide release is sustained by recruitment of rapidly diffusing secretory vesicles. *Proc. Natl. Acad. Sci. U.S.A.* 96:14577–14582, 1999.

Hellen, E., and D. Axelrod. Kinetics of epidermal growth factor/receptor binding on cells measured by total internal reflection/fluorescence recovery after photobleaching. *J. Fluorescence* 1:113–128, 1991.

Hellen, E. H., and D. Axelrod. Fluorescence emission at dielectric and metal-film interfaces. *J. Optic Soc. Am. B* 4:337–350, 1987.

Lagerholm, B. C., T. E. Starr, Z. N. Volovyk, and N. L. Thompson. Rebinding of IgE Fabs at haptenated planar membranes: Measurement by total internal reflection with fluorescence photobleaching recovery. *Biochemistry* 39:2042–2051, 2000.

Lang, T., I. Wacker, J. Steyer, C. Kaether, I. Wunderlich, T. Soldati, H.-H. Gerdes, and W. Almers. Ca^{2+}-triggered peptide secretion neurotechnique in single cells imaged with green fluorescent protein and evanescent-wave microscopy. *Neuron* 18:857–963, 1997.

Loerke D., B. Preitz, W. Stuhmer, and M. Oheim. Super-resolution measurements with evanescent-wave fluorescence excitation using variable beam incidence. *J. Biomed. Optics* 5:23–30, 2000.

McKiernan, A. M., R. C. MacDonald, R. I. MacDonald, and D. Axelrod. Cytoskeletal protein binding kinetics at planar phospholipid membranes. *Biophys. J.* 73:1987–1998, 1997.

Oheim, M., D. Loerke, W. Stuhmer, and R. H. Chow. The last few milliseconds in the life of a secretory granule. Docking, dynamics and fusion visualized by total internal reflection fluorescence microscopy (TIRFM). *Eur. Biophys. J.* 27:83–98, 1998.

Oheim, M., D. Loerke, W. Stuhmer, and R. H. Chow. Multiple stimulation-dependent processes regulate the size of the releasable pool of vesicles. *Eur. Biophys. J.* 28:91–101, 1999.

Omann, G. M., and D. Axelrod. Membrane proximal calcium transients in stimulated neutrophils seen by total internal reflection fluorescence. *Biophys. J.* 71:2885–2891, 1996.

Sako, Y., S. Miniguchi, and T. Yanagida. Single-molecule imaging of EGFR signaling on the surface of living cells. *Nat. Cell Biol.* 2:168–172, 2000.

Schmoranzer, J., M. Goulian, D. Axelrod, and S. M. Simon. Imaging constitutive exocytosis with total internal reflection microscopy. *J. Cell Biol.* 149:23–31, 2000.

Steyer, J. A., and W. Almers. Tracking single secretory granules in live chromaffin cells by evanescent-field fluorescence microscopy. *Biophys. J.* 76:2262–2271, 1999.

Sund, S. E., J. A. Swanson, and D. Axelrod. Cell membrane orientation visualized by polarized total internal reflection fluorescence. *Biophys. J.* 77:2266–2283, 1999.

Thompson, N. L., and A. G. Palmer. Model cell membranes on planar substrates. *Comments Mol. Cell. Biophys.* 5:39–56, 1988.

Todd, I., J. S. Mellor, and D. Gingell. Mapping of cell-glass contacts of *Dictyostelium amoebae* by total internal reflection aqueous fluorescence overcomes a basic ambiguity of interference reflection microscopy. *J. Cell Sci.* 89:107–114, 1988.

Vale, R. D., T. Funatsu, D. W. Pierce, L. Romberg, Y. Harada, and T. Yanagida. Direct observation of single kinesin molecules moving along microtubules. *Nature* 380:451–453, 1996.

Wang, M. D., and D. Axelrod. Time-lapse total internal reflection fluorescence video of acetylcholine receptor cluster formation on myotubes. *Dev. Dynam.* 201:29–40, 1994.

22

Laser Traps in Cell Biology and Biophysics

WILLIAM H. GUILFORD

1. INTRODUCTION

Steven Block of Stanford University, a leader in biological applications of laser traps, often describes these traps as "the closest thing science has devised to the 'tractor beam'" of Star Trek fame. This is not far from the truth. While they only work at the level of microscopic particles (rather than spacecraft), laser traps capture and hold objects in three-dimensional space using only a beam of light.

In its simplest incarnation, a laser trap is a beam of laser light brought to a diffraction-limited focus through a microscope objective. At the focal point, a translucent particle or cell may be captured, held, and manipulated. A carefully chosen wavelength of the laser can trap living cells with little or no ill effect.

Applications of laser traps in biology have been diverse, ranging from the simple manipulation and placement of individual cells (Townes-Anderson et al., 1998) to measuring the mechanics of single macromolecules (Dupuis et al., 1997). This chapter will not extensively review previous biological applications of laser traps, nor will it provide detailed instructions on the construction of complex systems, such as generating multiple traps, or combinations of traps with total internal reflection fluorescence microscopy (TIRF) or other imaging techniques. Numerous articles and books have been published on these subjects (e.g., Svoboda and Block, 1994; Fällman and Axner, 1997; Sheetz, 1998).

The chapter's purpose is to provide sufficient knowledge of laser traps to show when the technique is appropriate for certain applications and to point out common problems and their solutions. It will first review the physics and fundamental limitations of laser traps in cell biology and then cover the basics of laser trap construction and component selection, emphasizing points that often cause problems for the first-time trap builder. Finally, common problems encountered while trapping single cells will be discussed and solutions recommended.

2. The Physics of Trapping

Although many experienced microscopists know that a laser trap "traps," few know *how* it does so. There are both ray-optics and electromagnetic theories to explain the trapping effect (Svoboda and Block, 1994). The ray-optics theory explains reasonably well the trapping of objects greater than one wavelength in diameter, such as a typical sized cell. Perhaps more important to the average user, the ray-optics model of trapping is easily explained to visitors and students.

First, recall that photons have momentum, just as any other particle. Imagine a sphere with a ray of light passing exactly through the center, as shown in Figure 22–1. The ray is not disturbed from its path. If the ray passes through the sphere slightly off center, however, the light path is refracted in the direction of the sphere displacement, as shown. Due to conservation of momentum, if the sphere refracts the photons to one side (e.g., the right), there will be force acting on the bead in the opposite direction (left).

Envision this as a pair of billiard balls—one initially stationary ball representing the object being struck slightly off center by a second moving ball representing a photon. If the moving ball/photon is deflected to the right, the stationary ball experiences a force pushing it to the left.

Consequently, the bead tends to move left or right toward the position where the light is the brightest and best centered on the sphere. Similarly, if one were to bring the light in, not from directly above but from the side, this new ray would pull the sphere up or down to where the light is brightest. Herein lies the trick to

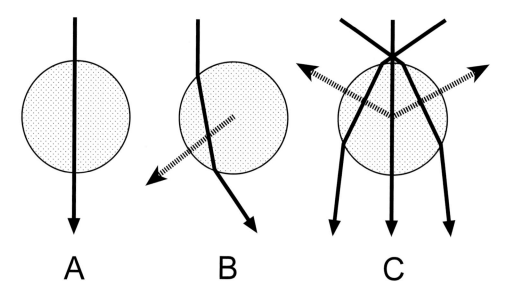

Figure 22–1 Ray-optics theory of laser trapping. *A,* A ray of light passing directly through the center of a sphere is undisturbed from its path. *B,* If the ray enters off center, it is refracted. Conservation of momentum requires that a force act on the sphere in the direction opposite the refracted ray. *C,* If rays enter symmetrically and from steep angles, the net force pulls the sphere up and toward the center of the focal point.

creating a stable three-dimensional trap. A microscope objective is used to focus a laser beam from very steep angles. Thus, the sphere is drawn not only laterally to the focal point, but vertically to the focus.

The principle of operation of such laser traps or "tweezers" differs somewhat from that used to trap atoms, for which the Nobel Prize in physics was recently awarded. Atom traps effectively use the Doppler effect to slow atoms that are moving toward a light source. A discussion of this fascinating technology and its relationship to laser tweezers is given by Chu (1991).

3. "RULES OF THUMB" FOR TRAPPING

The single most common question asked of laser trappers is "will laser trapping work in my application?" One should carefully consider each of the following issues in deciding whether a laser trap is appropriate for a given experiment. (Experienced trappers will note that some generalizations have been made.)

3.1 How Big is the Object to be Trapped?

Is the object a sizeable fraction of the wavelength of light? Small objects are more difficult to trap than large objects. Objects less than 0.1 or 0.2 λ (e.g., pinocytic vesicles) are generally not "trappable." As the size of the object approaches 0.5 λ, however, trapping becomes much easier. See Figure 5 in Svoboda and Block (1994) for a descriptive graph. Users may be tempted to simply use a shorter wavelength laser to trap smaller objects, but rapid photodamage is the typical outcome of this method (see Section 4.4).

3.2 What Outside Forces Act on the Object?

Will some outside force be acting on the trapped object, and, if so, how large a force? A trapped cell may exert traction on a substrate or be in a moving fluid, which creates hydrodynamic drag. Microspheres may have bound motor proteins that displace them from trap center. Whatever the origin of the force, bear in mind that typical laser traps are very weak and are seldom able to withstand forces over 50 piconewtons (pN). For comparison, a single myosin molecule generate forces in the range of 1–4 pN (Guilford et al., 1997), while the net generated by crawling macrophages in vitro are on the order of 1 nN, or 1000 pN (Guilford et al., 1995).

3.3 How Thick is the Specimen?

How thick is the object being trapped or the tissue in which it is found? While individual cells heat only slightly in a laser trap (see Section 6), thicker specimens will absorb a sufficient amount of laser power to be damaged or even incinerated. Thus, in vivo laser trapping is generally impossible. In general, near-infrared (near-IR) lasers are used in biological laser trapping, as most cells absorb very little energy at these wavelengths. There is, however, no substitute for a thin specimen.

3.4 Is the Refractive Index Appropriate?

Is the refractive index of the object significantly higher than that of the surrounding medium? Consider the above ray-optics explanation of trapping. If photons were bent *opposite* the direction shown, the net force would drive the object away from the laser focus, not toward it. This is the result when the object has a lower refractive index than the medium, and a laser fountain is created instead of a laser trap. Fortunately, this is seldom a problem in cell biology. Nevertheless, one should avoid situations in which the difference in refractive index is very small (e.g., a liposome containing pure water suspended in pure water).

3.5 How Fast Do Things Move?

How fast must the object move with or within the laser trap? When a cell is dragged through a fluid, even slowly, the viscous forces acting on a cell are large relative to the strength of the trap. Therefore, it is not practical to move cells hither and yon quickly or over large distances. The force acting on a spherical object as it moves through solution is given by the Stokes' equation, $3\pi d\eta v$, where d is the diameter of the sphere, η is the viscosity of the fluid, and v is the velocity of the fluid relative to the sphere. Similarly, if one is observing the movement of an object within the trap, such as in motor protein biomechanics, bandwidths greater than 1 kHz are seldom achieved. This is because motions of microspheres within traps are critically dampened.

3.6 Can the Trap Reach the Specimen?

It is sometimes difficult to form a stable trap at the location of the specimen. When using high numerical aperture (NA) oil-immersion objectives, a stable trap is only formed to a depth of around 15 μm into the aqueous solution beyond the coverslip! This is obviously a major limitation of the technique. While this problem may be circumvented by using water-immersion objectives (NA $\approx$ 1.2), they do not trap as stably as oil immersion.

4. Trap Assembly and Component Choice

4.1 Basic Laser Trap

A basic laser trap consists of four essential components: a laser, a beam expander, a high NA microscope objective, and a means of aligning the expanded beam to the objective. A common arrangement of optics is show in Figure 22–2. A laser beam is expanded to a diameter that will fill the back-aperture of the microscope objective to be used. The expanded beam is reflected off a pair of mirrors that are used to align the beam to the objective. Finally, the objective focuses the laser beam at the specimen plane where the trap is formed.

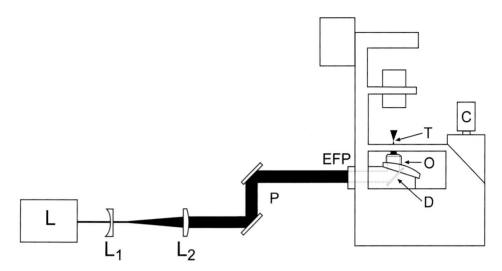

Figure 22–2 Schematic of a simple laser trap. The beam from a laser (L) expanded by a planoconcave (L$_1$) and a planoconvex (L$_2$) lens. The expanded beam reflects off two mirrors forming a periscope (P) and enters the epifluorescence port (EFP) of an inverted microscope. A dichroic mirror (D) reflects the beam to a high-numerical aperture objective (O) that focuses the beam to form a trap (T). This trap could be displayed on a monitor using a camera (C).

4.2 Objective Selection

The first choice one should make when designing a laser trap is the microscope objective. The first criterion is whether an oil- or water-immersion objective is appropriate. As discussed earlier, if the specimen is (or can be) within a few microns of the coverslip, then an oil-immersion objective is preferable. If a coverslip cannot be used, or if the specimen is greater that 10 or 15 μm below the coverslip, then a water-immersion objective is necessary.

The limitation is that light passing from oil into the glass coverslip and then into water introduces severe spherical aberration that worsens the further into the solution one focuses. As spherical aberration dominates, the trap fails. Passing from water into glass and back into water, however, introduces very little spherical aberration, and diffraction-limited focusing is achieved no matter how deep into the aqueous medium one focuses.

The second criterion is the NA. In general, the higher the NA, the more stable the laser trap. This is because high NA objectives bring light to the specimen from steeper angles, creating forces that pull the object up into the laser trap. These forces are needed to dominate the radiation pressure on the trapped object resulting from reflected light. At low NA, radiation pressure dominates and pushes objects away from the laser beam.

Unfortunately, high NA objectives tend to pass significantly less light in the near-IR than do low NA objectives. An NA 1.3 PlanNeoFluor objective (or its equivalent) from any manufacturer may pass over 60% of light at wavelengths near 1064 nm, while an NA 1.4 PlanApo may pass less than 30%. Thus, if laser powers will be especially high, a slightly lower NA may be appropriate.

To take advantage of the full NA of an objective, the laser beam must fill the back aperture, which is the clear diameter of the final lens visible from the back end of the objective. One major manufacturer uses very large diameter objectives to achieve a greater working distance at the expense of a very large back-aperture. The laser beam entering the backs of these objectives must be proportionally larger in diameter. This requires a larger clearance for the laser beam entering the microscope, a larger beam expander, and larger apertures for the optics in general. Be aware of the back-aperture diameter before purchasing an objective for laser trapping.

4.3 Microscope Selection

Some have elected to dispense with the microscope altogether and have built their laser traps on optical rails. Most modern microscopes are, however, perfectly acceptable for building a laser trap. There are four major points that should be taken into consideration when choosing a microscope for laser trapping.

4.3.1 Introducing the laser beam

Every manufacturer offers a "bottom port" as an option for introducing a laser beam. This may, however, conflict with imaging needs. Many investigators introduce the laser through modifications of the epifluorescence assembly or replacements of the differential interference contrast analyzer holder. Obviously imaging needs must be considered in conjunction with introducing the laser beam and must be decided on a case by case basis.

4.3.2 Inverted versus upright

Laser traps may be based on either upright or inverted microscopes. The deciding factors in choosing one or the other are usually (1) which surface, top or bottom, will have the coverslip and (2) how mechanically stable must the trap be? When "high-resolution" work is being done, such as the study of the mechanics of single proteins, mechanical stability is of overriding importance, and inverted microscopes are preferred. For most cell manipulation, stability is not a critical matter.

4.3.3 Infinity correction

All new microscopes (since the mid-1990s) use infinity corrected optics (see Chapter 4). This is particularly convenient for developing laser traps, as a collimated laser beam will be properly focused in the specimen plane. Older microscopes work perfectly well for laser trapping except that the laser beam entering the objective will *not* be collimated, but rather diverge at an angle determined by the manufacturer of the objective. This angle is sometimes difficult to achieve, and alignment is sometimes more challenging.

4.4 Laser Selection

The heart of the laser trap, and typically its most expensive single component, is the laser light source. Many laser technologies are appropriate for laser trapping, but one should consider six criteria in choosing a laser for trapping: wavelength, power, mode, beam quality, pointing stability, and polarization.

4.4.1 Wavelength

For trapping living cells or other biological materials, it is best to choose a wavelength of light at which the material absorbs little to minimize heating and photodamage. Generally, near-IR light in the range of 800–1100 nm is best (Svoboda and Block, 1994). There have been more detailed studies of the effects of laser light on living cells that suggest that a careful choice of wavelength within this range may have significant consequences for cell viability (see Section 6).

4.4.2 Power

The laser should emit continuously (CW) and deliver at least 0.1 W at the specimen plane. The optics will, however, significantly attenuate the laser beam. A good rule of thumb for choosing laser power is to use two to four times the required power at the specimen plane. Thus, a power of at least 0.5 W is advisable. Lasers with continuously variable power are preferable.

4.4.3 Mode

The "mode" of a laser is usually a reference to its transverse electromagnetic mode (TEM). TEM defines the distribution of energy across the width of the laser beam. To achieve stable and predictable trap performance, the energy distribution should be radially symmetric. Thus, most laser traps are based on a TEM_{00} laser, which has a Gaussian energy distribution. The beam is "brightest" at the center and trails off to zero near the edges.

There are other TEMs, including the TEM_{01} or "donut mode," which (as the name implies) has a donut-shaped energy distribution of zero at the center and bright near the edges. While TEM_{01} has some benefits over TEM_{00}, such lasers are more difficult to obtain and are quite expensive.

4.4.4 Beam quality

"Beam quality" refers qualitatively to how closely the measured energy distribution of the laser beam conforms to the primary TEM. Beam quality in TEM_{00} lasers is assessed quantitatively by the M^2 estimator. When there are modes and/or "noise" present in the laser beam besides the Gaussian primary, the laser beam cannot be as tightly focused as would be expected. The spot size becomes M times larger than expected, and the energy density becomes M_2 less than expected.

This is of critical importance in several areas of microscopy where focused laser beams are used as a light source. Stable laser trapping requires a steep gradient of energy at the focal point. Using lasers with M^2 values of greater than 1.1 or 1.2 may noticeably degrade the stability of a trap by reducing this energy gradient.

4.4.5 Pointing stability

Laser beams "wander." The mean angular change that the laser beam sweeps through over a given period is a quantification of the pointing stability and is usually given in μrad. A angular change of θ in the laser beam displaces the trap by θf_{eff} at the specimen plane, where f_{eff} is the effective focal length of the objective. Common values for f_{eff} and θ are on the order of 1 mm and 50 μrad, respectively, giving the laser trap a "wander" of 50 nm. For simple cell manipulation and placement, this is generally not an issue. In contrast, 50 nm of wander is unacceptable for molecular mechanics studies. Values for f_{eff} are not commonly published, but can be obtained from microscope dealers.

4.4.6 Polarization

Lasers are sold that emit linear, circular, or randomly polarized light. For a simple, single laser trap, polarization is of little concern. In applications where polarized light is needed, the various polarization states of lasers are easily interconverted.

4.5 Beam Expanders

Between the laser and the microscope is the beam expander. How many "fold" the beam needs be expanded is given by (d_0/d_1), where d_1 is the diameter of the laser beam, and d_0 is the diameter of the back-aperture of the objective. Laser beam diameters are normally given as the $1/e^2$ diameter, or the edge-to-edge distance where the brightness of the beam drops to 13.5% of the peak. Beam expansions of approximately $\times 10$ are typical. If one is fortunate in the degree of expansion that is needed, pre-assembled beam expanders are available from many optics vendors in common sizes (e.g., $\times 5$, $\times 10$, $\times 20$). Otherwise, an expander must be assembled from a pair of lenses.

There are two basic forms of beam expanders: Keplerian and Galilean. A Keplerian expander uses a pair of planoconvex lenses, as shown in Figure 22–3. A Galilean expander uses a plano- or biconcave lens as the first lens of the expander and a planoconvex as the second lens. The first lens must always have a shorter focal length than the second lens. The degree of expansion (n) is given by (f_2/f_1), where f_1 and f_2 are the focal lengths of the first and second lenses, respectively. When the lenses are placed a distance $(f_1 + f_2)$ apart, a collimated beam entering the expander will leave collimated and n-fold larger in diameter. Note that a concave lens has a negative focal length. Thus, Galilean expanders tend to be somewhat shorter than Keplerian expanders are.

Consideration should be given to choice of optics. Longer focal lengths and larger diameters make it easier to achieve true diffraction-limited performance. When possible, lenses should be used with antireflection coatings specific to the wavelength of the laser.

4.6 Assembly and Alignment

4.6.1 Safety

Lasers used in trapping are generally Class IV, which come with many safety requirements. The laser trap should be in a separate room, or at least in an area closed off by an appropriate laser safety curtain. Signs should be posted in the area indicating a danger from visible and/or invisible laser radiation. Entryways

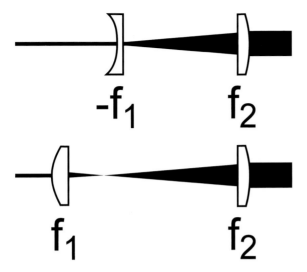

FIGURE 22–3 A Galilean beam expander (top) and a Keplerian beam expander (bottom). f_1 and f_2 are the focal lengths of the first and second lenses, respectively. The focal length of the first lens is negative in a Galilean beam expander, which reduces the overall length of the expander.

should be interlocked with the laser so that opening any door to the area will cut the laser power.

To illustrate the importance of safety, a 0.5 W beam, unexpanded yet unfocused, is sufficient to burn skin in only a few seconds. Even a 3 mW laser pointer, when directed into the eye, has an energy density on the retina 10,000 times greater than a 100 W light bulb held directly against the cornea! Brief exposure to a scattered portion of a laser beam used in trapping can result in permanent retinal damage. Thus, laser goggles should always be worn and should have optical densities of 5 or greater at the emission wavelength of the laser.

Finally, never look through the eyepieces of the microscope with the laser in operation. Remove the binocular head from any microscope being used for laser trapping to discourage the practice. Use a charge-coupled device (CCD) camera to observe the specimen and/or beam.

4.6.2 Special equipment

There are several special pieces of equipment one may need before attempting to assemble a laser trap. Infrared cards are absolutely necessary to make the IR beam visible to the naked eye. Wherever the laser beam strikes these cards, it appears as a bright spot. These are available from several vendors (Edmund Scientific, New Focus), in both reflective and transparent versions, to be viewed from the front or the back, respectively. A laser power meter is frequently useful, although certainly not necessary for assembling a simple laser trap. A CCD camera attached to the microscope is a valuable tool for assessing the quality of the alignment. One may use it to observe the IR light reflected from a glass surface, thus visualizing the trap itself.

4.6.3 Assembly

The following are steps to assemble a simple laser trap:

1. To assemble a simple laser trap, first position the microscope on an appropriate table. An air table with a drilled and tapped optical breadboard top is preferable. It is also preferable that the microscope be fastened directly to the table to prevent its being inadvertently moved.

2. It may be necessary to install an appropriate dichroic mirror somewhere in the microscope that will either pass or reflect the laser light (Fig. 22–2). Deciding which dichroic, or whether one is even necessary, depends on what imaging modalities are being used and how the beam is entering the microscope. Consult the microscope technical representative for help with these decisions.

3. Position the laser on the table, leaving adequate room for the beam expander and any mirrors needed to direct the beam into the microscope. Fasten the laser securely to the table. Install the periscope and/or mirrors that will be used to align the beam to the microscope, and raise it to the appropriate height. Estimate the initial positions of the mirrors.

4. The system is now ready for rough alignment. The goal is for the laser beam to proceed straight and centered into the back of the microscope objective. Begin by turning the objective turret to an open hole (i.e., no objective). Turn the laser on at a very low power (e.g., 100 mW), and use an IR card to follow the beam as it reflects off the mirrors and into the microscope. Adjust the mirrors so that the laser passes simultaneously through the center of the laser port (the point of entry into the microscope) and through the center of the objective hole.

5. Once rough alignment is complete, insert a low-power objective (around 10×) into the turret, and focus onto the surface of a glass slide. Use a CCD camera attached to the microscope to look for IR light reflected from the glass. The focal point will appear as a bright spot surrounded by interference rings. If the spot still cannot be located, try focusing onto the surface of an IR card. Move the IR card around and watch for a diffuse bright spot. Adjust the mirrors to try to bring the spot into the middle of the field of view.

6. Once the focal point has been located and centered, focus up and down and watch the shape of the spot. The spot will expand and contract symmetrically as one goes through focus (Fig. 22–4) if the trap is aligned. If it "smears" in one dimension or the other, adjust the mirrors on the offending axis *simultaneously* and in *opposite directions*. The beam should remain in the field of view, and, with patience, the "smearing" will lessen. If the brightness of the spot suddenly drops, the beam may have been steered off the edge of a mirror or the objective. Adjust the mirror positions and return to step 5.

7. Install the first lens of the beam expander (or the entire beam expander if purchased as a pre-assembled unit) with the curved surface facing

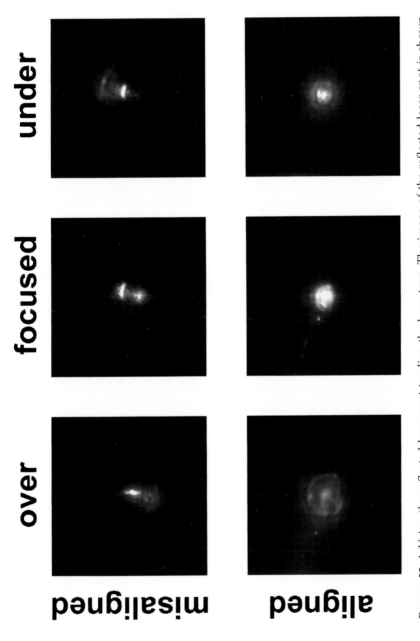

FIGURE 22-4 Using the reflected laser spot to align the laser trap. The image of the reflected laser spot is shown as it passes through focus (left to right) before (top) and after (bottom) alignment. Notice that the spot in the aligned case expands but remains roughly concentric through focus. When misaligned, the spot skews through focus.

the laser. Make sure that the beam passes through the lens dead center. Install the second lens at an appropriate distance from the first, with the curved surface facing the microscope. Adjust the expander so that the expanded beam passes simultaneously through the centers of the laser port and the objective. If the expander is assembled from individual lenses, the first lens may have to be on an XY translating stage to allow adjustment.

8. Repeat step 6, but instead of adjusting the mirrors to align the trap, adjust the beam expander.

9. Install the high NA objective, and couple it to a coverslip. Repeat step 8, adjusting the mirrors so that the reflected spot is well centered in the field of view and expands/contracts symmetrically as one passes through focus. This generally requires significant adjustments of the mirrors and may take some patience. Mark the location of the trap on the video monitor.

10. If all has gone well, the system should now be ready for trapping. A suspension of 1–2 μm polystyrene beads in water is ideal to test the trap. If the suspension is too concentrated, a small vortex will form where the trap should be. Dilute the beads and try again. Remember that with oil-immersion objectives a trap only forms within 15 μm of the coverslip, so working near the surface is vital. It is also worth noting that a laser trap cannot normally displace beads stuck to the glass surface.

11. Finally, it is generally best to cover the path of the laser beam with an acrylic or metal box. This keeps the optics clean, protects them from accidental bumps, and prevents air currents from making the trap jiggle.

5. TRICKS AND TROUBLESHOOTING

5.1 Difficulty Trapping

When the instrument fails to trap, the instinctual reaction is to increase the laser power. Although this sometimes helps, it is rarely the source of the trouble. The most common source of trap failure is misalignment. Optics occasionally get bumped, which ruins the alignment. Some traps have been known to lose alignment seasonally when the ambient temperature changes. In either case, the best diagnostic is to reflect the laser beam off a glass surface, as above, and check for symmetry.

Another common problem that prevents a laser trap from working is the degree of beam expansion. Even *slight* underfilling of the back-aperture will prevent a stable laser trap from forming. This is equivalent to a large reduction in NA. Ensure that the diameter of the beam is equal to, or even overfills, the back-aperture. Overfilling degrades trap performance only to the extent that some energy is lost to the aperture itself. It is better to overfill significantly, rather than to underfill even marginally.

If the trapped object appears out of focus, then the trap is forming at a different focal plane than the image. To alleviate this problem, adjust the separation of the lenses in the beam expander (pre-assembled expanders are usually adjustable) until the object comes into focus. This will sometimes degrade trap performance. If it does, one may have to adjust the degree of beam expansion or simply strike a balance between image quality and trap quality.

5.2 Cell Sticking

The single most common problem in trapping is the cell or microsphere sticking to the coverslip. Beads in water generally do not stick. Cells or beads in an electrolyte/protein solution will, however, become rapidly adherent. Adherent beads or cells cannot normally be dislodged using a laser trap. Unfortunately, there is no perfect solution to this problem. Siliconizing the coverslip before use will help reduce sticking. Another approach is to coat the coverslip with 0.5 mg/ml bovine serum albumin (BSA) for 1 or 2 minutes and include BSA in the working medium. The best advice for novice trappers, however, is to work *quickly* and with very dilute suspensions of cells or beads so that they simply do not have time to settle and adhere.

5.3 Microscope Stages

Using a laser trap to manipulate cells requires very slow stage movements, as discussed earlier. The standard mechanical stages provided with microscopes do not generally give fine enough control over specimen movement to allow successful manipulations with a laser trap. Stages based on pneumatics, stepper motors, and piezo-electrics have been used with laser traps to obtain fine control. A replacement stage is usually a significant investment and should be budgeted from the beginning.

6. Effects of Trapping on Cells

No discussion of laser traps in cell biology would be complete without some mention of the effects of trapping on cell physiology and growth. The impression is sometimes given that trapping has little or no effect on the cell. While trapping is a relatively gentle technique, it does have consequences for cell health.

There have been two key studies of the effects of laser traps on cell physiology and viability. In the first, cell temperature increased by approximately 1°C/100 mW trapping power at 1064 nm, but did not alter cytoplasmic pH or DNA structure (Liu et al., 1996). Loss of cell viability was observed, however, after trapping for more than 2 min at 300 mW. Therefore, applications that require long periods of confinement or negligible temperature increases may not be able to make use of a laser trap.

A second study demonstrated that the cloning efficiency of Chinese hamster ovary cells after trapping decreased with both laser power and time spent in the laser trap (Liang et al., 1996). Furthermore, cloning efficiency varied widely with wavelength, being maximal at 950–990 nm, providing an excellent guideline for laser selection. This study also highlights the importance of both working quickly and using the lowest possible powers when manipulating cells.

The author thanks David Warshaw, who first introduced him to the laser trap, and all the members of the Motility Subgroup of the Biophysical Society for their many laser trap innovations from which he has benefited. The author thanks Laura La Bonte for helpful suggestions on the manuscript and the National Institute for Arthritis and Musculoskeletal Diseases for their support (grant AR45604).

REFERENCES

Chu, S. Laser manipulation of atoms and particles. *Science* 253:861–866, 1991.

Dupuis, D. E., W. H. Guilford, J. Wu, and D. M. Warshaw. Actin filament mechanics in the laser trap. *J. Muscle Res. Cell Motil.* 18:17–30, 1996.

Fällman, E., and O. Axner. Design for fully steerable dual trap optical tweezers. *Appl. Opt.* 36:2107–2113, 1997.

Guilford, W. H., D. E. Dupuis, G. Kennedy, J. Wu, J. B. Patlak, and D. M. Warshaw. Smooth and skeletal muscle myosins produce similar unitary forces and displacements in the laser trap. *Biophys. J.* 72 (3):1006–1021, 1997.

Guilford, W. H, R. W. Gore, and R. C. Lantz. The locomotive forces produced by single leukocytes in vivo and in vitro. *Am. J. Physiol.* 268:C1308–1312, 1995.

Liang, H., K. T. Vu, P. Krishnan, T. C. Trang, D. Shin, S. Kimel, and M. W. Berns. Wavelength dependence of cell cloning efficiency after optical trapping. *Biophys. J.* 70:1529–1533, 1996.

Liu, Y., G. J. Sonek, M. W. Berns, and B. J. Tromberg. Physiological monitoring of optically trapped cells: Assessing the effects of confinement by 1064-nm laser tweezers using microfluorometry. *Biophys. J.* 71:2158–2167, 1996.

Sheetz, M. P., editor. Laser tweezers in cell biology. *Methods. Cell Biol.* 55, 1998.

Svoboda, K., and S. M. Block. Biological applications of optical forces. *Annu. Rev. Biophys. Biomol. Struct.* 23:247–285, 1994.

Townes-Anderson, E., R. S. St. Jules, D. M. Sherry, J. Lichtenberger, and M. Hassanain. Micromanipulation of retinal neurons by optical tweezers. *Mol. Vis.* 4:12–17, 1998.

23

Bioluminescence Imaging of Gene Expression in Living Cells and Tissues

Michael E. Geusz

1. Introduction

Biomedical research requires a better understanding of how genes are regulated. Unfortunately, most assays of gene expression disrupt cell integrity and provide information only at specific points in time. Numerous replicates are needed at each time point to show the temporal pattern of gene activity. Rapidly changing patterns in the induction of immediate-early genes and other cell responses following cell stimulation are not easily described by current methods. Recent studies have shown, however, that continuous monitoring of gene activity is possible by imaging bioluminescent reporter gene expression in live cells. Temporal data acquired in this way are being used to answer questions concerning development, signal transduction, hormone secretion, infection, circadian rhythms, and other processes. This chapter will describe current imaging using bioluminescent gene products, particularly those in live mammalian cells.

2. The Advantages of Bioluminescence Imaging

Live bioluminescence assays have several unique features that allow sustained imaging studies lasting hours or days:

1. No phototoxic damage: Bioluminescence does not cause the phototoxic damage associated with fluorescence-based assays in which radical oxygen species generated by the excitation light limit the practical duration of experiments. The extent of phototoxic injury from probes based on the popular green fluorescent protein (GFP) reporter genes has not yet been fully addressed, although red-shifted GFP variants excited and imaged at longer wavelengths and two-photon excitation techniques should minimize toxicity and autofluorescence artifacts (see Chapter 9–10).
2. No autofluorescence: Cell autofluorescence produces background noise that can limit signal discrimination with fluorescence techniques.

In contrast, only the ultraweak luminescence found in some cells and phosphorescent decay contribute to background with bioluminescence.

3. No bleaching: There is no signal loss comparable to the bleaching of fluorophores by excitation light, although other substrate and enzyme-dependent factors may affect bioluminescence.

4. Compatible with photoreception: Bioluminescence probably does not interfere with light-dependent signaling pathways and may be useful in retinal and photoreceptor studies because far fewer photons are involved than with fluorescence.

Two chemically distinct enzymes, beetle luciferase and bacterial luciferase, are commonly used in bioluminescence imaging with live cells. This chapter focuses on a beetle luciferase from the North American firefly *Photinus pyralis* (EC 1.13.12.7) and, to some extent, on a luciferase from the sea pansy *Renilla reniformis* (EC 1.13.12.5), an anthozoan coelenterate (Lorenz et al., 1991). Several reviews of the product of the bacterial luciferase gene *lux* are available (Billard and DuBow, 1998). The *lux* enzyme (EC 1.14.14.3) is not as effective as firefly luciferase in mammalian cells due to thermal denaturation (Pazzagli et al., 1992).

3. FIREFLY LUCIFERASE

3.1 Properties of the Light-Emitting Reaction

The essential reaction of *Photinus* firefly luciferase, as well as that of click beetle (*Pyrophorus*) and similar luciferases, is the ATP-dependent oxidation of the dihydro form of D-luciferin (4,5-dihydro-2-[6-hydroxy-2-benzothiazolyl]-4-thiazolecarboxylic acid) in the presence of Mg^{2+} (Fig. 23–1). The reaction consists of several steps that generate yellow-green light at 560 nm by adenylating and exciting luciferin with a high quantum efficiency (0.88) (DeLuca and McElroy, 1978; Brolin and Wetermark, 1992) before liberating adenosine monophosphate (AMP) and CO_2. Details of the light-emitting reaction have been extensively described elsewhere (Gould and Subramani, 1988; Brolin and Wetermark, 1992).

The slow release of enzyme-bound oxyluciferin while substrate levels are saturating maintains most of the available luciferase molecules in an occupied state, yielding low photon emission rates but also minimizing luciferin consumption (Aflalo, 1991; Brolin and Wetermark, 1992). Consequently, luciferase reactions can

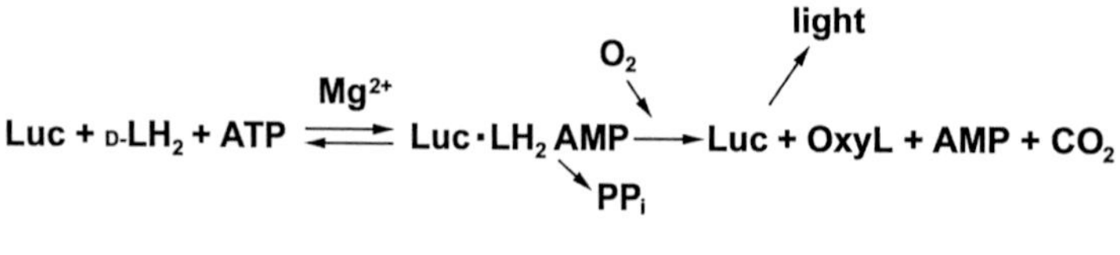

FIGURE 23–1 The luciferase–luciferin reaction. In its simplest form, the luciferase light-emitting reaction involves adenylation of luciferin (D-LH$_2$) and release of pyrophosphate followed by oxidation of luciferin, bioluminescence, and CO_2 generation. Oxyluciferin: OxyL, luciferase: Luc. (After DeLuca and McElroy, 1978).

provide a sustained glow for many hours while consuming little luciferin, an attractive feature for long-term imaging. Renewed or increased gene expression is marked by an increase in light emission from newly made luciferase.

The kinetics of luciferase in cell-free systems are complex, and its behavior inside living cells is poorly understood. Luciferase activity appears to be regulated by both substrates, adenosine triphosphate (ATP) and luciferin (Lembert and Idahl, 1995). In commercial luciferase assays, coenzyme A is included to produce a sustained glow rather than a brief flash of light, allowing more time for recording the light signal (Jones et al., 1995). A similar effect may occur in living cells due to the approximately 0.01 mM coenzyme A level in the cytosol (Kennedy et al., 1999). Notably, bovine serum albumin also stimulates the luciferase reaction (Lembert and Idahl, 1995), suggesting that interaction between luciferin and cytoplasmic proteins may significantly affect the kinetics. Because of the low activity of luciferase—0.1 U/mg (Brolin and Wetermark, 1992)—detectors sensitive enough to quantify individual photons are often needed, and considerations for this low photon flux are presented below.

3.2 Beetle Luciferin

Sufficient luciferin should be applied so that its concentration does not limit expected luciferase levels. Luciferin permeates the plasma membrane at physiological pH despite its negative charge and appears to distribute itself according to a Nernstian gradient. Due to this effect, a 20–40-fold lower concentration was estimated inside cardiomyocytes relative to outside (Gandelman et al., 1994). Luciferin applied externally at 0.1–1 mM is likely adequate for enzyme activity, although the distribution of luciferin within cells has not been shown directly. All beetle luciferases use the same luciferin, which is typically dissolved in pure water to prepare a stock solution (e.g., 100 mM) and then frozen as aliquots. Luciferin is quite stable, although it must be protected from light at all times, and the sodium and potassium salts are readily soluble. Luciferin and coelenterazine, the substrate for *Renilla* luciferase, are well tolerated by cells. Incidentally, the fluorescence of luciferin in near-ultraviolet light must be considered when imaging with some fluorescent probes will follow luciferase imaging.

4. Aequorin, a Ca^{2+} Photoprotein

The photoprotein aequorin is another very useful bioluminescent molecule for assaying activity in living cells (Jones et al., 1999). In a reaction found in several marine organisms, aequorin produces blue light upon binding Ca^{2+}, although green light is generated in vivo through fluorescence resonance energy transfer from aequorin to GFP. Normally, aequorin is present in a complex with coelenterazine (Shimomura and Johnson, 1975), and cells transfected with the aequorin gene can be used to assay intracellular Ca^{2+} responses or gene expression when coelenterazine is in the culture medium. Different coelenterazine molecules can provide altered reaction rates and Ca^{2+} affinities (Shimomura et al., 1988). Many principles of optimizing light capture described here apply equally to luciferase and aequorin bioluminescence as well as to chemiluminescent probes based on luminol and lucigenin that are commonly used in enzymatic detection of molecules.

5. IMAGING AND DETECTION SYSTEMS

5.1 Photomultiplier Tubes

Imaging a dim subject typically requires sacrificing spatial information for the pixel intensity needed to maintain a high signal-to-noise (S/N) ratio. In this way, cameras with "binning" capabilities collect light captured in several adjacent pixels, summing the signal to produce an overall brighter image comprising fewer pixels. Entirely photometric measurements (those without imaging) essentially bin light into a single pixel that captures the entire signal. Photometry is commonly preferred to imaging when acquisition speed or signal dynamics are critical, and photomultiplier tubes (PMTs)—vacuum tubes that convert photons to current pulses—are perhaps the best photometric detectors for bioluminescence. Although they can detect single-photon events, their intrinsic noise and high-voltage power supplies do constrain system designs, particularly when multiple channels are needed. The PMT can be cooled to lower the inherent "dark current" noise, but moisture condensation must be prevented.

To maximize recordings with a single detector, some systems move multiple samples continuously past a PMT or camera. One early design collected natural bioluminescence from the marine dinoflagellate *Gonyaulax polyedra* and recorded circadian rhythms in glow from several cultures sequentially (Broda et al., 1986). More recently, a circular turntable was used for automatic imaging of hundreds of cyanobacteria cultures expressing circadian rhythms in bioluminescence from *lux* reporter genes (Kondo and Ishiura, 1994). A luminometer that automatically positions PMTs above multiwell sample plates (Packard Instrument Co., Meriden, CT) has been used to record circadian rhythms in bioluminescence from *Drosophila* (Brandes et al., 1996; Plautz et al., 1997a,b) and *Arabidopsis* (Millar and Kay, 1996). All of these designs provide high sample throughput, clearly compensating for the prolonged duration of long-term and circadian recordings.

The HC-135 photon counting module from Hamamatsu (Bridgewater, NJ) is a relatively inexpensive device used for live-cell luciferase recordings. Figure 23–2 shows cell responses using the HC-135 to record from cells of a transgenic mouse in which the human c-*fos* promoter drives firefly luciferase expression (*fos::luc*) (Geusz et al., 1997).

5.2 Camera Designs

Two main types of cameras provide detectors suitable for bioluminescence imaging—intensified charge-coupled device (ICCD) cameras used with photon-counting software and cooled charged-coupled device (CCD) cameras. Cooling reduces noise due to thermal motion that accumulates over time in the sensor,

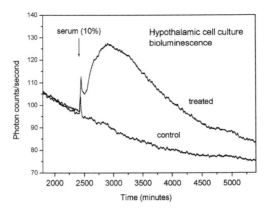

Figure 23–2 Photometric assay of luciferase reporter gene activity in live cells. Cell cultures from the hypothalamus of *fos::luc* mice were maintained in serum-free medium until the time shown when serum, a known stimulus for c-Fos production, was introduced for 15 min. The identical control culture was given medium alone, and luciferin was present at 1 mM at all times. Both treatments produced a transient signal increase followed by a large increase only from the treated culture. Methods are described in detail elsewhere (Geusz et al., 1997).

thereby allowing longer exposures and brighter images. In some cases, ICCD cameras can also be cooled. The ICCD camera allows images to be viewed in real time as pixels accumulate signal on a computer screen.

5.2.1 Cooled CCD cameras

The most sensitive cooled CCD camera contains a reservoir of liquid nitrogen (LN) that along with a heater circuit maintains the sensor temperature near −90°C. Thermal noise accumulating in the pixel wells as charge is reduced by half for about every 7°C decrease in temperature. Without cooling, this noise would mask the image during the particularly long (10–60 min) exposures used to collect bioluminescence. Proper gloves and face protection must be worn while handling LN, and it should only be used in an open, well-ventilated room.

Well-established manufacturers of LN-cooled cameras include Roper Scientific (Tucson, AZ), Hamamatsu, Micro Photonics (Allentown, PA), and PixelVision (Tigard, OR). Many of these designs were first developed for astronomy, which also has a need to observe a self-illuminated bright object on a dark background. To maximize sensitivity, the silicon CCD sensor is back-thinned by etching and then illuminated from the reverse side instead of imaging through the sensor's semiconductor gates. These steps increase quantum efficiency to over 80% at the maximum sensitivity, which is typically near the blue-to-green transition. Cooled CCD cameras provide excellent sensitivity, low noise, and megapixel spatial resolution. Some camera makers have achieved cooling to −70°C (Micro Luminetics, Inc., Los Angeles, CA) and −90°C (L.O.T.-Oriel GmbH & Co.) with thermoelectric (Peltier) cooling rather than LN.

The pixels, or "wells," of a cooled CCD camera are designed to accumulate charge (electrons) until they can be read at the end of the exposure. Amplifiers that read the charge are often the limiting source of noise in faint bioluminescence images, and this "read noise" increases when the sensor is read at high speed. Thus, many cooled CCD cameras require several seconds to pass data to the computer, which then displays the image. At high light intensities, "shot noise"—uncertainty in the photon flux—becomes the limiting noise, while read noise then has only a negligible effect on the total system noise. See Chapter 3 for further discussion of noise in CCD cameras.

Radiation events, or "cosmic rays," are apparent in CCD images when exposure times are longer than 10 min. These excessively bright areas, typically consisting of only one or two pixels, are generated by external radiation effects in the atmosphere or events within the silicon. Color Figure 23–3 shows cosmic ray noise in a raw image of a brain slice culture from the *fos/luc* mouse that can be removed by taking the lowest pixel values from two sequential images. Binning increases the likelihood of losing data to a cosmic ray event because several pixels can be ruined at once; but binning increases the S/N ratio by producing only one unit of read noise for all pixels of a bin rather than one for each pixel.

5.2.2 ICCD cameras

ICCD cameras in photon counting mode have been useful in several studies with live cells expressing luciferase. Each photon captured by the photocathode of the ICCD camera is converted to electrons by amplification through a narrow cylinder-shaped PMT, resulting in a cluster of charge at the opposite end that produces a flash of light at a specific position on a phosphor screen. A standard non-cooled CCD camera images the flashes. The "microchannel plate" of the intensifier consists of thousands of PMTs that form a two-dimensional array detector. Software is often applied to the video image to determine whether each flash is likely due to an incident photon or to thermal noise within the intensifier. Background noise is a constant presence and a limitation in intensified camera images, although extreme sensitivity and the ability to capture images in real time are distinct advantages. ICCD camera systems for photon counting often cost twice as much as those based on LN-cooled CCDs, generally have fewer pixels, miss photons incident between the microchannels, and can be destroyed by excessive light exposure during operation because of the high voltage applied across the microchannel plate.

A similar but unique camera uses a resistive anode to detect the position of the charge delivered by a stack of microchannel plates (Photek Inc, Tuscon, AZ). The x- and y-coordinates of each photon event are sent to a computer, which creates the image in real time. This position-sensitive detector provides very sensitive and fast detection of very low-intensity images.

6. OPTICAL CONSIDERATIONS

Standard microscope objectives and larger lenses, such as those in 35 mm cameras and video systems, are also used to record luciferase and aequorin bioluminescence. Relay lenses, mirrors, or prisms between the primary lens and the camera are avoided to prevent light loss at glass surfaces. The camera can be

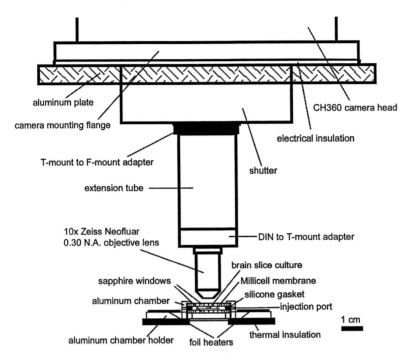

FIGURE 23–4 A simple system for long-term imaging of firefly luciferase in brain slices and dispersed cells. A microscope objective is coupled to the shutter of an LN-cooled CCD camera (Photometrics CH360). An aluminum plate mounted on steel rods supports the camera and prevents LN spills on the preparation. Plexiglas insulation between the camera and the plate provides electrical isolation to reduce noise. The culture chamber is manipulated using an upright microscope with all optics removed (not shown). The entire system is on a vibration-damping table in a darkroom with cables from the camera passing through a wall to a computer. The microscope can be removed for macroscopic imaging with a 35 mm camera lens.

placed at a port directly below the stage of inverted microscopes, as long as LN-type cameras can be adequately accessed for filling.

Older microscope objectives with fixed tube lengths can also be coupled directly to the camera with a DIN-to-C mount and spacers (Edmund Scientific, Barrington, NJ) (Fig. 23–4). Suitable images are possible even when the distance between the lens and the camera is less than the tube length. Distances shorter than the tube length reduce image magnification accordingly.

As with fluorescence microscopy, high numerical aperture (NA) lenses capture more bioluminescent light. Unlike fluorescence, however, the signal does not strengthen as excitation light is concentrated at high NA, and higher magnification, which usually includes higher NA, may not always improve image quality. At higher magnification, light is distributed over more of the sensor, lowering the S/N ratio at each pixel; furthermore, additional lens elements used in high-power objectives can filter out a substantial amount of the signal due to light losses at each of the many glass-to-air interfaces (Inoué and Spring, 1997). Using the highest NA at a given magnification is preferred because the light-gathering power of the lens equals the square of the NA divided by the square of the magnification (Inoué and Spring, 1997). Consequently, images with the highest S/N ratio often can be obtained at ×10 or ×20, which also provide large depth of focus.

When attached directly to the camera, the lens is most easily focused by moving the sample rather than the lens or camera. Proper focus is particularly critical for bioluminescence imaging because of the rapidly declining intensity of the already faint signal with distance from the focal plane. A vibration-damping table equipped with air-controlled pistons below a heavy steel surface is important to minimize any movement of the cells. Many cells, of course, are able to move on their own, change shape, and divide, thereby blurring images during long integration times. Methods must be used that fix cells or tissues as well as possible while also maintaining their viability and minimizing exposure times.

7. IMAGING CHAMBERS

The optimal chamber design for bioluminescence and other low-light imaging experiments should have good optical access, temperature control, sterility, fluid access, and a small internal volume to minimize luciferin usage. A Sykes-Moore type of chemotaxis chamber (Bellco Glass Inc., Vineland, NJ) is a good device for imaging brain slice explants maintained in stationary organotypic culture that have been made using the interface culture technique of Stoppini et al. (1991). In this method, a porous, translucent Teflon membrane (Millicell-CM, Millipore, Bedford, MA) supports a brain slice tissue explant on a disk of agarose gel mixed with culture medium, serum, and a buffered salt solution. Phenol red is omitted from all culture media to prevent signal loss. Slices can also be placed on the membrane alone within a thin film of medium to maximize gas exchange. Synthetic sapphire (Edmunds Scientific) is used for the windows of the chamber because of its high thermal conduction, preventing condensation inside the chamber window. This configuration allows imaging of individual cells within slices (Chandler et al., 1999). Individual dispersed cells can also be imaged by replacing the sapphire window with cell cultures grown on glass coverslips, which are imaged inverted with medium filling the chamber. Ports can be created for exchanging medium by inserting 22 gauge syringe needles. Open chambers are useful for imaging cells or tissues on an inverted microscope. In this case, the Teflon membrane can be placed over the tissue to hold it in place.

A temperature controller that provides proportional heating (e.g., CellMicro Systems, Virginia Beach, VA) ensures cell viability as well as optimal temperature for the luciferase reaction. Purified luciferase unfolds near 37°C (White et al., 1996), so the optimal compromise between luciferase activity and mammalian cell metabolism may be near 30°C, although 37°C is also used.

Imaging in complete darkness is essential to avoid background light that can quickly saturate the detector. In addition, potentially phosphorescent materials near the preparation, including certain plastics in white multiwell plates, must be shielded from direct sustained fluorescent light or sunlight before imaging. Finally, cells from some organisms have a high endogenous luminescence, while others can generate a faint background signal or ultraweak bioluminescence (Zhuravlev, 1990) that is below the sensitivity of most cameras.

8. BIOLOGICAL APPLICATIONS

8.1 The Development of Imaging Techniques Using Live Cells

Firefly luciferase was used in transgenic mice in 1988 as a reporter gene to study promoter activity, although cells were not imaged (DiLella et al., 1988). Transgenic mice were also made with a prolactin-luciferase fusion gene, again relying on measurements made with cell lysates (Crenshaw et al., 1989). At the same time, substantial improvements were made in luciferase cell lysate assays for measuring gene expression (Brasier et al., 1989). A direct comparison was made between *luc* and the chloramphenicol acetyltransferase (CAT) reporter gene by placing both under control of the interleukin-2 promoter (Williams et al., 1989). Luciferase was shown to be a far more sensitive reporter gene than the much more widely used CAT, with a very linear response and extremely accurate detection (10^{-20} moles) of gene product (de Wet et al., 1987; Pazzagli et al., 1992). Equally important, studies demonstrated that luciferase has a much shorter half-life than CAT—1–2 h in most cells at 37°C (Day et al., 1998; Nunez et al., 1998)—allowing detection of rapidly changing gene activity.

Only a few studies have used firefly luciferase for imaging live cells, in contrast to numerous studies using luciferase in cell lysates. Gene constructs coding for fusion proteins consisting of luciferase combined with a normal cell protein have provided even greater versatility. Photon-counting ICCD cameras were available for early studies of mammalian cells expressing luciferase. Later experiments began to use LN-cooled cameras, and techniques progressed to single-cell imaging and whole-organism studies.

Gene expression was imaged in single cells within a population of cells using luciferase in two recent studies. A gonadotropin-releasing hormone (GnRH) cell line expressing luciferase under control of the GnRH promoter showed pulsatile luminescence in live cells that may represent a previously unknown rapid regulatory switching of gene expression (Nunez et al., 1998). In the second study, a pituitary cell line containing the prolactin promoter upstream from *luc* showed surprising heterogeneity in individual cell expression levels in response to expression-inducing agents, although that response was not observed in the entire cell population (Takasuka et al., 1998). These phenomena are probably not unique to these cell lines but also occur in normal tissue; similar oscillations and independent signal fluctuations were observed in brain slice cells from a transgenic mouse containing the promoter/enhancer of the cytomegalovirus major immediate-early gene 1 upstream from *luc* (Chandler et al., 1999).

Early experiments used viral promoters to generate high *luc* expression (White et al., 1995). More recently, luciferase expressed by viral pathogens in the organs of live rats was imaged through the skin with a photon-counting ICCD system (Contag et al., 1997). This approach, described as a way to monitor disease progression, uses

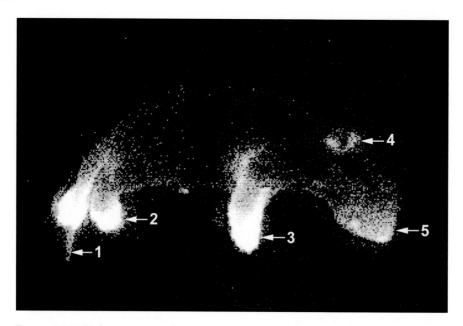

FIGURE 23–5 Bioluminescence from an intact neonatal *fos::luc* mouse. A newborn transgenic mouse anesthetized with sodium pentobarbital and imaged for 10 min with a photon-counting CCD camera system (Hamamatsu VIM) equipped with a 50 mm Nikkor f/1. 1 lens. Expression is high in the tail (1), paws (2, 3), ears (4), and snout (5). Signal strength is suitable for whole-animal screening of expression that could obviate the need for more invasive testing with many more animals.

less rats than standard assays that require biopsies to determine how the infection progresses, and results are obtained faster and with less statistical variation. Similarly, transgenic mice were shown to express the *fos::luc* reporter gene in the skin, brain, and other organs (Geusz et al., 1997), and bioluminescence can be imaged in intact anesthetized mice (Fig. 23–5).

8.2 Imaging Circadian Rhythms in Bioluminescence

Benefits of long-term luciferase imaging are perhaps most evident in studies of circadian rhythms. These near-24 hour oscillations in the physiology and behavior of organisms are spontaneously generated in cells of particular tissues and are best understood in the nervous system, where they control subordinate oscillators throughout the body. Recently, much progress has been made in describing the molecular basis of the rhythm's central cycle in *Drosophila*, mice and in the mold *Neurospora* (Dunlap, 1999). Mutations affecting the period of circadian rhythms were imaged in *Arabidopsis* (Millar et al., 1995) using *luc*, and aequorin was used in transgenic plants to record circadian rhythms in intracellular Ca^{2+} (Johnson et al., 1995). Live *Drosophila* expressing *luc* controlled by the circadian pacemaker gene *per* were imaged within different regions of the fly (Plautz et al., 1997a,b).

Long-term cultures of brain slice explants from *fos::luc* transgenic mice showed circadian rhythms when recorded with the HC-135 photon-counting module

(Geusz et al., 1997). The same device was used to record circadian rhythms from neural, liver, lung, and muscle cultures from transgenic rats containing a fusion protein made from the mammalian *per* gene and *luc* (Yamazaki et al., 2000). This study showed how the effects of jet lag could be due to the differential ability of oscillators in the body to re-establish correct phase relationships with each other. Images of the *fos::luc* signal in the major circadian pacemaker, the suprachiasmatic nucleus (SCN), of brain slices may reveal the cell types or areas of the SCN responsible for generating rhythms (Color Fig. 23–3).

8.3 Intracellular ATP Measurements

Despite the well-established use of luciferase to assay ATP in solution, attempts to perform the same measurement in living cells have been difficult because of the low Km for ATP (63 μM) (Lemasters and Hackenbrock, 1977; DeLuca et al., 1979). Luciferase assays are particularly effective at ATP concentrations below 1 μM (Brolin and Wetermark, 1992), which is 1000 times lower than normal cytosolic levels. Three recombinant luciferase molecules that target the enzyme to different cell compartments had unexpected Km values in the low mM range and responded to ATP changes at these locations (Kennedy et al., 1999). Large declines in intracellular ATP levels, around 10%, in response to mitochondrial poisons have been monitored with photon counting and luciferase injected into rat hepatocytes (Koop and Cobbold, 1993). Impending cell death and loss of ATP should be considered whenever a precipitous decline in the luciferase signal is observed.

8.4 Dual Reporter Genes

It is often useful to rapidly screen cell cultures or tissues to locate cells expressing a particular gene. One approach is to introduce by transfection a dual-reporter gene coding for a fusion protein composed of firefly luciferase and GFP (Day et al., 1998). Long-term imaging can be performed by applying luciferin after using fluorescence to locate transfected cells. In a related procedure, two bioluminescent reporter genes were used in primary islet β-cells that were co-injected with a plasmid containing the upstream flanking sequence of the pyruvate kinase gene driving *luc* and a plasmid containing the gene for *Renilla* luciferase under the control of the CMV immediate-early gene 1 enhancer/promoter (Kennedy et al., 1997). Because the two bioluminescence enzymes use different substrates, *luc* signals could be normalized to those from *Renilla* luciferase to provide an accurate measure of gene induction.

8.5 Bioluminescence Resonance Energy Transfer

Bioluminescence resonance energy transfer (BRET) is a powerful technique that can be combined with imaging to provide information on molecular interactions (Xu et al., 1999). As in fluorescence resonance energy transfer (see Chapter 15–17), an excited donor molecule delivers energy to an acceptor molecule only when they are within 10 to 100 Å of each other, causing a spectral shift in the emission that can be accurately described by the ratio of light intensity at two wavelengths. Using a bioluminescent donor molecule such as *Renilla* luciferase avoids any phototoxic damage due to fluorescence excitation and prevents inadvertent excitation of the acceptor molecule by the excitation light. *Renilla* luciferase has been used effectively with an enhanced yellow fluorescent protein as acceptor (Xu et al., 1999).

8.6 Thermal Stability of Luciferase

Experiments have not been limited to the North American firefly. Sequence homologies and similarities between *Luciola cruciata*, the Japanese firefly, and *Photinus* luciferase were described quite early (Masuda et al., 1989). Modifications of the *Luciola* gene produced a protein with greater stability at $37°C$ than *Photinus* or the native *Luciola* luciferase (Kajiyama and Nakano, 1993). Normal thermal instability may be preferred when rapid changes in gene expression are of interest (Leach and Webster, 1986). A longer half-life, which may be optimal in some systems in order to provide longer detection time, may become possible with further bioengineering of luciferase.

9. Conclusion

Current techniques for imaging bioluminescence enable continuous measurements of gene expression in individual live cells. A simple objective lens and LN-cooled CCD camera system can be used for long-term bioluminescence imaging. This experimental technique is still in its infancy but offers many possibilities.

References

Aflalo, C. Biologically localized firefly luciferase: A tool to study cellular processes. *Int. Rev. Cytol.* 130:269–323, 1991.

Billard, P., and M. S. DuBow. Bioluminescence-based assays for detection and characterization of bacteria and chemicals in clinical laboratories. *Clin. Biochem.* 31:1–14, 1998.

Brandes, C., J. D. Plautz, R. Stanewsky, C. F. Jamison, M. Straume, K. V. Wood, S. A. Kay, and J. C. Hall. Novel features of *Drosophila* period transcription revealed by real-time luciferase reporting. *Neuron* 16:687–692, 1996.

Brasier, A. R., J. E. Tate, and J. F. Habener. Optimized use of the firefly luciferase assay as a reporter gene in mammalian cell lines. *Biotechniques* 7:1116–1122, 1989.

Broda, H., V. D. Gooch, W. Taylor, N. Aiuto, and J. W. Hastings. Acquisition of circadian bioluminescence data in *Gonyaulax* and an effect of the measurement procedure on the period of the rhythm. *J. Biol. Rhythms* 1:251–263, 1986.

Brolin, S. E., and G. Wetermark. *Bioluminescence Analysis*. New York: VCH Publishers, Inc., 1992, pp. 47–56.

Chandler, T. R., L. A. Sigworth, and M. E. Geusz. A luciferase reporter gene reveals expression in cells of the dorsal suprachiasmatic nucleus. *Soc. Neurosci. Abstr.* 25:1371, 1999.

Contag, C. H., S. D. Spilman, P. R. Contag, M. Oshiro, B. Eames, P. Dennery, D. K. Stevenson, and D. A. Benaron. Visualizing gene expression in living mammals using a bioluminescent reporter. *Photochem. Photobiol.* 66:523–531, 1997.

Crenshaw, E. B. D., K. Kalla, D. M. Simmons, L. W. Swanson, and M. G. Rosenfeld. Cell-specific expression of the prolactin gene in transgenic mice is controlled by synergistic interactions between promoter and enhancer elements. *Genes Dev.* 3:959–972, 1989.

Day, R. N., M. Kawecki, and D. Berry. Dual-function reporter protein for analysis of gene expression in living cells. *Biotechniques* 25:848–850, 852–844, 856, 1998.

DeLuca, M., and W. D. McElroy. Purification and properties of firefly luciferase. *Methods Enzymol.* 57:3–15, 1978.

DeLuca, M., J. Wannlund, and W. D. McElroy. Factors affecting the kinetics of light emission from crude and purified firefly luciferase. *Anal. Biochem.* 95:194–198, 1979.

de Wet, J. R., K. V. Wood, M. DeLuca, D. R. Helinski, and S. Subramani. Firefly luciferase gene: Structure and expression in mammalian cells. *Mol. Cell. Biol.* 7:725–737, 1987.

DiLella, A. G., D. A. Hope, H. Chen, M. Trumbauer, R. J. Schwartz, and R. G. Smith. Utility of firefly luciferase as a reporter gene for promoter activity in transgenic mice. *Nucleic Acids Res.* 16:4159, 1988.

Dunlap, J. C. Molecular bases for circadian clocks. *Cell* 96:271–290, 1999.

Gandelman, O., I. Allue, K. Bowers, and P. Cobbold. Cytoplasmic factors that affect the intensity and stability of bioluminescence from firefly luciferase in living mammalian cells. *J. Biolumin. Chemilumin.* 9:363–371, 1994.

Geusz, M. E., C. Fletcher, G. D. Block, M. Straume, N. G. Copeland, N. A. Jenkins, S. A. Kay, and R. N. Day. Long-term monitoring of circadian rhythms in c-*fos* gene expression from suprachiasmatic nucleus cultures. *Curr. Biol.* 7:758–766, 1997.

Gould, S. J., and S. Subramani. Firefly luciferase as a tool in molecular and cell biology. *Anal. Biochem.* 175:5–13, 1988.

Inoué, S., and K. R. Spring. *Video Microscopy: The Fundamentals*. New York: Plenum Press, 1997, 741 pp.

Johnson, C. H., M. R. Knight, T. Kondo, P. Masson, J. Sedbrook, A. Haley, and A. Trewavas. Circadian oscillations of cytosolic and chloroplastic free calcium in plants. *Science* 269:1863–1865, 1995.

Jones, D. P., Sherf, B. A., Wood, K. V. Luciferase Assay System Vendor Comparison. *Promega Notes Magazine* 54:20, 1995.

Jones, K., F. Hibbert, and M. Keenan. Glowing jellyfish, luminescence and a molecule called coelenterazine. *Trends Biotechnol.* 17:477–481, 1999.

Kajiyama, N., and E. Nakano. Thermostabilization of firefly luciferase by a single amino acid substitution at position 217. *Biochemistry* 32:13795–13799, 1993.

Kennedy, H. J., A. E. Pouli, E. K. Ainscow, L. S. Jouaville, R. Rizzuto, and G. A. Rutter. Glucose generates sub-plasma membrane ATP microdomains in single islet beta-cells. Potential role for strategically located mitochondria. *J. Biol. Chem.* 274:13281–13291, 1999.

Kennedy, H. J., B. Viollet, I. Rafiq, A. Kahn, and G. A. Rutter. Upstream stimulatory factor-2 (USF2) activity is required for glucose stimulation of L-pyruvate kinase promoter activity in single living islet beta-cells. *J. Biol. Chem.* 272:20636–20640, 1997.

Kondo, T., and M. Ishiura. Circadian rhythms of cyanobacteria: Monitoring the biological clocks of individual colonies by bioluminescence. *J. Bacteriol.* 176:1881–1885, 1994.

Koop, A., and P. H. Cobbold. Continuous bioluminescent monitoring of cytoplasmic ATP in single isolated rat hepatocytes during metabolic poisoning. *Biochem. J.* 295:165–170, 1993.

Leach, F. R., and J. J. Webster. Commercially available firefly luciferase reagents. *Methods Enzymol.* 133:51–70, 1986.

Lemasters, J. J., and C. R. Hackenbrock. Kinetics of product inhibition during firefly luciferase luminescence. *Biochemistry* 16:445–447, 1977.

Lembert, N., and L. A. Idahl. Regulatory effects of ATP and luciferin on firefly luciferase activity. *Biochem. J.* 305:929–933, 1995.

Lorenz, W. W., R. O. McCann, M. Longiaru, and M. J. Cormier. Isolation and expression of a cDNA encoding *Renilla reniformis* luciferase. *Proc. Natl. Acad. Sci. U.S.A.* 88:4438–4442, 1991.

Masuda, T., H. Tatsumi, and E. Nakano. Cloning and sequence analysis of cDNA for luciferase of a Japanese firefly, *Luciola cruciata. Gene* 77:265–270, 1989.

Millar, A. J., I. A. Carre, C. A. Strayer, N. H. Chua, and S. A. Kay. Circadian clock mutants in *Arabidopsis* identified by luciferase imaging. *Science* 267:1161–1163, 1995.

Millar, A. J., and S. A. Kay. Integration of circadian and phototransduction pathways in the network controlling CAB gene transcription in *Arabidopsis. Proc. Natl. Acad. Sci. U.S.A.* 93:15491–15496, 1996.

Nunez, L., W. J. Faught, and L. S. Frawley. Episodic gonadotropin-releasing hormone gene expression revealed by dynamic monitoring of luciferase reporter activity in single, living neurons. *Proc Natl Acad Sci U. S. A.* 95:9648–9653, 1998.

Pazzagli, M., J. H. Devine, D. O. Peterson, and T. O. Baldwin. Use of bacterial and firefly luciferases as reporter genes in DEAE–dextran–mediated transfection of mammalian cells. *Anal. Biochem.* 204:315–323, 1992.

Plautz, J. D., M. Kaneko, J. C. Hall, and S. A. Kay. Independent photoreceptive circadian clocks throughout *Drosophila. Science* 278:1632–1635, 1997a.

Plautz, J. D., M. Straume, R. Stanewsky, C. F. Jamison, C. Brandes, H. B. Dowse, J. C. Hall, and S. A. Kay. Quantitative analysis of *Drosophila* period gene transcription in living animals. *J. Biol. Rhythms* 12:204–217, 1997b.

Shimomura, O., and F. H. Johnson. Chemical nature of bioluminescence systems in coelenterates. *Proc. Natl. Acad. Sci. U.S.A.* 72:1546–1549, 1975.

Shimomura, O., B. Musicki, and Y. Kishi. Semi-synthetic aequorin. An improved tool for the measurement of calcium ion concentration. *Biochem. J.* 251:405–410, 1988.

Stoppini, L., P. A. Buchs, and D. Muller. A simple method for organotypic cultures of nervous tissue. *J. Neurosci. Methods* 37:173–182, 1991.

Takasuka, N., M. R. White, C. D. Wood, W. R. Robertson, and J. R. Davis. Dynamic changes in prolactin promoter activation in individual living lactotrophic cells. *Endocrinology* 139:1361–1368, 1998.

White, M. R., M. Masuko, L. Amet, G. Elliott, M. Braddock, A. J. Kingsman, and S. M. Kingsman. Real-time analysis of the transcriptional regulation of HIV and hCMV promoters in single mammalian cells. *J. Cell. Sci.* 108:441–455, 1995.

White, P. J., D. J. Squirrell, P. Arnaud, C. R. Lowe, and J. A. Murray. Improved thermostability of the North American firefly luciferase: Saturation mutagenesis at position 354. *Biochem. J.* 319:343–350, 1996.

Williams, T. M., J. E. Burlein, S. Ogden, L. J. Kricka, and J. A. Kant. Advantages of firefly luciferase as a reporter gene: Application to the interleukin-2 gene promoter. *Anal. Biochem.* 176:28–32, 1989.

Xu, Y., D. W. Piston, and C. H. Johnson. A bioluminescence resonance energy transfer (BRET) system: Application to interacting circadian clock proteins. *Proc. Natl. Acad. Sci. U.S.A.* 96:151–156, 1999.

Yamazaki, S. et al. Resetting central and peripheral oscillators in transgenic rats. *Science* 288:682–685, 2000.

Zhuravlev, A. I. Spontaneous superlow chemiluminescence and creation of quantum biology. In: *Biological Luminescence*, edited by B. Jezowska-Trzebiatowska, B. Kochel, J. Slawinski, and W. Strek. Teaneck, NJ: World Scientific, 1990, pp. 19–48.

24

Imaging Living Cells and Mapping Their Surface Molecules with the Atomic Force Microscope

MUHAMMED GAD AND ATSUSHI IKAI

1. INTRODUCTION

In 1986 Gerd Binnig and Heinrich Roher shared the Nobel Prize in Physics for inventing the scanning tunneling microscope (STM) and discovering that it can image individual atoms with unprecedented resolution (Binnig et al., 1982). This novel type of microscopy is based on the quantum phenomenon that electrons can tunnel through a narrow insulating gap between two conductors.

Several scanning probe microscopes (SPMs) have been developed in the past few years, such as the atomic force microscope (Binnig et al., 1986), the photon scanning tunneling microscope (Ferrell et al., 1991), the scanning ion-conductance microscope (Hansma et al., 1989), the thermal profiler microscope (Williams and Wickramasinghe, 1986), and the magnetic force microscope (Martin and Wickramasinghe, 1987). These scanning probe microscopes use contrast formation mechanisms other than electron tunneling. The use of these microscopes has become routine practice in obtaining high-resolution images of insulating surfaces.

This chapter will explain the instrumentation involved in atomic force microscopy (AFM) imaging and the details of two biological applications. The first one is a cellular level application that shows the potential of AFM to study the dynamics of living cells at high resolution. A novel immobilization method using agar gel is presented. The yeast *Saccharomyces cereviciae* was used as a model plant cell for this study. The second application is one on a molecular level that shows the possibility of mapping a particular molecular species among a heterogeneous population of molecules on a living cell surface. For this purpose, cell surface mannan of the same yeast *S. cereviciae* was mapped using a concanavalin functionalized AFM tip, as will be shown later.

2. METHODOLOGY AND INSTRUMENTATION

2.1 Why AFM Is a Good Candidate for Studying Biological Materials

The resolution achieved by AFM is not limited by the wavelength of radiation used to form the image, but rather is determined by the finest tip diameter and tip–sample gap spacing that can be achieved. Because it is possible for AFM to image both conductors and insulators, it became a routine practice to image biological materials without metal coating. In effect, sample preparation was reduced to a minimum. Moreover, real-time three-dimensional images can easily be obtained (Binnig et al., 1986), even under fluids (Drake et al., 1989; Hansma et al., 1992). AFM can be also used as a nanoindentor to measure the microelastic properties of various materials, including biological ones (Shroff et al., 1994). Another unique capability of AFM is that it can be used to measure specific as well as nonspecific forces between the tip and the sample.

2.2 Atomic Force Microscopy Substrates

Because AFM images the surface of a sample, the sample should be supported on a substrate, such as mica or glass. Mica is an atomically flat, cleavable aluminum silicate crystal, which is useful for imaging molecules less than a few nanometers high. Although glass, with a typical surface roughness of 1–2 nm, is generally not used for imaging thin molecules such as DNA at high resolution, it serves as a good substrate for imaging thicker biological samples like cells. Both glass and mica have negative surface charge, which must be taken into account when developing conditions for binding the sample to the substrates. An ultraflat gold surface was also used for immobilizing biomolecules by chemisorption of thiol groups that may be introduced to the molecule of interest (Martin et al., 1993). It is possible to image living cells under culture conditions while being held in a gel matrix (Gad and Ikai, 1995).

2.3 Description of the Atomic Force Microscope Instrumentation

Whatever the origins of the force, all force microscopes have five essential components (Fig. 24–1):

1. A sharp tip mounted to a soft cantilever spring
2. A way of sensing the cantilever deflections
3. A feedback system to monitor and control the deflection (or, in other words, the interaction force between the atoms of the tip and the atoms of the sample)
4. A mechanical scanning system (usually piezoelectric) that moves the sample with respect to the tip in a raster pattern
5. A display system that converts the measured data into an image

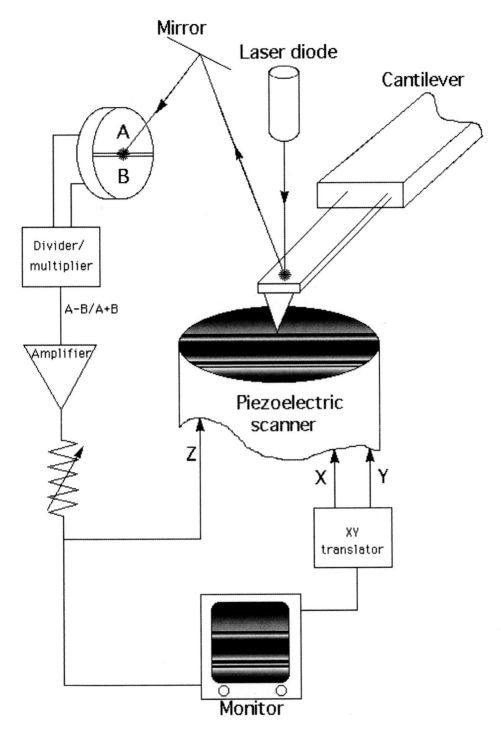

FIGURE 24–1 Schematic presentation showing the main components of an atomic force microscope.

Commercially available cantilevers are microfabricated from silicon, silicon oxide, or silicon nitride using photolithographic techniques (Albrecht and Quate, 1988). Typical lateral dimensions are about 100 μm with a thickness on the order of 1 μm and a width ranging from 27 to 30 μm. This geometry gives spring constants in the range of 0.1–1 N/m and resonant frequencies of 10–100 kHz.

The second important component of AFM is the sensor that detects the cantilever's deflections. Ideally, the sensor should have subangstrom sensitivity and should exert negligible forces on the cantilever. Optical detection schemes are the most commonly used ones with AFM. They are divided into two basic types: interferometry and beam deflection (Martin and Wickramasinghe, 1987; Meyer and Amer, 1988). Both of these methods are capable of measuring cantilever deflections about 0.1 Å. In a typical beam-deflection AFM, light from a laser diode is reflected from the back side of a mirror-like cantilever. The direction of the reflected light beam is sensed with a position-sensitive (bisected) photodetector. The amount of the beam light that is detected by the upper part of the detector is defined as A, while that of the lower part is defined as B. Thus, the tension between the tip and the sample can be recognized as the change of $(A - B)/(A + B)$. The sample to be imaged is mounted on the piezoelectric scanner, and the feedback gain is added to the piezo to keep $(A - B)/(A + B)$ constant. The signals are amplified and transmitted to the feedback loop of the microscope. Finally, the signal is converted by the display system into an image.

2.4 Operation Modes of Atomic Force Microscopes

Atomic force microscopes have three major operation modes: contact, noncontact, and tapping modes. In this chapter we focus mainly on the contact mode. Four types of measurements can be carried out by contact AFM:

1. Topography
2. Force measurement
3. Friction forces
4. Force modulation

The first two measurements are considered in details in this chapter.

2.4.1 Contact mode topography measurement

In contact mode, the interacting forces between the tip and the sample are very short-range repulsive forces, which make it possible to achieve atomic resolution (Fig. 24–1). The contact mode can be operated in two different ways.

The first is constant force mode (height mode), in which the cantilever deflection is maintained constant and the image contrast is due to z variations of the piezoelectric scanner. The height can be measured accurately from the resulting image. The second mode is the constant height mode in which the sample height is kept constant and the cantilever deflections are recorded. Thus, the image contrast is now due to the local force variations and is detected by the variation in the signal of the photodetector.

In the contact mode AFM the tension between the tip and the sample ranges from 10 to 100 nN. This force is applied to a small area of about 50 nm and causes a pressure of 100–1000 MPa, which is high enough to induce indentations on soft biological samples. By scanning under liquids, the capillary forces are removed, thus decreasing the load force to a few nanonewtons.

2.4.2 Force measurement

The capability of AFM to generate force versus distance curves proved to be a useful tool for measuring forces in the range of piconewtons. In the biological field, there is an increased interest to make use of that capability to understand biological processes that are largely governed by intermolecular forces (Florin et al., 1994; Radmacher et al., 1994; Dammer et al., 1996; Gad et al., 1997).

In this chapter we discuss two biological applications of cellular imaging and molecular level mapping of cell surface molecules. The first application is a method that addresses the possibility of studying cellular level dynamics using AFM.

Drake et al. (1989) presented the first dynamic study by AFM for the clotting process of the human blood protein fibrinogen that clearly showed the potential of AFM in the study of in situ biological processes. There have, however, been some limitations that have restricted the use of AFM to investigate real-time biological processes under physiological conditions. First, a properly fixed sample is a prerequisite for successful imaging. Second, a sample having a maximum height less than 1 μm is recommended. Third, tip sample affinity should not exceed a certain limit for a stable measurement.

Although immobilizing the sample has always been the major difficulty for imaging globular and nonadhesive samples under fluids, many successful methods have been developed to modify the substrates to allow the sample to absorb them physically or chemically. Many of these methods have proved to be suitable for biological applications (Bezanilla et al., 1993; Butt et al., 1990; Hansma et al., 1993; Hegner et al., 1993; Thundat et al., 1992; Yang et al., 1994; Henderson, 1994; Häberle et al., 1992). Only a few cases of imaging of live plant cells have, however, been reported thus far (Butt et al., 1990; Hörber et al., 1992; Kasas and Ikai, 1995). This is probably because there is a question as to the cell's proper fixation on the substrate and to the maintenance of stable scanning over tall objects.

Plant cells usually have very rigid and chemically inert walls that make up their shapes and protect them from rupturing. As a result, there is no tendency for them to spread on the substrate as animal cells do. The contact area between a plant cell and the substrate is much smaller than in the case of animal cells. Such situations make it difficult for them to withstand the lateral forces exerted by the AFM tip during the scanning operation. There is also the possibility that they might detach themselves from the substrate even if they are covalently immobilized to it while growing. This is due to the fact that the surface layer of the cell wall, in some microbial cells, peels off the cell surface as the cell expands (Cooper, 1991).

3. Applications

3.1 Atomic Force Microscope Imaging of the Dynamics of Yeast Cells' Growth

A method for immobilizing living haploid cells of *S. cerevisiae* on agar surface without using any chemicals has been developed.

3.1.1 Cell culture

A haploid strain of *S. cerevisiae* X2180-1A from the Yeast Genetic Stock Center at Berkeley was used in this study. They were grown in 10 ml of a medium containing 1% yeast extract (Difco), 2% polypeptone, and 2% glucose at 25°C with reciprocal shaking at 150 strokes/min.

3.1.2 Sample preparation

The cells were harvested in the logarithmic phase of growth by centrifugation at 2150g for 5 min, washed once with distilled water at 20°C, and suspended in 0.25 ml of the same fluid. Five microliters of a highly concentrated cell suspension was deposited on a clean circular coverglass of 15 mm in diameter. About 200 μl of 3% molten agar at 45°–50°C was dropped over the deposited cells by a micropipette. Another piece of coverglass was quickly put on the agar before it hardened, to create a flat surface almost parallel to the lower one.

On addition of the molten agar, the droplet of the cell suspension spread almost homogeneously over the coverglass. The agar was allowed to completely solidify for 30 min, and then the sample was inverted upside down. The lower coverglass was gently pulled parallel to the agar surface, while the agar disk was held between the index and the thumb fingers and bent slowly away from the coverslip facing the cells. Most of the yeast cells were found to be localized on that side of the agar disk because the high viscosity of molten agar prevented them from diffusing deep into the agar before it solidified. Weakly captured or uncaptured cells were washed off with deionized water. A control sample was prepared by fixing the yeast cells in 2% glutaraldehyde for 30 min, washing them with distilled water, and immobilizing them on agar, as described above.

3.1.3 Imaging

The cell-free side of the agar disk was dried carefully on a sheet of tissue paper and was put on the metal disk of the AFM without using glue. The liquid cell of AFM was set up, and a small volume of liquid culture medium was injected into the liquid cell to form a thin water interface between the agar surface and the liquid cell.

Images were obtained by contact mode AFM. We used sharpened silicon nitride tips with a spring constant of 0.06 N/m and a J scanner (165 μm maximal scan size). It was possible to image the same area continuously for more than 7 h with a time lapse between each subsequent image of 6 min. This was possible because additional culture medium was injected at intervals of the scanning to substitute the evaporating water.

Room temperature was kept at 22°C during imaging. Scanning parameters were as follows: Scan size was 20 μm, scan angle was 35°, scan rate was

0.702 Hz, integral gain was 0.266, proportional gain was 0.318, and set point was -1.95 V.

We recorded the error signal images to show the fine details on the surface of the cells (Putman et al., 1992; Neagu et al., 1994). In this mode, the feedback loop compensates for the large cantilever deflections by keeping the sample height constant, resulting in the height image. The remaining deflection signal was amplified and recorded, giving the error signal image.

Incubating some of the agar disks overnight in the same culture broth mentioned above checked the viability of the immobilized cells prepared (as described above). We noticed thick cell coverage on the agar surface after incubation. In addition, on examining the sample surface after imaging for several hours, it was possible to notice the formation of minute colonies.

Typical results with real-time images of live yeast cells are presented in Color Figure 24–2A (frames a, b, c, d, e, and f), which shows that some of these cells are increasing in height and width (cell numbers 1 and 3). The corresponding height mode images are shown in Color Figure 24–2B (frames a, b, c, d, e, and f).

It was possible to ascertain that the rapid height changes (Color Fig. 24–2A, B) probably resulted from the growth activity of the immobilized cells and are not due to the slow dissolution, the mechanical scraping of the agar surface, or shifting of the positions of cells within the agar matrix. One line of evidence supporting the above assertion came from the study of vertical sections of the cells and the measurement of their height change using AFM software.

In Color Figure 24–2B, the height change of five representative cells numbered from 1 to 5 was plotted against time. The results (Fig. 24–3) show that while the heights of the cells numbered 1 to 3 increased by 440 nm, or more than 1 μm, those of cells 4 and 5 remained almost constant throughout the scanning period of 36 min. A slight increase in the height of cell number 4 might have been induced by the growth of the neighboring cell, which could be either the upper cell in Color Figure 24–2B frame a or cell number 2 that emerged from the hole shown in the same figure. We do not know whether cells 4 and 5 were dormant or dead, but the very fact that there were such cells that did not show any noticeable increase in their heights attested to our tentative conclusion that the agar surface was stable against the force exerted by the scanning tip. Also, there was no continuous dissolution of agar to a significant degree during the scanning period. It can be clearly seen in Color Figure 24–2B that cell number 2 grew faster than the adjacent cell number 1, although the agar matrix surrounding them was exposed to the same scanning force. Another important observation that supports our theory of cellular growth is that the cells were becoming taller compared with the agar level at different rates (Fig. 24–3).

The results of the control experiment performed on chemically fixed cells as described in the previous section were quite distinct from the results on live cells given above. We imaged the fixed cells for several hours. Throughout the scanning time, none of the imaged cells showed rapid changes in their height and width or detached from the substrate, except for during the initial 90 min when there was a very slow increase in their heights. The total height change during this 90 min period was roughly equivalent to that observed within one scan of 6 min

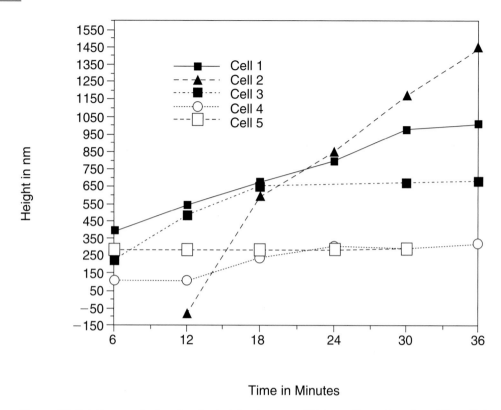

Time in Minutes

FIGURE 24–3 A line graph showing the height changes of living cells against time. The cells' numbers indicated correspond to the representative cells shown in Figure 24–4 and Color Figure 24–5.

in the case of live cells. This may be due to the removal of the very fine agar layer covering the periphery of the exposed cell surface. Another possibility that may explain this phenomenon is the shrinkage of the agar matrix as a result of the increasing osmotic pressure due to the evaporation of water. Furthermore, we tried to scrape the agar surface, by increasing the force applied by the AFM tip, hence the friction force, to the maximum setting. Surprisingly, none of the cells detached from the agar surface, indicating the firmness of adhesive force between agar and yeast cells when they are not alive.

The method presented here partly solved the height problem in AFM scanning, but as the cells grew in volume it became impossible to continue imaging them due to an inherent defect of AFM known as tip–sample convolution effect. This effect leads to the generation of a pyramidal shape artifact that is actually the tip shape and not the real sample topography.

3.2 Mapping a Living Cell Surface

We next explain molecular level application to show the potential of AFM in the study of cell surface molecules. In biology, specific intermolecular recognition is the basis for many biochemical processes. Such recognition processes involve a multitude of weak noncovalent bonds (e.g., electrostatic, van der Waals, and/or hydrogen bonds) or hydrophobic interactions between geometrically complementary surfaces. These interactions are highly specific and can be very strong, as in the molecular recognition between receptors and their ligands.

Several techniques have been developed to measure intermolecular forces. The optical trapping technique is one of them and is very sensitive to small forces, so its use has been limited to certain special samples and to measurements less than 10 pN (Ashkin et al., 1987; Kuo and Sheetz, 1993; Svoboda et al., 1993). Other techniques, such as the pipette suction and the surface force apparatus, are also sensitive, but lack a spatial resolution (Evans et al., 1991; Israelachvili, 1992). After the invention of AFM, a new precise technique was introduced that can probe surfaces in physiological environments with a high spatial resolution and can sense forces down to the piconewton range. Promising results were obtained from the biotin/streptavidin model systems, complementary strands of DNA, cell adhesion proteoglycans, and specific antigen--antibody interactions (Florin et al., 1994; Dammer et al., 1995, 1996; Hinterdorfer et al., 1996). All these measurements utilize the same basic idea of functionalizing the tip and the substrate with the complementary (or the same) type of molecules of interest by simple adsorption or by covalent immobilization using cross-linkers.

In our study, we have tried to use the force volume mode of AFM to map the distribution of a particular species of polysaccharides on a living microbial cell surface. We carried out a direct measurement of binding force between a functionalized tip (carrying specific lectin molecules) and its corresponding ligand (sugar) on a microbial cell surface. Such a probe should be capable of identifying its target molecules among a heterogeneous population of molecules on the surface, thus helping us to draw a map of the distribution of the molecules of interest. This is unlike the previous studies in which only preselected complementary molecules were used to carry out the measurements. The resulting force curves showed that binding events occurred presumably between the receptor protein on the tip and its sugar ligand on the cell surface.

For this purpose, a lectin from *Canavalia ensiformis,* namely, concanavalin A (conA) of molecular weight 102,300 that can bind to mannose residues of the yeast cell wall mannan, was used. The protein was derivatized with long-chain succinimidyl 6-[3'-(2-pyridyldithio)-propionamido] hexanoate (LC-SPDP) to introduce free sulfhydryl groups by reduction immediately before carrying out the measurement. The derivatized protein was first cross-linked to a gold-coated cantilever via Au-S bonds in a buffer solution. Yeast cells (*S. cerevisiae*) were immobilized in almost a monolayer form on a glass substrate that was previously coated with conA. The results indicated that a force in the order of 50–150 pN was needed to rupture the bonds formed between a single pair of ligand–receptor molecules. An interesting process was observed in which the sandwiched ligand–receptor complex was extended up to 500–600 nm or longer before the tip was freed. This type of force curve was only obtained when we used a covalently immobilized protein to the tip. We concluded that this behavior demonstrates a mechanical pulling of certain cell wall components, most likely mannoprotein polymers.

3.2.1 Sample preparation

The same yeast strain, *S. cerevisiae* X2180-1A, that was used in the immobilization method was cultured under the same condition, as described in Section 3.1.1. ConA was purchased from Sigma (St. Louis, MO) as a highly purified, essentially salt-free, lyophilized powder. It was reconstituted into a tetramer in phosphate-buffered saline (PBS) at pH 7 and was derivatized with LC-SPDP as previously described by Mitsuda et al. (1995). The derivatized protein was reduced with dithiothreitol (DTT) to make thiol groups available for the reaction with a gold-coated tip immediately before AFM measurement, so cross-linking between protein molecules would be avoided. The excess DTT was removed by passing the solution of the thiolated protein through a column of sephadex G-25 (Pharmacia, Uppsala, Sweden).

The average number of thiol groups that were introduced ranged from four to six as determined by spectrophotometric methods (Mitsuda et al., 1995). The reduced protein solution was concentrated by passing it through a Vivapore 10 ml concentrator from Vivascience (Lincoln, England). Before the derivatized protein was used, its biological activity was tested by a simple agglutination test on a slide glass. All the chemicals used in this experiment were of analytical grade.

3.2.2 Immobilization of cells

Cells were harvested by centrifugation and washed twice with distilled water. A 15 mm clean coverglass was coated with conA by applying 20 μl of 1 mg/ml native protein solution for 15 min. The coated substrate was gently washed with PBS buffer solution several times. Fifty microliters of a highly concentrated cell suspension was applied on a coated coverglass and was rotated slowly for a few seconds. The cells were then allowed to agglutinate for 15 min. Finally, the sample was washed with buffer and was left to partially dry for 10 min. This treatment allowed the cells to be imaged in a stable environment under PBS during AFM measurements.

3.2.3. Tip functionalization

Because the AFM used in this study did not have a combined light microscope to locate the cells, a bare gold-coated tip scanned the sample, and a large cell surface was zoomed into. The same tip was functionalized by disassembling the liquid cell as a whole from the optical head while holding the cantilever to minimize its displacement. Care was taken not to touch the spring that holds the cantilever in its housing in the liquid cell. It was then washed by injecting PBS buffer through the inlet rubber tubing of the liquid cell. Finally, it was functionalized by immersing it in a droplet of the SH-derivatized protein solution in a Petri dish by laying the liquid cell on a support of rolled vinyl tape. After 15 min of incubation at room temperature, the unbound protein was washed away by injecting more PBS, as described before. A model of the set-up configuration is shown in Figure 24–4a.

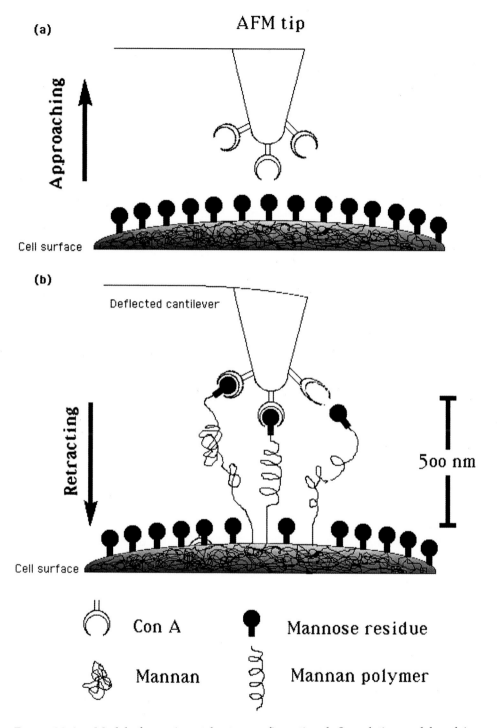

FIGURE 24–4 *a*, Model of experimental setup configuration. *b*, Speculative model explaining the stretching phenomenon of mannan.

3.2.4 Atomic force microscopy imaging of immobilized yeast cells

A coverglass on which the yeast cells were immobilized was glued on a metal disk of AFM. The liquid cell of the microscope was filled with a limited volume of PBS so that a thin water interface forms between the sample surface and the liquid cell. Gold-coated cantilevers with a spring constant of 0.025 N/m were used in this study. Force volume measurements were performed at 1 μm Z-scan range as gently as possible at 0.5 Hz to secure enough contact time for a ligand and receptor pair to recognize each other.

In the force volume mode, force curves were recorded while a square area of the sample was raster scanned under the tip. The maximum upward cantilever deflection was kept at 10 nm using the relative trigger mode to minimize the contact area between the tip and the cell surface. The scanned area was 3 μm $\times$ 3 μm, and the amount of data collected was 16 $\times$ 16 pixel for each scan.

The functionalized tip was lowered to the center of one of the immobilized cell surfaces. We were able to distinguish the cell surface in the low-resolution height image of the force volume mode.

A control experiment was performed using a bare gold-coated tip. There was only a weak adhesion between the tip and the sample. It is difficult to speculate the origin of such weak adhesions because the measurement was carried out on a heterogeneous surface that contained several classes of macromolecules.

Color Figure 24–5a shows a typical force volume data frame. The upper left image represents a height image for the area of interest. The upper right image represents a map of the elastic properties of the same area, also known as the *force volume image*. It is identified as a map of the cantilever deflection at each point on the sample surface at a constant *z* position of the sample (Laney et al., 1997). This image will not be considered in this chapter. The lower right plot represents the window where force curves of each data point in the height image are displayed. The plot in Color Figure 24–5a shows 60 superimposed curves. It can be noted that the tip frequently traveled a long distance with a continued downward deflection, suggesting that parts of the sample were pulled away from the cell surface. It is possible that the receptor ligand bonds pulled out several strands of mannoprotein polymer.

Unfolding conA does not seem to be contributing factor, because conA is a relatively low-molecular-weight tetrameric protein of MW = 102,300 that would never be extended to such a long distance if stretched. The shape of the negative cantilever deflection in a representative curve, like the one shown in Color Figure 24–5b, suggests that mannan polymers may be stuffed as random coils on the cell surface. Figure 24–4b shows a speculative model for such a phenomenon. Experiments with the surface force apparatus (Helm et al., 1991) and with pipette suction techniques (Evans et al., 1991) have revealed that lipids can be pulled out of a membrane by means of receptor–ligand bonds.

We have considered two types of pull-off points for each force curve. The first is the maximum deflection, and the second is the final pull-off point. Data collected from the maximum deflections were used to construct a map for the sugar distribution on the cell surface, while the last pull-off point was used to estimate the binding force between a single pair of conA and mannose residue as described previously by Florin et al. (1994) and Dammer et al. (1995). Because it is still difficult to hang a single small molecule like conA from the tip, we tried to limit the number of the active protein molecules by injecting 0.25% mannose solution into the liquid cell to block most of the available active sites, hoping to obtain single jump-off events. We noticed that the adhesion force frequencies between a conA functionalized tip and yeast cell wall mannan was reduced due to the reason mentioned above. The binding force was estimated to be in the order of 50–150 pN.

Eight surface graphs constructed from the maximum force values measured at each point of a force map for an area of 9 μm^2 are shown in Color Figure 24–6. A pair of two adjacent graphs represents subsequent scans of the same area in different experiments. The distribution of force in the two subsequent frames suggests that molecular recognition events can be reproduced to a large extent, indicating that conA functionalized tips are actually mapping the distribution of mannan on the cell surface. After several force map recordings, force profiles may change due to the following possibilities.

The first possibility is the inactivation of proteins on the tip due to the mechanical stress they were exposed to during force measurement cycles. The second possibility is a piezo scanner shift in the scan area, and a third is cell movement. Because mannan polymers may be floating as random coils on the cell surface, their movement may contribute to an increase in the frequency of the measured force within a small local area of several nanometers. This is due to their repeated interactions with conA on the tip as a result of their motion.

4. CONCLUSION

The first method presented in this chapter helped to image the dynamics of yeast cells that could be interpreted as their growth and budding processes. Because of the flexibility of the agar matrix, the cells were allowed to have a certain degree of freedom to expand within it. It was also possible to obtain high-resolution images for distinct structures on the cell surface (e.g., the bud and birth scars, image not shown). We believe that this method can be applied to other plant cells, as well as to nonliving bulky objects.

The second method showed that a functionalized AFM tip could be used to map the distribution of a molecular species on chemically heterogeneous surfaces like living cell surfaces. From our data we found that mannan is not uniformly distributed on some of the studied areas. Autoradiography and gold markers were previously used to study the location of mannan on the yeast cell. They showed that mannan was uniformly distributed on the outer layer of the cell wall (Farkas et al., 1974; Horisberger and Vonlanthen, 1977). Such contradictions may be due to the resolution differences in the above-mentioned techniques.

Although this technique has obvious applications in other fields, there are technical improvements that need to be dealt with in the future to achieve more accurate measurements. The most important is to develop easy methods to determine spring constants of cantilevers and immobilize a single molecule to the AFM tip. In the future, we think that data of kinetic importance can be obtained by studying the time dependence of specific molecular recognition events at increasing scan speeds. This will also require a precise technique to immobilize the molecules in a particular orientation at the tip surface to expose their active sites. Furthermore, it will be necessary to compare the numbers obtained from the energetics of the reaction and to study the effect of external parameters, such as temperature and pH.

References

Albrecht, T. R., and C. F. Quate. Atomic resolution with the atomic force microscope on conductors and nonconductors. *J. Vac. Sci. Technol. A* 6:271–274, 1988.

Ashkin, A., J. M. Dziedzic, and T. Yamane. Optical trapping and manipulation of single cells using infrared laser beam. *Nature* 330:769–771, 1987.

Bezanilla, M., C. J. Bustamante, and H. G. Hansma. Improved visualization of DNA in aqueous buffer with the atomic force microscope. *Scanning Microsc.* 7(4):1145–1148, 1993.

Binnig, G., C. F. Quate, and C. H. Gerber. Atomic force microscopy. *Phys. Rev. Lett.* 56:930–933, 1986.

Butt, H. J., E. K. Wolff, S. A. C. Gould, B. Dixon Northern, C. M. Peterson, and P. K. Hansma. Imaging cells with the atomic force microscope. *J. Struct. Biol.* 105:54–61, 1990.

Cooper, S. *Bacterial Growth and Division.* San Diego: Academic Press, 1991, p. 359.

Dammer, U., O. Popescu, P. Wagner, D. Anselmetti, H. J. Güntherodt, and G. Misevic. Binding strength between cell adhesion proteoglycans measured by atomic force microscopy. *Science* 267:1173–1175, 1995.

Dammer, U., M. Hegner, D. Anselmetti, P. Wagner, M. Dreier, W. Huber, and H. J. Güntherodt. Specific antigen/antibody interaction measured by force microscopy. *Biophys. J.* 70:2437–2441, 1996.

Drake, B., C. B. Prater, A. L. Weisenhorn, S. A. C. Gould, T. R. Albrecht, C. F. Quate, D. S. Cannell, H. G. Hansma, and P. K. Hansma. Imaging crystals, polymers and processes in water with the atomic force microscope. *Science* 243:1586–1989, 1989.

Evans, E., D. Berk, and A. Leung. Detachment of agglutinin-bonded red blood cells. Forces to rupture molecular point attachment. *Biophys. J.* 59:838–848, 1991.

Farkas, V., J. Kovařík, A. Kosinová, and S. Bauer. Autoradiographic study of mannan incorporation into the growing cell walls of *Saccharomyces cerevisiae J. Bacteriol.* 117:265–269, 1974.

Ferrell, T. L., J. P. Goundonnet, R. C. Reddick, S. L. Sharp, and R. J. Warmack. The photon scanning tunneling microscopy. *J. Vac. Sci. Technol.* B9:525–530, 1991.

Florin, E. L., V. T. Moy, and H. E. Gaub. Adhesion forces between individual ligand receptor pairs. *Science* 264:415–417, 1994.

Gad, M., and A. Ikai. Method for immobilizing microbial cells on gel surface for dynamic AFM studies. *Biophys. J.* 69:2226–2233, 1995.

Gad, M., A. Itoh, A. Ikai. Mapping cell wall polysaccharides of living microbial cells using atomic force microscopy. *Cell Biol. Inter.* 21(11):697–706, 1997.

Häberle, W., J. K. H. Hörber, F. Ohnesorge, D. P. E. Smith, and G. Binnig. In situ investigation of single living cell infected by viruses. *Ultramicroscopy* 42–44:1161–1167, 1992.

Hansma, H. G., M. Benzallina, F. Zenhausern, M. Adrian, and R. L. Sinsheimer. Atomic force microscopy of DNA in aqueous solutions. *Nucleic Acids Res.* 21(3):505–512, 1993.

Hansma, P. K., B. Drake, O. Marti, S. A. Goul, and C. B. Prater. The scanning ion-conductance microscope. *Science* 243:641–643, 1989.

Hansma, H., J. Vesenka, C. Siegerist, G. Kelderman, H. Morret, R. L. Sinsheimer, V. Elings, C. Bustamante, and P. K. Hansma. Reproducible imaging and dissection of plasmid DNA under liquids with atomic force microscope. *Science* 256:1180–1184, 1992.

Hegner, M., P. Wagner, and G. Semenza. Immobilizing DNA on gold via thiol modification for atomic force microscopy in buffer solution. *FEBS* 336:452–456, 1993.

Helm, C., W. Knoll, and J. Israelachivili. Measurement of ligand-receptor interactions. *Proc. Nat. Aca. Sci.* 88:8169–8173, 1991.

Henderson, E. Imaging of living cells by atomic-force microscopy. *Prog. Surf. Sci.* 46:39–60, 1994.

Hinterdorfer P., W. Baumgartner, H. J. Gruber, K. Schilcher, H. Schindler. Detection and localization of individual antibody–antigen recognition events by atomic force microscopy. *Proc. Natl. Acad. Sci. U.S.A.* 93:3477–3481, 1996.

Hörber, J. K. H., W. Häberle, F. Ohnesorge, G. Binnig, H. G. Liebich, C. P. Czerny, H. Mahnel, and A. Mayr. Investigation of living cells in the nanometer regime with the scanning force microscope. *Scanning Microsc.* 6:919–930, 1992.

Horisberger, M., and M. Vonlanthen. Location of mannan and chitin on thin sections of budding yeasts with gold markers. *Arch. Microbiol.* 115:1–7, 1977.

Israelachvili, J. N. *Intermolecular and Surface Forces.* London: Academic Press, 1992.

Kasas, S., and A. Ikai. A method for anchoring round shaped cells for atomic force microscope imaging. *Biophys. J.* 68:1678–1680, 1995.

Kuo, S. C., and M. P. Sheetz. Force of single kinesin molecules measured with optical tweezers. *Science* 260:232–234, 1993.

Laney D. E., R. A. Garcia, S. M. Parsons, and H. G. Hansma. Changes in the elastic properties of cholinergic synaptic vesicles as measured by atomic force microscopy. *Biophys. J.* 72:806–813, 1997.

Martin, H., P. Wagner, and G. Semenza. Immobilizing DNA on gold via thiol modification for atomic force microscopy imaging in buffer solutions. *FEBS Lett.* 336:452–456, 1993.

Martin, Y., and H. K. Wickramasinghe. Magnetic imaging by "force microscopy" with 1000 Å resolution. *Appl. Phys. Lett.* 50:1455–1458, 1987.

Mitsuda, S., T. Nakagawa, H. Nakazato, and A. Ikai. Receptor-linked antigen delivery system. Importance of 2-macroglobulin in the development of peptide vaccine. *Biochem. Biophys. Res. Commun.* 216:339–405, 1995.

Neagu, C., K. O. Van Der Werf, C. A. J. Putman, Y. M. Kraan, B. G. De Grooth, N. F. Van Hulst, and J. Greve. Analysis of immunolabeled cells by atomic force microscopy, optical microscopy, and flow cytometry. *J. Struct. Biol.* 112:32–40, 1994.

Putman, C. A. J., K. O. van der Werf, B. G. deGrooth, N. F. van Hulst, J. Greve, and P. K. Hansma. A new imaging mode in atomic force microscopy based on the error signal. *Proc. SPIE* 1639:198–204, 1992.

Radmacher, M., J. P. Cleveland, M. Frtiz, H. G. Hansma, and P. K. Hansma. Mapping interaction forces with the atomic force microscope. *Biophys. J.* 66:2159–2165, 1994.

Shroff, S. G., D. R. Saner, and R. Lal. Atomic force microscopy of arterial cells: Local viscoelastic mechanical-properties and imaging of cytoskeleton. *Biophys. J.* 66:278a, 1994.

Svoboda, K., C. F. Schmidt, B. J. Schnapp, and S. M. Block. Direct observation of kinesin stepping by optical trapping interferometry. *Nature* 365:721–727, 1993.

Thundat, T., D. P. Allison, R. J. Warmack, G. M. Brown, K. B. Jacobson, J. J. Schrick, and T. L. Ferrel. Atomic force microscopy of DNA on mica and chemically modified mica. *Scanning Microsc.* 6:911–918, 1992.

Williams, C. C., and H. K. Wickramasinghe. Scanning thermal profiler. *Appl. Phys. Lett.* 49:1587–1589, 1986.

Yang, J., J. Mou, and Z. Shao. Molecular resolution atomic force microscopy of soluble proteins in solution. *Biochim Biophys. Acta* 1199:105–114, 1994.

Index